Geometry and Thermodynamics

Common Problems of Quasi-Crystals, Liquid Crystals, and Incommensurate Systems

NATO ASI Series

Advanced Science Institutes Series

A series presenting the results of activities sponsored by the NATO Science Committee, which aims at the dissemination of advanced scientific and technological knowledge, with a view to strengthening links between scientific communities.

The series is published by an international board of publishers in conjunction with the NATO Scientific Affairs Division

A	**Life Sciences**	Plenum Publishing Corporation
B	**Physics**	New York and London
C	**Mathematical**	Kluwer Academic Publishers
	and Physical Sciences	Dordrecht, Boston, and London
D	**Behavioral and Social Sciences**	
E	**Applied Sciences**	
F	**Computer and Systems Sciences**	Springer-Verlag
G	**Ecological Sciences**	Berlin, Heidelberg, New York, London,
H	**Cell Biology**	Paris, and Tokyo

Recent Volumes in this Series

Volume 227—Dynamics of Polyatomic Van der Waals Complexes
edited by Nadine Halberstadt and Kenneth C. Janda

Volume 228—Hadrons and Hadronic Matter
edited by D. Vautherin, F. Lenz, and J. W. Negele

Volume 229—Geometry and Thermodynamics: Common Problems of
Quasi-Crystals, Liquid Crystals, and Incommensurate Systems
edited by J.-C. Tolédano

Volume 230—Quantum Mechanics in Curved Space-Time
edited by Jürgen Audretsch and Venzo de Sabbata

Volume 231—Electronic Properties of Multilayers and
Low-Dimensional Semiconductor Structures
edited by J. M. Chamberlain, L. Eaves, and J.-C. Portal

Volume 232—New Frontiers in Quantum Electrodynamics and Quantum Optics
edited by A. O. Barut

Volume 233—Radiative Corrections: Results and Perspectives
edited by N. Dombey and F. Boudjema

Volume 234—Constructive Quantum Field Theory II
edited by G. Velo and A. S. Wightman

Series B: Physics

GEOMETRY AND THERMODYNAMICS

Common Problems of Quasi-Crystals, Liquid Crystals, and Incommensurate Systems

Edited by

J.-C. Tolédano

National Center of Telecommunication Studies
Bagneux, France

Springer Science+Business Media, LLC

Proceedings of a NATO Advanced Research Workshop on
From Geometry to Thermodynamics: Common Problems of
Quasi-Crystals, Liquid Crystals, and Incommensurate Insulators,
held September 4-8, 1989,
in Preveza, Greece

Library of Congress Cataloging-in-Publication Data

NATO Advanced Research Workshop on From Geometry to Thermodynamics
  (1989 Preveza, Greece)
    Geometry and thermodynamics   common problems of quasi-crystals,
  liquid crystals, and incommensurate systems / edited by J.-C.
  Tolédano.
        p.   cm. -- (NATO ASI series. Series B, Physics , v. 229)
    Proceedings of a NATO Advanced Research Workshop on From Geometry
  to Thermodynamics, held 9/4-8/89, Preveza, Greece.
    Includes bibliographical references and index.
    ISBN 978-1-4613-6702-4   ISBN 978-1-4615-3816-5 (eBook)
    DOI 10.1007/978-1-4615-3816-5
    1. Crystals--Defects--Congresses. 2. Liquid crystals--Congresses.
  3. Molecular structure--Congresses.   I. Tolédano, Jean-Claude.
  II. North Atlantic Treaty Organization. Scientific Affairs
  Division. III. Title. IV. Series.
  QD921.N377 1989
  548--dc20   90-43309
                                                     CIP

© 1990 Springer Science+Business Media New York
Originally published by Plenum Press, New York in 1990
Softcover reprint of the hardcover 1st edition 1990

PREFACE

Distinct scientific communities are usually involved in the three fields of quasi-crystals, of liquid crystals, and of systems having modulated crystalline structures. However, in recent years, there has been a growing feeling that a number of common problems were encountered in the three fields. These comprise the need to recur to "exotic" spaces for describing the type of order of the atomic or molecular configurations of these systems (Euclidian "superspaces" of dimensions greater than 3, or 4-dimensional curved spaces); the recognition that one has to deal with geometrically frustrated systems, and also the occurence of specific excitations (static or dynamic) resulting from the continuous degeneracies of the stable structures considered.

In the view of discussing these problems, a NATO-Advance Research Workshop has assembled in Preveza (Greece), in september 1989, 50 experts of the three considered fields (with an equal proportion of theorists and experimentalists). 35 hours of conferences and discussions have led to a more detailed evaluation of the similarities and of the differences in the approaches implemented in the studies of the three types of systems. The papers contained in this NATO-series book provide the substance of this workshop. The reader will find three types of papers. Some very short papers giving the main ideas stated on a subject. Papers comprising 8-10 pages which stick closely to the contents of the talks presented. Longer papers providing more extensively the background and results relative to a given topic.

It is worth summarizing the principal outputs of the workshop.

"Exotic" spaces have appeared to possess a different status in the different systems. In modulated structures, Euclidian superspaces constitute a tool of mere geometrical conveniency for the determination of the crystal structure, and it is possible to solve the structures without recurring to them. By contrast they seem unavoidable in discussing the structure of quasi-crystals, and many a physical argumentation is based on their consideration. A different objective is pursued in the case of liquid crystals: that of estimating, with the help of curved spaces, the energy associated to the frustration of the actual 3D-structure. This difference between a geometrical point of view in the first two types of systems, and an energetic one in the last type, must be attenuated. Indeed, it is also possible to develop a unified theoretical framework which describes physical excitations (electrons, phonons) in quasi-crystals and in modulated crystals on the basis of the superspace description. Experimentally, the testing of these unified theories has not reached the same level of maturity in quasi-crystals (for which little has been done aside from the structural aspect) and in the simpler modulated structures (in which few questions remain to clarify, e.g. the existence of optical activity).

The "phason" excitation is another interesting point of comparison between liquid crystals and modulated crystals. Its observation seems more generally favourable in liquid crystals due to the lack of strong pinning centers. Such an observation has even been achieved, in liquid crystals, by means of light scattering, while the same technique has been tried unsuccessfully, up to now, in the other type of system. On the other hand, the existence of a gap in the phason spectrum is still a subject of controversy. Its systematic detection by NMR techniques relies on an interpretative framework whose validation has to be refined. It has been noted that there may be a difference of nature of the phason in the two types of systems due to the fact that there is no reference-frame for the sliding of the phase in liquid crystals, and also, that in the latter systems, there is a possible coupling to hydrodynamics degrees of freedom.

In a number of substances with modulated crystal structures, reliable and accurate atomic configurations have been determined. By contrast, a central problem in quasi-crystals, and in newly discovered liquid crystal phases (e.g. blue phases) is the delicate interpretation, in the form of a detailed structural description, of the available diffraction data and of the electron microscopic observations. Many questions relative to these structures are still open. One of the promising ways to answering them is to investigate theoretically and experimentally their mechanism of formation and of growth.

J.C. Tolédano

CONTENTS

STRUCTURE AND GEOMETRY

MODELS FOR STABILITY AND GROWTH

DYNAMICS AND PHASONS

HIERARCHIC STRUCTURE

Alan L. Mackay

Department of Crystallography, Birkbeck College
University of London, Malet Street, London WC1E 7HX

ABSTRACT

We are concerned with *structuration*, the ways in which complex structures are built from simpler components, and especially with *levels of organisation*, primarily from a geometrical point of view, with the objective of developing a *generalised crystallography* extending from the atomic level, beyond regular crystals, towards biological structures which contain fibres, membranes and other sub-structures, some informational. Some levels may be random, some crystalline and some may be regular in different ways Materials characterised as *textures* must now be specified in more detail. The quasi-crystal affair has provided the stimulus for a fundamental re-estimation of orthodox crystallography.

"If this is the best of all possible worlds, then what the others may be like hardly bears thinking about" — Voltaire, *(Candide)*, (1758).

Newton, having determined *"the motions of the planets, the comets, the Moon and the sea"*, was unfortunately unable to determine the remaining structure of the world from the same propositions because: *"I suspect that they may all depend upon certain forces by which the particles of bodies, by some causes hitherto unknown, are either mutually impelled towards one another, and cohere in regular figures, or are repelled and receed from one another. Those forces being unknown, philosophers have hitherto attempted the search of Nature in vain; but I hope the principles here laid down will afford some light either to this or some truer method of philosophy"*. Preface to the *"Principia"*, (1687). [1]

"It will be found that everything depends on the composition of the forces with which the particles of matter act upon one another; and from these forces, as a matter of fact, all phenomena of Nature take their origin". R. J. Bošković. *"Theoria Philosophiae Naturalis"*, Venice, (1763). (sec. 1.5).

"A number of quite independent lines of argument converge toward the assertion that there is often a limit to the complexity of systems". Richard Levins, (*"Towards a Theoretical Biology"*, **3**, 73-88, (1970)).

---

[1](qu. Nature, **329**, 772, 29 Oct. 1987).

Geometry and Thermodynamics
Edited by J.-C. Tolédano
Plenum Press, New York, 1990

"Physics is simple only when analysed locally". Misner, Thorpe and Wheeler, (*"Gravitation"*)

Quasi-crystals are only part of the revolution in thinking about structure brought about by the computer and computer graphics[2] [8]. I want to direct your attention to the "Generalised Crystallography" first promulgated by J. D. Bernal [2] about 25 years ago. X-ray crystal structure analysis has been so successful that people have forgotten that most materials are not in fact crystalline. Discussions current in the 1930s can now be picked up again at a new level.

As you will see, I am trying to collect together many rather disparate parts of a general science of spatial structure and, since it is very difficult to organise them into a linear sequence, have begun to use the Hypertext system to handle them as a strongly connected network. This corresponds better, I believe, to the way ideas are organised in the brain. We can think only about a few items at a time, so that In order to organise our thinking about structures, we have to seize on the key feature of hierarchy.

1. THE BLIND WATCHMAKER

The Blind Watchmaker [4], besides making life, also makes inanimate materials.

My comments are based on the parable of the two watchmakers presented by H. A. Simon in his book [20] *The Sciences of the Artificial.*

A watch has 100 parts. The first watchmaker assembles each part in turn but on each occasion when he is interrupted, the watch falls to pieces and he has to start again from the beginning. The second watchmaker, subject to the same interruptions, has broken down the construction into 10 subunits each of ten parts. He makes subunits and then assembles 10 of these to make a watch. Needless to say the probability of his completing a watch is enormously greater than that of the first watchmaker. The difference is due to *hierarchisation.* We should also note that the watchmaker has some kind of description of a watch in terms other than a completed watch itself.

Hierarchisation is one of the basic principles of constructing anything. This is well-recognised by protein crystallographers [17] who see primary, secondary, tertiary and quaternary levels of structure in proteins[3]. Protein structures are all built of domains in this way. The domains are seen as an "implicit structural consequence of the folding process" [17]. Inorganic structures are hierarchised by seeing coordination polyhedra which are then linked into larger features like, threads, sheets and frameworks. Molecular compounds are usually seen as having only two levels, the molecules themselves, where atoms are linked by covalent bonds which are much stronger than the forces between molecules, and solutions or crystals of molecules.

Liquids and amorphous materials present special problems because it is difficult to find distinct levels.

---

[2]Many colour pictures were shown in the lecture which thus corresponded only roughly with the present text.

[3]primary structure is the sequence of amino-acids; secondary structure is the alpha-helices and beta-sheets produced by favoured angles between units: tertiary structure is the way in which sheets and helices twist and turn to make globules: quaternary structure is the composition of globules to make the whole protein.

The key feature is that in a hierarchic structure the rule of composition at each level has to stop and to give way to a new rule for the composition of the next higher level. Orthodox crystals have only one rule. The classical symmetry operations do not explain how to stop. The considerations of facial energy introduced by G. V. Wulf and developed by Donnay and Harker, which explain external morphology, are the next step and after that questions of grain boundaries. N. Rivier [16] has connected grain boundaries with quasi-crystals, suggesting that, between coincidence site lattices and completely general grain boundaries, quasi-crystal interfaces form a preferred type.

2. LEVELS OF ORGANIZATION

At a given level there may be a particular kind of structure, *crystalline, random, textured, curved manifold.*

A good example is opal. Here, amorphous hydrated silica forms a random network; layers of such network form spheres like hailstones; these spheres crystallise (to give the layers with spacings in the optical wavelength range so prized in precious opal); regions of crystalline and non-crystalline opal form rock.

Hans Zocher has demonstrated crystals of crystals with $\beta - FeOOH$ and tungstic acid crystals. Sadanaga, Takeuchi and Morimoto [18] have shown that minerals with complex crystal strucrues can be seen as built of fragments of simpler structures. A. D. Wadsley and others have greatly simplified the crystal chemistry of complex oxides with this outlook which has been amply confirmed by high-resolution electron microscopy, which shows simple coordination polyhedra coupled in a complicated ways not easily to be discosed by X-ray diffraction.

A regular hexagonal packing may occur at the scale of the Giants' Causeway, although through physical processes which are quite different from the process which produce hexagonal graphite. Physical forces are not scale-independent.

3. IS THERE A LIMIT TO COMPLEXITY

Hierarchisation also takes place in the inorganic world and complex structures are built out of simpler in several levels [18]. There are limits to complexity. For example we may ask our data bank: what is the largest number of different elements which occurs in a crystal structure (where each atom is in a distinct site without solid solution)? The answer is about seven or eight.

We may ask also about the structures which have the largest unit cells. They are almost all proteins, or polytypes, that is, modulations of simple structues or complex framework silicates, that is, recognisably hierarchic structures. An example is *paulingite*, which is cubic with a unit cell dimension of about 35Åcontaing a thousand or so atoms, as many as a protein. How much information is necessary to control the formation of such a structure and where is it or, where are the genes for paulingite? Cellular automata begin to give us a clue.

I guess that there is a natural limit to complexity to to exceed this it is necessary to use special structures for the storage of programme. This is characteristic of life, but we see synthetic systems, computers and the search is on, under the stimulation of the ideas of Graham Cairns-Smith, for inorganic systems which may store arbitrary information controlling the reproduction of other components.

The development of cellular automata have perhaps changed our ideas about complexity in showing the richness of patterns which can appear from very simple local rules. In the bonding of atoms we have local rules, but with a stored programme, more arbitrary rules can be implemented. As an extreme, the Mandelbrot set is an infinitely complex structure emerging from a simple equation, more picturesquely than the digits of *pi* emerge. However, it is a mathematical and not a physical structure and takes no account of the natural objects, such as atoms, which are available for making structures. It has no scale and no upper or lower limits.

In a crystal the same rule takes one an enormous distance but characteristically in biological objects, only a few units can be added at the same level before the rules of combination change.

The decimal numbers used for counting are an example of perfect hierarchisation. After counting from 0 to 9 we must carry one to the next most significant place and begin again [4] . The Turkish army is organised like this. A corporal *Onbaşi* is (literally) the head of ten; a captain *Yüzbaşı* is the head of a hundred and a major *Bımbaşi* is the head of a thousand. As has been known from antiquity this is the way to control large numbers. If they act in unison the amplitudes of their individual efforts add and the total intensity is the square of this (as explained by F. W. Lanchester [11]) . If they act as individuals the intensities of their efforts add[5].

As Simon points out with respect to opening the combination lock of a safe, if we can separate a structure into its hierarchic levels we can analyse its behaviour enormously more easily (and the safe can be cracked)[6].

Nothing is totally separable. EPR (Einstein, Podolsky and Rosen), Bell and Aspect show that everything is connected to everything else, albeit weakly, and that levels in a heirarchy are not completely separable. Equivalence must be replaced by quasi-equivalence.

If the span of levels is large, for example 10, then the levels are more clearly separable than if the span is only two.

One of the most interesting features of the Penrose tiling / Quasi-crystal affair is that hierarchic structures have been introduced into formal crystallography [12]. This begins to resemble the renormalisation procedures which connect interatomic forces with whole-body phenomena such as phase transitions.

LOGICAL PARADOXES

The problem is that the levels in the hierarchy of inorganic and particularly metallic structures are not very clearly separable. Metals are almost all close-packed assemblies of atoms each with about 12 neighbours[7].

---

[4]The ordinary binary numbers are not the best way for machines to count because more than one digit may change at a time leading to uncertainty. The proper counting sequence is the Gray code. e.g. Nature, **340**, 514, (17 Aug 1989).

[5]Just these tactics were used intuitively by Octavian to defeat Mark Anthony in the naval battle of Actium which was fought just off Preveza on 2nd. September 31 BC (2019 years ago) — One of the decisive battles of the world.

[6]Richard Feynman has told you how to do it

[7]There was an interesting study by Stuart Kauffman in 1969, in which he showed that cellular automata with high coordination numbers had much longer cycles of repetition than those with low coordination

He writes: "The cyclic behaviour of the network now is equivalent to logical contradiction in its circuitry. We suggest that spontaneous persistent activity in deterministic discrete systems is the equivalenct of self-contradiction in the networks" [9]. That is, attempts to resolve frustrations repeat cyclically.

The Buergers vector is a measure of the frustration in a static, spatial structure.

It is also evident that in setting up some of the various logical paradoxes which have confounded philosophers, we are in fact constructing cellular automata of this type. An example of this type of frustrated structure is the card which carries on one side the words *"The statement on the other side of this card is true"* and on the reverse side the statement *"The statement on the other side of this card is false"*. We should not be surprised if linguistic structure reflects the structure of the real world. Linguistic philosophers have made very heavy weather of such paradoxes[8].

A similar situation can be found in spin glasses.

5. TWO DIMENSIONAL STRUCTURES AND DIFFERENTIAL GEOMETRY

The inorganic way to look at a structure is to consider the network of interactions between point atoms, attaching appropriate force functions to pairwise or multiple interactions.

A more biological way is to regard atoms as force centres and to allocate domains to them. The simplest type of domain is the Voronoi polyhedron and many more complex criteria for boundaries can be devised. The next simplest is the radical plane division and, with Coulomb forces between charged ions, equipotential surfaces appear dividing up space. Surfaces are structures orthogonal to lines of force.

The characteristic shapes of protein and other molecules which interact with great specificity are largely discussed in terms of potential contours. These are of intense interest in drug design.

Further up in scale we can consider membranes made of lipid bilayers which define inside and outside and give the possibility of cell structure characteristic of living systems. These membranes are effectively two-dimensional manifolds. We may consider surface tension either as energy per unit area over the surface or as force per unit length along the perimeter. Similarly, we may find the integrated Gaussian curvature either by summing over the area or along the perimeter.

Membranes may have elastic properties. F. C. Frank [5] has given general expressions for the energy of a nematic. Membranes, in addition to bend, splay and twist energies, may have some of the properties of a rigid medium. Gradually we can introduce force considerations into the geometry. Surface tension (and more complex expressions for energy) represent a way of carrying the interatomic forces seen at one level, to the handling of structures at the level of membranes where the individual atoms are not resolved.

Two-dimensional structures are of particular interest because, living as we do in three-dimensional space we have one more dimension and can sit back and see the two-dimensional structures embedded in the three. The topology of two-dimensional manifolds can be described exactly by specifying how many handles they have. We can make physical models

---

numbers (2)

[8]Such a frustrated structure is the basis of the children's game where an odd number of children stand in pairs in a ring The odd one out stands in front of a pair and the rear member of the pair is displaced.

or projections of non-Euclidean two-dimensional manifolds.

Examples of two-dimensional structures are the curved silicate layers which form cylinders of *imogolite* and the curved layers of graphite in similar tubules. Graphite layers also may form doubly curved sheets (of positive Gaussian curvature) as Kroto [10] has demonstrated. Certain kinds of sea-weed form sheets with negative Gaussian curvature and various lipids form layers which are periodic minimal surfaces with negative Gaussian curvature. Many protein components form tubes or spherulites rather than crystals. One example of tubes is furnished by a degradation product of haemoglobin.

The mathematics of differential geometry have now entered crystallography in a serious way as may be seen, for example, in the work of Stephen Hyde [1] and his colleagues.

6. THREE DIMENSIONAL MANIFOLDS

Three and higher dimensional manifolds are much more complicated. It is not even know whether there is an algorithm for deciding whether two three-dimensional manifolds are the same, but it is known that there is no such algorithm in more than three dimensions. The 3-D surface of a 4-D hypersphere provide a useful model of a space of non-Euclidean metric and has been much used.

In addition to real structures, manifolds of various dimensions occur in the description of variables, such as angles, in mechanical systems [21]. We see some promise here for the use of such space for handling the space of the six parameters necessary for the general docking problem (of fitting two molecules together).

7. TEXTURES

The symmetries of textures have been classified on a statistical basis. The study of textures is now at an interesting stage of relating statistical features to the inter-relationships of the individual units. Recent developments in liquid crystals, Penrose tilings and in techniques of electron microscopy have directed concern to the details of the arrangements of sub-units. Quasi-crystals are, in a sense, the most highly ordered textures known.

CENTRO-SYMMETRY

Centro-symmetry is, in a sense, one of the most general textures. A crystal structure, which has a centre of symmetry, has only centro-symmetrical (cosine) terms in its Fourier transform. The measured amplitudes of the Fourier components thus correspond to phases of either 0 or π, and the structure is far more readily solved by the direct methods of Karle and Hauptmann, since the information to be restored is far less [9].

---

[9]The German *Enigma* coding machine was made centro-symmetrical for administrative convenience (coding and decoding procedures were then the same) — in a setting where typing in A gave B, then typing in B would return A. This weakness led to an entry to the decypherment by M. Rejewski, J. Różycki and H. Zygalski, and later others at Bletchley Park, which perhaps utimately led to military defeat of the Axis. Probably no crystallographers were involved on either side. *Enigma*, Władysław Kozaczuk, (in Polish, 1976)(in English, 1984)

8. CONCLUSION

I regret that I will never be able to apprehend the astonishing book on Gravitation [14] by Misner, Thorpe and Wheeler. However, dipping into it, from back to front, I suspect that it contains many clues as to how solid state physics might develop, working from local order outwards, instead of from infinite perfect crystals as the primary structures. They start from the proposition that "physics is simple only when analysed locally". Newton himself concentrated on the more accessible aspects of the universe and regrettably did not look closely into the consequences of interatomic forces. I recommend more knowledgable people to try whether they can extract more.

I think that we are in a fascinating period where the prevailing paradigm is thinking about the structure of matter is no longer the X-ray crystal structure analysis of orthodox crystallography but has moved to cellular automata and the consequences of the computer in all its generality [8] as a tool, as a concept and as a parallel to living processes. Instead of conceptually infinite crystals, we look for layers of complexity, the type of order changing from level to level.

REFERENCES

[1] S. Andersson, S. T. Hyde and H. G. von Schnering, "The intrinsic curvature of solids", Z. f. Krist., **168**, 1-17, (1984).

[2] J. D. Bernal and C. H. Carlisle, "The range of generalised crystallography", Sov. Phys. Crystallography, **13**, 811-831, (1969).

[3] J. M. Burgers, "The emergence of patterns of order", Bul. Amer. Math. Soc., **69**, No.1, 1-25, (Jan. 1963). (Lecture given 20.1.59).

[4] R. Dawkins, *The Blind Watchmaker*, 1986. Penguin, 1988.

[5] F. C. Frank, Proc. Farad. Soc. (1958)

[6] S. Hachisu, and S. Yoshimura, "Optical demonstration of crystalline superstructure in binary mixtures of latex globules". Nature, **283**, 188-189, (1980).

[7] S. Harren, T.C.Lowe, R.J.Asaro and A. Needleman, "Analysis of large-strain shear in rate-dependent face-centered cubic polycrystals", Phil. Trans. R.S., **328**. 443-500, (1989).

[8] R. Herken (ed.), *The Universal Turing Machine*, Oxford Science Publications, 1988. pp. 657.

[9] S. Kauffman, "Metabolic stability and epigenesis in randomly constructed genetic nets". J. Theor. Biol., **22**, 437-467, (1969).

[10] H. W. Kroto and K. McKay, "The formation of quasi-icosahedral spiral shell carbon particles", Nature, **331**, 328-331, (1988).

[11] F. W. Lanchester, *Aircraft in Warfare*, Constable, London, 1916.

[12] A. L. Mackay, "Regular structures need not be crystalline", Phys. Bull., 495-497, (Nov. 1976).

[13] S. Mann and B. R. Heywood, "Crystal engineering at interfaces", Chem. in Brit., 698-700, (July, 1989)

[14] C. W. Misner, K. S. Thorne and J. H. Wheeler, *Gravitation*, W. H. Freeman, San Francisco, 1970.

[15] H. Pattee (ed.), *Hierarchy Theory*, Braziller, N.Y., 1973.

[16] N. Rivier, "Quasicrystals and Grain Boundaries". Proc. 3rd. Int. Meet. on Quasicrystals...Vista Hermosa, Mexico, June 1989.

[17] G. D. Rose,"Hierarchic organisation of domains in globular proteins", J. Mol. Biol., **134** 447-470, (1979).

[18] R. Sadanaga, Y. Takeuchi and N. Morimoto, "Complex structure of minerals", Rec. Progr. of Natural Sciences in Japan, **3** 141-206, (1978).

[19] A. H. Schoen, "Infinite periodic minimal surfaces without self-intersection", NASA Technical note, C-98 (1969).

[20] H. A. Simon, *The Sciences of the Artificial*, MIT Press, Cambridge, Mass., 1969. pp.90-93.

[21] W. P. Thurston and J. R. Weeks, "Three-dimensional manifolds", Sci. Amer., **251**, No.1, 94-106, (July, 1984).

[22] L. L. Whyte, A.G. Wilson and D. Wilson (eds.), *Hierarchical Structures*, Amer. Elsevier, New York, 1969.

[23] A. Winfree, *The Geometry of Biological Time*, Springer, Berlin, (1980).

[24] S. Wolfram, *Theory and application of cellular automata*, World Scientific, Singapore, (1986).

THE STRUCTURE OF QUASICRYSTALS :

FROM DIFFRACTION PATTERNS TO ATOM POSITIONS

Christian Janot[+], Jean-Marie Dubois[*] and Marc de Boissieu[+*]

+ Institut Laue-Langevin
  156X, 38042 Grenoble Cedex, France
* LSG2M, Ecole des Mines
  Parc de Saurupt, 54042 Nancy Cedex, France

ABSTRACT

As for periodic crystals, it is possible to make a crystallographic approach to the atomic structure of quasicrystals. Patterson's functions, possibly for partial correlations, and sometimes atomic density distributions may be extracted from diffraction patterns. These data are usefully complemented by the direct determinations of pair (or preferably partial pair) distribution functions which take into account the "in-between strong Bragg peaks" scattering. A good example is the specification of the structure of Al-Mn like quasicrystals for which atomic positions cannot be strictly described in terms of a canonical 3-dimensional Penrose lattice and require generalized $A3_{perp}$ volumes in 6-dim with components in the physical space. Insight into the structure of AlLiCu and AlCuFe quasicrystals have also been obtained.

INTRODUCTION

For periodic crystals, the structure is completely specified when both the unit cell (or the Bravais lattice) and the positions of atoms in this unit cell have been determined. The so-called direct methods of crystallography are the usual way to extract this structural information from diffraction data.

Basically, quasicrystals must be treated the same way : positions (indexing) of the diffraction peaks are related to the geometrical characteristics of the quasiperiodic framework while distribution of the scattered intensity should reveal where the atoms are located. The first step, i.e. the successful description of the quasiperiodic lattice, has been achieved with use of a variety of different schemes for generating them[1] : space tiling by two rhombohedral cells with matching rules, inflation-deflation procedure, multigrid or dual methods, strip-projection[2] or cut-projection approaches.[3] The latter, in particular, drives quasicrystal back to crystallography by showing that any 3-dim quasi-periodic lattice has actually hidden translation invariances which can be recovered if the structure is properly described in a higher-dimensional space. For instance, the archetypical icosahedral quasicrystals with $m\bar{3}5$ point-group symmetries cannot be periodic in 3-dim but there are three possible six-dimensional icosahedral periodic arrangements, i.e. the primitive (P), face-centred (F) and body-centred (I) cubic Bravais lattices. Each of them

Geometry and Thermodynamics
Edited by J.-C. Tolédano
Plenum Press, New York, 1990

corresponds to well defined indexing (positions and extinction rules) of the diffraction peaks[4]. The second step, i.e. saying where the atoms are, is intrinsically more difficult for a quasicrystal than for a crystal. A perfect quasi-periodic structure, without any disorder, still has an infinite number of sites in 3-dim which are not exactly equivalent. There are also practical difficulties to be overcome, related to the fairly low level of information that can be extracted from diffraction patterns of quasicrystals, e.g. only "strong" Bragg peaks (small Q_{perp}) are actually measured and each of them contains mixed contributions from both geometrical and chemical features of the structure. For that purpose, direct methods have been somewhat bypassed or at least complemented by modelling approaches.[5-9] For more complete reviews see for instance refs.[10,11].

The purpose of this paper is to present a brief report on mainly neutron diffraction approaches to the structure of the most popular families of quasicrystals, namely Al-Mn like, Al-Li-Cu and Al-Fe-Cu systems. As experimental details have been extensively reported elsewhere, this paper is focussed on the structural features that have been ascertained so far.

A GENERAL SCHEME OF DIFFRACTION APPROACH TO QUASICRYSTAL STRUCTURE

The best way to analyse the relation between diffraction data and the structure of a quasicrystal is probably to work out the problem within the so-called cut mehod[9].

In the cut method, an icosahedral quasiperiodic arrangement of atoms in a 3-dim space $R3_{par}$, taken as an example, corresponds to a periodic arrangement of 3-dim hypersurfaces, or atomic shells $A3_{perp}$ in 6-dim space R6. These atomic shells intersect the 3-dim real world hyperplane at the atom positions. For physical obvious reasons, the $A3_{perp}$ cannot intersect each other and they have to be invariant under the operations of the point group symmetry (120 in the case of $m\bar{3}5$) ; they do not have to be hyperplanes. For each type (or family) of atomic sites in 3-dim there is one $A3_{perp}$ shell whose relative volume is directly related to the corresponding relative atomic 3-dim density. In a simplistic idealistic monoatomic icosahedral quasicrystal, with a single site at the origin of the 6-dim structure, and triacontahedral $A3_{perp}$ the 3-dim atomic density is a distribution of Dirac functions at the vertex positions of a three dimensional Penrose tiling (3DPT). The volume $V(A3_{perp})$ of $A3_{perp}$ is equal to n_3a^6 in which a is the 6-dim lattice parameter and n_3 the 3-dim atomic density. Correspondence rules also exist between the reciprocal spaces $R6^*$, $R3^*_{par}$ and $R3^*_{perp}$ associated with R6, $R3_{par}$ and $R3_{perp}$, respectively ($R3_{perp}$ is the 3-dim space complementary to $R3_{par}$ into R_6). These reciprocal spaces contain the Fourier transforms (FT) of the densities and it is easy to demonstrate that the FT, $F(Q_{par})$ in $R3^*_{par}$ is the projection onto this very reciprocal subspace of the FT, $F(Q_6)$, in $R6^*$; $F(Q_6)$ in turn is a distribution of δ-functions modulated by $G(Q_{perp})$, the FT of $A3_{perp}$ (Q_{par} and Q_{perp} are the projections of Q_6 onto $R3^*_{par}$ and $R3^*_{perp}$, respectively). Such a correspondence scheme is illustrated in Figure 1, using a 2-dim → 1-dim simplification. The points of interest for an experimental appproach to quasicrystal structures may then be summarized as follows :
- There is a one-to-one correspondence between Q_6 and Q_{par} which generates a six integer indexing of diffraction peaks measured at Q_{par} in $R3^*_{par}$ and allows to determine the 6-dim Bravais lattice from diffraction data.
- Intensities $|F(Q_{par})|^2$ measured at Q_{par} in diffraction data are also the intensities $|F(Q_6)|^2$ that would correspond to a "6-dim diffraction experiment".
- The diffraction pattern in $R3^*_{par}$ is a very dense set of peaks whose intensity is a decreasing function of Q_{perp} (see Fig. 1).
- The direct FT of these measured $F(Q_6)$, or at least of $|F(Q_6)|^2$, gives the

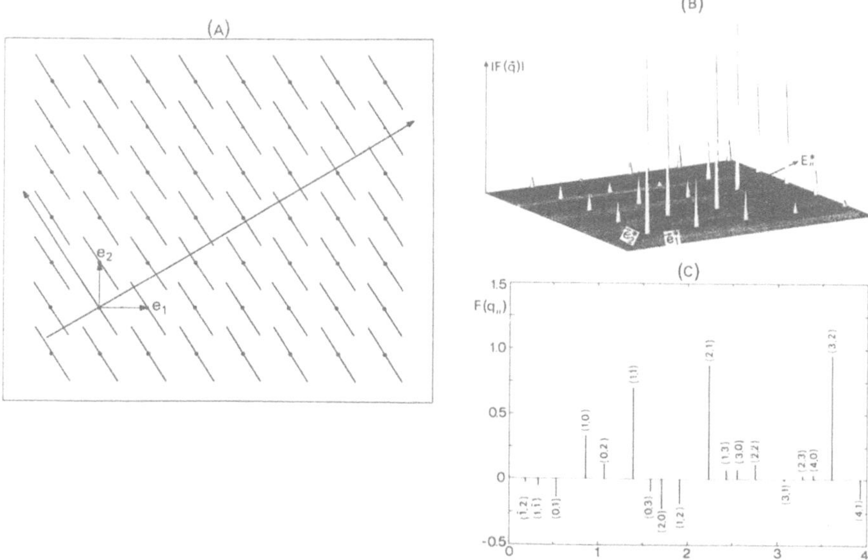

Fig.1   Illustration of the high dimensional crystallography methods. From diffraction data, integrated intensities of the strong peaks are obtained. They give at least amplitude of the structure factors and two integer indexing (c); this 2-integer indexing allows the high dimensional diffraction pattern to be reconstructed (b); Fourier transform of this pattern gives the high dimensional periodic structure (a) which, when cut by our physical space, determines atom positions.

6-dim structure (sites positions <u>and</u> $A3_{perp}$ functions), or at least the corresponding 6-dim Patterson function.
- The 3-dim cut of this 6-dim structure by $R3_{par}$ results in physical atom positions.

This is of course oversimplifying real life. A first, unavoidable, complication comes from the actual quasicrystal being not monoatomic nor single site systems which means that the $F(Q_{par})$ or $F(Q_6)$ are sums of several terms, with proper phase shifts between them. The FT of these quantities gives all right the lattice point in 6-dim but only <u>combinations</u> of the $A3_{perp}$ atomic shells. This complication can be somewhat <u>overcome by</u> contrast variation techniques in neutron diffraction, as extensively explained elsewhere[12-15]. Basically, isotopic or isomorphous substitutions allow to vary the weight of one, or several, atomic species into the scattered signal. By measuring several diffraction patterns weighted differently, it is possible to calculate what would be the diffraction patterns if each atomic species were scattering alone, i.e., so-called <u>partial</u> structure factor. The problem can then be treated as the super-position of several monoatomic structures. This is obviously the strong point of the methods as phase reconstruction procedure can then be attempted in 6-dim rather like for a classical 3-dim crystallography problem.

The second complication is due to experimental truncation effects in Q_{perp} space. As explained in details elsewhere[16], the diffraction pattern of a quasicrystal (in 3-dim) is singularly continuous but the strongest diffraction peaks only, corresponding to relatively small $|Q_{perp}|$ values, are actually measured (typically $|Q_{perp}| \lesssim 1$, in $2\pi/a$ units ; this range contains 283 independent reflexions of which only 60 are actually measured; had $|Q_{perp}|$ be measured up to 3, 5951 independent reflexions would have been concerned). Thus, the FT of the experimentally truncated $G(Q_{perp})$

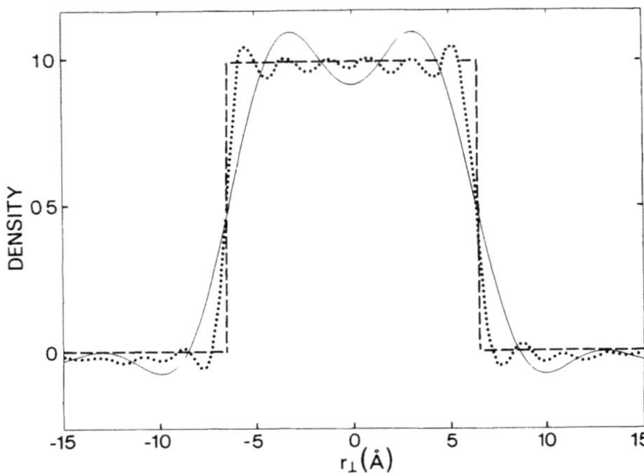

Fig.2　Illustration of the Q_{perp} truncation effects. The density profile in $R3_{perp}$ of a spherical $A3_{perp}$ function (dashed line) is compared to profiles calculated by Fourier transforms of the $G(Q_{perp})$ function when cut at $Q_{perp} = 1$ (full line) and $Q_{perp} = 3$ (dotted line). Ripples, depressions and boarder smearing are observed.

functions are somewhat able to give proper volumes but not the exact shapes of the $A3_{perp}$ atomis shells[16], as illustrated in Figure 2.

This truncation effects, along with the fact that the $A3_{perp}$ calculated from experimental $G(Q_{perp})$ may be somewhat noisy, prevent the atom positions from being completely specified[17] ; practically, background cut-off and shape smearing for $A3_{perp}(exp)$ results in rather bad reconstruction of the diffraction patterns when the determined structure is back Fourier transformed in reciprocal space. Thus, <u>final modelling is not completely unavoidable</u>[17], in which a further experimental approach may help, namely the so-called direct space methods (DSM). In DSM, currently employed for non-crystalline materials, pair distribution functions (PDF) and, preferably, partial pair distribution functions (PPDF) can be obtained[14,15,18]. Diffraction patterns are measured over a large Q range and a continuous FT of the whole (corrected) scattering signal gives the averaged, isotropically regrouped probability of atomic pairs as a functions of pair distances. Contrast variation with neutron diffraction allows the PPDF to be determined from convenient data sets. The negative point of such a procedure is that angular information is obviously lost. But the whole (diffuse) signal out of the strong peaks is reintroduced into the FT and thus contribute to the PDF or/and PPDF. Thus, any proposed structure of a quasicrystal has, interestingly, to be confronted to measured PDF or PPDF. As an example, Figure 3 shows measured PPDF compared to the ones deduced from the known crystalline structure for the β-Al_9SiMn_3 phase. Contrast has been produced on the Mn sites by σ-FeCr substitutions as for the quasicrystal samples[12]. This example validates both DSM for ordered structure and the formal consideration of the alloys as pseudo-binary compounds for the three PPDF extraction procedure.

Finally, a last complication arises when powder rather than single-grain samples are used in diffraction measurements. Different Q_{par} (or Q_6) vectors attached to non-equivalent Bragg reflexions (with different intensities then) may contribute to powder diffraction peaks having the same Q modulus value. Without further assumption or experimental evidence, the total intensity of such a global peak can only be shared uniformly between the contributing non-equivalent reflexions, thus inducing erroneous occupation factors of the atomic sites. The whole procedures that may go from diffraction data to quasicrystal structure is summarized in the following flow-chart :

12

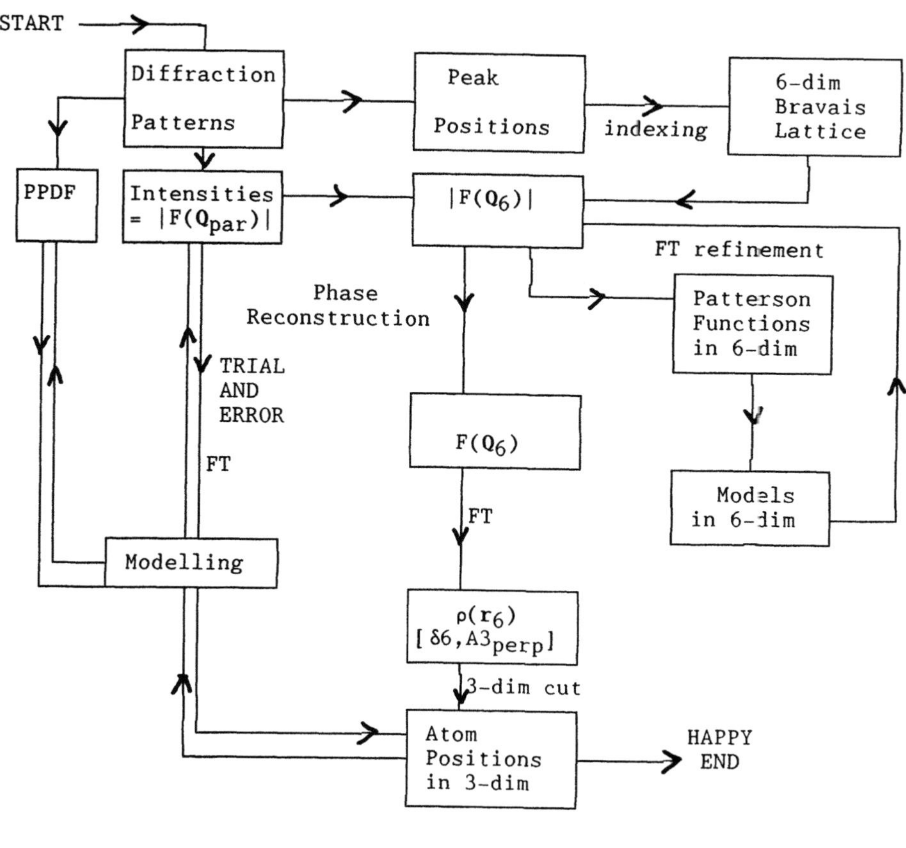

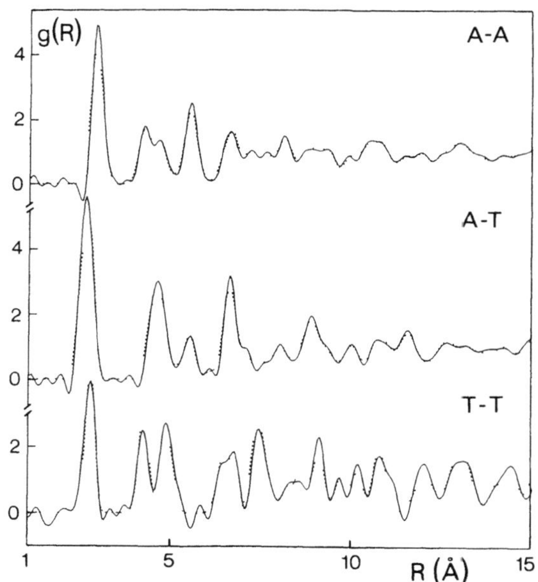

Fig.3 The partial pair distribution functions of the crystalline β-Al9SiMn3 phase : directly measured (—) and calculated from crystallography structure (...).

The rest of the paper will be devoted to a summary of the results of this procedure in the cases of Al-Mn, AlLiCu and AlFeCu quasicrystals.

EXAMPLES OF STRUCTURE SPECIFICATIONS FROM DIFFRACTION APPROACHES

The three major families of quasicrystals, namely AlMn(Si), AlLiCu and AlFeCu systems, have been investigated by neutron or X-rays diffraction. Up to a certain extent, they are quite illustrative of the various advantages or/and limits attached to the method. The AlMn(Si) system might be the more "dirty" quasicrystal but its basic structure happened to be relatively simple and its diffraction study has been pushed up to rather detailed aspects. Conversely, AlCuFe is considered as the more "perfect" quasicrystal but its structure seems to present chemical ordering effects and is far from achievement. Finally, AlLiCu quasicrystals have benefited from both isotopic substitution methods and single-crystal like approaches and the specification of their structure is presently in progress.

AlMn(Si) quasicrystals

Details of the investigation of the structure of $Al_{74}Si_5Mn_{21}$ quasicrystal, using powder neutron diffraction with contrast variation of the Mn site, can be found in previous publications[12-14,17]. In summary, this is the only system for which it has been possible to determine the partial structure factors, F_{Al} and F_{Mn}, with their amplitudes and phases, directly from diffraction data and thus to work out completely the 6-dim cut scheme. The 6-dim structure is quite simple and is related to an icosahedral (cubic) primitive lattice with a parameter $a = 6.5$ Å. There is only one site associated with the Mn atoms, at the origin of the 6-dim cube. The corresponding $A3_{perp}(Mn)$ is roughly spherical with smeared borders. There are two Al sites, one also at the origin and the other at the body centre of the 6-dim cube. $A3_{perp}(Al_O)$ is a distorted empty spherical shell and $A3_{perp}(Al_{BC})$ a very smeared rather weak spherical distribution.

Thus, the 6-dim structure is perfectly defined through such a scheme and it is possible to generate atom positions in 3-dim by a pertinent physical cut of this 6-dim structure. Without repeating the details already reported elsewhere[12,13,17], the cut 3-dim structure can be described as a 3DPT (4.6 Å edge) with Mn atom sites at the vertices, Al sites dividing $(1/\tau)$ the long diagonals of the rhombohedron faces (both prolate and oblate) and other Al sites at 2.95 and 6.77 Å from a vertex site along the triad axis of the prolate rhombohedra (the 6.77 Å site comes from the BC atomic volume in 6-dim). Of course, the 6-dim structure cannot generally induce a single perfectly defined decoration of the 3DPT. However, the 3-dim structure may be described in terms of an "average decoration", the atomic sites as reported here being only partially occupied. Noteworthy, there is no mid-edge Al sites contrary to what was formerly suggested in models based on Mackay icosahedron clusters[19] (MI) since the inner small Al icosahedron is not found and is replaced by a half occupied pentagonal dodecahedron. Fragments of filled Al/Mn small icosahedra are also found and the outer shells (large Mn icosahedron and large Al icosidodecahedron) of the MI are still observed.

However, such a structure, although describing correctly at least 85% of the atom positions, is too simple and, probably due to smearing truncation effects and/or background cut-off of the experimental $A3_{perp}$ profiles, fails to reproduce correctly the diffracted intensities when back Fourier transformed into reciprocal space[17]. Further modellings through reshapings attempts of the $A3_{perp}$ functions do not result in any better agreement[17], as long as these $A3_{perp}$ hypervolumes are kept into the $R3_{perp}$ complementary space, without any components (or contribution) to physical space. Actually, that might be the critical point, as suggested by DSM analysis and a more careful scrutiny of the 6-dim experimental structure.

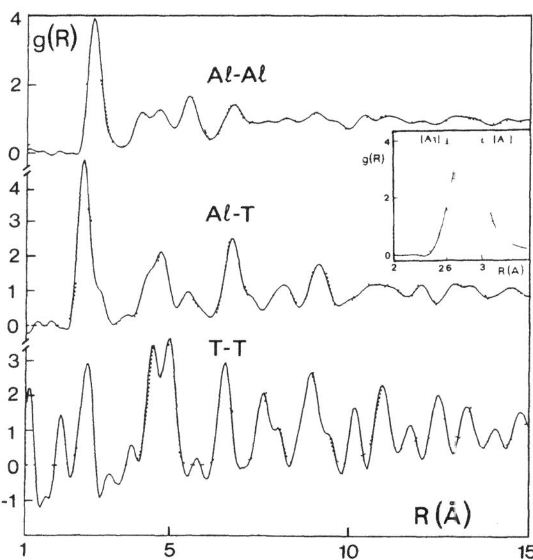

Fig.4 The partial pair distribution functions of the icosahedral $Al_{74}Si_5Mn_{21}$ phase : directly measured by DSM (—) and calculated from quasicrystallography approach (---). Details of the first Al-Al peaks are shown in the insert (weaker broadening of the calculated curve).

The PPDF shown in Figure 4, comparing DSM data with calculated curve obtained from the quasicrystallography approach, emphasize that beyond an overall qualitative agreement, there are significant discrepancies especially for the shortest pair distances.

In the quasi crystallography description the nearest neighbour Al-Al distance has in fact two main components : one at 2.59 Å comes from two Al_O sites on a three-fold axis, the other at 2.99 Å comes also from two Al_O sites but on a two-fold axis ; a third minor component at 2.85 Å is an Al_O-Al_{BC} pair. The Al-Al distribution obtained from DSM is more compact and centred on 2.82 Å. Thus, the actual structure seems to depart from the ideal average structure, the Al-Al shortest distances being expanded along three-fold axes and contracted along two-fold axes. This expansion-contraction modification is of the order of 0.2 Å. A similar trend is also visible on the Al-Mn PPDF which shows a drastic increase of the three-fold axis pairs (2.59 Å) to the disadvantage of the two-fold axis pairs (2.99 Å). The Mn-Mn pairs are less accurately determined and huge truncation oscillations have unfortunate screening consequences. However, a perfect Mn Penrose sublattice would give five-fold pairs at 4.6 Å (edges of the rhombohedra) and two-fold pairs at 4.85 Å (face short diagonals) ; the DSM experimental pair distances would rather be 4.5 and 4.99 Å, respectively. Some too short unphysical pair distances show up, as expected from spherical approximation for the $A3_{perp}$ volumes. A flatening around the five-fold axes, especially for the Al_O shell, suppresses easily these spurious distances.

In principle, theses changes in pair distances as introduced from examination of the PPDF scheme, can be interpreted in the frame of the 6-dim cut approach. This is illustrated in Figure 5 which shows a 2-dim rational cut of only a part of the 6-dim structure, namely the empty spherical shell $A3_{perp}(Al_O)$. Two-fold axes, one in our physical space and the other in complementary space, are represented. On the left hand side of the map, the structure is kept "perfect" and Al-Al distances are equal to L

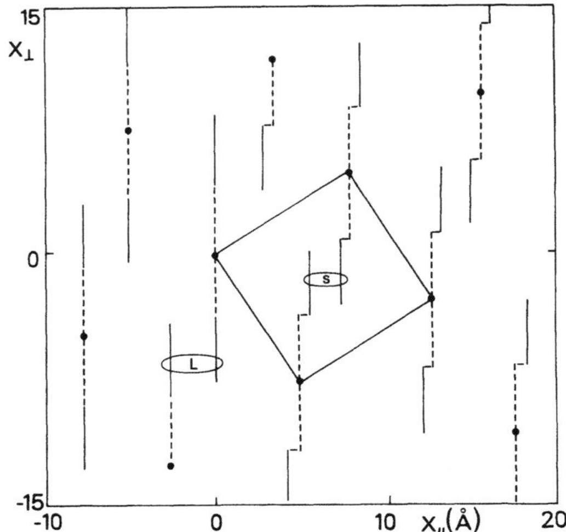

Fig.5 Illustration of simple parallel components which maintain icosahedral symmetry but modify atom distance in a particular direction
(here L becomes S along a two-fold axis).

along the physical two-fold axis. From the middle to the right hand side of
the figures, the $A3_{perp}(Al_0)$ have been given a small component parallel to
the physical two-fold axis, rightward for the top part and leftward for the
bottom part of $A3_{perp}$, with the result that now the Al-Al distances are
S < L. Another opposite modification of the $A3_{perp}(Al_0)$ can simultaneously
enlarge the Al-Al pair distances along the physical three-fold axes.
Actually one way to induce such parallel components into the Al-Al sublattice of the quasicrystal might be "digging" 30 spherical holes, centered
on the two-fold axes, into the $A3_{perp}(Al_0)$ atomic shell, and shifting
conveniently the corresponding density pieces by a r_{par} vector[20]. These
parallel components have to be only of the order of 0.2 Å if they are to be
consistent with the PPDF. They are going to influence mostly the rather

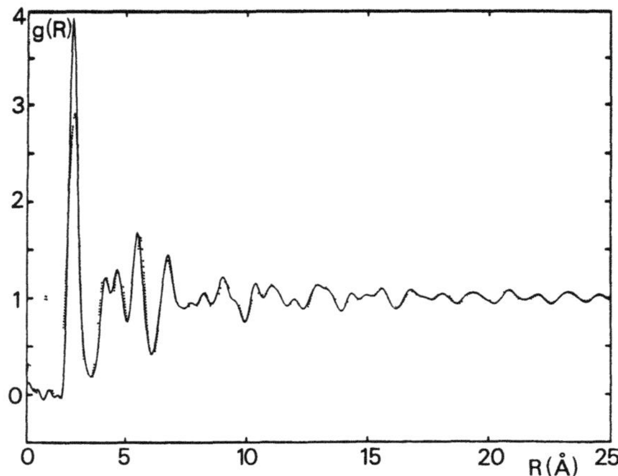

Fig.6 Influence of parallel components on calculated PPDF (crystallography
approach)(---) compared to DSM data. Note improvement with respect
to the curves shown in Figure 4.

16

weak diffraction peaks, as it can easily be realised by calculating
$G(Q_{perp})$, the FT of the modified $A3_{perp}$. But, as such, they result into
noticeable improvements of the agreement between DSM and "quasicrystallography" PPDF. This is illustrated in Figure 6, in the case of the Al-Al
partial. When scrutinized carefully, the 6-dim density maps obtained from
FT of the partial structure factors also show evidences for the existence
of these parallel components of the $A3_{perp}$ hypersurfaces. This is
illustrated in Figure 7 where a 2-dim cut of the experimental 6-dim
manganese substructure is compared to its calculated counterpart using a
basic spherical $A3_{perp}$ volume with a 0.5 Å parallel shift of 12 small
spheres centred on the five-fold axes[20] ; the antisymmetrical features
exhibited in both cases obviously obey to similar trends. Thus, the
structural information related to parallel components of the $A3_{perp}$ volumes
is actually contained into the 6-dim cut crystallography approach.
Unfortunately this information is somewhat smeared out by the disastrous
truncation effects, and to a less extent, by background cut-off. Noteworthy, parallel components into the $A3_{perp}$ volumes do not induce any
atomic disorder ; the icosahedral symmetry is still respected, diffraction
peaks are still sharp (Bragg-like) and diffuse scattering is not produced.
The wavy background which shows up in the base line of some of the diffraction pattern must have a different origin and is indeed the signature
of a true disorder. In that respect space fluctuations of the $A3_{perp}$
parallel components might contribute to this disorder. Finally, it is also
interesting to point out that the main structural consequence of the
parallel components as exemplified in the present paper, is to modify
(increase or decrease) the radii of the successive icosahedral atomic
shells[20].

Conclusively, it is already rather clear that a 3-dim description of
the structure in terms of a decorated 3DPT with 4.6 Å edge is over-
simplified and may be not very adapted anymore. The parallel components of
the $A3_{perp}$ volumes induce quite large distorsions of the rhombohedra. For
instance, the Al 3-dim subnetwork cut into the 6-dim structure contains
chains of corner shared tetrahedra. These tetrahedra, with their

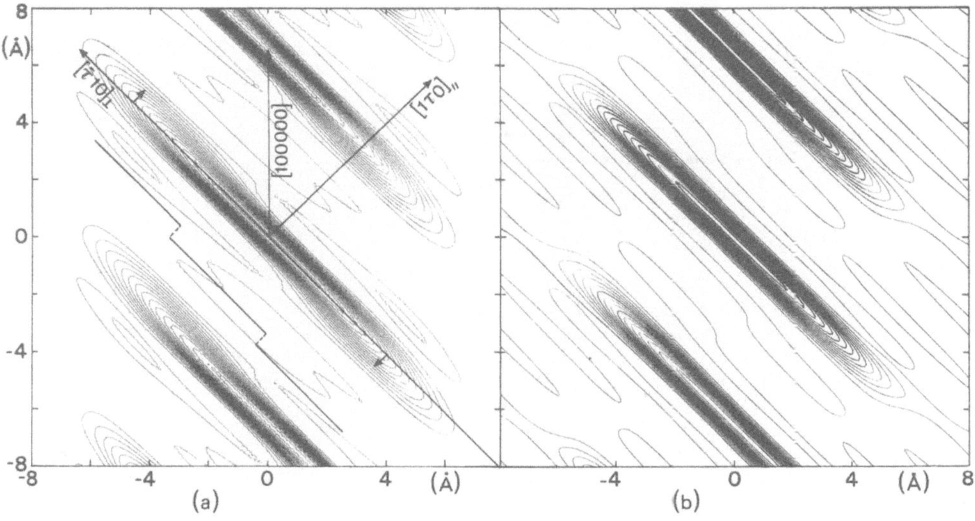

Fig.7 Fourier transforms of the F_{Mn} partial structure factor ; (a)
calculated with a $A3_{perp}$ volume based on a sphere with 12 smaller
spheres shifted along parallel five-fold axes[20] ; (b) as deduced
from diffraction data. The maps are 2-dim cut of the 6-dim structures containing five-fold axes (one $A5_{par}$ and one $A5_{perp}$).

incriminated 2.6 and 3 Å edges are not regular in the 3DPT description but relax toward regularity when the so-called parallel components are introduced, resulting also into a diameter contraction of the Al icosi-dodecahedra. Thus, a description of the structure in terms of statistical clusters, also derivable from a 6-dim cut scheme, might be more fruitful. Noteworthy, these parallel parts of the $A3_{perp}$ volumes are reminiscent of the commensurate displacive modulations, as introduced by Janssen[11] in order to stabilize 3DPT structures. Such displacive modulations have actually been observed when simulating structure relaxation[11] or growth process[21,22] for 2-dim quasi-crystals. Chemical modulations related to phason disorder can also be generated in a similar way by substituting a small fraction of a given $A3_{perp}$ with the same volume taken from another $A3_{perp}$ (for instance Al $\leftrightarrow$ Mn substitution). P. Bak[23] had also previously mentioned that detailed structures of quasicrystals might be difficult to be specified completely, due to complicated shapes of the $A3_{perp}$ volumes including components into the physical space. A complete modelling of the Al(Si)Mn structure is in progress.

T2-AlLiCu quasicrystals

The second major quasicrystalline system, namely AlLiCu alloys, might be the best candidate for X-rays and neutron studies. Actually, it has been the first quasicrystal to be grown into single grain approaching a millimeter across[24] and consequently raising the prospect for single crystal like investigations. On the other hand, Li and Cu have convenient isotopes for useful neutron experiments to be carried out and partial structure factors to be determined, as already done for the Al(Si)Mn system.

Single crystal like approaches have mostly illustrated that the major recorded reflections only exhibits 2-fold, 3-fold or 5-fold symmetries (for a more complete review see ref.[15]). Four circle single crystal diffraction scans, using both X-rays and neutrons, exhibits peaks which are mostly

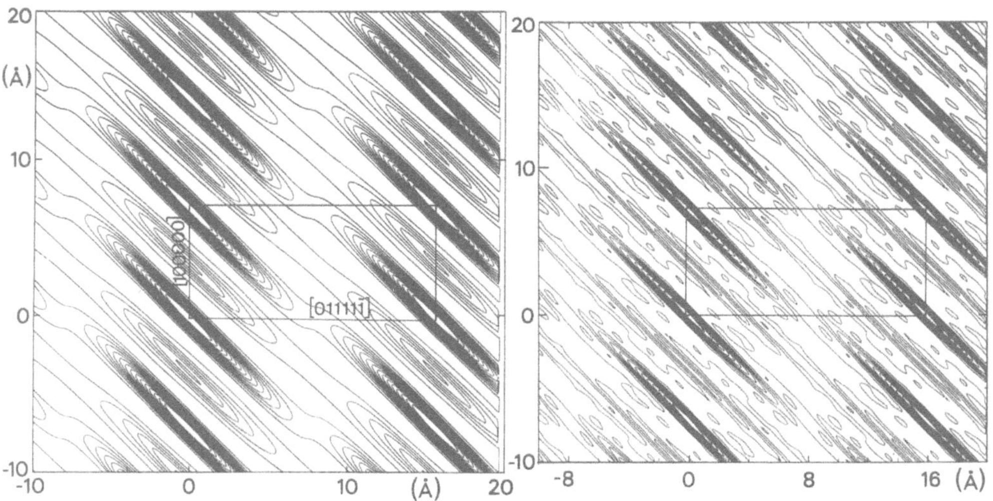

Fig.8 A 2-dim map of 6-dim Patterson functions calculated by FT of X-rays (a) and neutron (b) four-circle diffraction data. The BC features on the (b) map actually appear with a negative weight, due to the negative sign of the neutron scattering length of natural Li. Details are given in Figure 9.

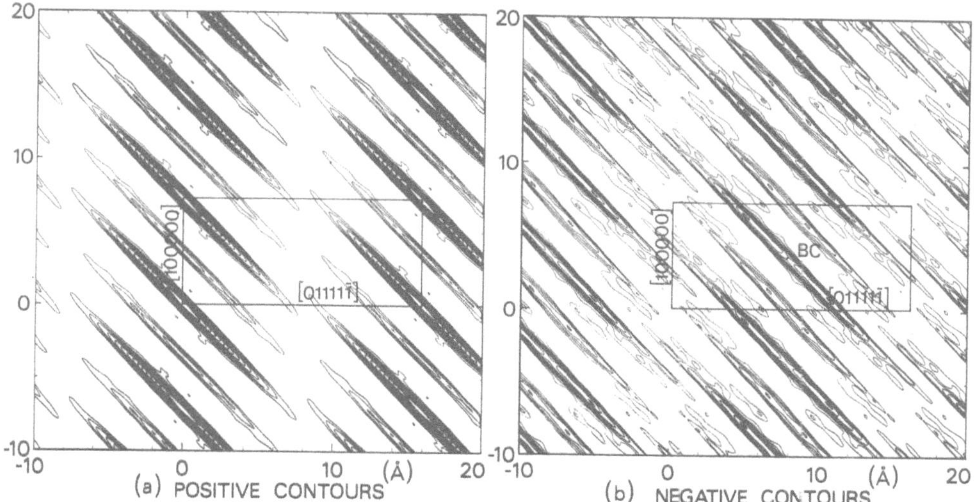

Fig.9 Five-fold axes maps of the Patterson function calculated with
4-circle neutron data. (a) reproduces the positive features (A-A and
Li-Li correlations) and (b) the negative features (A-Li correlations) of the map shown in Figure 8.

indexable with six indices that belong to a primitive icosahedral (6-dim
cubic) Bravais lattice. It is thus possible to use them for reconstructing
the corresponding 6-dim diffraction patterns and Fourier transform them
into their 6-dim Patterson functions. These Patterson functions are shown
in Figures 8 and 9, restricted of course to 2-dim maps. Single grain
measurements allow to account properly for possible degenerate reflections.
The weak point, at this stage, is that the chemical features of the structure cannot be analysed completely. Some insight into the 6-dim structure
can nonetheless be extracted from the comparison of the X-ray and neutron
Patterson function, due to the much higher relative cross-section of
lithium with neutron than with X-ray. Thus, it can be said that Al-Cu atoms
have origin and middle-edge sites into the 6-dim elementary cube while Li
atoms have body-centre sites. Again the $A3_{perp}$ volumes are roughly
spherical into the $R3_{perp}$ complementary space but with parallel components
as explained in the previous section.

Going further into the quasicrystallography analysis would require the
determination of the partial structure factors, or at least the partial
Patterson functions. This is actually in progress. So far, the isotopic
contrasts on Cu (2 isotopes with positive scattering length for neutrons)
and on Li (2 isotopes with opposite signs of their neutron scattering
lengths) have been used only in a DSM approach[15]. Using, first, alloys with
a mixture of $^6Li/^7Li$ isotopes in such a proportion that the lithium scattering cross-section becomes zero, and with natural Cu, ^{63}Cu and ^{65}Cu, it
has been possible to verify that Al/Cu order is very weak or inexistant.
This is illustrated in Figure 10 showing that the PDF are almost insensi-
tive to copper isotopic contrast. The differences are not significant
enough for expecting sensible PPDF calculations.

The next step has been to measure PDF with alloys prepared with
different $^6Li/^7Li$ mixtures. In principle, contrast changes in neutron
diffraction produced by the Li isotopic substitution would allow to
determine the A-A, A-Li and Li-Li partial correlations (PPDF)(with A
standing for the average Al-Cu atom). Actually the Li-Li partial is very
poorly weighted into the PDF and its calculation results in a function with
unphysical negative contributions. At least, A-A and A-Li partials have

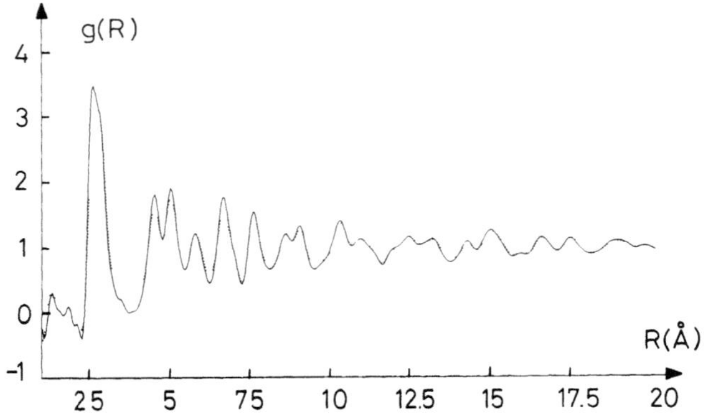

Fig.10 Measured PDF of the AlLiCu quasicrystal when prepared with ^{65}Cu (——)
and ^{63}Cu (---) isotopes and with a zero-scatterer ^{6}Li/^{7}Li mixture.

been obtained[15]. Their comparisons to their counterparts in the related bcc
R-phase (Figure 11) suggests that short range orders have common features
in both materials.

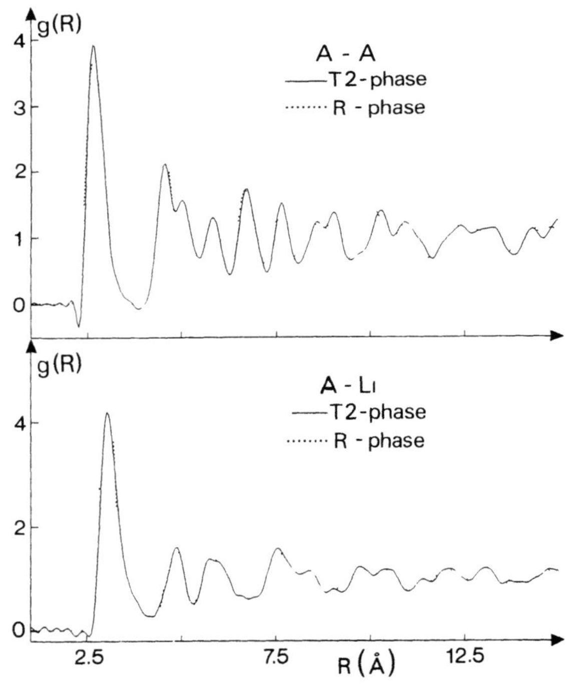

Fig.11 Measured PPDF of the T2 quasicrystal (full line) compared to those
calculated for the R-crystal (dashed line).

Using the R-phase structure as an approximant, tentative modelling of the T2-quasicrystal have been proposed. One of them [7] is a decoration procedure using a τ^3 inflated 3DPT and convenient clusters and shells of atoms taken out of the R-phase structure to generate atomic order and matching rules. Another one[25] is a phase reconstruction procedure using relations between the 6-dim structures of the R and T2-phases. Clearly, more definitive conclusions are expected from the current determination of the partial structure factors[26].

AlCuFe quasicrystals

The discovery by Tsai, Inoue and Masumoto[27] of a high quality new stable icosahedral phase in the ternary AlCuFe system has opened a wide variety of possible structural studies. The $Al_{65}Cu_{20}Fe_{15}$ alloy has been identified as a Face Centered icosahedral phase[28] issued from a chemical ordering of a primitive 6-dim hypercubic lattice. A typical aspect of this system is that the icosahedral phase forms by peritectic reaction from a monoclinic $Al_{13}Fe_4(Cu?)$ primary phase[29]. On the other hand quality and even basic structure seem to be strongly influenced by detail in the thermal history of the samples. Depending on annealing temperatures, cubic, monoclinic and icosahedral phases can nucleate ; their further stabilization are then related to the growth of phase domains up to minimum sizes[30]. When conveniently prepared, this quasicrystal shows very sharp peaks in neutron diffraction, with the existence of features revealing chemical ordering (Figure 12).

Some aspects of the structure are quite visible on the 6-dim Patterson function (PF) that can be calculated using these neutron diffraction data. A 2-dim map of this PF is shown in Figure 13, along with the density profile of the different $A3_{perp}$ volumes. A similar approach has been proposed by Gähler[31] using X-ray diffraction data.

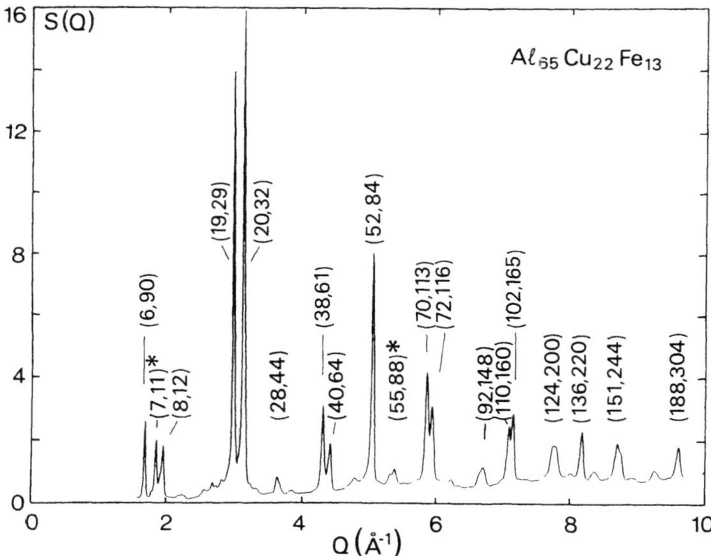

Fig.12 Interference function measured with an AlCuFe quasicrystal. Peaks labelled (*) reveal chemical order (data have actually been collected up to $Q \simeq 30$ Å^{-1}).

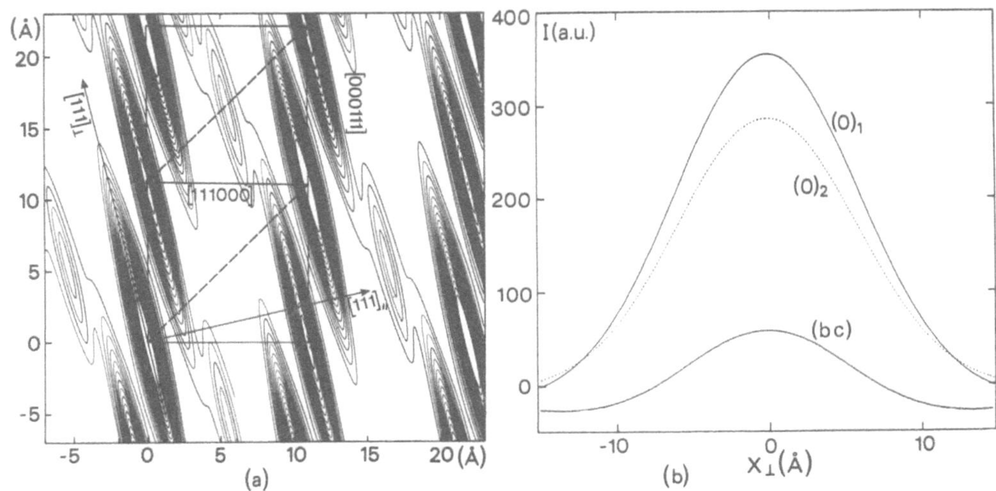

Fig.13 Patterson function (a) and density profiles of the $A3_{perp}$ volumes (b) obtained with neutron diffraction[29]. $(0)_1$ and $(0)_2$ refer to odd and even origin sites, respectively ; bc is the weakly occupied body-centre site.

The 6-dim cubic superstructure which is consistent with these data can be derived from the one proposed for the Al(Si)Mn quasicrystal, within the following changes :

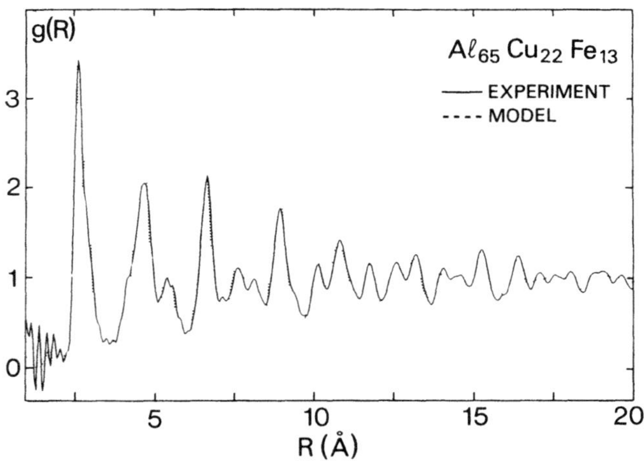

Fig.14 PDF of the AlCuFe quasicrystal as deduced from (FT) data shown in Figure 11 (—) compared to a calculated PDF as explained in text (---).

22

- the cubic unit cell parameter has to be taken twice as large and thus contains 2^6 elementary 6-dim cubes of the primitive structure,
- in this unit cell the $A3_{perp}$ (Al_{BC}) actually exist only on every other BC site,
- the odd "origin" sites are occupied by a spherical Cu core surrounded by an Al shell,
- the even "origin" sites have a small empty core surrounded by a narrow Fe shell and then an Al external shell.

The PDF calculated with this model structure is compared to the experimental one in the Figure 14. Possible "parallel component effects" on the shortest pair distances are less visible for this system than for AlSiMn. Long range order, site distribution and domain sizes are probably well organized in this AlCuFe quasicrystal. The complete specification of the structure directly from diffraction data might be however quite a messy business, even with the available contrast effects on Fe and Cu, due to the truly ternary aspect of the structure.

CONCLUSION

The 6-dim crystallography scheme is definitely a valuable tool for deriving the basic features of icosahedral quasicrystals. For the simple systems and with the use of contrast variation effects, average structures are obtained which can be refined through a last step modelling and comparison to partial pair distribution functions. The $A3_{perp}$ atomic volume in the 6-dim structure have to be somewhat generalized and exhibit components into the physical space. This makes the description of quasi-crystal structure on 3DPT basis less realistic than suggested earlier.

ACKNOWLEDGEMENTS

We gratefully acknowlege D. PRICE and J. RICHARDSON for their definitive contributions to our time-of-flight diffraction measurements. We are also grateful to the ILL and IPNS (Argonne National Laboratory) for generous allocation of neutron beam time.

REFERENCES

1. Ch. Janot and J.M. Dubois, J. Non Cryst. Sol. 106, 193 (1988).
2. M. Duneau and A. Katz, Phys. Rev. Lett. 54, 2688 (1985).
3. P. Bak, Phys. Rev. Lett. 54, 1517 (1985).
4. J.W. Cahn, D. Shechtman, and D. Gratias, J. Mater. Res. 1, 13 (1986)
5. M. Audier, and P. Guyot, in "Quasicrystalline Materials" p. 181 (1988), Ed. by Ch. Janot and J.M. Dubois (World Scientific, Singapore).
6. M. Audier, and P. Guyot, Philos. Mag. Lett. 58, 17 (1988)
7. P. Guyot, M. Audier and M. de Boissieu, Third International Meeting on Quasicrystals, Mexico, May 29th – June 2nd, 1989 – Preprint.
8. J.W. Cahn, D. Gratias and B. Mozer, J. Phys.(Paris) 49, 1225 (1988).
9. M. Duneau and C. Oguey, J. phys.(Paris) 50, 135 (1989)
10. Ch. Janot and J.M. Dubois, J. Phys. F: Metal Physics 18, 2303 (1988)
11. T. Janssen, Phys. Rep. 168, 55 (1988)
12. a) Ch. Janot, M. de Boissieu, J.M. Dubois and J. Pannetier, J. Phys.: Condensed Matter 1, 1029 (1989).
    b) Ch. Janot, J.M. Dubois, J. Pannetier, M. de Boissieu and R. Fruchart in : "Quasicrystalline Materials" p. 107, Ch. Janot and J.M. Dubois, Eds., World Scientific, Singapore (1988).
13. Ch. Janot, J. Pannetier, J.M. Dubois and M. de Boissieu, Phys. Rev. Lett. 62, 450 (1989).
14. J.M. Dubois and Ch. Janot, J. Phys.(Paris) 48, 1981 (1987).

15. M. de Boissieu, Ch. Janot, J.M. Dubois, M. Audier and B. Dubost, J. Phys.(Paris) $\underline{50}$, 1689 (1989).

16. M. de Boissieu, Ch. Janot and J.M. Dubois, Europhys. Lett. $\underline{7}$, 593 (1988).

17. Ch. Janot, M. de Boissieu and J.M. Dubois, Anniversary Adriatico Research Conference on Quasicrystals, Trieste, July 4-7, 1989 (to be published by World Scientific, Singapore).

18. J.M. Dubois, M. de Boissieu, Ch. Janot, B. Dubost, R. Fruchart and M. Audier, J. Non Cryst. Sol. $\underline{106}$, 217 (1988).

19. P. Guyot and M. Audier, Philos. Mag. B $\underline{52}$, L15 (1985) ; M. Audier and P. Guyot, Philos. Mag. B $\underline{53}$, L43 (1986).

20. M. Duneau, private communication.

21. F. Nori, M. Ronchetti and V. Elser, Phys. Rev. lett. $\underline{61}$, 2774 (1988).

22. P.W. Leung, C.L. Henley and G.V. Chester, Phys. Rev. B $\underline{39}$, 446 (1989).

23. P. Bak, Scripta Met. $\underline{20}$, 1199 (1986).

24. B. Dubost, J.M. Lang, M. Tanaka, P. Sainfort and M. Audier, Nature $\underline{324}$, 48 (1986).

25. M. Jaric, Anniversary Adriatico Research Conference on Quasicrystals, Trieste, July 4-7, 1989 (to be published by World Scientific, Singapore).

26. M. de Boissieu, M. Audier, Ch. Janot, J.M. Dubois, P. Guyot and B. Dubost, Ibidem.

27. A.P. Tsai, A. Inoue and T. Masumoto, Japan J. Appl. Phys. $\underline{26}$, L1505 (1987)

28. S. Ebalard and F. Spaepen, J. Mater. Res. $\underline{4}$(1), 39 (1989).

29. C. Dong, M. de Boissieu, J.M. Dubois, J. Pannetier and Ch. Janot, J. Mater. Sc. Lett. $\underline{8}$, 827 (1989).

30. Ch. Janot, J.M. Richardson Jr., M. de Boissieu, D. Price and J.M. Dubois (to be published).

31. F. Gähler, 3ème Colloque Français sur les Quasicristaux, Nancy, 16-17 mars 1989 (E3 Abstract Booklet, J.M. Dubois, Ed.).

DETERMINATION OF QUASI–CRYSTAL STRUCTURES BY HIGHER DIMENSIONAL ANALYSIS

Walter Steurer

Institut für Kristallographie und Mineralogie
der Universität München, Theresienstraße 41
D–8000 München 2, F.R.G.

INTRODUCTION

Since the detection of the first quasi–crystalline phase in the system Al–Mn by Shechtman, Blech, Gratias & Cahn (1984) and independently by Ye, Wang & Kuo (1985) and Zhang, Ye & Kuo (1985) several hundred papers have been published presenting structural information about such phases. But, to date no really *quantitative* structure refinement, with a quality comparable to that of a conventional structure determination, has been performed. Table 1 gives a short list of more or less quantitative analyses of quasi–crystal structures (QCS) known to the author. For a review of the state–of–art of QCS analysis confer Steurer (1989b).

For the other large class of aperiodic crystals, the incommensurately modulated structures (IMS), there exist well established structure determination methods and refinement programs (c.f. Yamamoto, 1982). A few techniques, originally developed for IMS, as embedding in the $\mathbb{R}_n$ (cf. Janssen & Janner, 1987) or the nd Patterson analysis (Steurer, 1987) have become important tools for the analysis of QCS with diffraction techniques, too. Due to the close resemblance of IMS and QCS, the problems of structure determination of quasi–crystals will be discussed in close connection to similar problems of IMS, in the following.

What is now the special problem with quasi–crystal structures? A short comparison of the characteristics of IMS and QCS is made in Table 2. Both types of aperiodic crystals show diffraction patterns with sharp Bragg reflections (which may be accompanied by diffuse scattering), but only in the case of IMS there is an

Table 1. Some reliable refinements (rf), Patterson (Pa) and Fourier analyses (Fa) of quasi–crystal structures from powder (pd) or single crystal data (cd) on the basis of the 3d or nd description, respectively.

Year	Reference	phase	Method
1988	Gratias, Cahn & Mozer	$i - Al_{73}Mn_{21}Si_6$	6d pd Pa
	Cahn, Gratias & Mozer	$i - Al_{73}Mn_{21}Si_6$	6d pd rf
	Elswijk, deHosson et al.	$i - Al_6CuLi_3$	3d cd rf
1989	Janot, deBoissieu et al.	$i - Al_{74}Mn_{21}Si_5$	6d pd Fa
	Steurer	$d - Al_{78}Mn_{22}$	5d cd Pa

Geometry and Thermodynamics
Edited by J.-C. Tolédano
Plenum Press, New York, 1990

outstanding subset of reflections: the main reflections. From this subset the so called average structure can easily be determined with classical methods. Since most of the diffracted intensity is concentrated in the main reflections the actual IMS, calculated from all reflections, represents a relatively small deviation from this average structure. Thus, for small amplitudes of the modulation waves, the structure factor is nearly linearly dependent on the atomic shifts and a 'trial – and – error' structure refinement will often succeed.

Another consequence of the existence of an average structure is that IMS exhibit crystallographic symmetry only. In the 3d space an IMS can be characterized by its conventional periodic basic structure and the shape, amplitude and phase of its modulation function. The lack of main reflections and of an average structure for QCS allows non – crystallographic symmetry and renders a structure determination much more difficult. For a QCS description with a non – isometric nd unit cell special cuts of the nd supercrystal always give IMS (with their corresponding average structure) and the application of IMS determination methods becomes possible. This only makes sense, however, in the case of 1d quasicrystalline phases. In all other cases the main scattering power would be concentrated in the satellite reflections and the IMS would represent a large deviation from the basic structure.

For a long time, a more practical but nevertheless very important drag on successful determinations of QCS has been the difficulty to grow crystals large enough (about 0.1 mm diameter) for single – crystal X – ray diffraction studies. To

Table 2. Characteristics of aperiodic crystals

Incommensurately modulated structures (IMS)	Quasi – crystal structures (QCS)
Diffraction patterns with sharp Bragg reflections	
Main and satellite reflections $\Rightarrow$ average structure	One kind of reflections only $\Rightarrow$ no average structure
Crystallographic point symmetry	Non – crystallographic point symmetry ($m\overline{3}\overline{5}$, 10/mmm, ...)
Translational symmetry neither in direct nor in reci – procal space $\mathbb{R}_3$ (external space)	
Embedding the reciprocal quasilattice of rank $n > 3$ in the $\mathbb{R}_n$ (configurational space) leads to a periodic nd structure	
Simple nd unit cell with nd atoms which can be described in the 3d external subspace as usual (spheres) and in the orthogonal $(n-3)$d internal subspace in the following way:	
Connected $(n-3)$d hypersurfaces with the shape of the modulation function (sinusoidal, e.g.)	Disconnected $(n-3)$d hypersur – faces (bars, pentagons, triacon – tahedra, ...)
The real 3d aperiodic structure is obtained by cutting the nd crystal with the external (physical) space $\mathbb{R}_3$. Each point of the $(n-3)$d hypersurfaces in the internal space cor – responds to one atom in the physical space in an unique way.	
	special cuts of non – isometric nd crystal structures always give IMS

the author's knowledge appropriate single crystals have been available for the icosahedral (i) phase of Al – Cu – Li (e.g. Elswijk, de Hosson, van Smaalen & de Boer, 1988) and the decagonal (d) phase of Al – Mn (Steurer & Mayer, 1989), only. The reason is that most of the known quasi – crystals are metastable and accessible by rapid solidification methods, only. By conventional techniques the following icosahedral phases can be prepared: Al – Cu – Li (Saintfort & Dubost, 1986), crystallizing in rhombic triacontahedra, Al – Cu – Fe (Tsai, Inoue & Masumoto, 1988a) and Ga – Mg – Zn (Ohashi & Spaepen, 1987), forming pentagonal dodecahedra, Al – Cu – Ru and Al – Cu – Os (Tsai, Inoue & Masumoto, 1988b).

STRUCTURE ANALYSIS

The Aim

The aim of a complete analysis of a regular 3d periodic crystal structure is its full characterization by symmetry (space group), metrics (lattice parameters), atomic coordinates and thermal parameters, and the kind and degree of disorder present.

Aperiodic crystal structures may equivalently be described in two ways: Either in the $\mathbb{R}_3$ by a quasilattice decorated in a position dependent way or in the appropriate $\mathbb{R}_n$ by a periodic nd structure, respectively. In the first case the quasilattice (one out of an infinite number of possible ones) and the way the atoms occupy the quasilattice sites have to be given. In the second case the aperiodic structure can be represented by symmetry, metrics and content of the nd unit cell in analogy to regular structures. Additionally, however, essential structural information is contained in the shape of the nd atomic hypersurfaces. Thus, for the example of IMS, a nd atom may be sinusoidally shaped in the extradimension. In the case of QCS, the atoms may, for ideal quasi – crystals, look like bars (Fibonacci – sequence), pentagons (2d Penrose quasilattice) or triacontahedra (3d Penrose quasilattice) parallel to the internal space.

One important difference to regular structures is that the reflection intensities do not decrease isotropically in the nd reciprocal space with increasing length of the nd diffraction vectors $\mathbf{H} = \mathbf{H_E} + \mathbf{H_I}$. They fall off much heavier as a function of $|\mathbf{H_I}|$ (component parallel to the internal subspace) than of $|\mathbf{H_E}|$ (component parallel to the external subspace). This is not only a result of the 'geometrical' atomic form (or shape) factor, but also of a 'temperature' factor T_I parallel to the internal space. For IMS it can be interpreted to arise from (statical or dynamical) fluctuations of the phase of the modulation wave. It leads to a weakening of higher order satellites and the generation of diffuse scattering. Thus, in most cases first order satellite reflections can be observed and harmonic modulation functions refined, only. In the case of QCS, the 'temperature' factor T_I decribes additional atoms at interstitial sites of the 3d QCS indicating a kind of substitutional modulation or disorder, respectively. Displacive disorder, however, corresponds to an increased value of the external temperature factor T_E.

In both cases, space – and time – averaged nd atomic hypersurfaces can be derived from the observed Bragg reflections, only. This means, that resulting structure models have a much more statistical character than those for regular structures. A characteristic example for this fact is that for most anti – phase domain structures first order satellites can be observed only, corresponding to a sinusoidal instead to a square – wave modulation. This is the consequence of an averaging of domains of fluctuating size (Fujiwara, 1957). To extract the structural characteristics of a QCS, it may be necessary, therefore, to combine results of the different techniques which are listed in the following. Especially those are valuable for a vivid description which give average information in the 3d physical space (partial pair distribution functions, e.g.).

Local Isomorphism Classes

There exists an infinite number of different quasilattices which belong to different local isomorphism (LI) classes. If two infinite quasilattices are not super-posable but any finite region belonging to one quasilattice can be found in the other, then the quasilattices belong to the same LI class, otherwise to different ones (Socolar & Steinhardt, 1986; Levine & Steinhardt, 1986).

The diffraction patterns of vertex decorated quasilattices, e.g., only differ if they belong to different LI classes. The Bragg reflections occupy the same reciprocal lattice positions for any LI class, the intensities change weakly. A comparison of the diffraction pattern calculated for the most extreme cases of vertex decorated 2d general Penrose quasilattices with $\gamma = 0$ and $\gamma = 1/2$ can be found in Ishihara & Yamamoto (1988). This low sensitivity of intensities to structural changes of this kind has to be taken into account whenever the correctness of a 3d structure model is checked by a *qualitative* comparison of a diffraction pattern calculated on the basis of the kinematical theory and a electron diffraction pattern with its dynamical diffraction phenomena.

One big problem connected with all 3d analyses of QCS which has been neglected often is: not only the *decoration* of a given quasilattice has to be refined, but also the correct *quasilattice* has to be selected out of an infinite number of possible ones. Fully arbitrarily, the original Penrose tiling has always been used as quasilattice in the structure models suggested so far. In the case of the decagonal phase of $Al - Mn$ this assumption has been confirmed by the results of the 5d Patterson analysis, however (Steurer, 1989a). All these problems with the 3d decoration method, which become even more complicated in the case of disorderd QCS, can be overcome by the embedding method. This approach does not separate the generation of a quasilattice, its position dependent decoration with atoms and possible disorder effects.

It should be kept in mind that, contrary to the nd supercrystal, the 3d quasi-crystal exhibits neither rotational nor inversion symmetry in the direct space (though there exists such a point symmetry in the diffraction pattern). Generally, orientational order of the building units does appear, only. Exceptions are some special LI classes like the exceptional singular Penrose tiling.

Truncation effects

Truncation effects are a general problem for all structure analyses since they all base on inherently incomplete reflection data sets. Even if a subset with $\theta < 30^{\circ}$ (the usual range for a conventional single crystal data collection with $MoK\alpha$ X-ray radiation) is cut out of the hypothetically infinite accurate data set, series termination effects in Patterson and Fourier syntheses will lead to riples and smearing of the maxima. A practical example for the influence of data truncation on the 5d Patterson analysis of $d - Al_{78}Mn_{22}$ (Steurer, 1989a) is illustrated in Fig. 1. Maps have been calculated with one 'complete' and two truncated single crystal X-ray data sets.

Nearly all quasi-crystalline phases show a characteristic increase of the half-widths of high-$|H_I|$ reflections due to phason-strain. Thus, even stronger reflections may become unoberservable due to their diffuseness, especially in powder-diffractograms. Remarkable series termination effects in Fourier and Patterson maps are the consequence, but the results of a structure refinement may be biased less. The only way, therefore, to obtain atomic parameters with high accuracy is to perform a least-squares structure refinement. Supposition is that the 'real' shape of the atoms, which results from an averaging over the statistical time- and space-dependent deviations of the equivalent atoms from their equilibrium positions can be fitted by appropriate parameters (temperature factors, e.g.) starting from an idealized shape. The probability density function (PDF) of an ideal atom (the Fourier

transform of the temperature factors), decribing such an average atom, can be
calculated with the same accuracy and without series termination errors. Even very
complicated PDFs may be described using anharmonic temperature factors (cf.Zucker
& Schulz, 1982a,b).

Analogously, one may carry out the same procedure for the QCS determination. The structure refinement starts from ideal nd atoms (bars, pentagons, triacontahedra, ...) with chemical structure, and allows for symmetry adopted shape
variations by fitting a kind of anharmonic 'temperature' factors. After Fourier
transforming them to the nd PDF, the 3d PDF results along the section with the
physical space. Acting in this way, the spurious atoms generated using the nd
Fourier synthesis can be avoided.

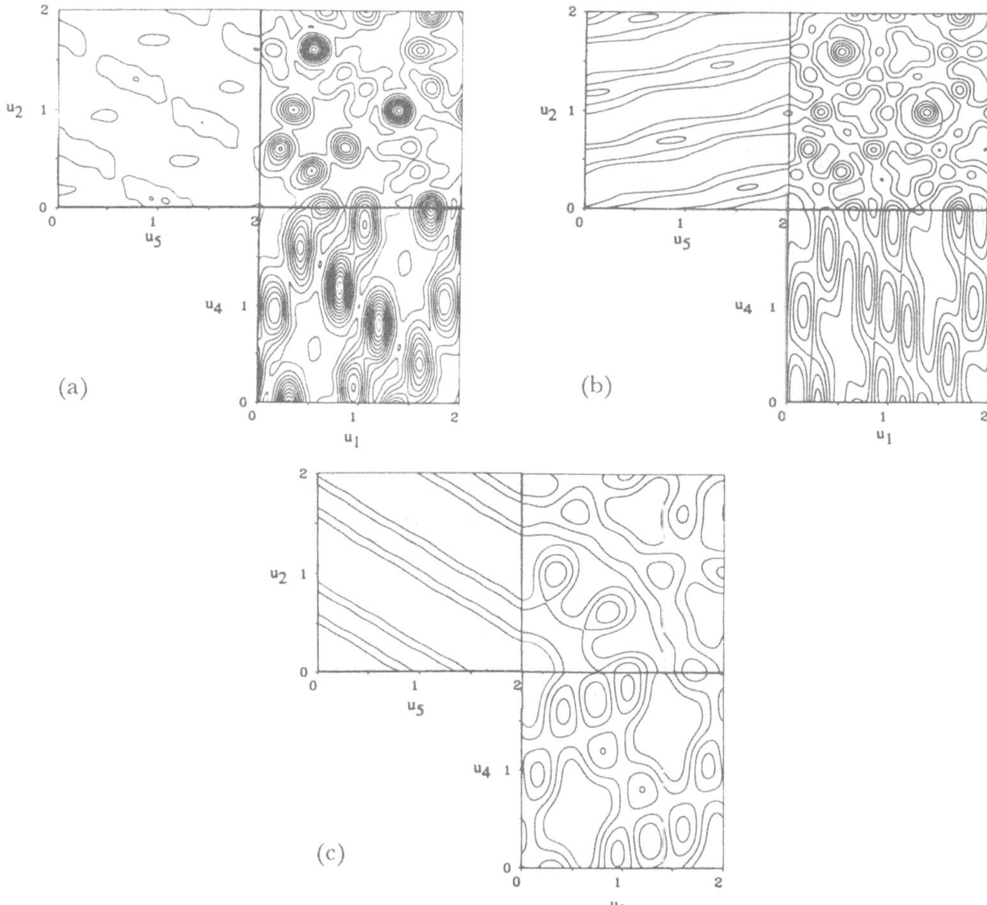

Fig. 1. Sections of 5d Patterson functions calculated from a single – crystal X – ray
data set of d – $Al_{78}Mn_{22}$ for $u_3 = 1/2$. The physical space section (11000) is
plotted (upper right) in combination with the sections (10010) (lower right)
and (01001) (upper left) including one external and one internal coordinate,
respectively. (a) shows the map calculated from the complete data set
$(0 < |H_E| < 1.4$ and $0 < |H_I| < 2.0)$, (b) for a truncation of the internal
$(0 < |H_I| < 0.25)$ and (c) of the external $(0 < |H_E| < 0.5)$ region of the reciprocal space. In (b) additional maxima appear in the (11000) section, and in
(c) peaks become broader or disappear.

Experimental techniques

Diffraction methods. For crystals which are well ordered over large volumes diffraction techniques are the methods of choice. The most powerful and best developed tools for quantitative structure determination are available for the interpretation of single – crystal X – ray data. Similar possibilities are offered by neutron – diffraction experiments which, additionally, allow the study of the dynamical properties of QCS by inelastic scattering (Suck, 1988). The main disadvantage is that millimeter – sized crystals are needed. X – ray and neutron powder diffraction methods have been used for the investigation of i – Al – Mn – Si (Gratias, Cahn & Mozer, 1988), e.g., but due to their limited resolution they are not optimal for quantitative structure determination of aperiodic crystals.

Electron diffraction allows a very quick imaging of zero layer reciprocal lattice planes from very small crystalline areas and is todays most used method for the investigation of QCS. Dynamical diffraction effects increase the intensities of weak satellite or quasi – crystal reflections and of the diffuse scattering. On the one hand this facilitates the detection of structural modulations and disorder but, on the other it makes quantitative calculations extremely difficult. Selected area electron diffraction (SAED) produces diffraction patterns from regions of about μm^2 whereas convergent beam electron diffraction (CBED) gives information from much smaller areas, and allows the determination of the point (and not only the Laue) symmetry.

Electron microscopy. Very impressive images of the characteristic features of IMS and QCS can be obtained by high resolution transmission electron microscopy (HRTEM). As examples may only be cited the studies of the IMS of labradorite by Nakajima, Morimoto & Kitamura (1977), of i – Al – Mn – Si by Li & Liu (1986) and of d – Al – Mn by Hiraga, Hirabayashi, Inoue & Masumoto (1987). Important for the study of of the real structure is the ability of HRTEM to image non – periodic features as crystal – defects, e.g.. HRTEM images always reflect projected structures. They do not allow the quantitative determination of atomic parameters.

Spectroscopy. Spectroscopical techniques have been used by many groups to elucidate the local structure of QCS, since they have the advantage to work with powder samples. Extended X – ray absorption fine structure (EXAFS) measurements of Ma & Stern (1988), e.g., yielded Mn – Al radial distribution functions for the cubic, icosahedral and decagonal phases in the system Al – Mn – Si. Mössbauer effect (ME) studies have been carried out on QCS containing Fe atoms by Koopmans, Schurer, van der Woude & Bronsveld (1987), and nuclear magnetic resonance (NMR) investigations have been performed for i – Al_6CuLi_3 in comparison to the cubic R phase of this system by Lee, White, Suits, Bancel & Heiney (1988), to quote only a few papers.

The structure determination methods

The solution of a crystal structure with diffraction methods is equivalent to the reconstruction of the phases of the structure factors F(H) which cannot be derived directly from the observed intensities I(H). The relationships between intensities, structure factors, Patterson function P(U) and electron density distribution $\varrho(R)$ is shown schematically in the following (the one – way connections are marked by single arrows)

$$I(H) \leftrightarrow F(H)F^*(H) \leftarrow F(H)$$
$$P(U) \leftarrow \varrho (R)$$

'Trial – and – Error' – method. The 'trial – and – error' – method has been the most used technique for the determination of IMS and QCS due to the lack of powerful direct methods for aperiodic structures. It works quite well for those IMS which have atoms modulated in phase approximately. In those cases it is possible to start the refinement with the known average structure and arbitrary small values for the amplitudes of the modulation functions, no model for the modulation is necessary. In the case of QCS there exists no average structure and therefore one needs a structural model to start a structure refinement. In most cases this has been a decorated 2d (Yamamoto & Ishihara, 1988, e.g.) or 3d Penrose quasilattice (Elswijk, de Hosson, van Smaalen & de Boer, 1988, e.g.). In the 3d description the type of decoration and quasilattice is fixed during the refinement, in the nd approach both the information about the quasilattice and the decoration is included in the shape of the nd atoms and, consequently, also can be refined.

Direct methods. This most popular approach for the solution of the phase problem has not been applied practically to IMS and QCS yet. Relationships between normalized structure factors are used to derive the phases. An important equation has been derived by Sayre (1952)

$$F(\mathbf{H}) = \Theta(\mathbf{H})/V \sum_{\mathbf{K}} F(\mathbf{K})F(\mathbf{H} - \mathbf{K})$$

$\Theta(\mathbf{H}) = f(\mathbf{H})/g(\mathbf{H})$
$f(\mathbf{H})$...atomic scattering factor (a.s.f.)
$g(\mathbf{H})$...a.s.f. for squared atoms

It is strictly valid for atoms which are identical, spherical and non – overlapping, only. It is not restricted by the assumption of positive electron densities. Hao, Liu and Fan (1987) proposed a modified Sayre's equation for IMS connecting the structure factors of satellites $F_s(\mathbf{H})$ with those of main reflections $F_m(\mathbf{H})$

$$F_s(\mathbf{H}) = 2\Theta(\mathbf{H})/V \sum_{\mathbf{K}} F_m(\mathbf{K})F_s(\mathbf{H} - \mathbf{K})$$

Using this equation, the authors determined successfully the phases of a centrosymmetric modulated test structure ($\gamma - Na_2CO_3$) which has nearly equally phased modulation functions with large similar amplitudes (easily to solve by 'trial – and – error – methods, too). The main problem with this method, always will be the calculation of $\Theta(\mathbf{H})$, which is a function of the a.s.f. and of the amplitudes of the modulation waves, especially for IMS with very different individual atomic modulation functions.

Li, Wang & Fan (1987) proposed the employment of an other modified Sayre's equation for QCS in the form

$$F(\mathbf{H})/S(\mathbf{H_I}) = \Theta(\mathbf{H})/V \sum_{\mathbf{K},\mathbf{K_I}} F(\mathbf{K})F(\mathbf{H} - \mathbf{K}) \,/[S(\mathbf{K_I})S(\mathbf{H_I} - \mathbf{K_I})]$$

The main problem with this approach is the calculation of $\Theta(\mathbf{H})$ and of $S(\mathbf{H_I})$. The latter is a kind of shape function of the nd atom. Consequently, the shape of the nd hypersurfaces and therewith the most important structural information about the QCS had to be known in advance before getting started. Experimentally, this method has not been tested by the authors yet.

Contrast variation. This method is the first quantitative phase determination for QCS followed by 3d and nd Fourier syntheses, respectively. Janot, de Boissieu, Dubois & Pannetier (1989) performed neutron powder diffraction experiments using isomorphous substitution for the Mn atoms. The intensities are split in the contributions

$$I(\mathbf{H}) = b_{Al}^2 |F_{Al}|^2 + b_T^2 |F_T|^2 + 2b_{Al}b_T |F_{Al}| |F_T| \cos\Delta\varphi$$

b...neutron scattering length

T...transition metal (Fe,Mn)

$$F_M = \sum_M \exp(iH.r_M)...\text{partial structure factor of atom M}$$

$\Delta\varphi$...phase difference between F_{Al} and F_T

The unknown quantities $|F_{Al}|$, $|F_T|$ and $|\Delta\varphi|$ can be determined for each reflection H by measuring at least three independent intensities $I(H,b_T)$. For centrosymmetric structures the $|\Delta\varphi|$ all are 0 or π. After the phase reconstruction procedure the partial structure factors are Fourier transformed to get the partial 3d or nd electron density distribution functions.

The main problem using this method of contrast variation with n-eutron – diffraction is that powder – data with their limited resolution (especially parallel to the internal space) have been used only. An other problem may be to find suitable elements for the isomorphic substitution. An X – ray single crystal λ – technique may probably overcome these problems.

A first step into this direction represents the X – ray powder – diffraction study of Bouchet – Fabre, Laridjani, Chenoufi & Dixmier (1988). At three different wave – lengths the diffractograms of i – Al – Cu – Fe alloys have been measured and partial pair distribution function evaluated. Principally, this technique can be used for phase determination as demonstrated above for neutron – diffraction.

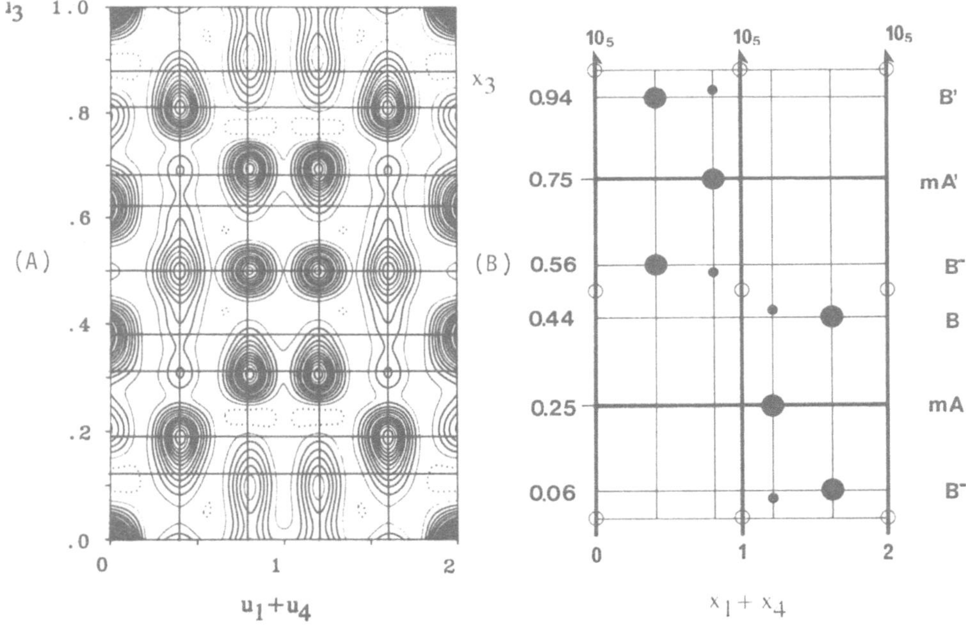

Fig. 2 (a) 5d – PF of d – Al – Mn showing the section parallel to (10110). The direction [10010] on the V – basis corresponds to the direction [11110] on the D – basis whereon the structure elements of the Penrose tiling are centered. (b) A model for the characteristic section of the 5d – unit cell of d – Al – Mn has been derived from the 5d – PF. The large/small circles are responsible for the large/small Patterson peaks. The symmetry elements in this plane are marked (o inversion center, | 10_5 axis and m reflection plane). The layers are denoted by A and B. The action of the screw axis is marked by the superscript ' and of the mirror plane by – .

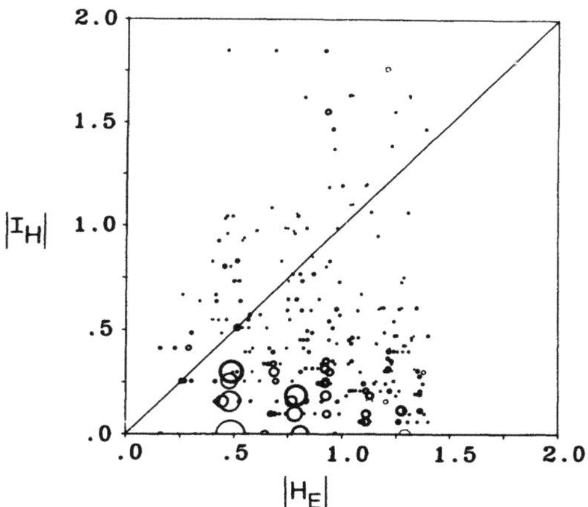

Fig. 3 Plot of the external component $|H_E|$ of the diffraction vector versus the internal one $|H_I|$ for all $|F| > 3\sigma(F)$. The radii of the circles are proportional to the $|F|$'s, $Q = 2\pi H$.

Patterson method. An approach for a model – free determination of aperiodic structures is the n – dimensional Patterson analysis first investigated by Steurer (1987). It works quite well even for more complicated IMS as the e – plagioclases (Steurer & Jagodzinski, 1988) and has also successfully been applied to QCS as i – Al – Mn – Si (Gratias, Cahn & Mozer, 1988) or d – Al – Mn (Steurer, 1989a). It is the extension of the well known 3d Patterson synthesis to higher dimensions. Consequently, the nd Patterson function (nd PF) corresponds to the weighted vector diagram between all nd atoms of the nd unit cell. The component of the nd PF in the 3d external space consists of the vector map between 'normal', more or less spherical atoms. The (n – 3d) internal component corresponds to a vector function between continous infinite atoms in the case of IMS, and between discrete extended elements (bars, pentagons, triacontahedra, ...) for QCS. The nd PF is inherently centrosymmetric and its unit cell is that of the nd structure. Due to the small number of atoms in the nd unit cell of QCS the nd Patterson map is much easier to interpret then the aperiodic 3d one.

What can we learn from such a higher dimensional PF? First, we get information about the occupation of the equipoints in the unit cell and can compare it with mathematical models (2d or 3d Penrose quasilattices, e.g.). Easily it can be recognized, whether the vertices only of a quasilattice are decorated or not since in that case additional sites in the nd unit cell would be occupied or the atoms would be enlarged parallel to the internal space V_I. Second, we get an idea about the amount of displacive disorder in both the external and the internal spaces studying the half – widths of the Patterson maxima in both spaces. Third, at least in some special cases, the LI class can be determined because the size, shape and distribution of the nd atoms in the nd unit cell depend on the LI class.

A practical example calculated from single – crystal X – ray diffraction data of d – Al – Mn are shown in Fig. 1 and 2. The main advantage of the nd PF is to use experimentally accessible diffraction intensities only for Fourier coefficients

$$P(U) = 1/V \sum_{H} I(H)\cos 2\pi H.U$$

U...vector in the nd Patterson space

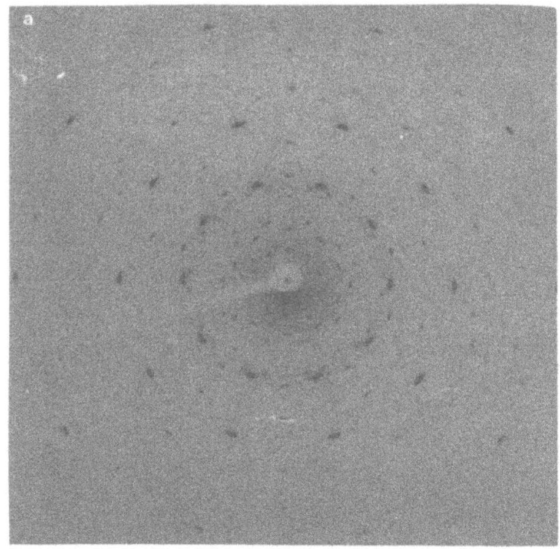
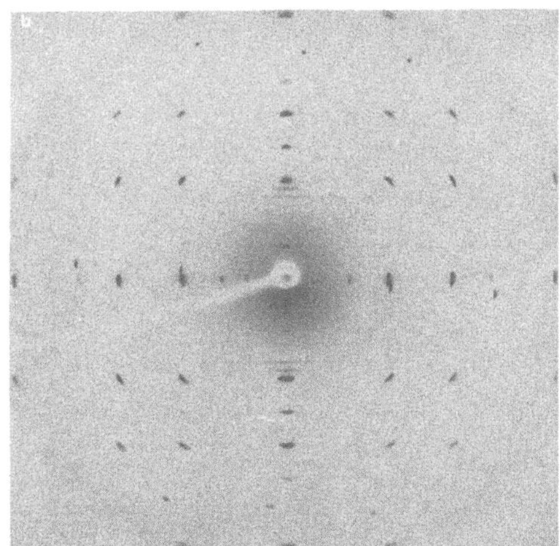

Fig 4 Monochromatic X – ray precession photographs of decagonal Al – Mn (a) Quasiperiodic reciprocal lattice plane, and (b) the one containing the tenfold axis The diffuse scattering is concentrated in layers perpendicular to the tenfold axis

Example:decagonal Al – Mn

A complete single – crystal X – ray data – set has been taken for a single crystal of composition $Al_{78(2)}Mn_{22(2)}$ and size 0.25x0.13x0.08 mm^3. Intensities could be oberved up to high internal components of the diffraction vectors (Fig. 3). The monochromatic X – ray diffraction patterns indicate a high degree of order, nevertheless the diffuse scatering concentrated in layers perpendicular to the tenfold axis must not be neglected (Fig. 4). The sharpness of the diffuse scattering parallel to the tenfold axis can be interpreted by long range ordered chain – like domains in that direction. Perpendicular to them, only short range order between them can be derived from the single maximum of the diffuse intensity distribution in the layers of the reciprocal lattice.

The results of the 5d Patterson analysis are: the structure consists of six non – equidistantly slightly puckered layers two of them in the asymmetric unit. Each layer represents a different decorated partial Penrose pattern. From the absence of a Patterson maximum in Fig. 2 it is possible to derive that the LI class is that of the original Penrose tiling. Disorder is present rather perpendicular to the layers than within. The exact shape of the atoms and corresponding therewith the correct structure of d – Al – Mn in the physical space cannot be obtained from the 5d Patterson synthesis That is the task of a 5d structure refinement.

The structures of the d – and the μ – phase seem to be closely related. μ – $MnAl_4$, spacegroup $P6_3$/mmc (Shoemaker, Keszler & Shoemaker,1989), has a translation period along the sixfold screw axis which corresponds to twice the translation period of the d – phase or directly to that of the diffuse superstructure, respectively. The arrangement of Patterson peaks in layers, leading to slightly puckered atomic layers, is comparable in both phases, too. μ – $MnAl_4$ has no complete MI in the structure.

Concluding remarks

Quasi – crystals do have lattices comparable to those which can be constructed by the strip – projection, or more general, the higher – dimensional embedding method. Special examples of such quasilattices are the 2d and 3d Penrose tilings. This is one of the few structural features known about quasi – crystal structures. An other one is that the basic structural building groups are all but complete Mackay icosahedra.

The results of last five years' structure determination, mostly performed on i – Al – Mn(– Si) and i – Al – Cu – Li do not allow the construction of ultimate structure models from the information scattered over hundreds of publications. To date, not a single quantitative analysis of a QCS, comparable in its quality to the determination of a regular crystal structure, has been published. The first steps on this way have been done by the application of higher dimensional Patterson methods and contrast variation techniques (isomorphous replacement). One big problem, the lack of single crystals suitable for X – ray structure analysis, has been overcome recently with the detection of the new copper based stable icosahedral phases. Now the way is open for accurate single – crystal X – ray and even neutron diffraction experiments, including inelastic scattering for the study of the dynamical properties of quasi – crystals.

REFERENCES

Bouchet – Fabre, B., Laridjani, M., Chenoufi, A., and Dixmier, J., 1987, X – ray study of the quasicrystalline structure of Al – Fe – Cu alloys, in: ”Quasicrystalline Materials” C.Janot, J.M.Dubois eds., World Scientific, Singapore.

Cahn, J.W., Gratias, D., and Mozer, B., 1988, Patterson Fourier analysis of the icosahedral (Al,Si) – Mn alloy, Phys. Rev. **B38**:1638 – 1642.

Elswijk, H.B., de Hosson, J.T.M., van Smaalen, S., and de Boer, J.L., 1988, Determination of the crystal structure of icosahedral Al – Cu – Li, Phys. Rev. **B38**:1681 – 1685.

Fujiwara, K., 1957, On the period of out – of – step of ordered alloys with antiphase – domain structure, J.Phys.Soc.Jpn.**12**:7 – 17.

Gratias, D., Cahn, J.W., and Mozer, B., 1988, Six – dimensional Fourier analysis of the icosahedral $Al_{73}Mn_{21}Si_6$ alloy, Phys. Rev. **B38**:1643 – 1646.

Hao, Q., Liu, Y. – W., and Fan, H. – F., 1987, Direct Methods in Superspace I. Preliminary Theory and Test on the Determination of Incommensurate Modulated Structures, Acta Crystallogr. **A43**:820 – 824.

Hiraga, K., Hirabayashi, M., Inoue, A., and Masumoto, T., 1987, High – resolution electron microscopy of Al – Mn – Si icosahedral and Al – Mn decagonal quasicrystals, J. Microscopy **146**:245 – 260.

Ishihara, K.N., and Yamamoto, A., 1988, Penrose Patterns and Related Structures. I.Superstructure and Generalized Penrose Patterns, Acta Crystallogr. **A44**:508 – 516.

Janot, C., de Boissieu, M., Dubois, J.M., and Pannetier, J., 1989, Icosahedral crystals: Neutron diffraction tells you where the atoms are, J. Phys.: Condens. Matter **1**:1029 – 1048.

Janssen, T., and Janner, A., 1987, Incommensurability in crystals, Advances in Physics **36**:519 – 624.

Koopmans, B., Schurer, P.J., van der Woude, F., and Bronsveld, B., 1987 X – ray diffraction and Mössbauer – effect study of the decagonal $Al_7(Mn_{1-x}Fe_x)_2$ alloy, Phys. Rev. **B35**:3005 – 3008.

Lee, C., White, D., Suits, B.H., Bancel, P.A., and Heiney, P.A., 1988, NMR study of Li in Al – Li – Cu icosahedral alloys, Phys. Rev. **B37**:9053 – 9056.

Levine, D., and Steinhardt, P.J., 1986, Quasicrystals. I.Definition and structure, Phys. Rev. **B34**:596 – 616.

Li, F.H., Wang, L.C., and Fan, H. – F., 1987, Possible approach to the phase problem in quasicrystals, Materials Science Forum **22 – 24**:397 – 408.

Li, F.H., and Liu, W., 1986, Combining High – Resolution Electron Microscopy and Electron Diffraction to Determine Quasicrystal Structure. Acta Crystallogr. **B42**:336 – 339.

Ma, Y., and Stern, E.A., 1988, Short – range structure of Al – Mn and Al – Mn – Si aperiodic alloys, Phys. Rev. **B38**:3754 – 3765.

Nakajima, Y., Morimoto, N., and Kitamura, M., 1977, The Superstructure of Plagioclase Feldspars, Phys. Chem. Minerals **1**:213 – 225.

Ohashi, W., and Spaepen, F., 1987, Stable Ga – Mg – Zn quasi – periodic crystals with pentagonal dodecahedral solidification morphology, Nature **330**:555 – 556.

Saintfort, P., and Dubost, B., 1986, The T2 Compound: a Stable Quasi – Crystal in the System Al – Li – Cu – (Mg)?, J. Phys. France **47**:C3 – 321 – 330.

Sayre, D., 1952, The Squaring Method: a New Method for Phase Determination, Acta Crystallogr. **5**:60 – 65.

Shechtman, D., Blech, I., Gratias, D., and Cahn, J.W., 1984, Metallic Phase with Long – Range Orientational Order and No Translational Symmetry, Phys. Rev. Lett. **53**:1951 – 1953.

Shoemaker, C.B., Keszler,D.A., and Shomaker,D.P., 1989, Structure of μ – $MnAl_4$ with Composition Close to that of Quasicrystal, Acta Crystallogr.**B45**:13 – 20.

Socolar, J.E.S., and Steinhardt, P.J., 1986, Quasicrystals. II.Unit – cell configurations, Phys. Rev. **B34**:617 – 647.

Steurer, W., 1987, (3 + 1) – dimensional Patterson and Fourier Methods for the Determination of One – Dimensionally Modulated Structures. Acta Crystallogr. **A43**:36 – 42.

Steurer, W., 1989a, Five – Dimensional Patterson Analysis of the Decagonal Phase of the System Al – Mn, Acta Crystallogr. **B45**:in the press.

Steurer, W., 1989b, The Structure of Quasicrystals, Z. Kristallogr.:in the press.

Steurer, W., and Jagodzinski, H., 1988, The Incommensurately Modulated Structure of an Andesine (An_{38}), Acta Crystallogr. **B44**:344 – 351.

Steurer, W., and Mayer, J., 1989, Single Crystal X – ray Study of the Decagonal

Phase of the System Al – Mn, Acta Crystallogr. **B45**:355 – 359.

Suck, J. – B., 1988, Atomic dynamics of rapidly quenched icosahedral alloys investigated by neutron inelastic scattering, in: "Quasicrystalline Materials" C.Janot, J.M.Dubois eds., World Scientific, Singapore.

Tsai, A. – P., Inoue, A., and Masumoto, T., 1988a, New stable quasicrystals in Al – Cu – M (M = V, Cr or Fe) systems, Trans. JIM **29**:521 – 524.

Tsai, A. – P., Inoue, A., and Masumoto, T., 1988b, New Stable Icosahedral Al – Cu – Ru and Al – Cu – Os Alloys, Jpn. J. Appl. Phys. **27**:L1587 – L1590.

Yamamoto, A., 1982, Structure Factor of Modulated Crystal Structures, Acta Crystallogr. **A38**:87 – 92.

Yamamoto, A., and Ishihara, K.N., 1988, Penrose Patterns and Related Structures. II. Decagonal Quasicrystals, Acta Crystallogr. **A44**:707 – 714.

Ye, H.Q., Wang, D.N., and Kuo, K.H., 1985, Fivefold symmetry in real and reciprocal spaces, Ultramicroscopy **16**:475 – 482.

Zhang, Z., Ye, H.Q., and Kuo, K.H., 1985, A new icosahedral phase withe m35 symmetry. Phil. Mag. **A52**:L49 – L52.

Zucker, U.H., and Schulz, H., 1982a, Statistical Approaches for the Treatment of Anharmonic Motion in Crystals. I. A Comparison of the Most Frequently Used Formalisms of Anharmonic Thermal Vibrations, Acta Crystallogr. **A38**: 563 – 568.

Zucker, U.H., and Schulz, H., 1982b, Statistical Approaches for the Treatment of Anharmonic Motion in Crystals. II. Anharmonic Thermal Vibrations and Effective Atomic Potentials in Fast Ionic Conductor Lithium Nitride (Li_3N), Acta Crystallogr. **A38**:568 – 576.

SIX-DIMENSIONAL ATOMS FOR A DECORATED THREE-DIMENSIONAL PENROSE TILING

Sander van Smaalen

Laboratory of Inorganic Chemistry
Materials Science Center
University of Groningen
Nijenborgh 16
NL 9747 AG Groningen
The Netherlands

INTRODUCTION

Detailed structural studies have been undertaken on only two quasi-crystalline materials of icosahedral symmetry. For one compound, Al_6CuLi_3, it has been possible to describe its structure as a decoration of a three-dimensional (3D) Penrose tiling.[1,2] Attempts to determine such a model for $Al_{73}Mn_{21}Si_6$ have not been succesful.[3] Instead, a six- dimensional model for its structure was proposed, which did give a good fit to the diffraction data.[4-6]

The 6D atoms comprising the structure in 6D space, as used by those authors, are spheres or spherical shells in perpendicular space, convoluted with an ordinary density in physical space. Later, this structure was confirmed by contrast-variation neutron diffraction experiments, which allowed for a direct experimental determination of the 6D structure and the constituting 6D atoms.[7] However, it was not possible to clearly establish whether or not the atoms have spherical symmetry. An alternative model was proposed, based on packing arguments, which lead to polyhedral shapes in perpendicular space, all of which did have icosahedral symmetry.[8]

In the above mentioned approaches, the perpendicular-space shape of the atoms in 6D space was derived from experimental information. Since such a procedure involves the fitting of in principle an infinite number of parameters to an experiment containing only limited information, a choice has to be made as to what parameters are allowed to vary in order to obtain the best fit to the experimental data. It seems therefore necessary to obtain some theoretical prescription for the shape of the 6D atoms, or at least to determine what aspects of the 6D shape can be determined from diffraction information.

Motivated by these considerations, as well by the fact that the structure of Al_6CuLi_3 is approximated very well by a decoration of a Penrose tiling, I present here the perpendicular-space shapes of the 6D atoms corresponding to decorations of a 3D Penrose tiling. It will be shown that both this shape and the position in 6D space depend on the decorating position in the 3D Penrose tiling. These results are then used to construct

Geometry and Thermodynamics
Edited by J.-C. Tolédano
Plenum Press, New York, 1990

a 6D structure model for the Al6CuLi3 quasicrystal, based on the earlier
determined decorated-Penrose tiling model.[2] Experimental evidence for this
model is given by the Patterson functions in 3D and in 6D space, as
calculated from single crystal X-ray diffraction data, and by a refinement
of the 6D structure model on these diffraction data.

FUNDAMENTAL SHAPES

Consider the usual Penrose tiling in 3D space, built from prolate
rhombohedra (PR) and oblate rhombohedra (OR). A decoration of a rhombohedron
is a filling of this body with atoms on specific positions (Fig. 1). A decoration of the Penrose tiling is defined as a Penrose tiling where each PR is
decorated identically, as well as that each OR is decorated identically.

A density function describing an "atom" in 6D space is defined by the
requirement that it gives the density function of an atom in physical space
when it is restricted to a particular 3D section of this 6D space. The quasi
periodic 3D density function of the structure is then obtained as the section of a 6D, periodic density function.[4-7,9] Because of this periodicity, a
structure is characterized by a finite number of atomic positions in the 6D
unit cell. Complicating factor is that the 6D-density function describing a
single atom is not known, contrary to the case for atoms in 3D, physical
space.

All vertices of the 3D Penrose tiling can be obtained as a 3D section
of a hypercubic lattice in 6D space, when each 6D lattice point is decorated
with a object which has the shape of a triacontahedron (TR) in perpendicular
space and is infinitely thin in physical space.[10-12] It immediataly follows
that a decoration of the 3D Penrose tiling with an atom on each vertex is
represented in 6D space by a 6D-atom, which is the convolution of an
ordinary density in physical space and the triacontahedron in perpendicular
space. The TR is called the perpendicular-space shape of this 6D-atom. An
identical atom is positioned on each node of the 6D hypercubic lattice. The
center of symmetry (midpoint) of this 6D atom is chosen to coincide with the
nodes of the 6D lattice, so that the lattice points coincide with inversion
centers of the 6D icosahedral space group. Following Ishihara and Shingu,[13]
Yamamoto and Hiraga,[6] and Gähler[14] we now derive the perpendicular-space
shapes for the other decorating positions.

Consider a decoration with an atom on the midpoints of all edges of the
3D Penrose tiling. Such a decoration can result from atoms in the 6D unit
cell positioned halfway the unit-cell vectors, for example: (0.5, 0, 0, 0,
0, 0). Both the perpendicular-space component of the position and the

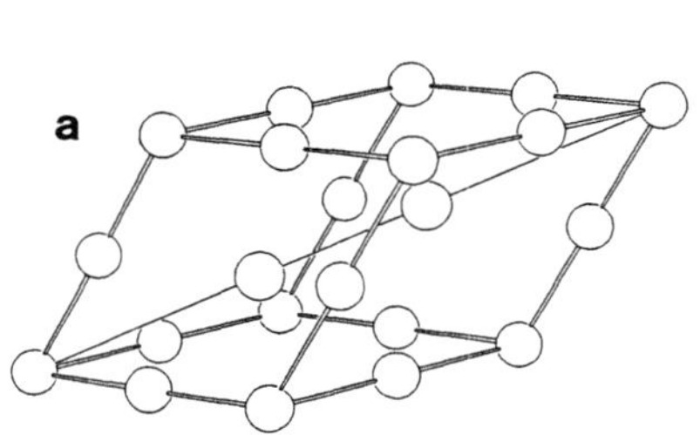
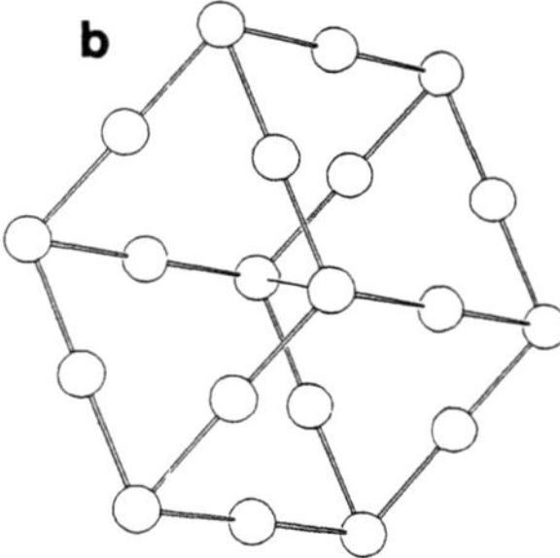

Fig. 1. The prolate rhombohedron (a) and oblate rhombohedron (b)
constituting the 3D Penrose tiling. The decorating sites shown are
the vertex position, the middle-edge position and two positions
(0.5±x) on the long body diagonal of the prolate rhombohedron.

perpendicular-space shape still have to be determined. An atom at the
midpoint of an edge is part of the decoration if the edge is part of the
Penrose tiling. An edge is part of the Penrose tiling if both end-points
(vertices) are part of the Penrose tiling. This latter condition can easily
be refrased in 6D space. For two vertices to be part of the Penrose tiling,
3D space must intersect with both triacontahedra centered on the corresponding lattice points in 6D space, respectively. As an example, take the origin
and the lattice point $\vec{e}1$. For physical space to intersect the triacontahedra
on both lattice points it must go through the intersection of a triacontahedron centered on the origin and one displaced over $\vec{e}_\perp^1$. This intersection
then defines the perpendicular-space shape for atoms decorating the mid-edge
positions of the Penrose tiling. The resulting shape is the rhombic
icosahedron (RI), a polyhedron with $\bar{5}m$ point symmetry and half the volume of
the triacontahedron. The perpendicular-space component of the position also
follows from this definition, and gives as position in the 6D unit cell
(0.5, 0, 0, 0, 0, 0). Due to the 6D icosahedral space group symmetry,
identical atoms are found halfway the other five lattice vectors of the 6D
unit cell.

Analogously, the perpendicular-space shapes for a decoration with atoms
on the mid-points of the faces follow as the intersection of three
triacontahedra, one on the origin, one displaced over $\vec{e}_\perp^1$ and one displaced
over $\vec{e}_\perp^2$, resulting in a rhombic dodecahedron (RD). The perpendicular-space
shape for an atom in the middle of a PR and a OR, respectively, is obtained
as the intersection of four triacontahedra in both cases. The first is
obtained from triacontahedra shifted over $\vec{e}_\perp^1$, $\vec{e}_\perp^2$, and $\vec{e}_\perp^3$, respectively,
defining a PR in perpendicular space. The second is obtained from
triacontahedra shifted over $\vec{e}_\perp^1$, $-\vec{e}_\perp^2$, and $-\vec{e}_\perp^4$, respectively, giving a OR in
perpendicular space. These shapes and the positions in the 6D unit cell are
summarized in the top-part of Table 1.

For a decoration of the vertices, one TR per 6D unit cell is positioned
on the nodes of the 6D hypercubic lattice. The site symmetry of this
position is $\bar{5}3m$, the full icosahedral point group. For the other decorating
positions 6D atoms of lower symmetry are found, which also are placed on
special positions of the 6D icosahedral space group of lower site symmetry
(Table 1). Alternatively, this means that the icosahedral space group
generates more than one of such atom per 6D unit cell. The number of atoms
per unit cell is given as the multiplicity in Table 1. The density of atoms
in physical space is proportional to the perpendicular-space volume of the
atoms, multiplied by the number of atoms per unit cel (multiplicity). It is
easy to check from Table 1 that this gives relative densities corresponding
to a decorated Penrose tiling.

The five perpendicular-space shapes, TR, RI, RD, PR and OR, derived in
this section can be considered as the fundamental shapes. They are
completely determined by the requirement of obtaining a decorated Penrose
tiling in 3D space. Furthermore, their positions in the 6D unit cell are
special positions of the 6D icosahedral space group, which do not allow for
any variation of positional parameters.

Deviations of the structure from the decorated 3D Penrose tiling can be
obtained when different perpendicular-space shapes are chosen. These
alternative shapes, however, still have to obey the site symmetry of their
respective positions.

GENERAL DECORATIONS

In the previous section the five fundamental perpendicular-space shapes
have been derived. The decorating positions considered thus far are special

Table 1. Perpendicular-space shapes and special positions of the 6D atoms. Shown are: the decorating position in the 3D Penrose tiling; The perpendicular-space shape of the corresponding atom in 6D; The special position with respect to the 6D unit cell; site symmetry; multiplicity (mult.); and the relative volume of the perpendicular-space shape ($v_\perp$). The special positions incorporate shifts along the vectors $\vec{r}(51) = (0.5, \alpha_1, \alpha_1, \alpha_1, \alpha_1, \alpha_1)$, $\vec{r}(52) = (\alpha_1, 0.5, \alpha_1, -\alpha_1, -\alpha_1, \alpha_1)$, $\vec{r}(53) = (\alpha_1, \alpha_1, 0.5, \alpha_1, -\alpha_1, -\alpha_1)$ and $\vec{r}(54) = (\alpha_1, -\alpha_1, \alpha_1, 0.5, \alpha_1, -\alpha_1)$, with $\alpha_1 = 1/(2\sqrt{5})$; along the vectors $\vec{r}(21) = (\alpha_2, \alpha_2, \alpha_3, 0, 0, \alpha_3)$ and $\vec{r}(22) = (\alpha_3, -\alpha_3, 0, \alpha_2, \alpha_2, 0)$, with $\alpha_2 = 1/(3-\tau)$ and $\alpha_3 = \alpha_2/\tau$; and along the vectors $\vec{r}(31) = (\alpha_4, \alpha_4, \alpha_4, \alpha_5, -\alpha_5, \alpha_5)$ and $\vec{r}(32) = (\alpha_6, -\alpha_6, -\alpha_5, -\alpha_6, \alpha_5, \alpha_5)$, with $\alpha_4 = 1/(1+((\tau-1)/(\tau+1))^2)$, $\alpha_5 = \alpha_4((\tau-1)/(\tau+1))$ and $\alpha_6 = \alpha_4((\tau-1)/(\tau+1))^2$; $\tau = (1+\sqrt{5})/2$. The values of the structural parameters x, y, z range from -0.5 tot 0.5.

Decorating position	Perp.-space shape	6D position	site symmetry	mult.	$v_\perp$
vertex	TR	$(0,0,0,0,0,0)$	$\bar{5}\bar{3}m$	1	1
mid-edge	RI	$(0.5,0,0,0,0,0)$	$\bar{5}m$	6	$\frac{1}{2}$
mid-face	RD	$(0.5,0.5,0,0,0,0)$	mmm	15	$\frac{1}{5}$
mid-PR	PR	$(0.5,0.5,0.5,0,0,0)$	$\bar{3}m$	10	$1/(10\tau)$
mid-OR	OR	$(0.5,-0.5,0,-0.5,0,0)$	$\bar{3}m$	10	$1/(10\tau^2)$
edge	$\frac{1}{2}$RI	$(0.5,0,0,0,0,0)$ $+ x\,\vec{r}(51)$	$5m$	12	$\frac{1}{4}$
face-diagonal	$\frac{1}{2}$RD	$(0.5,0.5,0,0,0,0)$ $+ x\,\vec{r}(21)$	$mm2$	30	$1/10$
face	$\frac{1}{4}$RD	$(0.5,0.5,0,0,0,0)$ $+ x\,\vec{r}(21) + y\,\vec{r}(22)$	m	60	$1/20$
PR-diagonal	$\frac{1}{2}$PR	$(0.5,0.5,0.5,0,0,0)$ $+ x\,\vec{r}(31)$	$3m$	20	$1/(20\tau)$
OR-diagonal	$\frac{1}{2}$OR	$(0.5,-0.5,0,-0.5,0,0)$ $+ x\,\vec{r}(32)$	$3m$	20	$1/(20\tau^2)$
PR-mirror	$\frac{1}{6}$PR	$(0.5,0.5,0.5,0,0,0)$ $+ x\,\vec{r}(53) + y\,\vec{r}(21)$	m	60	$1/(60\tau)$
OR-mirror	$\frac{1}{6}$OR	$(0.5,-0.5,0,-0.5,0,0)$ $+ x\,\vec{r}(54) + y\,\vec{r}(22)$	m	60	$1/(60\tau^2)$
PR	$\frac{1}{12}$PR	$(0.5,0.5,0.5,0,0,0)$ $+ x\,\vec{r}(51) + y\,\vec{r}(52) + z\,\vec{r}(53)$		120	$1/(120\tau)$
OR	$\frac{1}{12}$OR	$(0.5,-0.5,0,-0.5,0,0)$ $+ x\,\vec{r}(51) - y\,\vec{r}(52) - z\,\vec{r}(54)$		120	$1/(120\tau^2)$

ones: on the vertices, half way the edges, in the middle of the faces, or in the middle of either rhomb. However, more general decorations of the 3D Penrose tiling exist, for example, one atom in each PR, placed somewhere in the inner part of this body. We will now derive the 6D atoms corresponding to these special positions of lower symmetry.

First, consider a decoration of a pair of atoms placed symmetrically on an edge around its mid-point. By the same argument as in the previous section, the perpendicular-space shape is a RI. The perpendicular-space component of the 6D position must also be the same as for the mid-edge position. Any deviation from the original position should be accomplished by a shift in 6D-space parallel to physical space. Using the relation between the six basis vectors and their projections on perpendicular space and physical space,[11] shows that a shift along $\vec{e}_\parallel^{\,1}$, with zero component in perpendicular space, must be parallel to $\vec{r}(51) = (0.5, \alpha_1, \alpha_1, \alpha_1, \alpha_1, \alpha_1)$, with $\alpha_1 = 1/(2\sqrt{5})$. The single mid-edge position (e.g. on $\vec{e}_\parallel^{\,1}$) is replaced by two positions $(0.5\pm x)\vec{e}_\parallel^{\,1}$. In the 6D unit cell this corresponds to the two positions $(0.5, 0, 0, 0, 0, 0) \pm x\,\vec{r}(51)$. The site symmetry of this new special position in 6D is $5m$, a subgroup of the site symmetry group of the mid-edge postion ($\bar{5}m$).

As the site symmetry of this position is lower, it is allowed to use perpendicular-space shapes of lower symmetry, still preserving the icosa-hedral 6D space group symmetry of the structure. For example, the RI can be cut in two halves by a surface perpendicular to the 5-fold axis. One half is placed at $-x$ and the other half at $+x$. To retain the icosahedral symmetry of the structure, the body at $+x$ must be the image of that at $-x$ under the inversion operator. To ensure that each cell is decorated equivalently, the sum of the two halves must be exactly the RI, i.e., overlap and holes in the perpendicular space projection are not allowed. The symmetry of the cutting surface must be the site-symmetry of the original special position, in this case $\bar{5}m$. Analogously, perpendicular-space shapes for other decorating positions of lower symmetry can be derived. They are given in Table 1.

The restrictions on how to divide a fundamental perpendicular-space shape in a collection of bodies of lower symmetry, can be given as a set of three rules governing this cutting procedure. They are: (1) the sum of the parts must give exactly the original fundamental shape; (2) the parts must be each other image under the symmetry operators relating the respective positions; (3) The set of cutting surfaces must have point symmetry according to the site symmetry of the original special position of the fundamental shape, of which it is derived. As an example consider a general position in the PR. There are four cutting surfaces, dividing the PR in perpendicular space into twelve parts (Table 1). Three surfaces go through the 3-fold axis, the $\bar{3}m$ symmetry requires these surfacees to coincide with the mirror-planes; their shape is therefore a plane too. The fourth cutting surface is perpendicular to the 3-fold axis. The site symmetry requires it to go through the inversion center of the PR, and to have rotational symmetry according to the $3m$. However, within these limits there is still freedom in the choice of the exact shape of this cutting surfacee. This is one of the properties what makes these perpendicular-space shapes less fundamental than those defined in the previous section.

Obviously, using only half a RI as perpendicular-space shape of anatom on a position in the 6D unit cell, leads to the presence of this atom on only half the edges of the 3D Penrose tiling. Only one of the positions $(0.5\pm x)\,\vec{e}_\perp^{\,1}$ will be occupied. This still leads to a decorated Penrose tiling, since these two positions are indistinguishble. However, the internal $\bar{3}m$ symmetry of the rhombohedra in the Penrose tiling is lost. Apparantly, this does not destroy the overall (6D) icosahedral symmetry. The choice of one particular cutting surface for dividing the fundamental perpendicular-space

Table 2. 6D structure model for icosahedral Al6CuLi3.
Atoms are found on three 6D special positions. The values
for the occupation probability (P_μ) are those obtained with
the 6D refinement program. $\vec{r}(31)$ is defined in Table 1. The
value for x was found as 0.123 after refinement.

6D-atom	P_μ	position
(Al, TR)	0.70	(0, 0, 0, 0, 0, 0)
(Cu, TR)	0.19	(0, 0, 0, 0, 0, 0)
(Li, TR)	0.11	(0, 0, 0, 0, 0, 0)
(Al, RI)	0.75	(0.5, 0, 0, 0, 0, 0)
(Cu, RI)	0.11	(0.5, 0, 0, 0, 0, 0)
(Li, RI)	0.14	(0.5, 0, 0, 0, 0, 0)
(Al, PR)	0.16	(0.5, 0.5, 0.5, 0, 0, 0) $- x\ \vec{r}(31)$
(Cu, PR)	0.01	(0.5, 0.5, 0.5, 0, 0, 0) $- x\ \vec{r}(31)$
(Li, PR)	0.83	(0.5, 0.5, 0.5, 0, 0, 0) $- x\ \vec{r}(31)$

shape corresponds to precisely one distribution of atoms in physical space.
Although there is much freedom in the choice of the cutting surface,
corresponding to many different decorations of the Penrose tiling, the
distribution of the atoms in physical space is not random, but restricted by
the restrictions imposed on the cutting surface.

SIX-DIMENSIONAL STRUCTURE OF AL6CULi3

The structure of the icosahedral quasicrystal Al6CuLi3 has been
described previously as a decoration of the 3D Penrose tiling (Fig. 1).[1,2]
Refinement of such a structure model on single crystal X-ray diffraction
data lead to an excellent fit with an agreement factor (using unit weights
w) $R_F2 = \{\Sigma w(|F_o|-|F_c|)^2/(\Sigma w|F_o|^2)\}^{1/2} = 0.07.$[2] Direct experimental evidence
for this structure was found in the 3D Patterson function, calculated with
the diffraction data (Fig. 2).[15]

With the results of the previous sections, it is a straightforward
procedure to derive the 6D structure model corresponding to the decorated
Penrose model of ref.2. The result is given in Table 2. Note that each of
the three 6D special positions is disorderly occupied by three 6D atoms of a
different chemical kind. This is the same kind of disorder as is often
encountered in the 3D structure of 3D-periodic alloys.

Direct comparison of this 6D structure model can be done in two different ways again. First, the 6D Patterson function obtained from the X-ray
diffraction data is given (Figs. 3, 4).[16] Fig. 3 shows the perpendicular-
space shape of the origin peak in a 3-fold plane. Note the absence of any
intensity at the body-center of the cell. Fig. 4 shows the peak halfway the
lattice vector; it can be interpreted as due to an atom at the origin and
one on (0.5, 0, 0, 0, 0, 0). The peak on (0, 0.5, 0.5, 0, 0, 0) then is due
to a vector between two mid-edge atoms. No other peaks of significant
intensity were found.

The second comparison with experimental data is obtained by refinement
of the 6D structure on the diffraction data. The atomic form factor for a 6D

atom can be written as the product of the ordinary atomic form factor in physical space, $f_{\parallel}^{\mu}(|\vec{S}_{\parallel}|)$, and the Fourier transform of the perpendicular-space shape in perpendicular space, $f_{\perp}^{\mu}(\vec{S}_{\perp})$. Note that the former depends only on the modulus of the physical space component of the diffraction wave vector, while the latter depends also on the direction, but now of the perpendicular-space component. The complete expression for the structure factor is,

$$F(\vec{S}) = \sum_{\mu} f_{\parallel}^{\mu}(|\vec{S}_{\parallel}|)\, m_{\mu}\, P_{\mu}\, \exp(-\tfrac{1}{4}B_{\parallel}^{\mu}|S_{\parallel}|^{2})\, \sum_{R} f_{\perp}^{\mu}(\vec{S}_{\perp})\, \exp(2\pi i\vec{S}\cdot R\vec{r}^{\mu})$$

where the summation is over the independent atoms μ in the 6D unit cell and over the symmetry operators R. The multiplicity, m_{μ}, is as given in Table 1. The population, P_{μ}, is as given in Table 2. B^{μ} is the isotropic temperature factor.

A computer program for refinement of the 6D structure was written, incorporating an exact calculation of the perpendicular-space part of the atomic form factor using the rhombohedral dissection of these bodies.[13] Variation of the population parameters, with the same restrictions as in ref.2, and of the single positional parameter, lead to a weighted R-factor of $R_F2 = 0.05$ (0.07 with unit weights). The resulting parameters are given

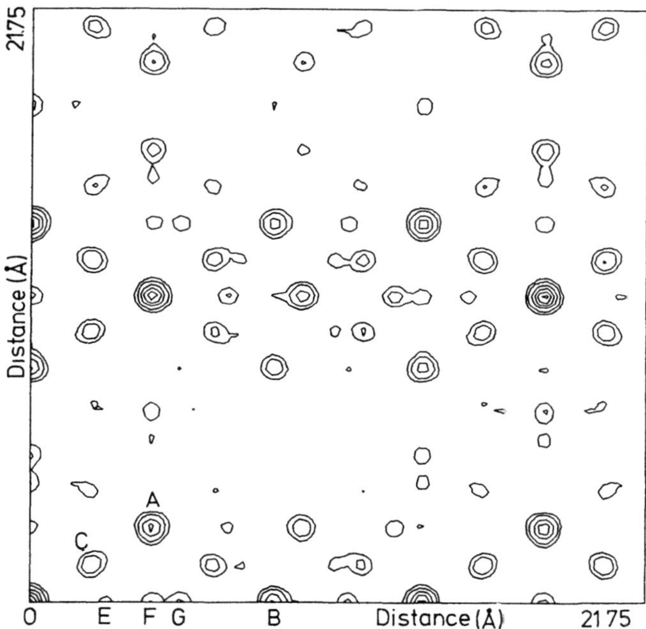

Fig. 2. Section of the 3D Patterson function for icosahedral Al6CuLi3, obtained by Fourier transformation of the measured X-ray diffraction intensities (after ref.15). The two coordinate axis are two-fold symmetry axis. The peaks O, A, B and G are obtained for atoms on the vertices of a Penrose tiling. The peaks C, E and F can be explained by assuming both the vertices and middle-edge positions to be decorated by atoms.

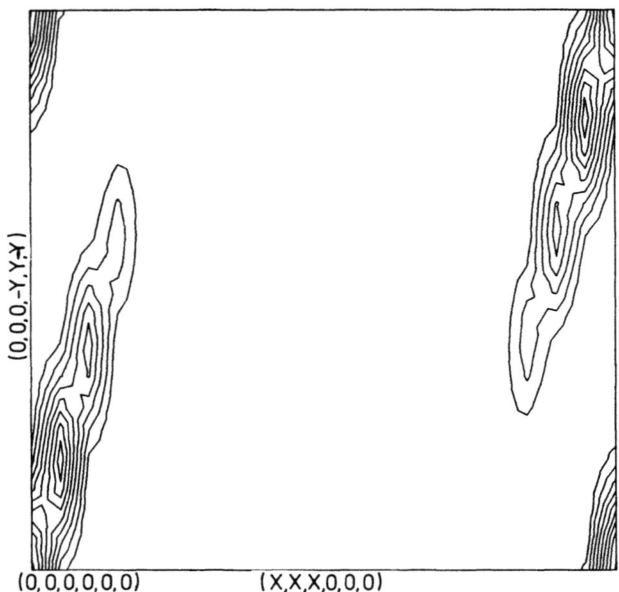

Fig. 3. Section of the 6D Patterson function for icosahedral Al₆CuLi₃,
obtained by Fourier transformation of the measured X-ray diffraction
intensities. Contour lines are drawn at intervals of 10% of the
maximum value. This section incorporates the origin. It intersects
with both perpendicular space and physical space in a line of 3-fold
symmetry. The values for x and y range from zero to one.

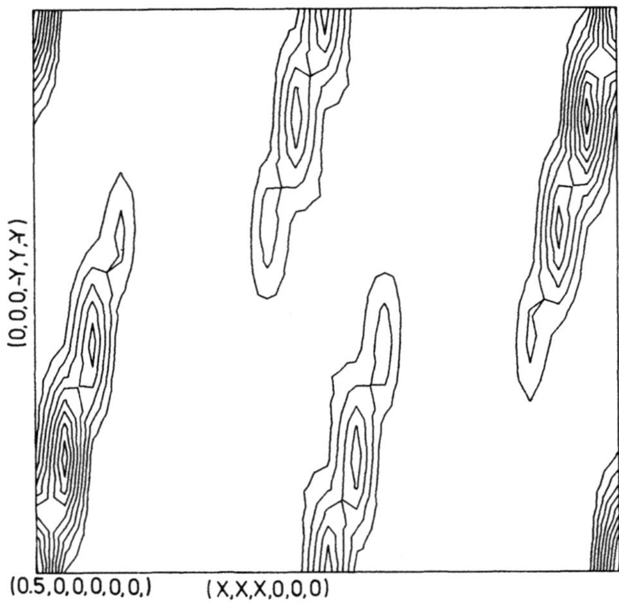

Fig. 4. The same as Fig.3, but now starting in the point (0.5,0,0,0,0,0).

46

in Table 2. They differ only marginally from those obtained in ref.2, as is
to be expected. This again gives experimental evidence for the correctness
of the 6D structure model presented here.

DISCUSSION AND CONCLUSIONS

In this paper the decorated 3D Penrose tiling as model for the
structure of icosahedral quasicrystals is considered. The 6D atoms and
their positions in the 6D unit cell were derived. This lead to the
definition of five fundamental perpendicular-space shapes and nine secundary
shapes (Table 1). The five fundamental shapes are completely determined by
the requirement that the 3D structure must be a decorated Penrose tiling. As
shown in the third section, the other perpendicular-space shapes may have
boundaries which are undetermined, even within the limits imposed by the
symmetry and the requirement of obtaining a decorated Penrose tiling in
physical space. This means that for those decorating positions, e.g. only
one of the two symmetry related positions on the long body diagonal of the
PR being occupied, there is freedom in the choice of which of the two sites
is occupied in each PR.

From the above analysis the conclusion is that no matching rules exist,
which take into account the orientation of the decorated PR and OR used for
building a 3D Penrose tiling. In order words, from the same pair of
decorated PR and OR, (infinitely) many inequivalent decorated 3D Penrose
tilings can be made. All these tilings are a 3D section of a 6D periodic
structure with icosahedral symmetry. However, the analysis did show that the
orientation of such decorated PR and OR is not entirely random; The 6D
icosahedral symmetry does impose some restrictions on the perpendicular-
space shape of these atoms, and therefore on the orientation of the
decorated PR and OR in physical space.

These results were applied to the structure of icosahedral Al_6CuLi_3,
for which the 3D structure model[1,2] was translated into a 6D structure
model. Experimental evidence for the latter model was given in the form of a
6D Patterson synthesis and a 6D refinement on single crystal X-ray
diffraction data.

ACKNOWLEDGEMENTS

W. Steurer is thanked for making his Patterson program available. This
research has been made possible by financial support form the Royal
Netherlands Academy of Arts and Science (KNAW).

REFERENCES

1. Y. Shen, S.J. Poon, W. Dmowski, T. Egami, and G.J. Shiflet, Structure of
   Al-Li-Cu Icosahedral Crystals and Penrose tiling, Phys. Rev. Lett.
   58:1440 (1987).
2. H.B. Elswijk, J.Th.M. de Hosson, S. van Smaalen and J.L. de Boer,
   Determination of the crystal structure of icosahedral Al-Cu-Li,
   Phys. Rev. B38:1681 (1988).
3. K.N. Knowles and W.M. Stobbs, Diffraction pattern simulations of
   quasiperiodic structures, Nature 323:313 (1986).
4. J.W. Cahn, D. Gratias and B. Mozer, A 6-D structural model for the
   icosahedral (Al, Si)-Mn quasicrystal, J. Phys. France 49:1225 (1988).
5. D. Gratias, J.W. Cahn and B. Mozer, Six-dimensional Fourier analysis of
   the icosahedral $Al_{73}Mn_{21}Si_6$ alloy, Phys. Rev. B38:1643 (1988).

6. A. Yamamoto and K. Hiraga, Structure of an icosahedral Al-Mn quasicrystal, _Phys_. _Rev_. B37:6207 (1988).

7. C. Janot, M. De Boissieu, J.M. Dubois and J. Pannetier, Icosahedral crystals: neutron diffraction tells you where the atoms are, _J_. _Phys_.: _Condens_. _Matter_ 1:1029 (1989).

8. M. Duneau and C. Oguey, Ideal AlMnSi quasicrystal: a structural model with icosahedral clusters, _J_. _Phys_. _France_ 50:135 (1989).

9   W. Steurer, Five-Dimensional Patterson-Analysis of the Decagonal Phase of the System Al-Mn, _preprint_ (1988).

10. P.A. Kalugin, A.Yu. Kitaev and L.S. levitov, $Al_{0.86}Mn_{0.14}$: a six-dimensional crystal, _JETP_ _Lett_. 41:145 (1985). [_Pis'ma_ _Zh_. _Eksp_. _Teor_. _Fiz_. 41:119 (1985).]

11. V. Elser, The diffraction pattern of projected structures, _Acta_ _Crystallogr_. A42: 36 (1986).

12. A. Katz and M. Duneau, Quasiperiodic patterns and icosahedral symmetry, _J_. _Phys_. _France_ 47:181 (1986).

13. K.N. Ishihara and P.H. Shingu, Calculation of the structure factor for three-dimensional Penrose tilting, _J_. _Phys_. _Soc_. _Japan_ 55:1795 (1986).

14. F. Gähler, Crystallography of dodecagonal quasicrystals, _in_: "Quasicrystalline Materials", Ch. Janot and J.M. Dubois, eds., World Scientific, Singapore (1988), pp. 272-284.

15. S. van Smaalen, Three-dimensional Patterson function for the Al_6CuLi_3 quasicrystal, _Phys_. _Rev_. B39:5850 (1989).

16. A modified version was used of the 6D Patterson synthesis program by W. Steurer.

METRICAL ASPECTS OF QUASICRYSTAL EMBEDDING IN SUPERSPACE

A.Janner

Institute for Theoretical Physics, University of Nijmegen
Toernooiveld, 6525 ED Nijmegen, The Netherlands

INTRODUCTION

The role played by the lattice for the atomic positions in a
normal crystal is taken over in a quasicrystal by a vector module M
spanned by all integral linear combinations of n basis vectors. The
integer n is the rank of M and is larger than the dimension m of the
physical space, which is also the dimension of the vector module.

The aim of this paper is to show how enflation rules appear as
scaling transformations leaving M invariant. So after embedding M on a
n-dimensional lattice Σ, scaling symmetries become crystallographic
point group transformations in superspace, and they can be combined
with lattice translational symmetries. As the lattice Σ is the same
which appears while considering the rotational symmetries of the
quasicrystal (also leaving M invariant), both rotational and scaling
transformations are compatible point group transformations of the
lattice in question. This means that they are expressible as elements
of $G\ell(n,\mathbb{Z})$, the group of n×n matrices with integral entries. It does
not imply, however, equivalence with n-dimensional orthogonal
transformations.

Previous investigations[1,2,3] have already shown that despite the
fact that scaling transformations being of infinite order do not leave
any Euclidean metric invariant, they possibly impose restrictions on
the Euclidean holohedry of the lattice Σ. In a number of cases studied
so far it is even so that the possibility of embedding a given vector
module M on a lattice with non-reducible holohedral point group depends
on the scaling properties of M. That interplay between rotational and
scaling symmetries also appears while considering enflation rules for

Geometry and Thermodynamics
Edited by J.-C. Tolédano
Plenum Press, New York, 1990

tiling, so that e.g. in an octagonal tiling one finds a specific
enflation rule, different from that one gets in a decagonal or in a
dodecagonal tiling.[3]

Despite all these observations, the structural relevance of the
rotational lattice symmetries one possibly gets in additional to those
leaving the underlying quasicrystal invariant, remains an open
question. As an example consider an icosahedral quasicrystal phase
embedded as 6-dimensional crystal structure with hypercubic lattice.
One then only considers the (super)space groups having as point group
the icosahedral one, regarding as physically irrelevant both the
additional hypercubic transformations, and those arising from a
golden-mean scaling invariance of M.

The author believes that crystallography arises from an interplay
between arithmetic and metric. Arithmetic here means integers (as
implied by $G\ell(n,\mathbb{Z})$ being called the arithmetic group) and metric is
normally speaking the Euclidean (n-dimensional) one. As we have
already said, embedded scaling symmetries are incompatible with
Euclidean metric. They can naturally be considered as affine
transformations, or as orthogonal ones but for a group of signature
lower than n, so that the superspace gets a multimetrical character:
Euclidean for the rotational point group and indefinite (we shall say
Minkowskian) for the scaling point group.

The justification of this (multimetrical) superspace approach has
to come from experiment. Accordingly, the present paper still has an
exploratory character. When two separate point groups are considered
one also gets two different groups of motions: one which includes
rotational symmetries and is a n-dimensional space group, and another
which we may call a scale group that includes the scaling symmetries.
Combining both type of transformations into a single point group one
then gets a scale/space group.[3,4] From the theoretical point of view the
interesting point is that there are more inequivalent scale/space
groups than one would get from space groups and scale groups considered
as separate entities.[3]

Independently of whether the two types of symmetries are
considered together or separately the quasicrystal is a given one, and
that requires compatibility relations for the pattern(s) representing
it in superspace. That compatibility has already been demonstrated on
the level of the lattice, but needs further investigation on the level
of the (higher dimensional) unit cell pattern, or equivalently on that
of the decoration of the quasicrystal tiling.

Note that a multimetrical superspace implies for a given lattice Σ
two different reciprocal lattices Σ_e^* and Σ_m^* (or correspondingly for a
given Σ^* two lattices Σ_e and Σ_m in direct space).

That is conceptually acceptable and does not exclude selection rules following from the two types of symmetries, nor multiple superspace patterns for a given physical structure. It represents, however, a complicating factor. It is, therefore, interesting to be aware that in the point-atom approximation scaling symmetries are, in some sense, expressible in terms of Euclidean superspace transformations. This is related with a minimal distance property among different atoms. The present paper suggests the even more fascinating possibility that these Euclidean symmetries are among the non-reducible ones of a high symmetric embedding (e.g. hypercubic in the icosahedral quasicrystal case). An example of such a case is given in terms of a decorated Fibonacci chain. Whether this possibility is verified by the atomic structure involved in real crystals is at the moment still an open question. Already now, it seems clear that scaling symmetry is only an approximate one and is broken by a continuous charge distribution or by a modulation. Nevertheless, it is an important ingredient in a crystallographic characterization of quasicrystals.

The structure of the paper is based on a multimetrical crystallographic approach (disregarding reflections). The first step is represented by the restrictions imposed on rotations and scalings by a (superspace) lattice translational symmetry. Conversely, these point group elements allow a distinction among lattices having different point symmetry which are geometrically inequivalent with respect to the appropriate orthogonal groups. The lattice being a given one in superspace, this implies compatibility relations and eventually one gets then a lattice classification according to crystallographic systems. Then centering has to be considered leading to the concept of holohedry based on arithmetic equivalency of the lattice symmetries involved. Admitted point groups are then, as usual, subgroups of these holohedries. Combination of point group transformations with lattice translations by means of group extension leads to the groups of rigid motions (space groups and their generalization) expressed in terms of a system of non-primitive (or intrinsic) translations. Finally, the symmetry of the quasicrystal is defined by the invariance of a suitably associated periodic pattern with respect to those groups of rigid motions. The pattern is expressed in terms of positions in superspace, classified as by Wyckoff according to site symmetry and multiplicity.

That all represents a well-known set of logical steps. Dealing correctly with a multimetrical superspace, however, gives rise to many problems. Furthermore, the procedure on how to connect a quasicrystal structure with a lattice periodic pattern is not unique, the various descriptions being a kind of spectral decomposition of a complex pattern according to different structural relations.

It is not possible, at present, to discuss properly all these aspects. Here only point-like pattern in direct superspace will be considered. The question of the associated atomic surfaces (or

equivalently of the windows) is disregarded and also that of the
structure factors.

LATTICE SYMMETRY OF QUASICRYSTALS

Incommensurate crystals are characterized by Fourier wave vectors
$\vec{k}$ generating a (reciprocal) vector module M^* of dimension m and rank
$n = m+d$.[5] This means that $\vec{k}$ is a vector of a m-dimensional reciprocal
space expressible as integral linear combination of a the n ones
$\vec{a}_1^*,\ldots,\vec{a}_n^*$ defining a basis of M^*:

$$\vec{k} = \sum_{i=1}^{n} h_i \vec{a}_i^* , \qquad \text{integers } h_i . \qquad (2.1)$$

In the normal superspace approach, M^* is embedded as a lattice Σ^*
in a n-dimensional Euclidean space (the superspace), in such a way that
by an orthogonal projection π one gets M^* back:[5]

$$M^* = \pi \, \Sigma^* . \qquad (2.2)$$

Quasicrystals share this property,[6] but in addition their basic atomic
positions $\vec{n}$ (disregarding thus possible modulations and continuous
charge densities) are expressible, up to initial values, as integral
linear combinations of n ones, $\vec{a}_1,\ldots,\vec{a}_n$ defining a basis of a (direct)
vector module M, of dimension m and rank n:

$$\vec{n} = \sum_{i=1}^{n} n_i \vec{a}_i , \qquad \text{integers } n_i . \qquad (2.3)$$

This allows, for quasicrystals and within the present
approximation, to consider an alternative superspace approach, by
embedding M on a lattice Σ in the n-dimensional superspace, in such a
way that one gets M back by orthogonal projection π:

$$M = \pi \, \Sigma . \qquad (2.4)$$

The two lattices Σ and Σ^* can be chosen to be reciprocal (making use
of the Euclidean nature of the superspace). Due to incommensurability,
both embedding procedures (in direct and in reciprocal space,
respectively) are 1-to-1 and fully equivalent, but of course different.
The reciprocal space approach[7,8] (denoted here as standard) is better
suited once dealing with a structure determination based on diffraction
pattern, the direct superspace approach makes the discussion of
properties of atomic positions more transparent, and will be adopted
here.

An atomic quasicrystal position is given in terms of an initial
position vector r_{sj} in the unit cell of the lattice Σ and a lattice
vector n_s

$$r_s(n_s,j) = r_{sj} + n_s \qquad (2.5)$$

with $n_s = \sum_{i=1}^{n} n_i a_{si}$ integers n_i. The corresponding position in the physical (or external) space is then:

$$r(\vec{n}, j) = \pi\, r_s(n_s, j) = \pi\, r_{sj} + \pi n_s = \vec{r}_j + \vec{n} \,. \tag{2.6}$$

What has been formulated is a necessary, but not a sufficient condition in the sense that it only indicates which are the *possible* positions. The *occupied* positions have also to be given: that can be done in terms of a density function $C_j(n_s)$ expressing the probability of occupancy of the crystallographic position $\vec{r}(\vec{n}, j)$ by a given atomic species. In the case that $C_j(n_s)$ is a characteristic function in superspace taking only the values 0 and 1, it corresponds to the so-called window within the cut-projection method,[9] and gives rise to atomic surfaces in the standard superspace approach.[6,10]

As well-known, the structure depends on the position of the window with respect to the physical space, i.e. on the internal coordinate of r_{sj}.[11] That is the reason why, for the characterization of a quasicrystal structure the position vector r_{sj} is relevant and not only its projection $\vec{r}_j$ in space.

From now on, we consider superspace atomic positions and restrict ourselves to the single atomic arrangement, i.e. to a single set of equivalent points (equivalent under an appropriate symmetry group) and disregard the occupation functions $C_j(n_s)$, focusing our attention on the admitted symmetry transformations and on their mutual relations.

A) Rotational Symmetries

Back now to the standard superspace approach. The Fourier module M^* is embedded on a Euclidean n-dimensional lattice Σ^*, defining a corresponding direct one Σ. In the present quasicrystal case one gets the positional vector module M by projection of Σ on the physical space.

Admitted rotational symmetries are elements R of $O(m)$ leaving M^* invariant which induce transformations R_S of $O(n)$ leaving Σ^* invariant, and thus Σ. As in fact $R_S \in O(m) \times O(n-m)$, R leaves also M invariant. One has:

$$RM = M \qquad \text{with} \qquad R\vec{a}_i = \sum_{j=1}^{n} \Gamma_{ji}(R)\vec{a}_j \tag{2.7a}$$

$$R_S\Sigma = \Sigma \qquad \text{with} \qquad R_s a_{si} = \sum_{j=1}^{n} \Gamma_{ji}(R_s)a_{sj} \tag{2.7b}$$

where $\Gamma(R) \equiv \Gamma(R_s)$ is an element of $G\ell(n, \mathbb{Z})$. The set of all such elements of $G\ell(n, \mathbb{Z})$ defines the rotational holohedry H_R of the quasicrystal:

$$H_R = \{\Gamma(R) \in G\ell(n,\mathbb{Z}) \mid RM = M, \quad R_s\Sigma = \Sigma\} \tag{2.8}$$

and $R \in O(m)$, $R_s \in O(m) \times O(n-m)$.

Note that the holohedry of M (defined by all $R \in O(m)$ leaving M invariant) can be larger: in the normal approach one implicitly assumes properties of the diffraction intensities such that the Bragg spots can be indexed. Also the holohedry of Σ is in general larger than H_R, as it possibly involves other elements of $O(n)$, not leaving the physical space invariant.

B) <u>Scaling Symmetries</u>

Well-known models of quasicrystals are characterized by a tiling admitting enflation (and deflation) rules. These linear transformations correspond to scalings (possibly combined with rotations also), which leave the underlying vector module M invariant:

$$SM = \lambda(S)M = M, \qquad S \in D(m) \tag{2.9}$$

with $\lambda(S)$ a real number different from zero and 1 and $D(m)$ the extended orthogonal group of V involving dilatations (homotheties) as well. (Actually, it may happen that only part of M is scaling invariant, a possibility not taken into account in (2.9), but for the present exposition that possibility is not an essential one.) From (2.9) one has:

$$S\vec{a}_i = \sum_{j=1}^{n} \Gamma_{ji}(S)\vec{a}_j \tag{2.10a}$$

with $\Gamma(S)$ element of $G\ell(n,\mathbb{Z})$, so S induces an affine transformation S_s in superspace leaving the lattice Σ invariant.

$$S_s\Sigma = \Sigma \quad \text{with} \quad S_s a_{si} = \sum_{j=1}^{n} \Gamma_{ji}(S_s)a_{sj} , \tag{2.10b}$$

where again $\Gamma(S) \equiv \Gamma(S_s)$. In general S_s is of infinite order, the orbit of a lattice point is not finite and no Euclidean metric is left invariant. As explained elsewhere,[1,2,3] in all the cases investigated so far there is, however, an indefinite metric tensor g_m such that $\Gamma(S)$ is either an automorph (+) or a negautomorph (−):

$$\tilde{\Gamma}(S)g_m\Gamma(S) = \pm g_m , \tag{2.11}$$

so that S_s appears as an element of the group $\hat{O}(m,d)$ which includes the orthogonal group of the quadratic form $(+,+,\ldots,+,-,\ldots,-)$ of dimension m+d and signature m and the negautomorphs of that form as well.

As an example for m=d=1, $\hat{O}(1,1)$ includes both Lorentz boosts and anti-Lorentz ones:

$$A = \begin{pmatrix} \cosh\chi & \sinh\chi \\ \sinh\chi & \cosh\chi \end{pmatrix} \quad \text{and} \quad N = \begin{pmatrix} \sinh\varphi & \cosh\varphi \\ \cosh\varphi & \sinh\varphi \end{pmatrix} . \tag{2.12}$$

The crystallographic restrictions on those scalings imply

$$2\cosh\chi = \mu \quad \text{and} \quad 2\sinh\varphi = \nu\,, \qquad \mu,\nu \text{ integers}\,,$$

respectively. For the small (positive) values of those integers one
gets scaling factors $\nu_\pm$ and $\mu_\pm$ appearing in known examples of
quasicrystals as illustrated in Table 1. Note that $\det N = -1$, whereas
$\det A = +1$, so that A defines a proper scale transformation and N an
improper one.

Table 1. Examples of quasicrystal scaling factors

Trace	Scaling factor(s)	Examples
$\nu = 1$	$\nu_+ = \frac{1+\sqrt{5}}{2} = \tau\,,\ \nu_- = -\frac{1}{\tau}$	Fibonacci chain (1D). Decagonal quasicrystal (2D). Face centered hypercubic icosahedral phase (3D). Body centered hypercubic icosahedral phase (3D).
$\nu = 2$	$\nu_+ = 1 + \sqrt{2} = \omega\,,\ \nu_- = -\frac{1}{\omega}$	Octagonal quasicrystal (2D).
$\nu = 4$	$\tau^3\,, -\frac{1}{\tau^3}$	Simple hypercubic icosahedral phase (3D).
$\mu = 3$	$\tau^2\,, \frac{1}{\tau^2}$	Penrose pattern (2D).
$\mu = 4$	$\mu_+ = 2 + \sqrt{3} = \xi\,,\ \mu_- = \frac{1}{\xi}$	Generalized Fibonacci chain (1D). Dodecagonal quasicrystal (2D).

In addition to the rotational holohedry we now define the scaling
holohedry of M as the appropriate subgroup H_S of $G\ell(n,\mathbb{Z})$:

$$H_S = \{\Gamma(S) \in G\ell(n,\mathbb{Z}) \mid SM = M, \qquad S_S\Sigma = \Sigma\} \qquad (2.13)$$

where $S \in D(m)$ and $S_S \in \hat{O}(m,d)$ with m+d=n. Note that d has to be large
enough in order to allow the scaling invariant part of M to be embedded
on the light cone of $O(m,d)$. (For example for the decagonal crystal
phase, the scaling invariant part of M is 2-dimensional, whereas in the
icosahedral phase it is 3-dimensional.)

The total holohedry of M is then the group H_{RS} generated by H_R and
H_S:

$$H_{RS}(M) = \{H_R, H_S\}\,. \qquad (2.14)$$

SYMMETRY OF LATTICES

In the previous section the restrictions imposed on rotations and
scalings by lattice translational symmetry have been considered.
Conversely, rotational and scaling symmetries impose restrictions on
the lattices left invariant. The Euclidean case needs to be considered
here only for fixing the idea: so for example a threefold rotational
symmetry implies in two dimensions that the lattice is hexagonal.

Analogous is the situation with respect to scaling invariance.
Let us again consider the two-dimensional case, and automorphs of the
form:

$$A_\mu = \begin{pmatrix} \cosh\chi & \sinh\chi \\ \sinh\chi & \cosh\chi \end{pmatrix} \quad \text{with} \quad 2\cosh\chi = \mu \text{ integer} \tag{3.1}$$

and a lattice Σ left invariant by A_μ (in the 2D Minkowskian plane with
orthonormal basis e_1, e_2 and metric tensor $g_{11} = 1$, $g_{22} = -1$, $g_{12} = 0$).
We can choose e_1 as one basis vector for Σ and we can consider as
primitive the lattice M_μ with basis

$$a_{s1} = e_1 \qquad\qquad a_{s2} = A_\mu e_1 \tag{3.2}$$

and call it (space-like) primitive. An alternative choice is $a_{s1} = e_2$
leading in the same way to a time-like basis. The isotropic case is
non-compatible with lattice invariancy with respect to A_μ. All other
(possibly inequivalent) cases can be obtained by the process of
centering. In the lattice basis (3.2) A_μ takes the form:

$$\Gamma(A_\mu) = \begin{pmatrix} 0 & \bar{1} \\ 1 & \mu \end{pmatrix}, \qquad\qquad \text{integer } \mu \tag{3.3a}$$

and the corresponding metric tensor left invariant is proportional to

$$g_\mu = \begin{pmatrix} 1 & \frac{\mu}{2} \\ \frac{\mu}{2} & 1 \end{pmatrix}. \tag{3.3b}$$

In the same way for a negautomorph N_ν we have

$$\Gamma(N_\nu) = \begin{pmatrix} 0 & 1 \\ 1 & \nu \end{pmatrix}, \qquad\qquad \text{integer } \nu \tag{3.4a}$$

and the metric tensor of a primitive lattice Λ_ν left invariant is
proportional to:

$$g_\nu = \begin{pmatrix} 1 & \frac{\nu}{2} \\ \frac{\nu}{2} & 1 \end{pmatrix}. \tag{3.4b}$$

As determinant and trace are invariant with respect to basis
transformations, M_μ and Λ_ν are representatives of the corresponding
Minkowskian system and each A_μ, N_ν determines different Minkowskian
crystal classes.

The derivation of the arithmetic crystal classes (and thus of the
classification of the scaling invariant lattices into Minkowskian
Bravais classes) is based on centering. That requires for each lattice
point a site symmetry A_μ and N_ν, respectively, and so one has to find
the lattices $M_{\mu,1}$ and $\Lambda_{\nu,1}$ of all points with site symmetry A_μ and N_ν,
respectively.

For $M_{\mu,1}$ that has already been done within the frame of
relativistic crystallography,[12,13,14] the result being that the index of
M_μ in $M_{\mu,1}$ is given by

$$I(M_\mu \subseteq M_{\mu,1}) = |\mu - 2|. \tag{3.5a}$$

In particular for $\mu = 3$, which characterizes a τ^2 invariant scaling, as indicated in Table 1, there is no additional centering, the index being 1.

The corresponding result for the negautomorph case is

$$I(\Lambda_\nu \subseteq \Lambda_{\nu,1}) = |\nu| \ . \tag{3.5b}$$

For the lowest value $\nu = 1$, which leads to a scaling by τ (see Table 1), there is also no additional centering. At this stage, it makes sense to restrict the admitted symmetries of the lattice Σ to those one gets from the holohedry of the underlying vector module M, expressed as group of integral matrices H_{RS} as in eq. (2.14). These are reducible affine transformations of the superspace leaving the physical space invariant. Of course for the holohedral point group H_{RS} of M one has:

$$H_{RS}\Sigma = \Sigma \tag{3.6}$$

as $H_{RS} \subset G\ell(n,\mathbb{Z})$, but H_{RS} expresses more than simply lattice basis transformations. Admitted point groups K_{RS} are then subgroups of H_{RS}, generated by a rotational point group K_R and a scaling point group K_S, subgroups of H_R and H_S defined in (2.3) and (2.13), respectively.

$$K_{RS} = \{K_R, K_S\} \qquad K_R \subseteq H_R \ , \ K_S \subseteq H_S \ . \tag{3.7}$$

SPACE GROUP, SCALE GROUP AND SCALE/SPACE GROUP

For a given point group K_{RS} as above, one can consider

(i) A group extension of $\mathbb{Z}^n$ by K_R which defines a n-dimensional space group G_R, subgroup of the Euclidean group E(n).
(ii) A group extension of $\mathbb{Z}^n$ by K_S which gives a scale group G_S subgroup of E(m,d), semi-direct product of the translation group T(n) and O(m,d) where n=m+d.
(iii) A group extension of $\mathbb{Z}^n$ by K_{RS} which yields a scale/space group G_{RS}.

Equivalent extensions are associated with the elements of a second cohomology group; and the unit element of it corresponds to a split extension (say a semi-direct product group structure). The key point is that there are cases where both $H_2(K_R, \mathbb{Z}^n)$ and $H_2(K_S, \mathbb{Z}^n)$ are trivial whereas $H_2(K_{RS}, \mathbb{Z}^n)$ not.[3] This implies that a scale/space group G_{RS} is not necessarily generated by G_R and G_S so that:

$$G_R \not\subset G_{RS} \quad \text{and} \quad G_S \not\subset G_{RS} \ , \ \text{in general.} \tag{4.1}$$

One has an analogous situation in Euclidean crystallography for the non-symmorphic space group case, because then its point group is not a subgroup of the space group. All these types of space groups G_R, G_S and G_{RS} can be made to appear, as subgroups of $IG\ell(n, \mathbb{R})$ the inhomogeneous linear group, semi-direct product of $\mathbb{R}^n$ by $G\ell(n, \mathbb{R})$ the n-dimensional linear group of $n \times n$ non-singular real matrices. Elements of this group can be written (in the Seitz notation) as $g = \{T|t\}$ with T the linear part and t the translational part of g. The action on the superspace is given by

$$g\, r_s = \{T|t\} r_s = T r_s + t \tag{4.2}$$

and one has a semi-direct product rule:

$$gg' = \{T|t\}\{T'|t'\} = \{TT'|t + Tt'\} . \tag{4.3}$$

Accordingly, in the groups G_R, G_S and G_{RS}, expressed in terms of their action on superspace, systems of non-primitive (or intrinsic) translations appear. So e.g. for an element g of G_{RS} the scale/space group one has:

$$g = \{T \,|\, n + v(T)\} \tag{4.4}$$

with $T \in K_{RS} \subset G\ell(n, \mathbb{Z})$, $n \in \mathbb{Z}^n$ and $v(T) \in \mathbb{R}^n$. The system of non-primitive translations $v(K_{RS})$ has elements which obey the rule of a crossed homomorphism:

$$v(TT') \equiv v(T) + Tv(T') \qquad (\text{modulo } \mathbb{Z}^n) \quad \text{for } T, T' \in K_{RS} \tag{4.5}$$

and is defined only modulo lattice translations and choice of the origin. Its equivalence class corresponds to an element of the first cohomology group

$$[v(K_{RS})] \in H_1(K_{RS}, \mathbb{R}^n/\mathbb{Z}^n) . \tag{4.6}$$

Analogous relations can be written for G_R and G_S. From a crystallographic point of view, relevant for defining the symmetry of a pattern is the action of the group on the superspace, so for classifying these symmetry groups one has to consider the elements of the group $H_1(K_{RS}, \mathbb{R}^n/\mathbb{Z}^n)$. The group structure, however, is fixed by the elements of the cohomology group $H_2(K_{RS}, \mathbb{Z}^n)$. In Euclidean crystallography both are related by a group isomorphism:

$$H_2(K_R, \mathbb{Z}^n) \simeq H_1(K_R, \mathbb{R}^n/\mathbb{Z}^n) . \tag{4.7}$$

One derives that relation from K_R being finite and $\mathbb{R}^n$ divisible. As in general K_S and K_{RS} are not finite, equation (4.7) is no more valid for the scale groups and the scale/space groups:

$$H_2(K_{RS}, \mathbb{Z}^n) \not\simeq H_1(K_{RS}, \mathbb{R}^n/\mathbb{Z}^n) , \qquad \text{in general.} \tag{4.8}$$

In dealing with crystallography of quasicrystals one has to take into account the two fundamental facts expressed by (4.1) and (4.8).

In what follows, we shall avoid those difficulties by restricting considerations to the symmorphic case, for which all systems of non-primitive translations are equivalent to zero:

$$v(K_{RS}) = v(K_R) = v(K_S) = 0 \qquad \text{implying} \quad G_{RS} = \{G_R, G_S\} \tag{4.9}$$

so that we can deal with scaling and rotational symmetries separately
without harm.

SOME PROPERTIES OF EQUIVALENT POSITIONS

We recall that in the present approach, and in the point-atom
approximation of a basis structure, atoms appear in superspace as
point-like objects at equivalent positions (2.5) of the appropriate
symmetry group G_R, G_S and/or G_{RS}, respectively. In physical space,
only a fraction of those atoms occurs and that according to the values
of the atomic occupation functions $C_j(n_S)$. The question of how one can
extract from a diffraction pattern the necessary information for
determining the occupation functions is not a trivial one, but will not
be discussed here.

Relevant for a given position is its multiplicity (i.e. the
number of equivalent points within a unit cell of the lattice) and its
site symmetry. No particular problems arise in this respect for G_R as
it is a Euclidean space group. For G_S the multiplicity of a general
position (which has site symmetry 1) is infinite. As we deal here with
a symmorphic case, K_S is a subgroup of G_S and the semigroup of the
enflation transformations has to leave the occupation functions
invariant. This implies that occupation of a general position is not
compatible with the requirement of a finite minimal atomic distance in
physical space: only special positions with finite multiplicity are
thus admissible.

Finite multiplicity occurs for positions having a sufficiently
large site symmetry. As illustration let us consider the m=d=1 case
and a G_S with primitive lattice M_μ having basis vectors a_{S1} and a_{S2} as
in (3.2) and scaling symmetry $K_S = \{A_\mu\} \simeq C_\infty$ (reflections are
disregarded).

The set of positions having site symmetry A_μ^k forms a lattice $M_{\mu,k}$.
The index of M_μ in $M_{\mu,k}$ is

$$I(M_\mu \subseteq M_{\mu,k}) = |p_{k+1}(\mu) - p_{k-1}(\mu) - 2| . \qquad (5.1)$$

where $p_k(\mu)$ is a solution of the recurrence relation in k:

$$p_{k+1}(\mu) = \mu\, p_k(\mu) - p_{k-1}(\mu) \qquad (5.2)$$

with initial values $p_0(\mu) = 0$ and $p_1(\mu) = 1$. For k=1 one gets from
(5.1) the special case indicated in (3.5a).

The multiplicity of a position belonging to $M_{\mu,k}$ is not for all
points the same and its determination requires a more detailed
analysis. In any case, the index (5.1) represents an upper value. As
an example used further on, we give for the case $\mu = 3$ a set of basis

vectors for the lattices $M_{3,k}$ expressed in terms of the basis (3.2) of M_μ.

For k = $2\ell+1$ one has:

$$a_{1,2\ell+1} = \frac{1}{p_{\ell+1}(3) + p_\ell(3)} \, a_{s1} \qquad a_{2,2\ell+1} = \frac{1}{p_{\ell+1}(3) + p_\ell(3)} \, a_{s2} \qquad (5.3a)$$

and for k = 2ℓ:

$$a_{1,2\ell} = \frac{1}{5p_\ell(3)} \, (-a_{s1} + a_{s2}) \qquad a_{2,2\ell} = \frac{1}{5p_\ell(3)} \, (-3a_{s1} + 2a_{s1}) \, . \qquad (5.3b)$$

The validity of (5.1) can be verified on the basis of the factorization expressions:

$$\Delta p_{2\ell+1}(\mu) - 2 = (\mu - 2)(p_\ell(\mu) + p_{\ell+1}(\mu))^2 \qquad (5.4a)$$

$$\Delta p_{2\ell}(\mu) - 2 = (\mu^2 - 4)p_\ell^2(\mu) \qquad (5.4b)$$

with

$$\Delta p_k(\mu) \overset{\text{def}}{=} p_{k+1}(\mu) - p_{k-1}(\mu) \, . \qquad (5.5)$$

EUCLIDISATION OF SCALING PROPERTIES

Crystallographic properties have to be independent from the description adopted. Nevertheless, there are descriptions which are better adapted than others for expressing given properties.

In the frame of a multimetrical superspace approach, it is natural to look for reference systems such that a given lattice gets its largest possible multimetrical holohedry. Indeed, as we are dealing with single objects in superspace (a lattice, a pattern and so on) a change in the Minkowskian reference system leaves the Minkowskian properties invariant but not the Euclidean ones, and conversely. Furthermore, it makes sense to examine whether a given quasicrystal module can be embedded on those high symmetric lattices and to discuss the structural meaning of that. Notice that the possibility of embedding on a lattice having a lower symmetry is always possible, but not always on one having a larger holohedry. That aspect has been discussed in several previous papers,[1,2,3,15] where it has been shown in which sense the possibility of embedding a Fibonacci chain on a square lattice is related to its golden-mean enflation invariance. Here, scaling symmetry has been expressed in terms of the Minkowskian holohedry of a 2-dimensional lattice, which in casu appears to be a square lattice from the Euclidean point of view. For investigating the generality of such situations, we introduce the concept of euclidisation.

By euclidisation of scaling properties we mean a Euclidean characterization of scaling properties expressed in terms of non-Euclidean symmetries in physical space and in superspace.

60

For fixing the ideas, consider a vector module M with scaling holohedry H_S generated by an automorph A_μ embedded on a primitive lattice M_μ as defined in Section 4. The 1-dimensional physical space V being on the light cone of the Minkowskian superspace, it is invariant with respect to $O(1,1)$ transformations, i.e. to the choice of (Minkowskian) reference systems, and so also the Minkowskian holohedry of M_μ is invariant, in the same sense as the Bravais class of a lattice remains the same by rotating the lattice. Different choices modify, however, the Euclidean properties of V and of M_μ. In V it simply corresponds to a change in the unit of length; for M_μ, and because of the interplay of contraction and dilatation, the Euclidean holohedry of the lattice is modified. We now try to express the Minkowskian isometric properties of M_μ by means of a Euclidean isometric lattice. In the present case we try to transform M_μ covariantly into a square or into a hexagonal lattice (these are the 2-dimensional Euclidean lattice isometric with respect to rotations).

Adopting an orthonormal basis e_1, e_2 with Minkowskian metric tensor $g_{11} = 1$, $g_{22} = -1$ and $g_{12} = 0$, a basis for M_μ can be chosen as

$$a_{S1}^{o} = e_1 = (1,0) \qquad a_{S2}^{o} = A_\mu e_1 = (\cosh\chi, \sinh\chi) \tag{6.1}$$

with $2\cosh\chi = \mu$. After a change of reference system the basis becomes:

$$a_{S1} = L\, a_{S1}^{o} = (\cosh\alpha, \sinh\alpha)$$

$$a_{S2} = L\, a_{S2}^{o} = (\cosh(\alpha + \chi), \sinh(\alpha + \chi)) \, . \tag{6.2}$$

Going over to a Euclidean metric tensor $g_{11} = 1$, $g_{22} = 1$ and $g_{12} = 0$, for the same basis vectors e_1, e_2 the vectors a_{S1} and a_{S2} get a Euclidean length and angle. So e.g.:

$$|a_{S1}|^2 = \cosh^2\alpha + \sinh^2\alpha.$$

Let us denote by a_1 and a_2 the isometrical basis of M_μ, keeping $a_{S1} = a_1$ as it is the lattice vector with the starting smallest Euclidean length. The possible second basis vector a_2 then has the general form:

$$a_2 = (a_{S2} - z a_{S1}), \qquad \text{integer } z. \tag{6.3}$$

The properties:

$$|a_1| = |a_2| \qquad \text{and} \qquad a_1 \cdot a_2 = |a_1||a_2|\cos\varphi \tag{6.4}$$

are then expressible in terms of the two equations:

$$(\mu - 2z)\cos\varphi = 2 - z(\mu - z) \tag{6.5a}$$

$$(1 + \tanh^2\alpha)(\mu - 2z - 2\cos\varphi) + 4\sqrt{\mu^2 - 4}\tanh\alpha = 0. \tag{6.5b}$$

In particular for the square lattice case one puts $\cos\varphi = 0$ and one gets the diophantine equation $2 = z(\mu - z)$ with solutions $\mu = 3$ and $z = 2$ or $z = 1$ implying the values:

$$\tanh\alpha = \pm(\sqrt{5} - 2) . \tag{6.6a}$$

Therefore the angle β made by a_{s1} with the physical space is

$$\tan\beta = \pm\frac{\sqrt{5} - 1}{2} \tag{6.6b}$$

the inverse of the golden ratio.

In the same way, in the case of an hexagonal lattice ($\cos\varphi = 1/2$) the only solution is $\mu = 4$ and $z = 3$ yielding

$$\tanh\alpha = \frac{1}{\sqrt{3}} \quad \text{and} \quad \tan\beta = 2 - \sqrt{3}. \tag{6.7}$$

Both results are consistent with the previous analysis.[1,2,3]

The remarkable fact is that the only primitive lattices M_μ admitting euclidisation are those for $\mu = 3$ and $\mu = 4$. These are of course not the only scaling invariant lattices admitting euclidisation, as the negautomorphic Λ_ν and all the centered lattices have not yet been investigated. But already now we can show that infinitely many lattices exist with the good properties. Consider in particular the lattice $M_{\mu,k}$ consisting of all points with site symmetry A_μ^k for $\mu = 3$ and arbitrarily power k and apply the same euclidisation procedure as above with $\tanh\alpha = \sqrt{5}-2$. The basis vectors (5.3) expressed in the square basis a_1, a_2 are given by:

$$a_{1,2\ell+1} = \frac{1}{p_{\ell+1}(3) + p_\ell(3)}\,a_1 \qquad a_{2,2\ell+1} = \frac{1}{p_{\ell+1}(3) + p_\ell(3)}\,a_2 \tag{6.8a}$$

and

$$a_{1,2\ell} = \frac{1}{5p_\ell(3)}\,(a_1 + 2a_2) \qquad a_{2,2\ell} = \frac{1}{5p_\ell(3)}\,(2a_1 - a_2). \tag{6.8b}$$

Accordingly, after euclidisation all the lattices $M_{3,k}$ become square lattices.

It follows that there are scaling invariant patterns having as symmetry the irreducible space group p4. From the multimetrical interplay among Wyckoff positions of the groups G_R, G_S and G_{RS}, a new group appears: the n-dimensional space group G, which is the Euclidean symmetry group of the pattern one gets from an appropriate embedding.

It is important to be aware that the structural implications of the pattern cannot be understood in terms of G only: so for example a four-fold rotation in superspace gets its meaning only when interpreted in terms of an equivalent transformation of G_{RS}, which expresses rotation and/or scaling symmetry of the quasicrystal structure.

As an illustrative example a centrosymmetric scaling invariant
decoration of a Fibonacci chain is considered (but in terms of
automorphs only). The groups involved are:

$$
\begin{aligned}
K_R &= \bar{1} & G_R &= p2 \\
K_S &= \{A_3\} \overset{\text{def}}{=} \hat{3} & G_S &= p\hat{3} \\
K_{RS} &= \hat{3}\,\bar{1} & G_{RS} &= p\hat{3}\,2 \\
K &= 4 & G &= p4
\end{aligned}
\tag{6.9}
$$

where $\hat{3}$ denotes the generator A_3 and where K and G represent the
euclidisation of K_{RS} and G_{RS}, respectively (also that of K_S and G_S).
A number of sets of equivalent positions with finite multiplicity and
high site symmetries are indicated in terms of multiplicity, Wyckoff
letter, site symmetry and coordinates. Their euclidisation requires
sometimes an appropriate grouping of those positions, and that is
possible in a straightforward way as indicated in Table 2.

Table 2. Scaling invariant positions in $G_{RS} = p\hat{3}\,2$ and
their G = p4 euclidisation.

$p\hat{3}2$ positions				p4 euclidisation		
Multiplicity	Wyckoff letter	Site symmetry	Coordinates	Multiplicity	Wyckoff letter	Site symmetry
1	1a	$\hat{3}2$	$0,0$	1	a	4
2	2a	$\hat{3}^2$	$\frac{1}{5},\frac{2}{5}$ $\frac{4}{5},\frac{3}{5}$			
2	2b	$\hat{3}^2$	$\frac{3}{5},\frac{1}{5}$ $\frac{2}{5},\frac{4}{5}$	4	d	1
3	3a	$\hat{3}^32$	$\frac{1}{2},\frac{1}{2}$	1	b	4
			$\frac{1}{2},0$ $0,\frac{1}{2}$	2	c	2
6	3b	$\hat{3}^3$	$\frac{1}{4},0$ $\frac{3}{4},0$ $\frac{1}{2},\frac{1}{4}$ $\frac{1}{2},\frac{3}{4}$ $\frac{1}{4},\frac{3}{4}$ $\frac{3}{4},\frac{1}{4}$			
6	3c	$\hat{3}^3$	$0,\frac{1}{4}$ $0,\frac{3}{4}$ $\frac{1}{4},\frac{1}{2}$ $\frac{3}{4},\frac{1}{2}$ $\frac{1}{4},\frac{1}{4}$ $\frac{3}{4},\frac{3}{4}$	12	d	1
4	4a	$\hat{3}^4$	$0,\frac{1}{3}$ $0,\frac{2}{3}$ $\frac{1}{3},\frac{1}{3}$ $\frac{2}{3},\frac{2}{3}$			
4	4b	$\hat{3}^4$	$\frac{1}{3},0$ $\frac{2}{3},0$ $\frac{1}{3},\frac{2}{3}$ $\frac{2}{3},\frac{1}{3}$	8	d	1

For explaining Table 2 some few examples are given. With respect to
the square lattice basis a_1, a_2 the matrices corresponding to A_3^k are:

$$\Gamma_Q(A_3^k) = \begin{pmatrix} 2p_k(3) - p_{k-1}(3) & p_k(3) \\ p_k(3) & p_{k+1}(3) - 2p_k(3) \end{pmatrix} \qquad (6.10)$$

and in particular for k = 1 and 2 one has:

$$\Gamma_Q(A_3) = \begin{pmatrix} 2 & 1 \\ 1 & 1 \end{pmatrix} \qquad \text{and} \qquad \Gamma_Q(A_3^2) = \begin{pmatrix} 5 & 3 \\ 3 & 2 \end{pmatrix} . \qquad (6.11)$$

One verifies that $(\frac{1}{2}, \frac{2}{5})$ of the Wyckoff position 2a is invariant (modulo
$\mathbb{Z}^2$) with respect to A_3^2

$$\begin{pmatrix} 5 & 3 \\ 3 & 2 \end{pmatrix} \begin{pmatrix} \frac{1}{5} \\ \frac{2}{5} \end{pmatrix} = \begin{pmatrix} \frac{11}{5} \\ \frac{7}{5} \end{pmatrix} \sim \begin{pmatrix} \frac{1}{5} \\ \frac{2}{5} \end{pmatrix}$$

and is transformed by $\hat{3}$ and by 2, the two generators of the point group
$\hat{3}2$, into $(\frac{4}{5}, \frac{3}{5})$.

$$\begin{pmatrix} 2 & 1 \\ 1 & 1 \end{pmatrix} \begin{pmatrix} \frac{1}{5} \\ \frac{2}{5} \end{pmatrix} = \begin{pmatrix} \frac{4}{5} \\ \frac{3}{5} \end{pmatrix} \qquad \begin{pmatrix} \bar{1} & 0 \\ 0 & \bar{1} \end{pmatrix} \begin{pmatrix} \frac{1}{5} \\ \frac{2}{5} \end{pmatrix} \sim \begin{pmatrix} \frac{4}{5} \\ \frac{3}{5} \end{pmatrix} .$$

The 1a position of $p\hat{3}2$ corresponds to the position a of p4. In most
cases more than one position having a given site symmetry is needed for
giving rise to a p4 invariant pattern: e.g. both positions 2a and 2b
together of the scale/space group symmetry $p\hat{3}2$ define a pattern having
the non-reducible space group symmetry p4. The converse situation also
occurs: so e.g. the single position 3a of $p\hat{3}2$ consists of the two
positions b and c of p4.

Clearly, what has been discussed here is not the end of the story,
but just a beginning. Negautomorphs can also be treated in a same way.
Whether the euclidisation as it has been approached here, can be
extended to other cases (the hexagonal one in particular) and to higher
dimensions has still to be investigated.

ACKNOWLEDGEMENTS

The final version has been written after the oral presentation of
the present paper. Intensive discussions with Dr. Ted Janssen and
Dr. André Katz on scaling invariance in quasicrystals greatly helped
the author in clarifying his point of view. Thanks are also expressed
to the Stichting voor Fundamenteel Onderzoek der Materie of the
Nederlandse Organisatie voor Zuiver Wetenschappelijk Onderzoek for
partial financial support.

REFERENCES

1. A. Janner, Crystallography of quasicrystals, _in_: "Fractals, Quasicrystals, Chaos, Knots and Algebraic Quantum Mechanics", A. Amann, L. Cederbaum and W. Gans, eds., Kluwer Academic Publ., Dordrecht (1988).

2. A. Janner, Symmetry of higher dimensional crystallography, _Phase Trans_. 16/17: 87-101 (1989).

3. A. Janner and T. Janssen, Alternative superspace embeddings of quasicrystals, _to appear in_: "Proceedings Third International Conference on Quasicrystals and Incommensurate Structures", Vista Hermosa, Mexico (1989).

4. T. Janssen, Symmetries of tilings and quasicrystals, _to appear in_: "Proceedings of the Adriatic Anniversary Research Conference on Quasicrystals", World Scientific, Trieste (1989).

5. A. Janner and T. Janssen, Symmetry of incommensurate crystal phases I, II, _Acta Cryst_. A36: 399-408, 408-415 (1980).

6. T. Janssen, Crystallography of quasicrystals, _Acta Cryst_. A42: 261-271 (1986).

7. P.M. de Wolff, Symmetry operations for displacively modulated structures, _Acta Cryst_. A33: 493-497 (1977).

8. A. Janner and T. Janssen, Symmetry of periodically disordered crystals, _Phys. Rev_. B15: 643-658 (1977).

9. A. Katz and M. Duneau, Quasiperiodic patterns and icosahedral symmetry, _J. de Physique_ 47: 181-196 (1986).

10. P. Bak, Icosahedral crystals from cuts in six-dimensional space, _Scr. Metall_. 20: 1199-1204 (1981).

11. N.G. de Bruijn, Algebraic theory of Penrose's non-periodic tilings of the plane, _Math. Proc_. A84: 39-66 (1981).

12. A. Janner and E. Ascher, Crystallography in two-dimensional metric spaces, _Zeits. für Krist_. 130: 277-303 (1969).

13. A. Janner and E. Ascher, Bravais classes of two-dimensional relativistic lattices, _Physica_ 45: 33-66 (1969).

14. A. Janner and E. Ascher, Relativistic crystallographic point groups, _Physica_ 45: 67-85 (1969).

15. A. Janner, Superspace embedding of 1-dimensional quasicrystals, _J. de Physique Colloque_ 47C3: 95-102 (1986).

Landau Theory and Direct Methods for Crystal Structure Analysis*

Alan L. Mackay

Department of Crystallography
Birkbeck College, (University of London)
Malet Street, London WC1E 7HX

Abstract

Fourier analysis has long been used in crystal structure analysis for constructing a "picture" of the structure. Each component sinusoidal density wave is determined by measuring one Bragg diffraction spot. The consequences of regarding these density waves as real and as having energetic interactions are examined.

Landau theory, originally developed (Landau, 1936) for examining the relative energies of possible crystal structures at a second-order phase transition, gives a phenomenological, *ad hoc*, expression for the free energy Φ of a crystal as the sum of sinusoidal density waves (of "matter", but unspecified in detail) superimposed on a background density ρ_0 representing some standard, higher temperature, higher symmetry, structure, perhaps even amorphous.

$$\rho(\underline{r}) = \rho_0(\underline{r}) + \sum_{\underline{g}} \rho_{\underline{g}} \exp(2\pi i \underline{g}.\underline{r}) \tag{1}$$

where g is a reciprocal lattice vector and ρ_g is complex, representing the amplitude and phase of a "matter" wave. As stated, for example, by Nelson and Halperin (1985), the free energy of the crystal can be expressed as:

$$\Phi = 1/2 \sum_{\underline{g}} K_g |\rho_g|^2 + \omega_3 \sum_{\underline{g}_1 + \underline{g}_2 + \underline{g}_3 = 0} \rho_{\underline{g}_1} \rho_{\underline{g}_2} \rho_{\underline{g}_3} + \omega_4 \sum_{\underline{g}_1 + \underline{g}_2 + \underline{g}_3 + \underline{g}_4 = 0} \rho_{\underline{g}_1} \rho_{\underline{g}_2} \rho_{\underline{g}_3} \rho_{\underline{g}_4} + \ldots \tag{2}$$

The first term represents the self-energy of the individual waves, the second the triplet interactions and the third the quadruplet interactions and so on. This expression is obtained by assuming that the free energy varies smoothly through the second order phase transition and that near that transition it can be expressed in

---

*based on a paper presented at the BCA meeting, London, 20 Nov. 1986

Geometry and Thermodynamics
Edited by J.-C. Tolédano
Plenum Press, New York, 1990

powers of $\rho(\underline{r})$, the deviation from the mean density. The algebraic results that $\int_0^1 \cos 2\pi h x . \cos 2\pi k x . dx = 0$, unless $h = k$ when it is $1/2$, and that

$$\int_0^1 \cos(2\pi \underline{g}_1 . \underline{r}) . \cos(2\pi \underline{g}_2 . \underline{r}) . \cos(2\pi \underline{g}_3 . \underline{r}) . |d\underline{r}| = 0 \tag{3}$$

unless $\underline{g}_1 + \underline{g}_2 + \underline{g}_3 = 0$, in which case it equals $1/4$, are used. The second result uses the identity:

$$4 \cos A . \cos B . \cos C = \cos(A+B+C) + \cos(-A+B+C) + \cos(A-B+C) + \cos(A+B-C) \tag{4}$$

It will be observed that the triplet, quadruplet, and further terms are just the structure invariants used in crystallography for the direct methods of phase determination developed by Karle (1986) and Hauptman (1986) from 1950 onwards.

$F_{\underline{h}}$ is the structure factor with phase $\phi_{\underline{h}}$, so that $F_{\underline{h}} = |F_{\underline{h}}| \exp(i\phi_{\underline{h}})$ and $F_{\underline{h}}$ and the electron density $\rho(\underline{r})$ are related by

$$F_{\underline{h}} = \int_V \rho(\underline{r}) \exp(2\pi \underline{h}.\underline{r}) . dV \tag{5}$$

where V is the volume of the unit cell, and

$$\rho(\underline{r}) = \frac{1}{V} \sum_{\underline{h}} F_{\underline{h}} \exp(-2\pi i \underline{h}.\underline{r}) \tag{6}$$

(Ladd and Palmer, 1980). The structure factors $F_{\underline{h}}$ are normalised to $E_{\underline{h}}$ by division by the average electron density for the same Bragg angle. $|E_{\underline{h}}|$ are obtainable experimentally, but the phases $\phi_{\underline{h}}$ are unknown.

"Direct methods" is a procedure for restoring the unobservable phases of the structure factors $F_{\underline{h}}$ and permitting the Fourier inversion to obtain the electron density distribution of the scattering crystal. The physical information used in the method is the assumption that the structure consists of non-overlapping atoms with non-negative electron density. Other methods of finding the phases require the measurement of structure amplitudes from several chemical modifications of the original crystal structure.

By considering the effect of a shift of origin it can be shown that, if $\underline{h}_1 + \underline{h}_2 + \underline{h}_3 = 0$ then $(\phi_{\underline{h}_1} + \phi_{\underline{h}_2} + \phi_{\underline{h}_3})$ is an invariant with respect to a shift of origin and similarly for quadruplets, etc. Thus, the product of three complex structure factors for which $\underline{h}_1 + \underline{h}_2 + \underline{h}_3 = 0$ is invariant with respect to a shift of origin:

$$F_{\underline{h}_1} . F_{\underline{h}_2} . F_{\underline{h}_3} = |F_{\underline{h}_1}||F_{\underline{h}_2}||F_{\underline{h}_3}| \exp i(\phi_{\underline{h}_1} + \phi_{\underline{h}_2} + \phi_{\underline{h}_3}) \tag{7}$$

It can be shown further that the probability of this invariant phase angle $\phi = (\phi_{\underline{h}_1} + \phi_{\underline{h}_2} + \phi_{\underline{h}_3})$ is a distribution centred on 0 deg. with a width which depends on the product $|E_{\underline{h}_1}||E_{\underline{h}_2}||E_{\underline{h}_3}|$, being narrower, the greater the product [Ladd and Palmer, 1980, p.161]. This result turns physically on the assumption of non-negative electron density.

The general condition that a Fourier series should represent a non-negative density can be expressed as a series of determinantal conditions $D_m \geq 0$, where

$$D_m = \begin{vmatrix} 1 & U_{12} & U_{13} & \cdot \\ U_{21} & 1 & U_{23} & \cdot \\ U_{31} & U_{32} & 1 & \cdot \\ \cdot & \cdot & \cdot & \end{vmatrix} \tag{8}$$

U_{ij} is the unitary structure factor $|E_{\underline{h}}| = N^{1/2}|U_{\underline{h}}|$.

The value of D_m depends solely upon structure invariants and is itself a structure invariant. "Among all sets of phases that are compatible with inequalities, the most probable set is that which leads to a maximum value of the determinant".

This is an alternative way of stating the result above. D_m depends on $m(m - 1)(m - 2)/6$ independent triplet invariants.

Structure invariants are defined in transform space. What structures do they represent in real space? With increasing order of the determinant they represent pictures of the structure with increasing resolution. Alexander and McTague (1978) showed that, for example, sets of 15 reciprocal lattice vectors parallel to the edges of an icosahedron, produced many triplets and corresponded to an icosahedral structure, perhaps later realised as a quasi-crystal, in real space.

X-ray crystal structure analysis by direct methods thus consists in measuring the structure factors $|F_{\underline{h}}|$ and in finding the phases $\phi_{\underline{h}}$ such that for triplets $|F_{\underline{h}_1}||F_{\underline{h}_2}||F_{\underline{h}_3}|$ for which, as this product is larger, the invariant $\phi = \phi_{\underline{h}_1} + \phi_{\underline{h}_2} + \phi_{\underline{h}_3}$ is the more probably closer to zero. When a consistent set of phases $\phi_{\underline{h}}$ has been recovered, the inverse Fourier transformation

$\rho(\underline{r}) = -\frac{1}{V}\sum_{\underline{h}}|F_{\underline{h}}|\exp(i\phi_{\underline{h}} - 2\pi\underline{h}.\underline{r})$ can be performed and the structure thus imaged as an electron density distribution. It should also be possible to use direct methods for the establishment of a self-consistent phase field for the amplitudes of the continuously distributed scattering from an amorphous material.

It has been pointed out (Mackay, 1986) that this procedure is such as to minimise Φ (if the coefficient ω_3 is negative) in the expression for the energy and thus, that the experimental crystal structure determination and the Landau minimisation of the energy, should give the same answer. Thus, the direct methods of crystal structure analysis, which operate in transform space, correspond to the geometrical packing of atoms in real space, the connection being the Karle-Hauptman determinants for $\rho(\underline{r}) \geq 0$.

Since the Landau theory considers in transform space the energetics of the structure (not recognising atomicity) it should be the equivalent of real space calculations of the minimum energy structure based on interatomic forces.

The problem, then, is what is the nature of the density waves which make up the crystal, and how do they interact? That is, is there a physical interpretation of Landau's *ad hoc* phenomenological treatment?

It is customary to make a "picture" of the electron density in a crystal by the *linear* superposition of the density waves measured by crystalllography but, the direct methods procedure outlined above, shows that the individual sine wave terms are not independent of each other and that they interact. Moreover, they interact in the same form as in the Landau theory, which implies that the crystal can be regarded, taking energy into account, as the *non-linear* superposition of electron density waves.

We must recognise that, although the X-ray method measures electron density, the electrons are everywhere accompanied by protons which preserve electrical neutrality. However, the whole arrangement could be repeated for neutron diffraction from a hypothetical structure containing only atoms having a positive scattering length for their scattering "matter".

We must ask whether it is possible to give a physical interpretation to the Ewald-Bertaut method of calculating the Madelung constant for Coulomb (and for van der

Waals (Williams, 1971)) interactions, since this shows that the energies of waves represented in reciprocal space can be calculated.

Parseval's theorem, $\int f(x).g(x).dx = \int trans(f).trans(g).dh$, permits the calculation of energy in either direct or Fourier transform spaces.

As used in his program WMIN, W. R. Busing calculates the Coulomb energy of a structure, both in direct space and in transform space. An error function is applied in real space and its transform (a Gaussian) in transform space to make both calculations converge better. The function used for transform space is:

$$\frac{1}{2\pi V Z} \sum_{\underline{h} \neq 0} |F_c(\underline{h})|^2 Q(\underline{h})^{-2} [\exp(-b^2)]$$

where $F_c(\underline{h})$ is the structure factor with the charge q_i replacing the usual X-ray expression for the atomic scattering factor.

$$F_c(\underline{h}) = \sum_{i}^{one\ cell} q_i \exp(2\pi i \underline{h}.\underline{r}_i)$$

$Q(\underline{h})$ is the reciprocal lattice vector representing the wave.

It is implied that, by adjusting the error function which is applied, the whole calculation could be carried out in transform space. The expression used is a simple one where the energy attributable to each wave is proportional to its intensity and thus it implies that these waves are independent of each other and add linearly as regards energy.

Is there here some contradiction or do the essential interactions and non-linearity discussed above arise from the non-Coulomb interactions?

References

[1] S.Alexander and J. McTague, Phys. Rev. Let., **41**, No.10, 702-705, (1978).

[2] F.Bertaut, J. Phys., **13**, 499-, (1952)

[3] W.L.Bragg, Z. f. Krist., **70**, 483-, (1929)

[4] W.R.Busing, "WMIN a computer program to model molecules and crystals in terms of potential energy functions", Oak Ridge National Laboratory, 1981. (ORNL-5747).

[5] P.P.Ewald, Ann. Physik, **64**, 253-, (1921)

[6] H. Hauptman, Angew. Chemie, **25**, 603-613, (1986) and Science, **233**, 178-183, (1986)

[7] J.Karle, Angew. Chemie, **25**, 614-629, (1986)

[8] M.F.C.Ladd and R.A.Palmer, (eds.), *Theory and Practice of Direct Methods in Crystallography*, Plenum Press, NY, 1980.

[9] A.L.Mackay, Science, **234**, 12, (3 Oct. 1986).

[10] D.R.Nelson and B.I.Halperin, Science, **229**, 233-238, (19 July 1985)

[11] J.C.Slater, *Insulators, Semiconductors and Metals*, McGraw-Hill, NY, (1967). pp. 215-220.

[12] D.E.Williams, Acta Cryst., **A27**, 452-455, (1971)

GEOMETRY OF FILMS OF AMPHIPHILE MOLECULES:
A CURVED SPACE APPROACH

J.-F. Sadoc and J. Charvolin

Laboratoire de Physique des Solides
Bât. 510, Université Paris-Sud
91405 Orsay, France

INTRODUCTION

Periodical organizations of two fluid media separated by interfaces are very common in liquid crystals where they most often occur under the forms of periodical stackings along one dimension of fluid layers of molecules . In lamellar phases of lyotropic liquid crystals , built by amphiphilic molecules in presence of water [1] , paraffinic bilayers of amphiphiles and polar layers of water are alternatively stacked with flat interfaces defined by the polar heads of the amphiphiles , as shown in fig. 1a . Besides these phases with periodicity along one dimension the phase diagrams of liquid crystalline systems may present other ordered phases with periodicities along two or three dimensions , curved interfaces and various topologies [1,2,3] . For instance consider the particular case of phases with cubic symmetry and bicontinuous topology . They are of the type shown in fig. 1b where a film of water without self-intersection separates two identical labyrinths of amphiphilic molecules . We wonder if the complex geometry of such structures could not rely upon the same principle than that which is easily discernable in the much simpler lamellar and smectic phases .

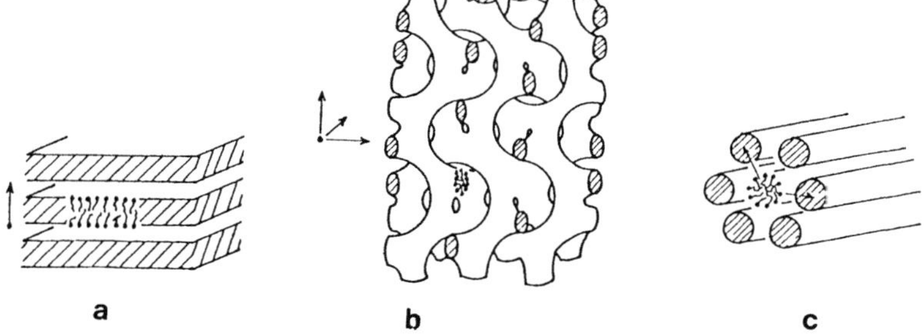

fig. 1- a) schematic representation of the lamellar phase of a lyotropic liquid crystal , hatched layers are the amphiphilic layers , they are separated by water layers.-b) schematic representation of a cubic phase with symmetry Ia3d , two interwoven but not connected labyrinths of amphiphilic molecules are separated by a film of water whose middle surface is at mid way between the two interfaces. -c)Hexagonal structure.

Geometry and Thermodynamics
Edited by J.-C. Tolédano
Plenum Press, New York, 1990

The latter are periodic organizations of flat interfaces at constant distances , could bicontinuous cubic phases be periodic organizations of curved interfaces at optimized distances , adjacent interfaces having the same concavity relative to the media they separate ?

We therefore formulate the problem of the description of these cubic structures in purely geometrical terms and reduced it to the search for geometrical elements, points, lines or surfaces, around which, or along which, the films may be organized in order to conciliate optimal distances and symmetric interfacial curvatures . We are obviously facing a frustration , between curvatures and distances , which does not exist in the case of bent parallel layers . This difference implies that the search for solutions in our case can not be done directly in the Euclidean space , as it is done in the latter case , but must go first through the search for solutions in a curved space . Once such solutions are found defects of rotation , or disclinations, are introduced in order to decrease the curvature of the curved space and come back to the Euclidean space . The structures obtained that way can therefore be considered as structures of disclinations . This procedure is currently used in condensed matter physics for solving cases of frustration , as for instance bi-dimensional tilings of regular polygons or tri-dimensional packings of regular polyhedra [4] .Its recent application for solving the curvature-thickness frustration in a liquid crystalline bilayer has provided solutions whose topologies are similar to those observed in liquid crystalline phases [5] .

PHYSICAL BASIS OF THE FRUSTRATION

We focus our attention on the well documented case of films built by amphiphilic molecules and water . In that case the simplest system of periodic films is the lamellar structure represented in fig. 1a . The distances between interfaces , measured along the normals to the interfaces , are constant . The interfaces are flat , or the lateral area per polar head and per paraffinic chains are equal , revealing equal equilibrium distances between heads and between chains in planes parallel to the interfaces . Finally the layers have isotropic properties in these planes .

There are several forces which fix the inter and intra layer distances in such a system : Van der Walls forces , electrostatic interactions between polar groups at the interfaces , forces created by water polarization and charge distribution in aqueous layers , hydrophobic interaction which prevent the presence of water in amphiphilic layers and steric forces associated with undulations of the layers [6] . It is reasonable to think that the role of the components normal to the interfaces is to define the relative positions of the interfaces , and therefore the thickness of the aqueous and amphiphilic layers , and that the role of the components parallel to the interfaces is to determine the interfacial curvature . Their variations are controlled by the thermodynamical variables which are the degree of hydration and the temperature . The normal components make the distances between interfaces vary but , as we consider homogeneous interfaces , these distances should remain constant when moving along the interfaces . The parallel components make the lateral distances between molecules vary , but not necessarily in concordance at different levels along the normals to the interfaces , so that variations of the parameters away from their values in the lamellar phase may induce differences in the lateral area at different levels of the bilayer and therefore interfacial curvatures . A coarse way to systematize the non concordant actions of these second components might be to say that the interactions between polar heads are mainly dominated by the degree of hydration whereas the behaviour of paraffinic chains is dominated by the temperature .

Owing to the symmetry of the amphiphilic bilayer with respect to its middle surface it is clear that the fact that two adjacent interfaces have different area than the middle surface between them , or have symmetric curvatures , is not compatible with constant distances between these interfaces if the lamellar structure is kept , as shown in fig. 3 . This is a situation where two physical forces oppose , i.e. a frustration .

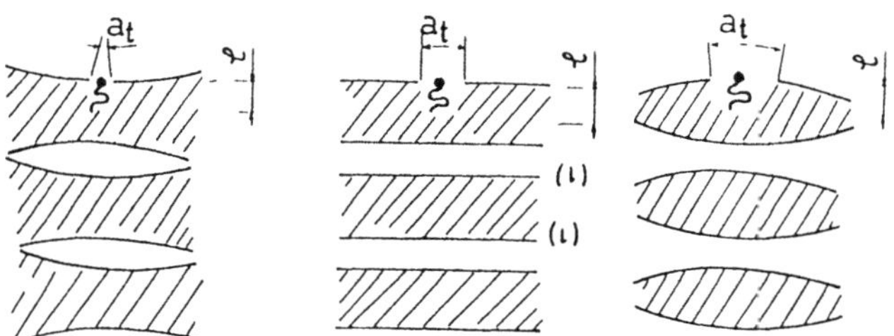

fig.2- Schematique representations of a periodic system of films with flat interfaces (Ccenter) : constant interfacial distance and zero curvature are compatible in R_3. The same with symmetrically curved interfaces (left and right). Constant interfacial distances and non zero curvatures are no longer compatible in R_3. and the system becomes frustrated.

THE FRUSTRATION AND ITS RELAXATION

<u>Frustration in R3</u>

So we consider the interplay between forces normal to the interfaces, which maintain constant distances between them, and forces parallel to the interfaces, which impose symmetric curvatures to adjacent interfaces of the film. The periodical stacking of the films can be preserved only if the structure is embedded in a space of adequate curvature. For instance the situation with flat interfaces is possible in the Euclidean space R_3, while the situations with curved interfaces are not, as the constant distances between their interfaces can not be maintained if they are symmetrically curved. In this latter case the system becomes frustrated if left in R_3. However the frustration can be relaxed if the system is transferred into curved spaces with positive Gaussian curvatures. This relaxation was obtained by applying the middle surface of the bilayer on one of the two possible surfaces separating S3 in two equivalent sub-spaces , the great sphere S2 or the spherical torus T2 , so that the two interfaces were applied on surfaces parallel to either S2 or T2 and with smaller area [7]. When T2 was used a structure with bicontinuous topology was obtained The structures created in such spaces are ideal structures, as they strictly conciliate interfacial. curvatures and distances, but have obviously no reality. The geometrical configurations possible for the real structures are obtained by mapping the curved spaces onto the Euclidean space R_3 by an homogeneous introduction of defects of rotation, or disclinations, which does not change the amphiphile/water volume and area ratio. The real structures can therefore be seen as structures of disclinations.

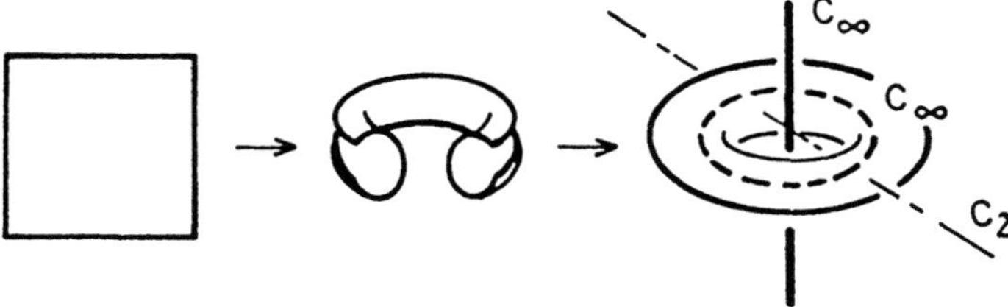

fig.3. The spherical torus can be built by identifications of the sides of a square sheet in S3 , as this is done in a curved space the sheet suffers no distorsion , the torus represented in this figure is a stereographic projection in R3 of the spherical torus in S3 , the two C∞ axes are identical orthogonal polar circles in S3

We then use the three types of spacial structures:

-The spherical torus T2 in S3

In this case the middle surface of the film is supported by the spherical torus and the two interfaces are supported by parallel torii at equal distances of one molecular length l from it. As represented in fig. 3,

-The cylinder S1*R1 in S2*R1

$S_2 * R_1$ can be obtained from S_3 by a disclination process but this aspect does not intervene here.

-The great sphere S2 in S3

In this case the middle surface of the film is supported by the great sphere and the two interfaces are supported by parallel spheres at equal distances of one molecular length l from it.

<u>Relaxation of the frustration : example of the spherical torus T2 in S3</u>

The curved space S3 can be described as the hypersphere in R4 , it is a finite space with positive Gaussian curvature . In this space the spherical torus T2 can be built by identification of opposite sides of a square sheet . T2 has two C∞ axes , which are great circles of S3 , and an infinity of C2 axes normal to its surface. T2 has therefore zero Gaussian curvature and admits the same tilings by regular polygons than the Euclidean plane ,which are {4,4} , {6,3} and {3,6} [8]. The area of T2 is maximal and that two tori parallel to T2 , at equal distances on either sides of it , have equal area smaller than that of T2 .

From the above properties of S3 and its family of parallel torii it appears that the set of a spherical torus T2 surrounded by two parallel torii at equal distances can be considered as a representation without frustration of the system of frustrated fluid films , as illustrated in fig. 6 . The periodicity in R3 , which is the repetition of the films when moving along geodesics normal to the films , is reproduced in S3 by the cyclic crossings of the spherical torus when moving along geodesics of S3 normal to T2 . Also the two sub-spaces separated by T2 are identical as are the two media separated by the film in the cell of the structure . Finally the frustration is obviously relaxed as the two interfaces of the film , which are supported by two torii parallel to T2 , have smaller area than the middle surface of the film supported by T2 . Once the system has been transfered into S2 possible structures are found by mapping S3 onto R3 . We shall limit ourselves to the process which maintains the bicontinuous topology of T2 . In the course of this process the positive curvature of S3 will be decreased to zero and the support of the film , the spherical torus with zero Gaussian curvature , will become an hyperbolic surface with negative Gaussian curvature .

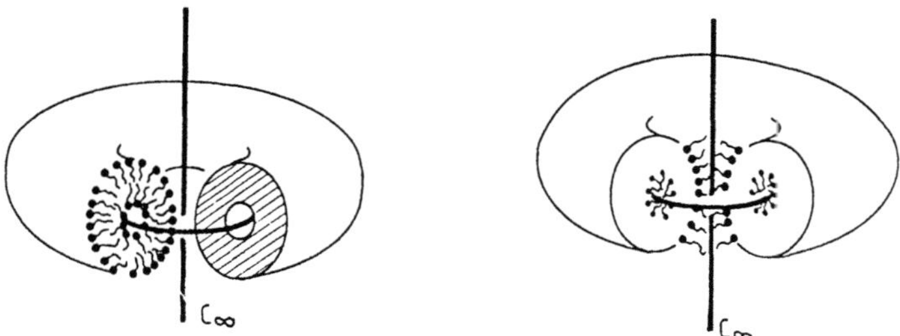

fig.4. Schematic representations of the relaxations of the two cases of frustrations ,corresponding to a film of water or to a film of oil, when the cells are transfered onto the spherical torus

DISCLINATIONS AND GENERATION OF HYPERBOLIC SURFACES

<u>The disclination process</u>

In a space with positive curvature the integration of the Gaussian curvature over the whole space is 4π revealing an angular defect [9] . To decrease the curvature of the space it is therefore necessary to fill in the defect and this is obtained by introducing defects of rotation , or disclinations . The disclinations are introduced following a Volterra process which respects the symmetries of the structure present in that space [10] and their density is related to the curvature to be suppressed [11] .

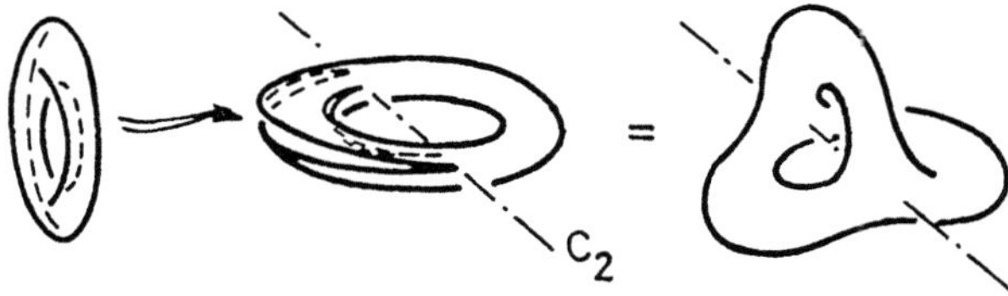

fig.5. a A -π disclination around a C2 axis of T2 in S3 , half the geodesic sphere bounded by C2 cuts T2 along two half geodesic lineswhich are orthogonal and bounded by the traces of C2 on the surface , the lips of the cut are separated (middle) , half a torus (left) is inserted in between the lips , a torus of higher genus is generated (right)

In our case the structure is the spherical torus T2 and we introduce disclinations around its C2 axes. A pictorial representation of the first step of the process is given in fig. 7 . The torus is cut along half a plane limited by a C2 axis , a rotation separates the two lips , half a torus is inserted and the system relaxes . It is necessary to introduce half a torus in order not to disrupt the continuity of the torus and to respect its C2 symmetry , this is a -π disclination [12] . Such an introduction of disclinations around axes normal to the surface preserves the bicontinuous topology of the torus , i.e. the surface still separates two identical sub-spaces , but its genus has been changed . Finally , as a disclination is a defect of rotation and concerns angles only, distances and area /volume ratio of the cell are preserved .

Each disclination axis pierces the surface at four points . At each of these points the effect of a -π-disclination on the element of surface surrounding the point is to introduce a supplementary amount of surface of area one half that of the element . As the disclinations have to be homogeneously introduced to flatten out S3 this element must be one of the tiles of a regular tesselation of the torus , moreover it must respect the symmetry operations which are related to the disclinations around C2 axes and therefore be a square of the {4,4} tiling . The -π disclination transforms the square into an hexagon without changing the coordinence of the vertices. The effect of the complete decurving of S3 on the torus is to transform its {4,4} tiling into a non planar {6,4} tiling , as suggested in fig. 9 . A surface with negative Gaussian curvature is generated.

fig.6.   A non- Euclidean {6,4} tiling of a hyperbolic surface (note that the angles of the hexagons in this picture are 2π/3 , if they were π/2 the hexagons would not be planar). The {6,4} tiling of the hyperbolic surface is represented in the Poincaré's model .

In principle we can therefore built three classes of surfaces : surfaces with orthoscheme triangles having the two sides of their right angle straight and a curved hypotenuse which are supported by one of Schœnflies' periodic skeleton of straight lines , surfaces with triangles having a straight hypotenuse only which are supported by another Schœnflies' skeleton and surfaces with triangles having all sides curved .

RELATION WITH INFINITE PERIODIC MINIMAL SURFACES

The two first classes of surfaces which are solutions to our problem are supported by skeletons of straight lines which can be organized periodically in R3 . Surfaces of this type have been studied by mathematicians under the condition that they are minimal or have zero mean curvature , i.e. $1/R_1 - 1/R_2 = 0$ where $-R_1, R_2$ are the radii of curvature , a condition which we have not introduced yet and whose necessity in our problem will be discussed now . The minimal property of the middle surface stems out from the symmetry of the cell , or the identity of the two sub-spaces separated by the film . It implies that the Gaussian curvatures of the two interfaces , at equal distances l on either sides of the midlle surface , are identical . If the middle surface has curvature radii $-R_1$, R_2 those of the two parallel surfaces are respectively $-(R_1-l)$, (R_2+l) and $-(R_1+l)$, (R_2-l) . The equality of their Gaussian curvatures $-(R_1-l)(R_2+l) = -(R_1+l)(R_2-l)$ imposes $R_1 = R_2$ and the middle surface is minimal . Following this the examination of the surfaces of the mathematicians will provide us with the solutions to our problem . Mathematicians first calculated finite minimal surfaces bounded by quadrangles of straight lines , such as those of Schœnflies , and showed that they could be used as fundamental regions for building infinite periodic minimal surfaces , or IPMS [12] . Their approach is summarized in [13], including further developments . There are two IPMS which are supported by the quadrangles of our two first classes and a third IPMS without straight lines , obtained from the two others by a geometrical transformation , which provides information about our third class: they are the Schwarz' F surface, the Schwarz' P surface and the Schoen's G surface.

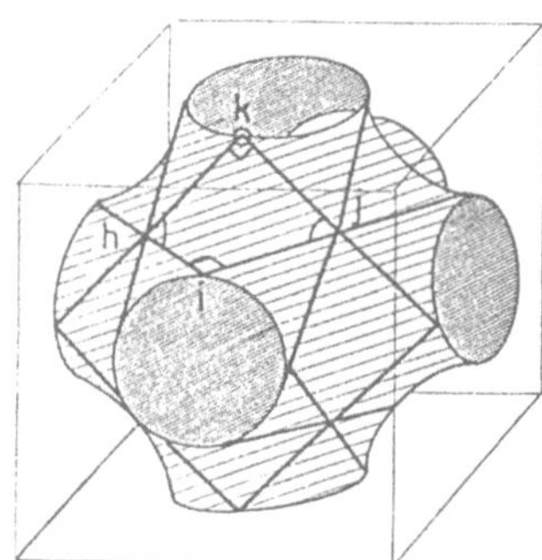

fig.7. The usual representation of the cell of surface P with the schoenflies quadrangles supporting it.

<u>Structure with space group Im3m</u>

It was proposed first to interpret electron micrographs of prolamellar bodies of the membrane system of a plant [18?] and was observed more recently in a ternary system by X-ray scattering [15] . The structure is represented in fig. 7 . The two congruent labyrinths , each with rods connected six by six at right angles , are those of surface P .

<u>Structure with space group Pn3</u>

It was proposed first for a phosphatidyl ethanolamine/water system , the lipid having been extracted from insects [14] . More recently it was identified in the glycerol monooleate/water system [17] . This is also the structure proposed for the phase formed by the polymeric material quoted above which was studied by electron microscopy [16] . The structure is represented in fig.8. The two congruent labyrinths , each with rods connected four by four at angles of 109°28' , are the labyrinths of surface F .

Structure with space group Ia3d

It was the first structure to be characterized with certainty [26?] . It is quite widespread and can be found in soap , phospholipid , detergent systems , hydrated or not [1,14] . The structure is represented in fig.1-b. The two oppositely congruent labyrinths , each with rods connected three by three at angles of 120° , are those of surface G and of our third class .

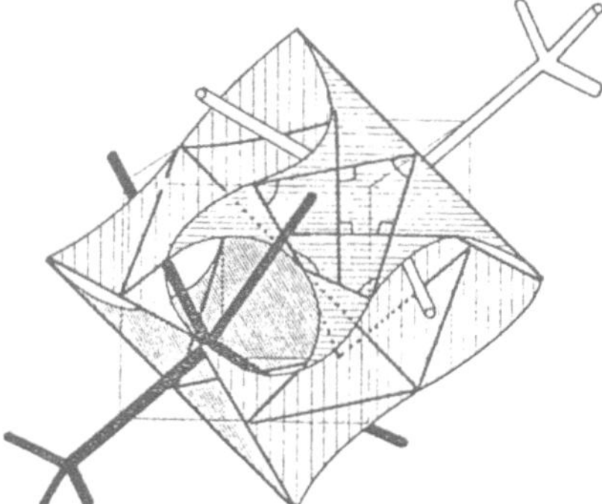

fig 8 The translation cell for the F surface and the two labyrinths separated by the surface.

Generation of the "micellar" topology [20]

The interplay , or frustration , of forces acting in a periodic system of fluid films built by amphiphilic molecules in presence of water can find ordered geometrical solutions within the frame of a topology of finite cells separated by a self-intersecting film .

A symmetric film , made of two facing interfaces , supported by a great sphere S2 of S3 is a representation of a frustrated lamellar structure whose frustration has been relaxed by the transfer into the curved space . This results from the fact that S2 separates S3 in two identical finite sub-spaces : the interfaces at equal distances from S2 have area smaller than that of S2 , so that the frustration is relaxed.

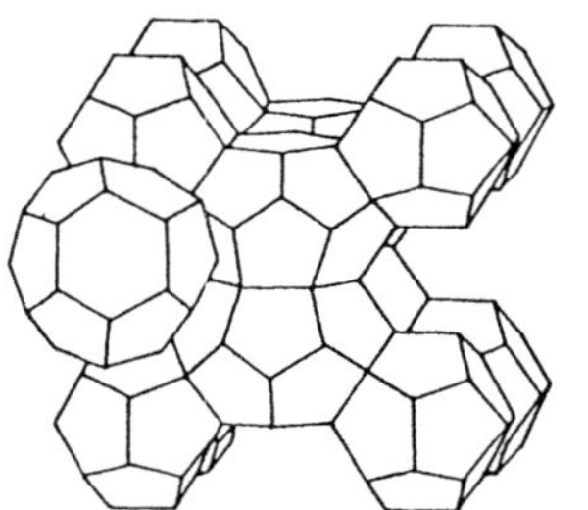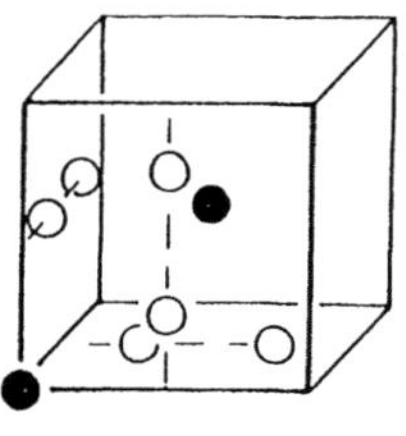

fig.9. -Aggregation of distorted 12-hedra and 14-hedra in type I structure , on the right the positions of the 12-hedra (O) and 14-hedra (O) in the cell of the Pm3n lattice .

These solutions can be described as periodic packings of polyhedral cells - slightly distorted dodecahedra , tetrakaidecahedra and hexakaidecahedra - certain of which are formally similar to type I and II structures of water clathrates . This description is in correct agreement with the results of known structural studies of cubic phases found in between hexagonal and micellar phases of some phase diagrams . This agreement , as the one previously obtained in the case of "bicontinuous" cubic phases [21] , exemplifies the fact that liquid crystalline polymorphism proceeds from the action of physical forces , existing in systems of films , within the frame of the geometrical constraints of our Euclidean space . If we suppress one force , for instance the interaction between two films , the order is lost , as in macroscopic soap bubble froths. Also it must not be forgotten that the structures found here are but possible geometrical configurations and that we do not discuss their physical stabilities relative to that of neighboring hexagonal and disordered micellar phases . For instance there are systems which do not present ordered micellar phases and in which the ordering of the micelles appears as short range fluctuations only [22] , and others which might exhibit ordered phases with more complex structures and fluctuations in the same region of the phase diagram [23] .

CONCLUSIONS

We have considered "bi-dimensional" objects which are fluid films organized by interfaces. Their structures result from the actions of forces normal to the interfaces , which control the periocity of their stackings , and forces parallel to the interfaces , which control the interfacial curvatures . These physical forces must exert their actions in Euclidean space R3 which has geometrical properties of its own . The structures are therefore to be seen as adaptations of physical forces to geometrical constraints . This is indeed the point of view of classical crystallography which has been mainly developed for "zero-dimensional" objects such as atoms and molecules . A typical example of the rigor of geometrical constraints in this case is provided by metallic atoms which tend to pack in a compact tetrahedral structure with tetrahedral interstices. As it is impossible to obtain a dense packing of regular tetrahedra in Euclidean space R3 different solutions are found : the classical f.c.c. or h.c.p. structures in which tetrahedral interstices are mixed with octahedral ones and a large number of structures which correspond to packings of distorted tetrahedra such as Franck and Kasper structures , amorphous metallic structures and quasi-crystals. This point of view should not be limited to point objects; films of molecules appear as good examples to investigate the crystallography of bi-dimensional objects .

We showed that perfect stackings of films with symmetrically curved interfaces can be built in two spaces with positive curvatures which are the hypersphere S_3 and the hypercylinder $S_2 * R_1$. There the middle surface of the film must be supported by the surfaces separating the spaces in two equal sub-spaces , the spherical torus T_2 and the great sphere S_2 in S_3 and the cylinder $S_1 * R_1$ in $S_2 * R_1$. The mapping of these two curved spaces onto the Euclidean space R_3 is obtained by the introduction of disclinations and the three surfaces are transformed into organizations of films having the topologies of liquid crystalline structures , the "bicontinuous" topology of cubic phases results from the use of the spherical torus T_2 , the "cylindrical" topology of the cylindrical phases results from the use of the cylinder $S_1 * R_1$ and the "micellar" topology of the micellar phases results from the use of the great sphere S_2.

REFERENCES

1 -Luzzati V. , *Biological Membranes* 1 (1968) 71 , edited by D. Chapman (Academic Press)
2 -Hardouin F. , Levelut A.M. , Achard M.F. and Sigaud G. , *J.Chim. Phys.* 80 (1983) 53
3 -Diele S. , Brand P. and Sackmann S. , *Mol. Cryst. Liq. Cryst.* 17 (1972) 163
  Tardieu A. and Billard J. , *J. Physique* 37 (1976) C3-79
  Etherington G. , Leadbetter A.J. , Wang X.J. , Gray G.W. and Tabakhsh A. , *Liquid Crystals* 1 (1986) 209
4 -Sadoc J.F. and Mosseri R. , *Phil. Mag.* B 45 (1982) 467
  Sadoc J.F. and Mosseri R. , *Pour la Science* , (jan. 1985)
    Mosseri R. in *Du cristal à l'amorphe* , proceedings of the Beg Rohu Summer School 1986, Editions de Physique , to be published
5-Sadoc J.F. and Charvolin J. , *J. Physique* 47 (1986) 683
6 -Israelachvili J. N. , *Intermolecular and Surface forces* , Academic Press (1985)
7 -Coxeter H.S.M. , *Regular Complex Polytopes* , Cambridge University Press (1974)
  -Hilbert D. and Cohn-Vossen S. , *Geometry and the Imagination* , Chelsea   Publishing Company , New York (1983)
  -Coxeter H.S.M. , *Introduction to Geometry* , John Wiley , New York (1961)
8-The Schlafli notation {p,q} means that the tiling is made of polygons having p edges that meet q by q at each vertex
9-Cartan E. , *Géométrie des espaces de Riemann* , page 189 , Gauthier Villars (1963)
    Misner C.W. , Thorne K.B. and Wheeler J.A. , *Gravitation* , Freeman Cy , San Fransisco (1970)
10-Friedel J. , *Proceedings of the sixth General Conference of the European Physical Society* , 25 , Prague (1984)
11-Sadoc J.F. and Mosseri R. , *J. Physique* 45 (1984) 1025
12-Schwarz H.A. , *Gesammelte Mathematische Abhandlungen* , Band 1 , Springer Verlag , Berlin (1890)
13-Schoen A.H. , *NASA Technical Note* D-5541 , (1970)
14-Tardieu A. , *Thesis* , Université Paris-Sud Orsay (1972)
15-Luzzati V. , Mariani P. and Gulik-Krzywicki T. , in *Physics of Amphiphilic Layers* , edited by D. Langevin and J. Meunier , Springer Verlag , to, be published
16-Thomas E.L. , Alward D.B. , Kinning D.J. , Martin D.C. , Handlin D.J. and   Fetters L.J. , *Macromolecules* 19 (1986) 2197
17-Bicontinuous cubic phases are observed in the vicinity of lamellar ones , cubic phases with different topology may exist far from lamellar phases, in the vicinity of micellar phases, see for instance
  Eriksson P.O. , Khan A. and Lindblom G. , *J. Phys. Chem.* 86 (1982) 387
18-Longley W. and Mac Intosh J. , *Nature* 303 (1983) 612
18-Gunning B.E.S. , *Protoplasma* 60 (1965) 111
26-Luzzati V. and Spegt P.A. , *Nature* 215 (1967) 701
20-Sadoc J.F. and Charvolin J. , *J. Physique* 47 683 (1986)
21 -Charvolin J. and Sadoc J.F. , *J. Physique* 48 1559 (1987)
22-Cabos C. , Delord P. and Martin J.C. , *J. Phys. Lett.* 40 , L-407 (1979)
23- Luzzati V. , private communication

GEOMETRICAL APPROACH OF BLUE PHASES

Brigitte PANSU and Elisabeth DUBOIS-VIOLETTE

Laboratoire de Physique des Solides
Bât. 510 Université de Paris Sud
91405 Orsay cedex, France

1. WHAT ARE BLUE PHASES ?

Blue phases are liquid crystalline phases which appear in chiral compounds between the isotropic and the cholesteric phase in a narrow range of temperature[1,2]. In the cholesteric phase, the periodicity which typically lies in the range of visible light wavelength is only due to orientational ordering. The mean orientation of the molecules or director n is constant inside planes perpendicular to one axis but twists when moving along this axis. This induces periodic properties of physical parameters such as the dielectric tensor in this direction with a period given by π/q where q is the cholesteric pitch. When this pitch is small enough, new phases may appear when temperature increases. These are the blue phases. There exist three blue phases : BPI, BPII and the blue fog. The two first ones are cubic phases with space group symmetry $O^8 = I4_132$ et $O^2 = P432$ as determined by optical scattering experiments. The third one is still mysterious[3].

The existence of phases with cubic symmetry has been interpreted not as due to the propagation at long distance of a local order but as the solution found by the system to a problem of geometrical frustration. Indeed, in the cholesteric phase, twist occurs in one particular direction perpendicular to the director. However, since the molecules rotate around their main axis, all the directions perpendicular to the director are equivalent and twist is expected to occur radially around the director. This natural local order in the molecular orientation cannot be extended at long distances. Starting from a given orientation at point A and moving from point A to a point B, the resulting orientation at point B obtained by following the double twist rule depends on the path to go from A to B. Therefore double twist can be performed only in an approximative way in finite regions such as the double twist tubes shown in fig. 1. In these tubes, the orientation twists radially from the axis of the tube but perfect double twist is performed only on this axis. Models of the blue phases [4] have been built by piling such cylinders in cubic arrays. The only geometrical constraint is that the orientations of the director at the surface of two adjacent cylinders fit together at the tangency point. But what happens in between the cylinders ? When moving on a closed loop around some 3 fold axis of the structure, there appears that the director n is transformed in - n. This shows the presence of defects which are disclinations of order s =- 1/2. To each stacking of cylinders is associated a cubic network of disclination lines. Such arrangements of defects are very similar to those encountered in some metallic or alloy phases and we would like to develop now this analogy.

Geometry and Thermodynamics
Edited by J.-C. Tolédano
Plenum Press, New York, 1990

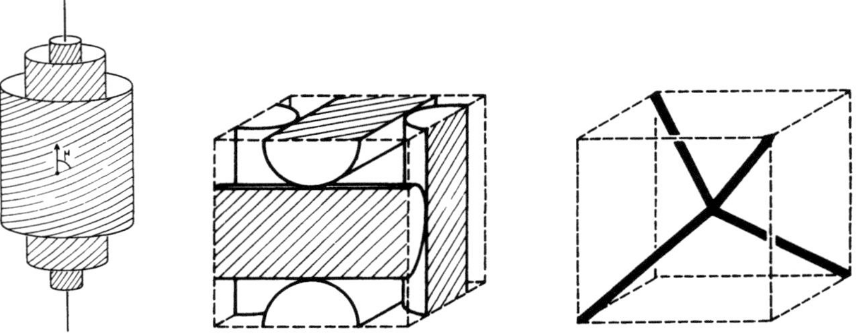

Fig. 1. Double twist model of blue phases and resulting array of disclination lines.

2. ANALOGY WITH METALLIC AND ALLOY PHASES

At present it is well established that complex structures found in some metals such as Mn, W, V... can be described as polytetrahedral structures reticulated by a periodic network of disclination lines [5,6]. In all these structures many coordination polyhedra are nearly icosahedral which shows that the system locally tends to the greatest compactness. As it is well known that the five fold symmetry of the icosahedron is incompatible with long range ordering, one recovers frustration as in the case of blue phases and a similar analysis of the structure can be done. The local geometrical constraint of icosahedral environment is more or less well satisfied for some sites surrounded by a slightly distorted icosahedron. But other coordination polyhedra can be seen as obtained by a disclination process. In this case the disclination order is s=- 1/5 since it consists in adding one fifth of space. To each cubic structure one can therefore associate an array of disclination lines like in the cubic blue phases.

As far as we know, there is no perfect correspondence between any networks of disclination lines in blue phases or in metallic phases. Nevertheless it is possible to find common sub lattices as for instance for BPII and the Laves phase of CuMg. Let us now describe these two phases. In the Laves phase, copper atoms are arranged like carbon atoms in a diamond structure. Magnesium atoms are in the free space of this structure. In the diamond structure, there are four tetrahedral holes in a cubic cell and one can put in these four holes tetrahedra made by four atoms of Mg. There are thus 16 atoms of Mg in a cell. The environment of one magnesium site is a slightly distorted icosahedron with a coordination number equal to 12 (6 Cu and 6 Mg). The coordination polyhedron for the copper site has 16 vertices (4 Cu and 12 Mg) and can be seen as obtained by a disclination procedure from an icosahedron. There are thus four defect half lines which cross at a cooper site. They build a periodic network of lines which is connected in the same way as bonds in a diamond structure.

In the double twist model of BPII, there are two shifted but intertwined such line structures. This specific defect array is generated by the arrangement of double twist cylinders shown in figure 1 . In this configuration infinite tubes lie along the three main directions of a cube without intersecting each other. As in all the model of blue phases, disclinations are present along some parts of the three-fold axes. Along the other part of these axes, the director can "escape" in the direction of the axis and one recovers a new double twist tube. Therefore such a network of disclination lines can be obtained by setting double twist cylinders in the same way as the dislocation array previously described, that is two shifted diamond bond structures.

In blue phases and in some metallic or alloy phases the linear defects seem therefore to play an important role. In both cases the cubic structure can be interpreted as resulting from the existence of disclinations and from their interactions. A better understanding of these phases could result from a fine estimation of the energy linked to the defect lines. The main difference between the two systems is that metals or alloys are described by the discrete positions of the atoms and their environment (coordination number and mean distance) when in blue phases, the order parameter is a director and the description needs to be continuous. On the other hand, the local symmetry imposes disclinations of order - 1/5 in the first case (icosahedron) and of order - 1/2 in the second case (director).

3. FRUSTRATION RELEASED IN CURVED SPACE

As previously said, blue phases and metallic phases are both characterized by geometrical frustration. In both cases, this frustration can be released in a curved space. This is well known for the metallic phases[7]. In blue phases, the double twist rule can be expressed with the mathematical notion of connection or parallel transport. The fact that the local orientational order cannot be extended at long distance is revealed by the non zero value of the curvature of this connection. This curvature is constant and depends on the square q of the pitch. In 1983, J. Sethna[8] has shown that it was possible to have a perfect double twisting director field in the hypersphere S^3 : when taking a sphere of radius R=q, the curvature of the sphere balances in a certain way the curvature of the connection. In a more precise language[9], there exists in S^3 a connection built in the same way as the one in R^3 and which has no curvature. This connection is associated to a given parallel transport which can be more easily described using the group structure of S^3 . Indeed in the same way as the circle S^1 is isomorphic to the group of unit complex numbers, S^3 is isomorphic to the group of unit quaternions (a quaternion is in 4D what a complex number is in 2D). Starting from one frame of three tangent vectors at point A, this group structure generates two frame fields. One is linked to right multiplication and the other one to left multiplication.

At this point one can notice that R^3 is also a group of translations and that the Euclidean frame i, j, k is built on this way ; but in this case the group is commutative and right translations are equivalent to left ones. In S^3 , one frame field was shown to be parallel for the right double twist connection (+q), and the other one for the left one (q). Each vector field which has constant components in the right frame field thus performs a perfect right double twist. This is analogous to the perfect nematic phase in R^3 . That is why the blue phase in S^3 can be seen as a nematic phase.

For coming back to our physical space R^3 , one needs to introduce orientational defects or disclinations. Such processes have been extensively described in metallic systems. In this case the perfect structure in S^3 are polytopes for instance the {335} one. The introduction of disclination lines can be done in various ways which always generate topological environments, those which are experimentally encountered. One could have thought that the same procedures would apply to blue phases but their continuous structure requires metrical and not only topological processes. At present only one defect has been introduced[10] and the next step will be much more difficult.

4. BLUE PHASES AND MINIMAL SURFACES

The second analogy we would like to develop now concerns soap films[11]. In some lyotropic systems, the amphiphilic molecules aggregate in small cylinders which are joined 3 (4 or 6) by 3 (4 or 6) and build two different but interlaced labyrinths. They are separated by a water film which is probably closely fitting a minimal surface. Cubic lyotropic phases are therefore well described by infinite periodic minimal surfaces. Now the space groups experimentally observed in blue phases can be deduced from those of soap films by removing all the mirror operations incompatible with chirality. Moreover in the model of blue phases previously described, one can set double twist tubes at the same place as the cylinders of amphiphilic molecules of the corresponding (same symmetry except mirrors) lyotropic phase. In blue phases, the organization of these tubes corresonds to two separate networks. If one lets the cylinders spread around their axes, one obtains a partition of the space into two equivalent domains. They are separated by an infinite surface whose topology is controlled by the tube network[12]. This surface is equivalent to the amphiphile film in lyotrops and can be described by infinite periodic minimal surfaces: the G surface for BPI and the F one for BPII.

The director field must satisfy as best as possible double twist near the surface. From symmetry arguments, the director field is either normal or tangent to the surface. The normal field is defined everywhere and is thus incompatible with the existence of disclination lines which cross the surface. Therefore only tangent fields need to be considered. With use of the connection previously mentioned, one can define a double twist energy involving the gradients of the director field and the pitch q. When neglecting the gradient in the direction of the normal to the surface, one obtains a surface energy which can be minimized. This local condition of minimum energy on the surface will determine a prefered orientation on the surface which will be further considered as a boundary condition. This is equivalent to neglect the influence of the boundary conditions at the center of the channels on the field near the surface. Straightforward computation leads to the following result: for each minimal surface, the director field which minimizes double twist surface energy lies along the asymptotic directions of the minimal surface. Since the mean curvature of the minimal surface vanishes, curvatures along the two principal directions are opposite. At 45° from the principal directions, there are the asymptotic directions along which curvature vanishes. There are two perpendicular sets of asymptotic directions: one is associated to right twist, the other one to left twist. Each set of asymptotic directions exhibits a singularity around the three fold axes where the Gauss curvature vanishes (all tangent directions are asymptotic) and it corresponds to a s=-1/2 disclination line.

5. CONCLUSION

All these approaches are geometrical and thus crystallographic. They tend to solve the fascinating problem of the orientation of the molecules in the cubic phases. Till now they do not really explain why these phases are thermodynamically stable. The purpose of this short paper is to show the analogy between blue phases and other cubic phases such as the Laves phases or the lyotropic phases. A better understanding of the interaction between the defects and their spatial ordering is now needed for both systems.

RERERENCES

1. V. A. Belyakov and V. E. Dmitrienko, The blue phase of liquid crystals, <u>Sov. Phys. Usp.</u>, 28:7 (1985).
2. H. Stegemeyer, Th. Blumel, K. Hiltrop, H. Onusseit and F. Porsch, Thermo-dynamic, structural and morphological studies on liquid crystalline blue phases, <u>Liq. Cryst.</u>, 1:3 (1986).
3. R. H. Hornreich in this book.
4. S. Meiboom, M. Sammon and D. W. Berreman, Lattice symmetry of the cholesteric blue phases, <u>Phys. Rev. A</u>, 28:3553 (1983).
5. F. C. Frank and J. S. Kasper, Complex alloy structures regarded as sphere packings, <u>Acta Crist.</u>, 11:184 (1958).
6. J. F. Sadoc, Periodic networks of disclination lines: application to metal structures, <u>J. Physique Lett.</u>, 44:707 (1983).

7. M. Kleman and J. F. Sadoc, A tentative description of the cristallography of amorphous solids, <u>J. Physique Lettres</u>, 40:569 (1979).

8. J. Sethna, Frustration, curvature, and defect lines in metallic glasses and the cholesteric blue phase, <u>Phys. Rev. B</u>, 31:6278 (1985).

9. E. Dubois-Violette and B. Pansu, Frustration and related topology of blue phases, <u>Mol. Cryst. Liq. Cryst.</u>, 165:151 (1988).

10. B. Pansu and E. Dubois-Violette, Disclination in the S^3 blue phase, <u>J. Physique</u>, 48:305 (1987).

11. J. F. Sadoc, in this book.

12. S. Andersson, S. T. Hyde, K. Larsson and S. Lidin, Minimal surfaces and structures: from inorganic and metal crystals to cell membranes and biopolymers, <u>Chem. Rev.</u>, 88: 221 (1988).

ELECTRON MICROSCOPY AND QUASICRYSTALS

P. Guyot and M. Audier

LTPCM.(UA CNRS n°29). Institut National Polytechnique de Grenoble
BP 75, 38402 Saint Martin d'Hères, France

INTRODUCTION

Since the discovery by D. Shechtman of the icosahedral AlMn phase
by electron diffraction, a number of investigation techniques commonly
used in materials science have been applied to the case of quasicrystals,
like X-rays and neutron diffraction and local spectroscopies. But indispu-
tably the understanding of their structure has been driven by electron
microscopy. The major reason for that is due to the fact that quasicrystals are generally of micronic size, when obtained for example by rapid
solidification, but also coexisting, at a more or less fine scale, with other
intermetallic crystalline compounds; this last point hampers evidently
characterization techniques with limited spatial resolution which sample
the alloys at a heterogeneous scale or which requires a large amount of
matter.

Different variants of electron microscopy have been extensively used:
transmission electron microscopy (TEM), analytical electron microscopy
(AEM), scanning electron microscopy (SEM) and to a much less extent
field-ion microscopy and tunneling microscopy, bringing informations on
several aspect, as growth morphology, structure, relationships with crystals and glasses, defects topology, phase transformations,...

An exhaustive scan of all these results is evidently beyond the scope
of this paper. We will rather limit ourselves to present characteristic and
as clear as possible results in the field, as well as data still in course of
development and analysis.

BASIC RESULTS: DIFFRACTION PATTERNS AND HIGH RESOLUTION IMAGING. VARIOUS CLASSES OF QUASICRYSTALS

The original selected area diffraction (SAD) patterns of Shechtman
et al. (1) on melt-spun AlMn alloys were consistent with the $m\bar{3}\bar{5}$ icosahedral group, with reflection positions indexed by 6 integers n_i, $\bar{q} = \sum_{i=1}^{6} n_i \bar{q}_i$,
with base vectors $\bar{q}_i$ pointing from the center towards six of the vertices
of an icosahedron. Cuts of the reciprocal space of the icosahedral (i)-phase,
perpendicularly to 2-, 3- and 5-fold axes are shown in Fig. 1 for the
AlCuFe system (2), which has been produced in an i-state of best quality.

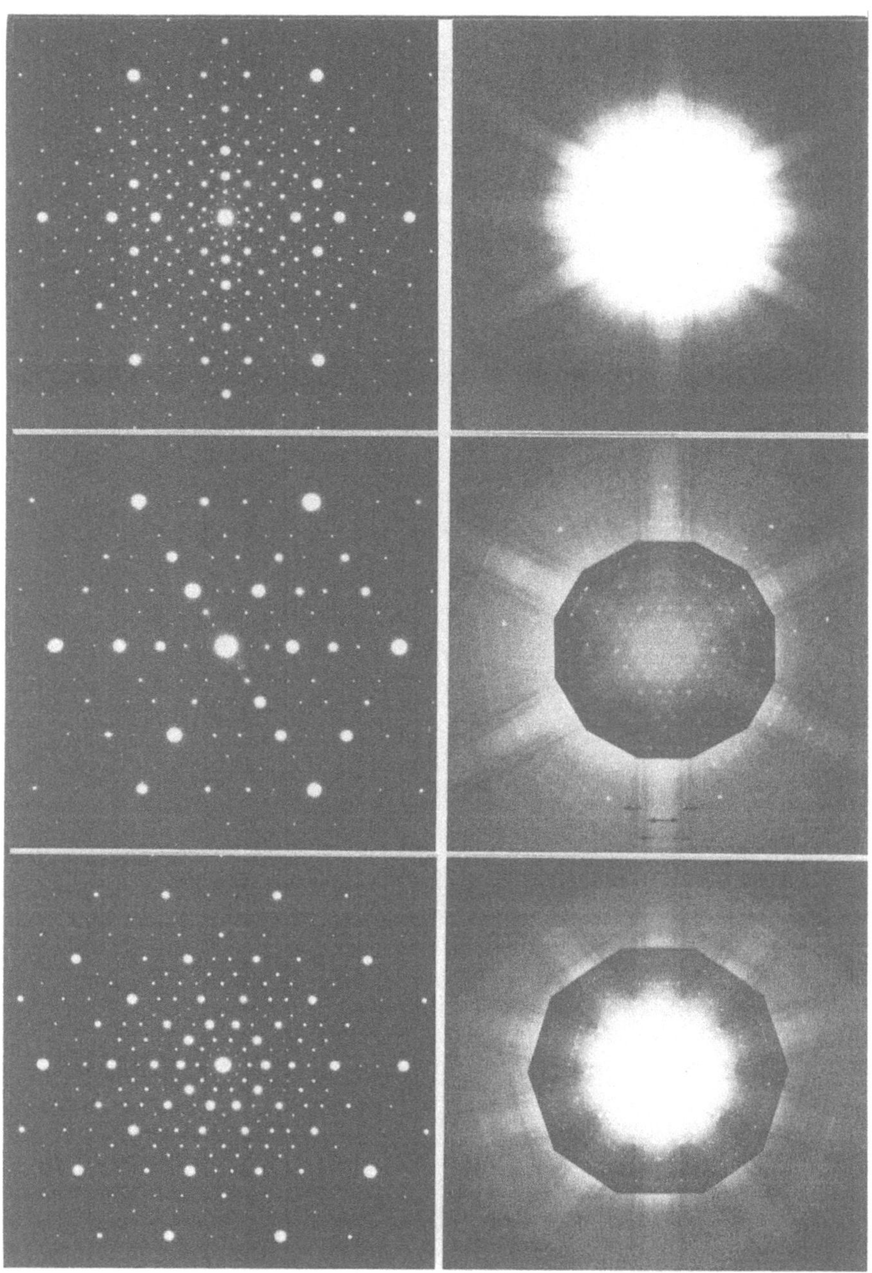

Fig. 1. SAD patterns of the AlFeCu icosahedral phase. Left column: 2, 3 and 5-fold orientation successively. Right column: patterns taken with a shorter camera length to reveal higher Laue zones and Kikuchi patterns (from ref. 2).

A close examination of the intensity distribution of diffraction spots in high order Laue zones and of Kikuchi side-bands prove that the symmetry is really icosahedral. A scaling factor τ ($\tau = (1+\sqrt{5})/2$) relies the spot positions along a systematic row, dedicating an incommensurate structure. The difference with conventional modulated crystals is the absence of a periodic support dressed with incommensurate satellites. Interference imaging, the so-called high resolution imaging (HRI) mode, with one example given in Fig. 2 for AlCuFe in the 5-fold orientation, exhibits a discrete distribution of white dots with orientational order but no translational periodicity. HRI 1985 works (3-5) already mentionned the quasiperiodic character of the white dots positions: Fibonacci sequences along rows, inflated pentagons for the 5-fold orientation,...

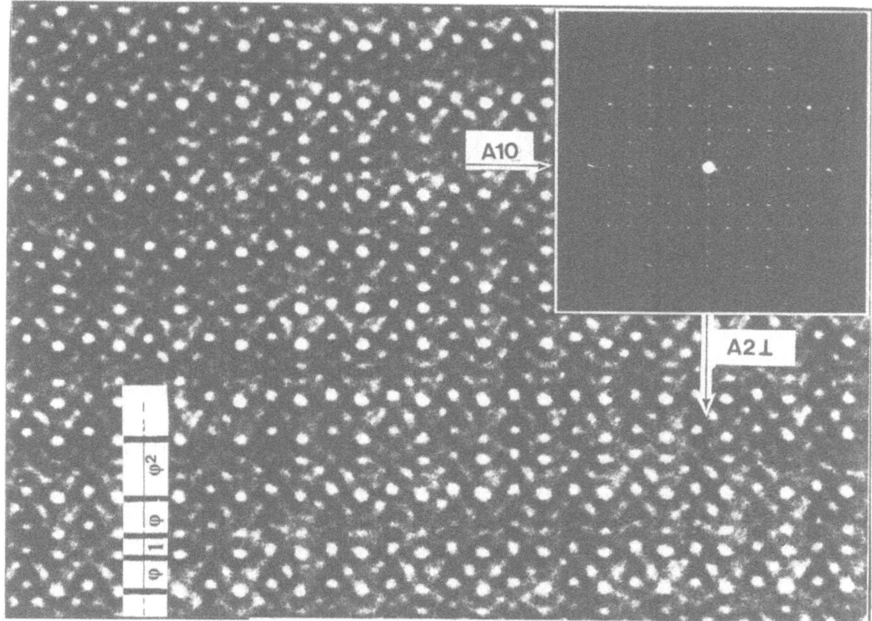

Fig. 2. High resolution electron micrograph of AlFeCu along a 5-fold symmetry axis.

The quasiperiodicity, instead to be 3 dim., as for i-phase, can be restricted in 2 dim. or 1 dim., the periodidicity being restored in 1 or 2 dim. respectively. Fig. 3 gives the example of the decagonal AlMn phase (7), which is a periodic tiling of layers with internal 10-fold symmetry (8): white dots layers of thickness 12.69Å are stacked periodically perpendicularly to the 10-fold axis, whereas rows of dots parallel to this axis are spaced

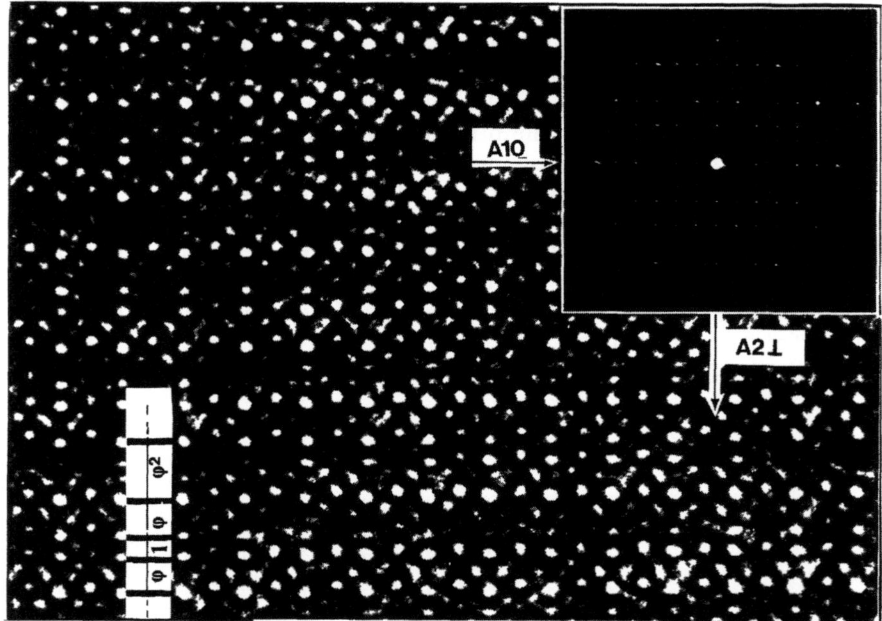

Fig. 3. Decagonal AlMn phase. The HRI is taken along a 2-fold axis perpendicular to the 10-fold axis. A periodic distribution of white dots is noticeable along the direction [A10], whereas it is quasiperiodic parallely to the 2-fold axis [A2⊥] (from ref. 8).

quasiperiodically. Similar polytypical structures with a 8 or 12-fold symmetry axis have also been reported in transition-metal alloys (9,10).

Classical incommensurate structures with a 1 dim-modulation, not rationally related to the basic structure, are known. This is the case of antiphase boundary modulated long period structures, where commensurate and incommensurate modulations are driven by composition variations (15,16). Some observations have reported the transformation upon annealing of a 2 dim-decagonal phase for AlNiSi and AlCuCo alloys, in 1 dim-quasicrystal (17): the periodicity along the 10-fold axis is conserved, while a periodic order appears along one perpendicular 2-fold axis, which was originally aperiodic, leaving therefore a residual quasiperiodicity along the third 2-fold orthogonal axis; the SAD patterns are however complexe, and no definitive conclusion in our opinion, may be established on this transition.

To close this part of quasiperiodic phases inventoried by electron microscopy, it is worthwile to notice the existence of metastable vacancy ordered phases in Al-transition metal alloys, also obtained by rapid solidification, which display in the CsCl structure a stacking of (111) planes according to a Fibonacci sequence (17,18). Such structures are periodic approximants of a 1 dim-quasicrystal at different levels of truncation of the Fibonacci sequence.

DIFFRACTION AND IMAGE CONTRAST CALCULATIONS

Diffraction properties of the i-quasilattice have been established, on the basis of kinematical scattering, by several techniques applied to a 6 dim. cubic lattice: cut and projection method (6) or section one (11-13). Kinematical scattering means simply Fourier transform, which gives for the amplitude of the Bragg peaks at $\bar{q}_{//}$:

$$F(\bar{q}_{//}) = \delta(\bar{q}_{//}) . G(\bar{q}_{\perp})$$

where $\bar{q} = \bar{q}_{//} + \bar{q}_{\perp}$, according to the now familiar notations in physical space ($//$) and orthogonal space ($\perp$); $G(\bar{q}_{\perp})$ is either the Fourier transform of the window function of the cut and projection technique, or the Fourier transform of the decorative function of the 6 dim-lattice nodes in the section approach. The remarkable properties of the diffraction patterns previously presented, once the icosahedral group has been applied to the hypercubic lattice, are then qualitatively reproduced.

So it is for the high resolution images within the framework of the projected potential approximation, used long ago for sufficiently thin crystals, where 2 dim. images are basically the projections along the electron beam parallel to crystallographic axes of the 3 dim. structure; the approach can be extented to quasicrystals, the existence of white dots in the image and their quasiperiodic arrangement being explained by the stacking along the projection direction of the lattice nodes of a hyper-cubic lattice (6,14).

Dynamical effects, important for electron scattering, have been so far insufficiently evaluated. A theoretical analysis of these effects has been estimated on the basis of a diffraction Hamiltonien built on a plane wave basis, and the consequence of the basis truncature has been studied (14), explaining for instance an apparent fringe periodicity on H.R. images taken along a 2-fold axis of the i-AlMnSi phase.

All these works have been however developed for monoatomic quasilattices (only one kind of atom per lattice site), and are too short to be applied to real (polyatomic) materials where chemical ordering has been furthermore proved to exist (see next paragraph). At the level of HRI simulations, chemical ordering complicates seriously the problem. Several authors made structure calculations for atomic cluster models (for example the Mackay icosahedron for AlMn, or more complex one for AlLiCu), either by embedding them in a 6-dim. hyperspace (20,21) or in a 3 dim. Ammann tiling (22,23). But image reconstruction was not made. Dynamical multislice image simulations of 6 dim. cluster model of icosahedral AlMnSi and AlLiCu are in progress (24). It seems for ultra thin foils and given focus conditions, that the calculated images are consistent with experimental ones, with details in the potential projection which may be characteristic of the icosahedral clusters.

RELATIONS WITH CRYSTALS. LOCAL ORDER.

The existence of local order in quasicrystals is inferred first fro
the observation of the coexistence of quasicrystals with crystals, whic
known structures contain local atomic arrangements of icosahedral
pseudo-icosahedral order (like the Frank-Kasper phases), according
precise epitaxy relationships.

Such a situation is met for example in AlMnSi (22). There, the ic
sahedral phase coexists with a cubic Pm$\bar{3}$ α phase, as shown in the H.
image of Fig. 4. The α structure is a cubic packing of Mackay icosahedr
face connected along $\langle 111 \rangle$ and edge connected along $\langle 100 \rangle$. The relati
orientation of the two structures follows the i-group _ c-subgroup rel
tion: namely the three $\langle 100 \rangle$ cubic axes are parallel to three orthogon
2-fold axes of the i-phase, whereas its 3-fold axes are parallel to $\langle 11$
and $\langle \bar{1}^2 01 \rangle$ and 5-fold axes are parallel to $\langle 1\bar{1}0 \rangle$. The icosahedral order
the α phase modulates its reflections and its more intense ones are near
in coincidence with those of icosahedral phase, for example $[006]_\alpha$
$[532]_\alpha$ and $[0/0\ 2/4\ 0/0]_i$. The cubic phase is clearly an approximant
the i-phase. The higher the crystal unit cell, the better is the approx
mation for a given space group. Other approximants with different spac
groups have been then proposed to explain diffraction pattern distortior

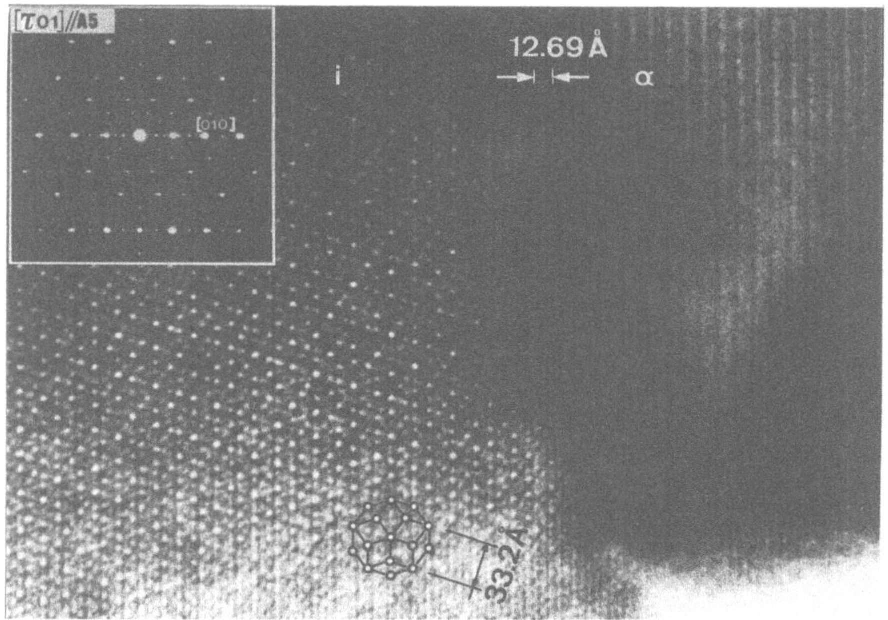

Fig. 4. HR image of coexisting i- and α-phase AlMnSi taken along a 5-fol
axis of the i-phase.

in a number of cases (25,26). Orientation relationships have been also found between icosahedral and decagonal structures in AlMn (7,8): the unique 10-fold axis of the decagonal phase coincides with one 5-fold axis of the i-phase, as well as two orthogonal 2-fold axes (Fig. 5)(8).

Besides EXAFS and neutron diffraction, there are also several electron microscopy experiments proving directly the existence of such an order Fig. 6 is an electron diffraction pattern obtained in fixed beam mode in a STE HB501 microscope, for i-AlMnSi. The electron beam, delivered

Fig. 5. Coexisting decagonal (T) and icosahedral (QC) phases in AlMn. HRI taken along a common 2-fold axis (from ref. 8).

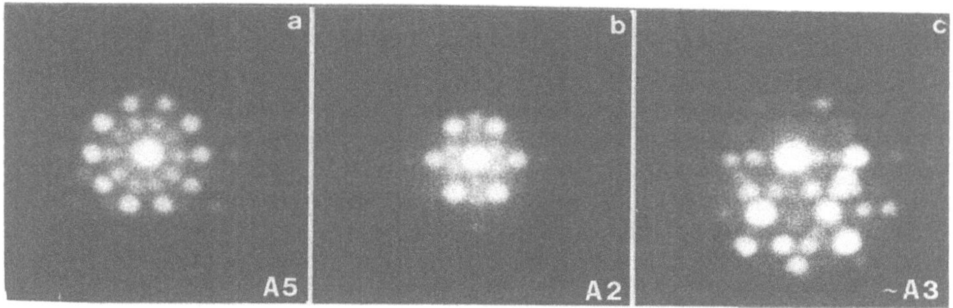

Fig. 6. Nanometric convergent beam diffraction patterns of i-AlMnSi, along zone axes parallel to 5, 2 and 3-fold axes.

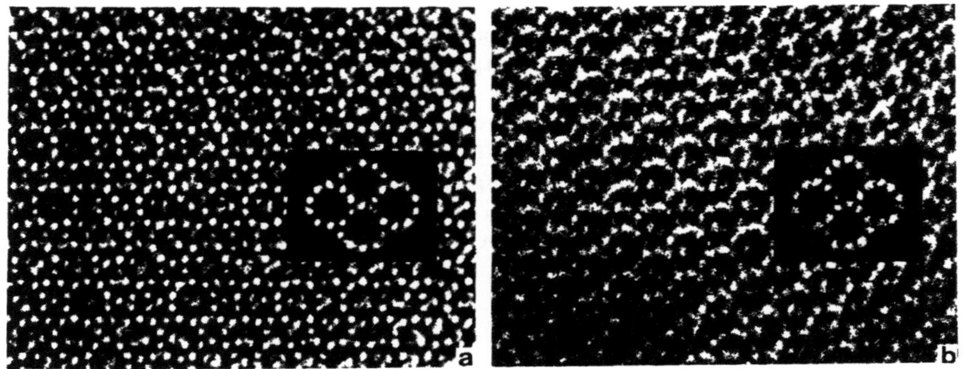

Fig. 7. HR image of i-AlFeCu (a) and of a rhombohedral approximant (b). The zone axis is 5-fold in a), and pseudo 5-fold in b) (from ref. 27).

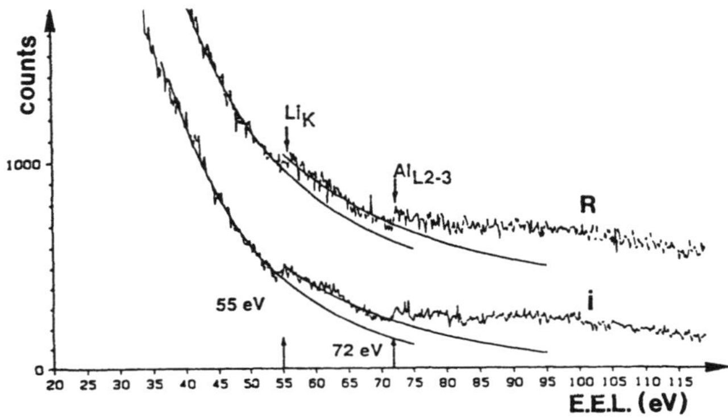

Fig. 8. Electron energy loss near edge structure at the Li_K and Al_{L2-3} edges in $R(Al_5CuLi_3)$ and $i(Al_6CuLi_3)$ phases (from ref. 29).

by a field emission gun, has a diameter smaller than 2nm, and the diffraction patterns are consequently issued from a region of equivalent lateral size. It is clear that these patterns have already, at this scale, all the characteristic features of the SAD ones.

Another example is shown in Fig. 7, which is a H.R. image of the AlFeCu i-phase on the left side, and a rhombohedral approximant on the right side (27). Both images are taken along 5-fold axis and pseudo 5-fold axis respectively. Four tangent rings of ten white dots are clearly distinguished, and confirm an i-local order common to quasicrystal and crystal.

Similar results have been obtained by analytical microscopy. Fig. 8 displays two electron energy loss spectra recorded in STEM with a nanometric beam, in i-AlLiCu and a cubic approximant, the R-phase (29). The near edge structures of Li_K and Al_{L2-3}, which results from topological and chemical short range order around Li and Al atoms, are again very similar.

The last example is borrowed to Shindo et al. (30), who performed EDX-rays analyses of i-AlFeCu in the electron microscope, parallely to high resolution imaging. As previously pointed out (see Fig. 1 and 2), the

AlFeCu may be obtained with a high degree of perfection. It appears that the characteristic X-rays spectra are sensitive to the channeling of the electrons: the I_{Fe}/I_{Al} and I_{Cu}/I_{Al} intensity ratios are larger for channeling along 5, 2 and 3-fold axes than in non-channeling conditions, i.e. when the incident electron beam propagates away from highly symmetrical directions. Such results indicate that an ordering between Fe and Cu atoms is unlikely, and that they occupy sites different from those occupied by Al atoms.

DEFECTS

We intend by defects, departures from the ground state. They include therefore a variety of origins: chemical disorder for ordered ground state and vice-versa, topological disorder (phonons, phasons, dislocations,...), as well as their coupling. In all cases, defects scatter electrons outside Bragg peaks of the ground state or broad and shift them.

In crystals, the treatment of such effects on the basis of dynamical scattering has been achieved for a number but relatively simple cases (simple structures and simple defects), but complexe cases, as intermetallic compounds, are still in course of development, even if, most of the time, very large computer time consuming in H.R. contrast calculations is the only obstacle.

Quasicrystals are much more complicated, since the perfect ground state itself has a theoretically discrete spectrum of Bragg reflections above an intensity threshold only, all weaker ones, which fill space densely, behaving as an intrinsic diffuse scattering. On the other hand we are even not sure to have identified in real materials a perfect quasicrystalline ground-state, as discussed previously.

However, there are a number of TEM data, both in reciprocal and direct spaces, proving incontestably that quasicrystals are most of the time highly defected.

Phonons, phasons, dislocations

The theory of phonons, phasons and dislocations in quasicrystals has been developed in 2 dim. within both descriptions of density wave and unit cell tiling (31). In the density wave description, the density writes:

$$\rho(\bar{r}) = \sum_{\bar{q} \in RL} \rho_{\bar{q}} \exp[i(\bar{q}.\bar{r} + \Phi_{\bar{q}})]$$

Distortions of $\rho(\bar{r})$ can be characterized by spatially varying phases $\Phi_{\bar{q}}(\bar{r})$:

$$\Phi_{\bar{q}_i}(\bar{r}) = \bar{q}_{i//} . \bar{u}(\bar{r}) + \bar{q}_{i\perp} . \bar{w}(\bar{r})$$

where the vector field $\bar{u}(\bar{r})$ represents phonons and $\bar{w}(\bar{r})$ represents phasons. Linear phonons and phasons fields produce in the diffraction patterns a shift of $\bar{q}_{//}$ proportional to $\bar{q}_{//}$ and $\bar{q}_{\perp}$, whereas non-linear fields introduce peak broardening. In real space, phonon strains result in bending of straight lines of density maxima or equivalently distort the unit cells of the Ammann tiling, phason strains shift the straight lines of density maxima and modify the arrangements of the unit cells. Phonon-phason coupling is in principle possible (32).

Dislocations include phonon and phason terms and are characterized by a 6 dim. Burgers vector $\bar{b}$ which measures the net displacement in $\bar{w}(\bar{b}_\perp)$ and $\bar{u}(\bar{b}_{//})$ around a circuit enclosing its core. The phason field makes a difference with respect to dislocations in crystals where only translation $\bar{u}(\bar{r})$ is involved.

Experimentally, the situation is as follows:

Except may be in AlFeCu, all i-quasicrystals contain a high density of phason strains, which respond to the above description, i.e. jags on rows of white dots in H.R. images, accompanied by shifts of the Bragg reflections in the corresponding directions of the diffraction patterns (33). One example is shown in the Fig. 9.

Static phonon strains may exist in real quasicrystals only if they are coupled with phasons. This may occur near dislocation cores. However, the detection of line (plane) bending, or eventually extra-lines for dislocations is very difficult in a material which already contains phasons. For example, the line counting method used by Hiraga and Hirabayashi (34) along a close circuit around a heavily strained region of AlMnSi (assigned to a dislocation) is not really convincing. In order to improve the visibility

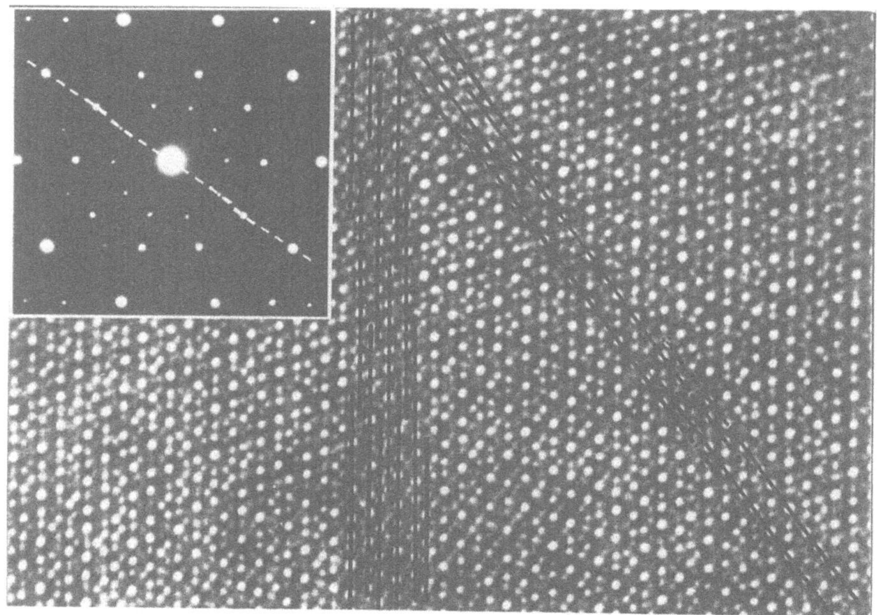

Fig. 9. HR image of AlMn along a 5-fold zone axis. The examination of the white dot rows along the projections of the 5-fold direction reveals jags giving a stripped texture of the image.

of phasons, phonons and dislocations, numerical image processing, i.e.
image digitizing, extraction of a given Fourier component $\bar{q}$ by masking in
the power spectrum, image reconstruction by back Fourier transform, has
been attempted (35,36). One example of such an image reconstruction is
given in Fig. 10 for a so-called dislocation lying parallel to a 2-fold
direction seen edge-on (36) in an i-AlFeCu quasicrystal. Phonon component
lattice bending is clearly seen in these pictures, as well as the presence of
additional fringes ending at the presumed dislocation core. A discussion of
the results has been given in the projected potential approximation for
a dislocation described in a 6 dim. lattice, as well as a partial determination of the Burger vector and a discussion of the difficulty to extend the
extinction rules well established in two-beam TEM of crystal dislocations.
If such approaches have the very merit to make in progress the defect
contrast understanding, one must be conscious of the artefacts known to
be introduced by direct filtering technique, and more generally of the
stringent conditions: foil thickness, specimen-beam relative orientation,
objective lens excitation,... imposed in lattice beam imaging to have a
one-to-one correspondence between object and image. They are also valid
for quasicrystals.

<u>Antiphase boundaries</u>
       The AlCuFe alloy has been recognized as forming a 6 dim. face-cen-
tered-hypercubic lattice (37), with superstructure reflections forbidden for
the primitive lattice (see extinction rules in ref. 38). As a consequence,
this system is able to show a domain structure with two kinds of domains.
Such a structure has been studied by standard dark field TEM (39), i e.
with primitive or superlattice reflections, revealing the existence of possible smooth but well defined antiphase boundaries.

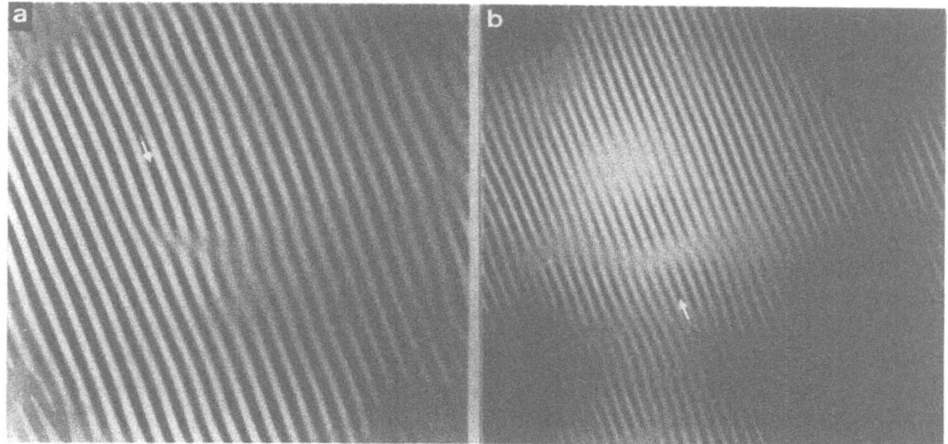

Fig. 10. Filtered images of a dislocation parallel to a 2-fold direction seen
edge-on in AlFeCu: a) filtering for $\bar{q}_1 = \pm[1/0\ 0/1\ 0/0]$, b) filtering for
$\bar{q}_2 = \pm[0/1\ 0/1\ 0/1] = \pm\bar{1}\bar{q}_1$. Arrows indicate additional fringes (from ref. 36).

Lastly, we present some results showing how the resolution power of TEM allows to discriminate between pure quasicrystals and textured crystalline domain structures with pseudo-quasicrystal symmetries. Such structures are not properly defected quasicrystals, although the existence of local translational order is a deviation from perfect quasicrystalline order. Their SAD patterns may look very close to those of perfect quasicrystals, with the addition of spot asterism or more or less diffuse scattering. In fact, we have seen that phasons shift the diffraction spots.

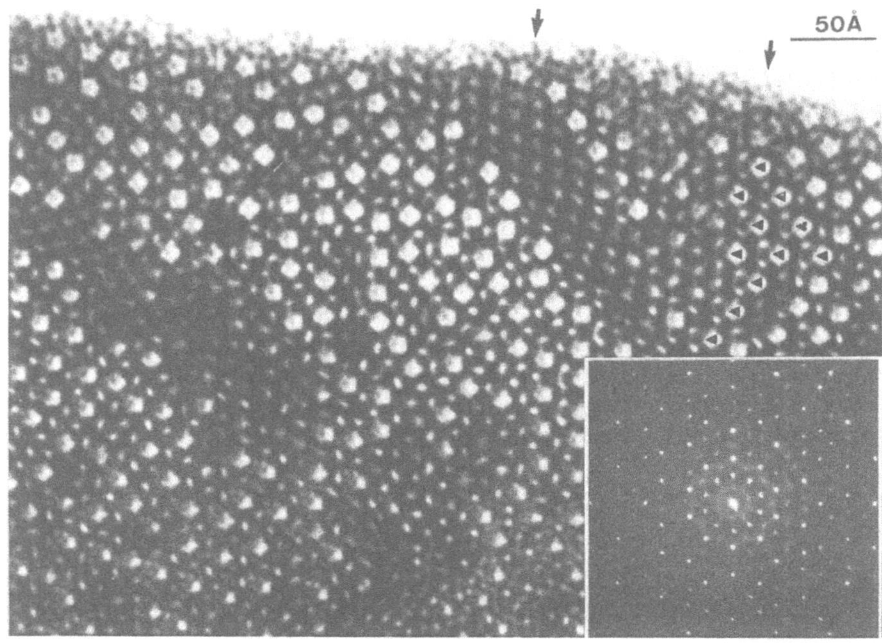

Fig. 11. HR image taken along the 10-fold axis of the decagonal Al₄Mn phase. Periodic domains are arrowed (from ref. 40).

When the phason field is linear in space, it rotates the section cut in a rational orientation and the quasicrystal is transformed in a (approximant) crystal: geometrically linear-phason field and crystal are equivalent. To be physically realistic such a transformation must involve a basic atomic motif which is able to propagate either periodically or aperiodically, and this can be observed in HR imaging. Several cases have been investigated, in AlMn (40), AlFeCu (28) and transition-metal alloys like VNiSi (41). Such structures appear to be microdomain, with a domain size in general of few nanometers. They are illustrated in Fig. 11 and 12.

Fig. 11 is a HR image of the decagonal AlMn phase along its 10-fold axis; five-branched stars, assumed to represent the icosahedral motif, are seen to be periodically distributed over lengths of 4 to 15 nm in some parts of the micrograph; arcs of diffuse scattering are distinguished on the SAD pattern.

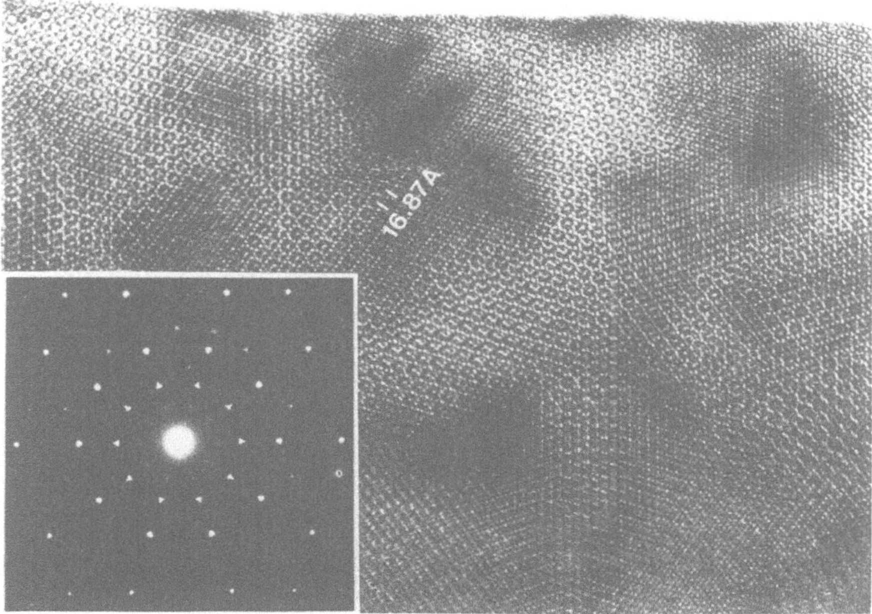

Fig. 12. Pseudo-icosahedral AlFeCu phase observed along a pseudo 5-fold axis (from ref. 28).

Fig. 12, for slowly solidified AlFeCu, is moreover clearer; it is constituted of grains of 10 to 20 nm in size, which exhibit a number of periodic features. The diffraction spots are splitted with a ($\top$) and ($\perp$) shapes alternate on successive rings. A carefull HR study has identified the Bravais lattice of the crystal as being trigonal with unit cell parameters a=3.21 nm and α=36° (28). In order for such microcrystalline aggregates to display an overall pseudo 5-fold symmetry, it is necessary to consider orientational relationships between domains. In AlFeCu such relative orientation has been checked to be relevant with a 20-branched stellate polyhedron composed of 20 prolate rhombohedra of the Ammann type; some of them are in twin relation, with twinning planes effectively observed, as shown in the Fig. 13. It is natural to presume that such a domain orientational order is inherited from the propagation of a local order common to quasicrystal and crystal. The above mentionned stellate polyhedron could play for example to that respect the role of seed.

TEM is also a beneficial technique to follow at a microscopic scale "in-situ", i.e. directly occuring into the microscope, structural transformations excited by electron irradiation or heating. Such experiments should help our understanding of the filiation between quasicrystals, glasses and crystals. We believed that a too small number of them have been realized.

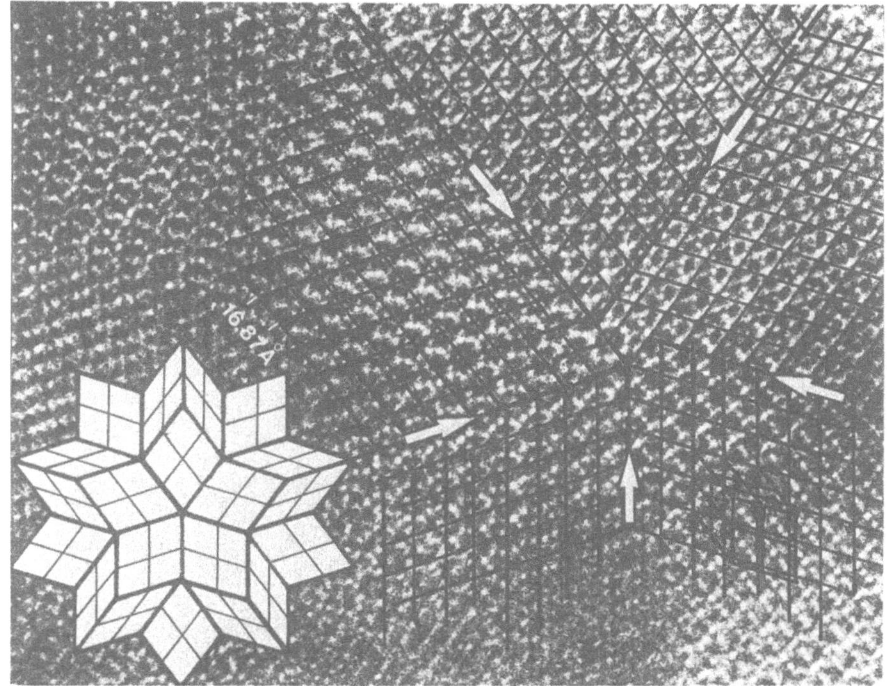

Fig. 13. Traces of {100} twinning planes observed in the AlFeCu specimen of Fig. 12. The twinning planes agree with those involved in the stellate polyhedron shown in inset (from ref. 28).

Fig. 14 shows coexisting amorphous and icosahedral phases in a AlMnSi melt-spun alloy (42), indicating a close correlation between first interatomic distances. So it is for the AlCrSi alloy shown in the Fig. 15. Upon in-situ heating, the i-particle does not grow into the a-matrix, which should have indicated a higher stability of the i-phase for apparent similar short range orders. In fact, both the i- and a-phases appear to crystallize in a hexagonal phase which cell parameters correspond to those of the μ-Al$_4$Mn phase. The last chosen example concerns the AlFeCu system. It is particularly interesting, for it has been recognized to give an i-quasicrystal with a minimun of defects (2). However making process (melt-spinning or

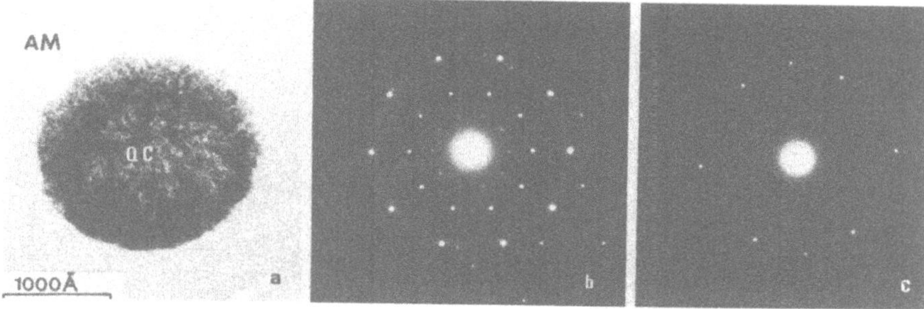

Fig. 14. Coexisting amorphous and icosahedral states in AlMnSi (a). 5-fold zone axis diffraction pattern of the i-particle in relation with the am.-ring (b). 2-fold zone axis diffraction pattern of the i-particle (c) (from ref. 42).

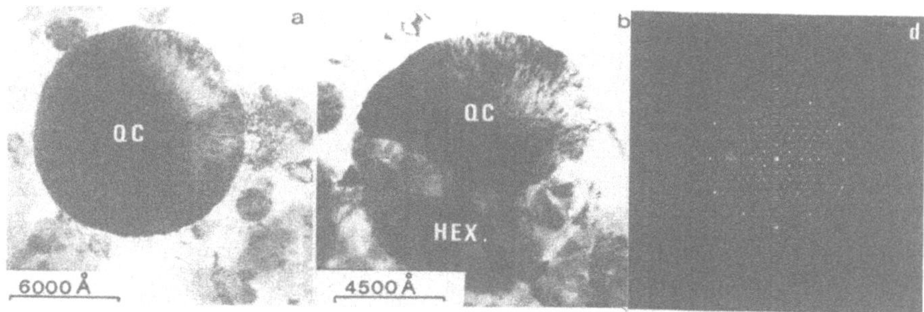

Fig. 15. "in situ" heating of AlCrSi: a) i-particle in an amorphous matrix which SAD pattern of [0001] zone axis is shown in c), (from ref. 42).

or slow solidification) and subsequent thermal heating has led to some controversy about the path to be followed to obtain such a perfection. We have shown that in the slowly solidified state, pseudo-icosahedral crystals of composition $Al_{63.5}Fe_{12.5}Cu_{24}$ are obtained (28), which characteristics have been previously described (micro-domain, trigonal). Heating transforms, at about 860°C, the micro-domain structure in the perfect i-quasicrystal· asterism and diffuse scattering disappear on the diffraction pattern, bright and dark field or HR images become homogeneous, as shown in Fig. 16. Furthermore, it appears that no other solid state phase transformation takes place on further heating up to the melting point (1050-1100°C), proving the thermodynamical stability of the i-phase at high temperatures. On the basis of such results, one expects a possible reverse transformation: icosahedral → trigonal, to occur below 860°C. By in-situ annealing at 650°C of the quenched i—state, an unexpected phenomenon has been discovered: a pair of stallites lying along different 5-fold directions appear around each reflection of the i-phase, Fig. 17. The origin of such modulation, of wave length ≥ 13 nm is under study.

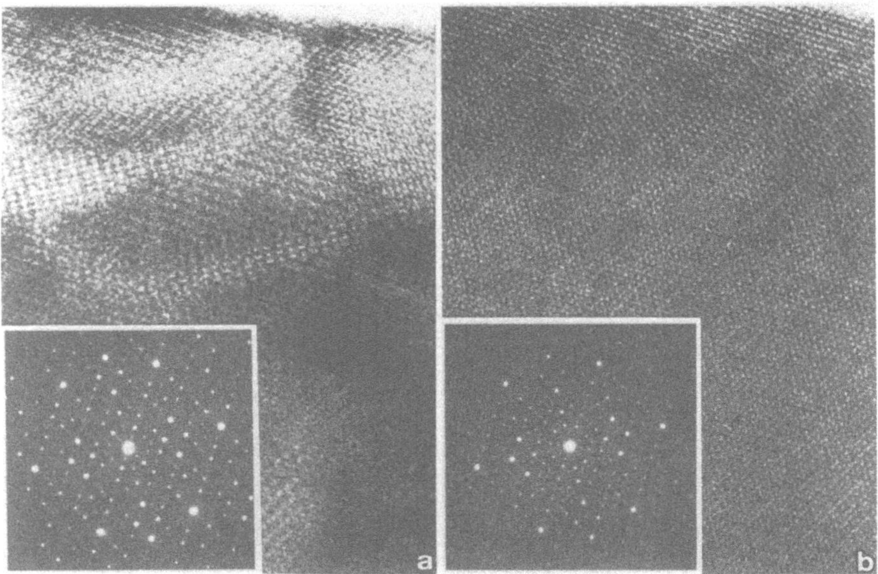

Fig. 16. HRI in the 2-fold orientation showing the transformation of the micro-domain structure of the trigonal AlFeCu phase (a) in perfect i-quasicrystal (b).

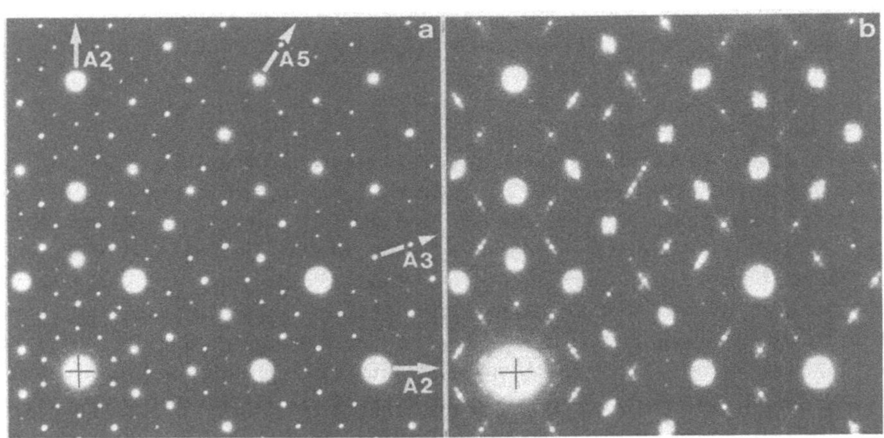

Fig. 17. 2-fold SAD patterns of i-AlFeCu before (a) and after (b) annealing at about 650°C.

104

The ultimate transformation, directly observed in the microscope, that we report, concerns the faceting of i-AlFeCu (43). The faceting ability of quasicrystals has been reported in various cases from solidification morphologies observed by scanning electron microscopy (44-46). We have confirmed this property by electron beam heating of an i-AlCuFe thin foil. Upon irradiation, holes form and grow from the specimen edge and they are faceted perpendicularly to the high symmetry directions (5, 3 and 2-fold); at the 3Å resolution of the microscope, they are planar and parallel to the bulk dense planes, as shown in Fig. 18.

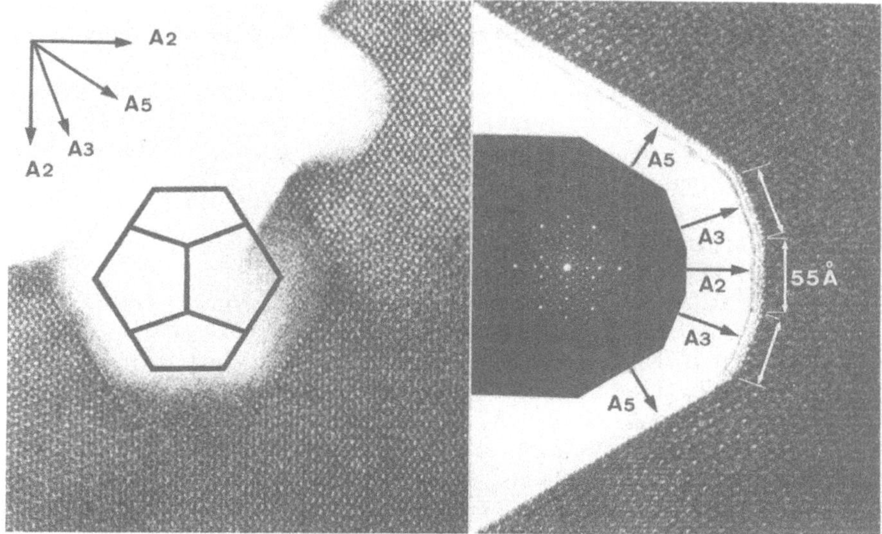

Fig. 18 Faceted hole edges in i-AlFeCu: a) HRI in 2-fold orientation where a dodecahedral hole is outlined and, b) HRI of small 3-fold and 2-fold facets joining two 5-fold facets, observed along a 2-fold zone axis

CONCLUSION

Besides to have been the discovering technique of quasicrystals, transmission electron microscopy, added to analytical microscopy, is a powerful tool to investigate their properties: symmetry, structure, topological and chemical order, defects, phase transformations,..., in a story similar to that of crystals initiated some fourty years ago. Among the large amount of obtained results, it appears that quasicrystals are generally included in multi-components microstructures, or anyhow are highly disordered (at a scale varying between the nanometer and the micron) with respect to theoretical quasiperiodic objects. Then, it comes out that electron microscopy controls are absolutely necessary for materials studied by other investigation techniques of easier quantification or more selective (like anomalous X-rays scattering or isomorphous and isotopic contrast for neutron scattering). High resolution electron microscopy is very helpful to elucidate the atomic structure; it needs, however, efforts in the field of dynamical image simulations, a hard task when one considers the difficulties already encountered in the case of large unit cell crystals, of which the quasicrystal is, in principle, the infinite limit.

REFERENCES

(1) D. Shechtman, I. Blech, D. Gratias and J.W. Cahn, Phys. Rev. Lett., 53, 1951 (1984).
(2) K. Hiraga, B.P. Zhang, M. Hirabayashi, A. Inoue and T. Masumoto, Jap. J. of Appl. Phys., 27, L951 (1988).
(3) K. Hiraga, M. Hirabayashi, A. Inoue and T. Masumoto, Sc. Rep. of Res. Inst., Tohoku Univ., A32, 2309 (1985).
(4) R. Portier, D. Shechtman, D. Gratias and J.W. Cahn, J. Microsc. Spectrosc. El., 10, 107 (1985).
(5) M. Audier and P. Guyot, Phil. Mag. B, 53, L43 (1986).
(6) A. Katz and M. Duneau, J. Phys., 47, 181 (1986).
(7) L. Bendersky, Phys. Rev. Lett., 55, 1461 (1985).
(8) P. Guyot and M. Audier, J. Microsc. Spectrosc. El., 10, 333 (1985).
(9) N. Wang, H. Chen and K. Kuo, Phys. Rev. Lett., 59, 1010 (1987).
(10) T. Ishimasa, H.U. Nissen and Y. Fukano, Phys. Rev. Lett., 55, 511 (1985).
(11) P. Bak, Phys. Rev. Lett., 54, 1517 (1985); Phys. Rev. B, 32, 5764 (1985).
(12) T. Janssen, Acta Crystallogr., A, 42, 261 (1986).
(13) T. Janssen and A. Janner, Adv. Phys, 36, 519 (1987).
(14) M. Cornier-Quiquandon, Thesis, Paris (1988).
(15) D. Broddin, G. Van Tanderloo, J. Van Landuyt, S. Amelinckx, R. Portier, M. Guymont and A. Loiseau, Phil. Mag. A, 54, 395 (1986).
(16) J. Van Landuyt, G. Van Tendeloo and S. Amelinckx, Mat. Sc. Forum, 22-24, 115 (1987).
(17) L.X. He, X.Z. Li, Z. Zhang and K.H. Kuo, Phys. Rev. Lett., 61, 1116 (1988).
(18) K. Chattopadhyay, S. Lele, N. Thangaraj and S. Ranganathan, Acta Metall., 35, 727 (1987).
(19) M. Van Sande, R. de Ridder, J. Van Landuyt and S. Amelinckx, Phys. Stat. Sol., a), 50, 587 (1978).
(20) J.W. Cahn and D. Gratias, J. de Phys. Coll. C3, 47, 415 (1986).
(21) A. Yamamoto and K. Hiraga, Phys.Rev. B, 37, 6207 (1988).
(22) P. Guyot, M. Audier and R. Lequette, J. de Phys. Coll. C3, 47, 389 (1986).
(23) P. Guyot, M. Audier and M. de Boissieu, Third Int. Meeting on Quasicrystals, Incommensurate Structures in Condensed Matter, Mexico (1989), to be published.
(24) M. Hirabayashi, K. Hiraga, D. Shindo and T.B. Williams, ibid (23).
(25) M. Audier, P. Sainfort and B. Dubost, Phil. Mag. B, 54, L105 (1986).
(26) K.M. Knowles, Quasicrystalline Materials, World Scientific Ed., 158 (1988).

(27) M. Audier and P. Guyot, Acta Metall., $\underline{36}$, 1321 (1988).

(28) M. Audier and P. Guyot, Quasicrystals AAR Conf., ICTP, Trieste (1989), to be published.

(29) P. Sainfort and P. Guyot, Scripta Metall., $\underline{21}$, 1517 (1987).

(30) D. Shindo, K. Hiraga, T. Williams, M. Hirabayashi, A. Inoue and T. Masumoto, Jap. J. of Appl. Phys., $\underline{28}$, L688 (1989).

(31) J.E. Socolar, T.C. Lubensky and P.J. Steinhardt, Phys. Rev. B, $\underline{34}$, 3345 (1986).

(32) P.L. Kalugin, A. Kitayev and L.S. Levitov, J. Phys. Lett. Paris, $\underline{46}$, 601 (1985).

(33) K. Hiraga and M. Hirabayashi, J. Electron Microsc., $\underline{36}$, 353 (1987).

(34) K. Hiraga and M. Hirabayashi, Jap. J. of Appl. Phys., $\underline{26}$, L155 (1987).

(35) D.N. Wang, T. Ishimasa, H.U. Nissen, S. Hovmoller and J. Rhyner, Phil. Mag. A, $\underline{58}$, 737 (1988).

(36) J. Devaud-Rzepski, M. Cornier-Quiquandon and D. Gratias, ibid (23).

(37) S. Ebalard and F. Spaepen, to appear in J. Mat. Res.

(38) J.W. Cahn, D. Shechtman and D. Gratias, J. Mat. Res., $\underline{1}$, 13 (1986).

(39) J. Devaud-Rezpski, A. Quivy, Y. Calvayrac, M. Cornier—Quiquandon and D. Gratias, Phil. Mag., to be published.

(40) M. Audier and P. Guyot, Suppl. to Trans. JIM, $\underline{29}$, 467 (1988).

(41) D.S Zhou, H.Q. Ye, D.X. Li and K.H. Kuo, Phys. Rev. Lett., $\underline{60}$, 2180 (1988).

(42) S. Garçon, DEA Metallurgie, Grenoble Univ. (1987).

(43) M. Audier, P. Guyot and Y. Brechet, submitted to Nature.

(44) O. Ohashi and F. Spaepen, Nature, $\underline{330}$, 555 (1987).

(45) B. Dubost, J.M. Lang, M. Tanaka, P. Sainfort and M. Audier, Nature, $\underline{324}$, 48 (1986).

(46) A.P. Tsai, A. Inoue and T. Matsumoto, Jap. J. of Appl. Phys., $\underline{26}$, L1505 (1987).

(47) K. Hiraga, M. Hirabayashi, A. Inoue and T. Matsumoto, J. of Microscopy, $\underline{146}$, $\underline{245}$ (1986).

ON THE DARK FIELD IMAGING BEHAVIOUR OF ICOSAHEDRAL PHASES IN RAPIDLY COOLED ALUMINIUM ALLOYS

K. M. Knowles and W.M. Stobbs

University of Cambridge
Dept. of Materials Science & Metallurgy
Pembroke Street, Cambridge CB2 3QZ, U.K.

INTRODUCTION

The recent report by Guryan *et al.* (1989) of an icosahedral phase in $Al_{65}Cu_{20}Ru_{15}$ for which the X-ray diffraction lines are much sharper than those from any other icosahedral phase discovered to date suggests that this phase is more 'perfect' than the other icosahedral phases. However, the nature of the disorder in these other icosahedral phases is still poorly understood. For example, some icosahedral phases, such as those in melt spun ribbons of $Al_{86}Mn_{14}$ and $Al_{90}V_{10}$ investigated by Knowles and Stobbs (1987), have very obvious mottling effects in dark field transmission electron microscope images, while the icosahedral phase in melt spun $Al_{73}Mn_{21}Si_6$ investigated by Calvayrac *et al.* (1989) exhibits clear bend contours just like crystalline materials and fails to exhibit mottling contrast even at large deviation parameters for the chosen diffracted beam, i.e. under essentially weak beam conditions. Quasicrystals in other alloy systems can exhibit other behaviour, such as diffuse scattering in reciprocal space (Mukhopadhyay *et al.*, 1987a, 1987b; Ebelard and Spaepen, 1989; Gibbons *et al.*, 1989) and/or clear thickness fringes in dark field at the edge of thin foils (Ishimasa *et al.*, 1988), but here we restrict our attention to a quantitative analysis of the differences in dark field imaging behaviour from one icosahedral system to another and to an discussion of whether this correlates with changes in the electron diffraction behaviour. We examine these two aspects with particular reference to our experimental results on three alloy systems which were made to be nominally $Al_{86}Mn_{14}$, $Al_{90}V_{10}$ and $Al_{74}Mn_{20}Si_6$.

EXPERIMENTAL METHOD

Melt-spun samples of the three alloys were prepared for transmission electron microscope examination by standard electropolishing procedures in either nitric acid/methanol solutions or perchloric acid/methanol solutions. These foils were subsequently examined in a Philips EM300 and a JEOL 2000FX electron microscope. $Al_{90}V_{10}$ samples were further examined in a JEOL 100CX electron microscope after three different heat treatments to determine the effect of ageing on the dark field contrast of the icosahedral phase in this alloy. The three different heat treatments chosen were: (a) 24hr at 300°C, (b) a differential scanning calorimeter run up to 500°C at 20°C/min, during which time the heat flow was carefully monitored and as a result of which a further heat treatment (c) of 1hr at 360°C in the differential scanning calorimeter was chosen, during which time the heat flow arising

Geometry and Thermodynamics
Edited by J.-C. Tolédano
Plenum Press, New York, 1990

from any phase transformation could be measured isothermally. The differential scanning calorimeter runs were made in a Perkin-Elmer 7 Series thermal analysis system.

EXPERIMENTAL RESULTS AND DISCUSSION

Our experimental results on the melt spun ribbons of $Al_{90}V_{10}$ and $Al_{86}Mn_{14}$ in their quenched state have been presented in a previous paper (Knowles and Stobbs, 1987). The pertinent experimental observations relevant to the behaviour of the dark field images in these two alloys are:

(1)   There are both coarse variations and fine variations in dark field contrast within the quasicrystalline particles in these two alloys. The coarse variations in contrast are on a ~ 0.1 μm scale. These arise from large scale compositional variations which occur as a result of macrosegregation within the melt at the solid-liquid interface. These coarse variations are also clearly seen in electron micrographs of $Mg_{32}(Al,Zn)_{49}$ (Chandra and Suryanarayana, 1988) and of $Al_{74}Mn_{20}Si_6$ and $Mg_{32}(Al-Zn-Cu)_{49}$ (Mukhopadhyay *et al.*, 1987b) and thus reflect the way in which the quasicrystal grows from the melt. These relatively large scale variations will be associated with slight changes in the lattice parameters of the quasicrystalline phase.

(2)   The fine scale variations in contrast are on a ~ 5 nm scale. The critical observation of this fine scale contrast is that it is speckle contrast which is not aperture-limited, even though it behaves in a similar manner to speckle contrast from amorphous materials, in that a small change in deviation parameter of the diffracted beam alters the areas which are strongly diffracting. This behaviour is explicable in terms of a fine scale chemical inhomogeneity which occurs as a fairly smooth modulation within the quasicrystalline phases. We have already suggested (Knowles and Stobbs, 1987) that this fine scale inhomogeneity might arise as a result of the way in which the atoms in a particular quasicrystalline phase decorate the underlying 'tiling', which could, for example, be the three-dimensional Penrose tiling, which is one of the geometries with the correct diffraction symmetries for icosahedral quasicrystals (Socolar and Steinhardt, 1986). Tilings such as this have a large number of vertices with different local coordination geometries, and we suggested that it is the way these vertices are distributed in three dimensions which provides the "driving force" for the chemical inhomogeneity.

(3)   There are subtle distinctions in the behaviour of the fine scale contrast in $Al_{90}V_{10}$ and $Al_{86}Mn_{14}$. The fine scale speckle contrast in weak beam dark field images of $Al_{90}V_{10}$ can have smaller dimensions than the fine scale contrast in $Al_{86}Mn_{14}$, and this we attribute to distinctions in the way in which the atoms decorate the underlying 'tiling'. There are two possible reasons for this: a substantial difference in the degree to which there are localised departures on a nanometre scale from the average chemical composition for the icosahedral phases in the two alloys and/or a fundamentally different way of decorating the 'tiling' because of the difference in size or ionicity of the manganese and vanadium atoms.

Our new experimental results were obtained with the aims of investigating the behaviour of the dark field contrast as a function of heat treatment in $Al_{90}V_{10}$ and of clarifying whether or not such contrast is seen in dark field images of an Al-Mn-Si alloy, in which the quasicrystalline phase is supposedly more 'perfect', in that the X-ray spectra from this alloy exhibit reasonably sharp peaks and the electron diffraction patterns exhibit many more reflections than other aluminium-based icosahedral phases, thus signifying the more dynamical nature of the scattering for this material in comparison with $Al_{90}V_{10}$ and $Al_{86}Mn_{14}$.

The effect of a 24hr heat treatment at 300°C on the imaging behaviour of the quasicrystalline phase in $Al_{90}V_{10}$ is shown in Fig. 1. It is clear that this phase has not transformed to any other phase and is therefore stable to this temperature. The dark field imaging behaviour of the icosahedral phase was indistinguishable from that seen from the as-spun alloy (Knowles and Stobbs, 1987).

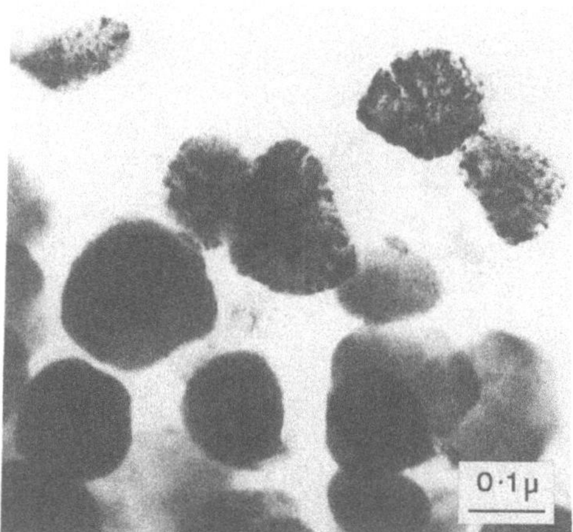

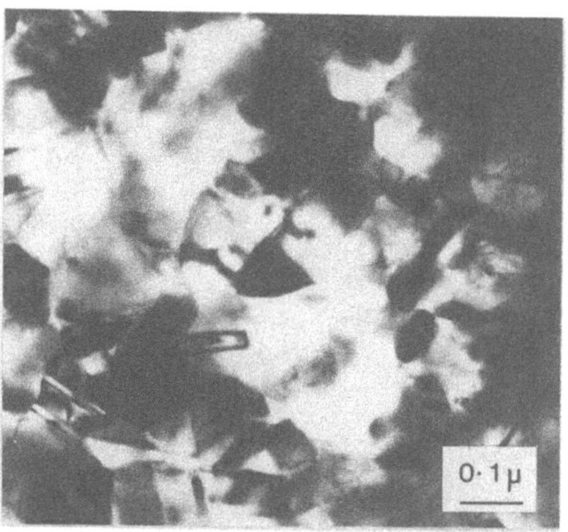

Fig. 1   Bright field electron micrograph of $Al_{90}V_{10}$ after a heat treatment of 24hr at 300°C.

Fig. 2   Bright field electron micrograph of $Al_{90}V_{10}$ after a DSC run up to at 500°C at 20°C per minute. The microcrystals are $Al_{10}V$.

As a result of this observation, a further series of heat treatments were carried out in a differential scanning calorimeter to assess the temperature at which any quasicrystalline $\rightarrow$ crystalline transformation would take place. A DSC run up to 500°C at a heating rate of 20°C produced the microstructure shown in Fig. 2, which consists of a fine arrangement of $Al_{10}V$ crystals, recognisable by their distinctive f.c.c. diffraction patterns. The heat flow from this DSC run was recorded (Fig. 3), together with DSC runs up to 550°C at different scan rates, such as those shown in Fig. 4.

A Kissinger analysis of the peak temperatures in these DSC runs as a function of scan rate (Kissinger, 1957; Henderson, 1979) gave an activation energy for the quasicrystalline $\rightarrow$ crystalline transformation of 183 ± 5 kJ mol^{-1}, which, when combined with an Avrami equation with a time exponent of 2, suggested that the volume fraction of crystalline phase after one hour at 360°C would be about a half. In practice this drastically overestimated the actual volume fraction of $Al_{10}V$ formed after one hour, but there were clear regions of $Al_{10}V$ formed between quasicrystalline particles, such as the one shown in Fig. 5(a). Again, it was significant that the dark field imaging behaviour in the retained quasicrystalline phase remained unaltered by this heat treatment (Figs. 5(b)-5(d)).

Turning now to to the behaviour of the quasicrystalline phases in the $Al_{74}Mn_{20}Si_6$ alloy, bright-field electron microscopy revealed that it had a coral reef-like contrast (Fig. 6), as seen for the $Al_{86}Mn_{14}$ alloy we had investigated previously (Knowles and Stobbs, 1987) and also found for the $Al_{74}Mn_{20}Si_6$ alloy investigated by Mukhopadhyay et al. (1987). In addition to the icosahedral phase, there were residual crystalline aluminium alloy régimes between the quasicrystalline grains.

The dark field imaging behaviour showed clear differences from the icosahedral phase in $Al_{86}Mn_{14}$. Series of dark field images as a function of deviation parameter, such as the one in Fig. 7, showed clear thickness fringes at moderate values of deviation parameter (e g. Figs. 7(b) and 7(c)),

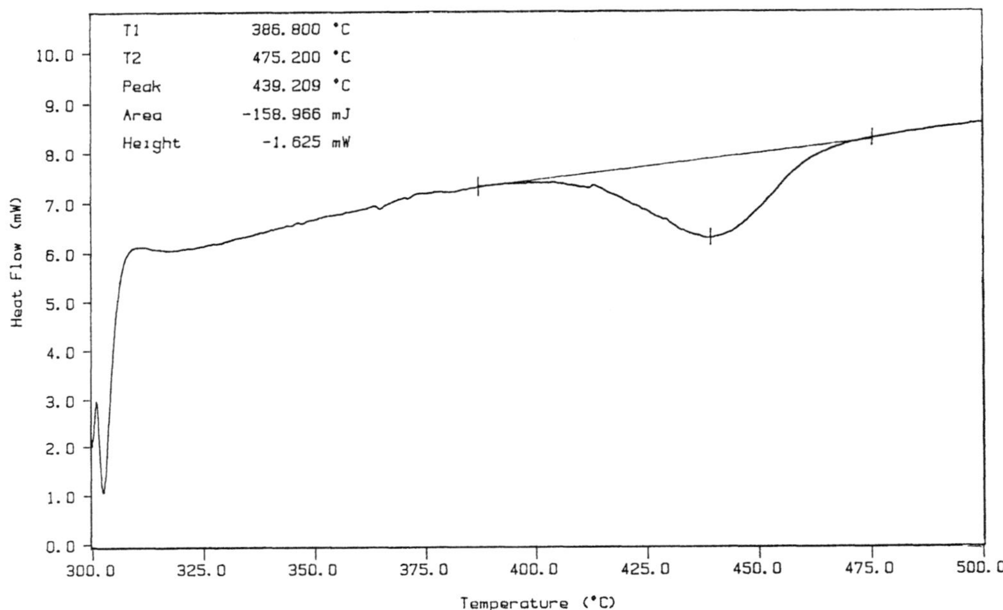

Fig. 3   DSC run of a sample as-received $Al_{90}V_{10}$ melt-spun strip conducted at a rate of 20°C per minute.

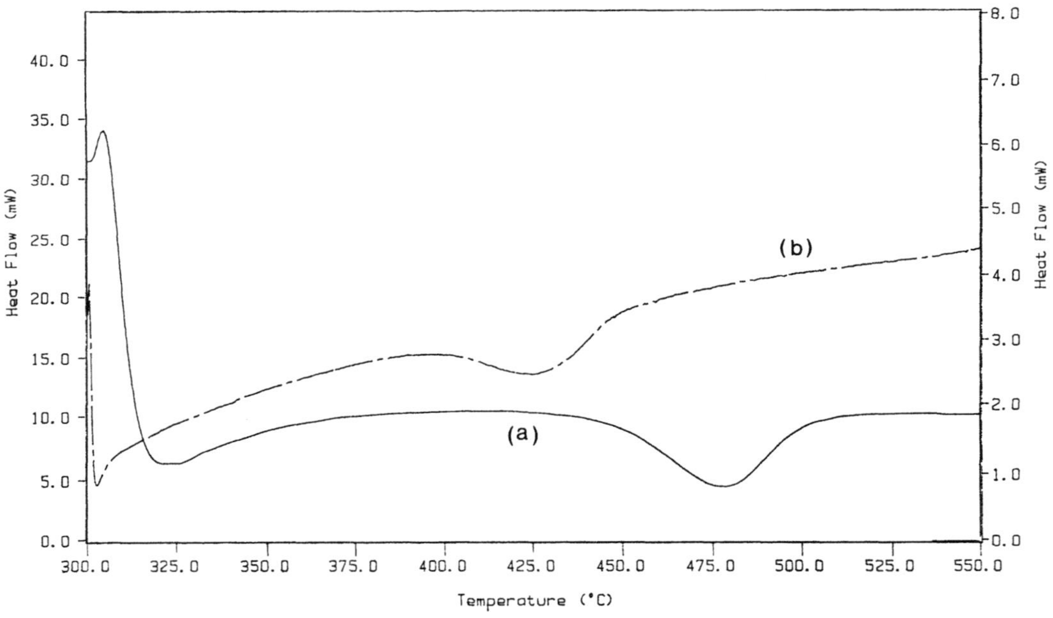

Fig. 4   DSC runs of samples of as-received $Al_{90}V_{10}$ melt-spun strip conducted at rates of (a) 100°C per minute and (b) 10°C per minute. The heat flow for curve (a) is shown on the left hand vertical axis and the heat flow for curve (b) is shown on the right hand vertical axis.

112

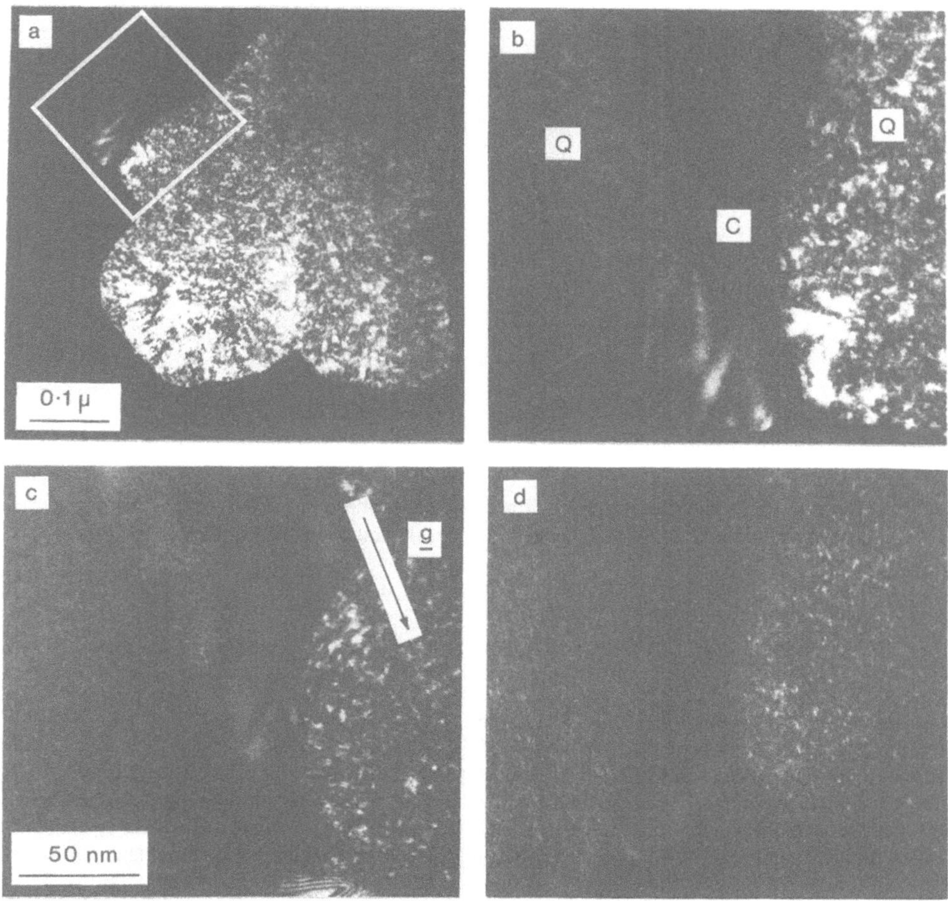

Fig. 5   Dark field electron micrographs from the quasicrystalline (Q) and crystalline (C) phases in a $Al_{90}V_{10}$ alloy heat treated at 360°C for 1hr. (a): a grain of icosahedral phase lighting up strongly in a $\mathbf{g} = 221001$ type diffracting vector in the Elser notation (Elser, 1985). (b), (c), (d): Enlargements of the area boxed in white in (a) showing the behaviour of the dark field images as a function of increasing deviation parameter. Note the imaging behaviour of the crystalline grain and the weakly diffracting quasicrystalline grain to the left of (b), (c) and (d).

as well as speckle contrast at high deviation parameters (e.g. Figs. 7(c) and 7(d)), whereas at very low values of deviation parameter the icosahedral phase exhibited more uniform contrast of a type similar to that seen for crystalline and decagonal phases.

It is relevant to compare the imaging behaviour of this icosahedral phase with the one investigated by Calvayrac *et al.* (1989). The composition of their alloy was $Al_{73}Mn_{21}Si_6$, whereas the one we investigated was nominally Al-33.7wt%Mn- 5wt%Si, corresponding to an alloy composition of 74.2at%Al, 20.0at%Mn and 5.8at%Si, and yet there are clear differences in the imaging behaviour of these two alloys. However, it must be remembered that the overall

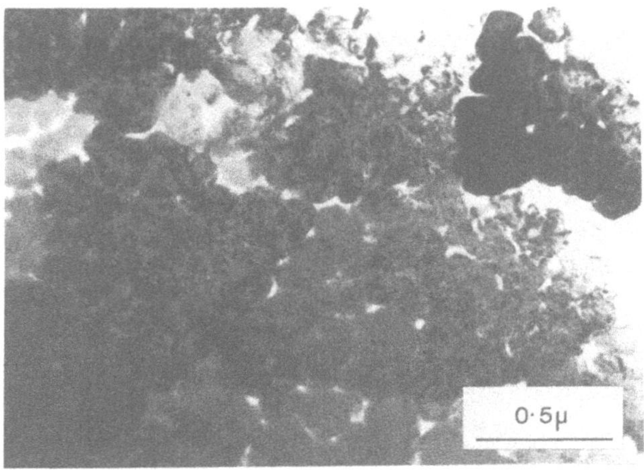

Fig. 6   Bright field electron micrograph of the icosahedral phase in $Al_{74}Mn_{20}Si_6$. The bright areas between and within individual 'fingers' of icosahedral phase are residual aluminium.

compositions are only approximate indications of the extent to which the icosahedral phases found (as a function of the particular heat treatment used) might differ in composition.Nevertheless, it is still significant that Calvayrac *et al.* (1989) did not see any speckle (or sparkle) contrast in their alloy even for large values of deviation parameter, whereas we do. The bend contours seen by Calvayrac *et al.* (1989) have an explanation entirely analagous to that for the thickness fringes that we see so clearly in this alloy and Ishimasa *et al.* (1988) see in $Al_{65}Cu_{20}Fe_{15}$, namely that the extinction distances for the strong reflections in these alloys are of the same order, or slightly larger, than those found in crystalline materials. By comparison, the extinction distances for the strong reflections in the icosahedral phases in both $Al_{90}V_{10}$ and $Al_{86}Mn_{14}$ are very large, which is why thickness fringes are not seen in these phases.

The reason for this difference in behaviour between our Al-Mn-Si alloy and that of Calvayrac *et al.* (1989) is tied in with the reason for the speckle contrast. Suppose we had an underlying 'tiling' such as a three-dimensional Penrose tiling, in which the 'average' atom description in terms of a smooth atomic occupancy were valid. In this case, there would, by definition, be no chemical modulation, and the contributions to a given diffraction spot in an electron diffraction pattern would add as coherently as the geometrically imposed constraints of the underlying 'tiling' allowed. This would then give rise to dynamical electron scattering behaviour similar to that found routinely in crystalline materials. Furthermore, there would be no fine scale speckle in weak beam dark field images because of the chemical homogeneity of the material.

However, if there were a chemical inhomogeneity, the effect would be to prevent the monotonic additions of amplitude to a particular diffraction spot through the thickness of a TEM specimen. This would cause speckle contrast in the way we describe, with changes in the regions showing high contrast as the deviation parameter is changed arising from changes in the projectional geometry of the specimen. The necessary coherent summing of amplitudes to a particular diffraction spot, together with changes in phase differences arising partially from the changes in separation of the strongly scattering regions on tilting, will cause rapid changes of contrast with deviation parameter. The partial randomisation of the phase additions would also have the effect of increasing the extinction distance for the particular beam. This would in turn have the effect of making thickness fringes more difficult to observe as the degree of inhomogeneity increased.

114

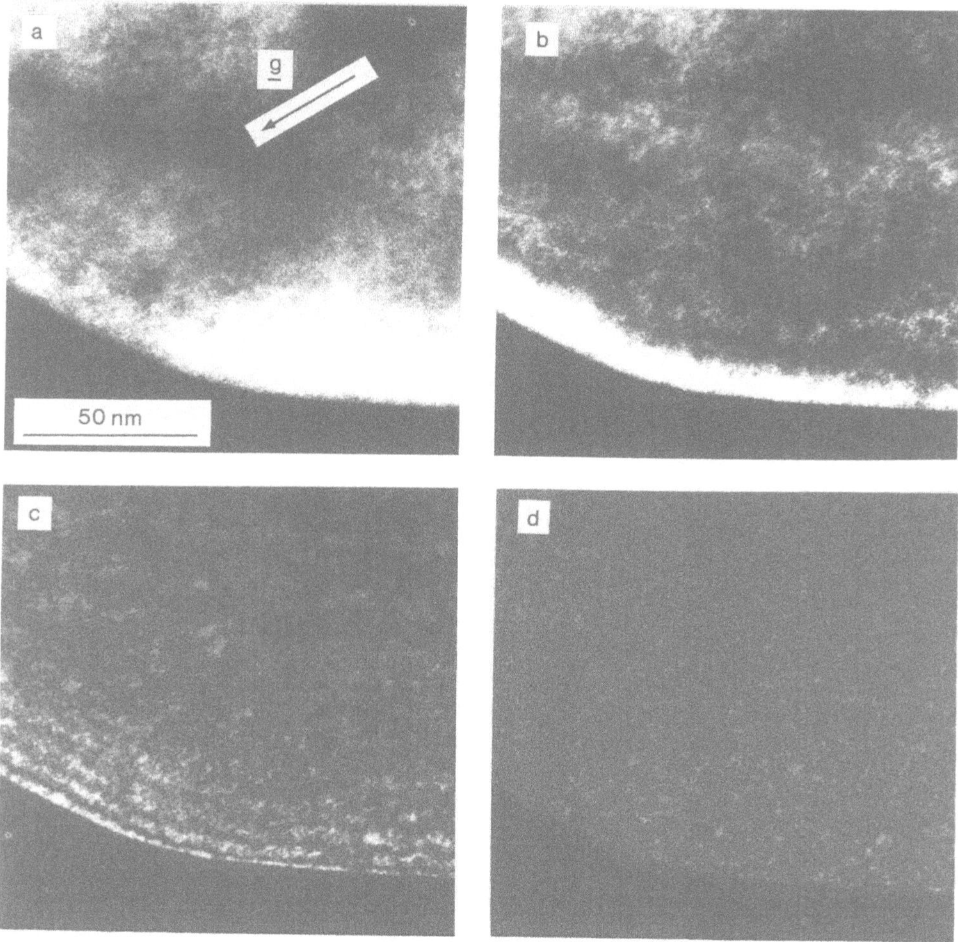

Fig. 7   (a) - (d) are dark field electron micrographs from a **g** = 221001 type diffracting vector in as-spun $Al_{74}Mn_{20}Si_6$ ribbon with the deviation parameter increasing from (a) to (d). Note the thickness fringes in (b) and (c) and the speckle contrast in (c) and (d).

We thus attribute the differences between the alloy we investigated and the alloy investigated by Calvayrac *et al* (1989) to large and localised departures from the mean chemical composition for our material. This would result in differences in the way in which atoms populate the underlying 'tiling' of the quasicrystal: in the one case (theirs) there is a virtually uniform composition at the atomic level whereas, in the other case (ours), small differences in the overall composition have led to chemical inhomogeneities at the near-atomic level sufficent to cause speckle contrast, albeit at very large deviation parameters, coupled with the occurrence of thickness fringes. By comparison, the chemical inhomogeneities in $Al_{86}Mn_{14}$ and $Al_{90}V_{10}$ are much larger, which is why the speckle contrast for these materials is so much more obvious. We also note that this explanation also accounts for the experimental results of Mukhopadhyay *et al.* (1987) on $Al_{74}Mn_{20}Si_6$: we deduce that their alloy was of yet a third composition, slightly different from the ones that we and Calvayrac *et al.* (1989) have investigated, and one which from the behaviour of the bright field image in Fig. 1(a) of their paper we would expect to exhibit strong speckle contrast, but not to exhibit thickness fringes so readily.

Finally, it is relevant to discuss the nature of the diffuse intensity seen in the electron diffraction patterns from some alloys, such as $Mg_{32}(Al\text{-}Zn\text{-}Cu)_{49}$ (Mukhopadhyay *et al*, 1987) and Ti_2Mn (Gibbons *et al.*, 1989). It is evident that in these alloys there are ordering phenomena which have no direct bearing on our discussion on the dark field speckle contrast, but which will nevertheless affect the character of the diffracted intensity distribution from a thin foil because of the nature of the ordering.

CONCLUSIONS

We have shown that the differences in dark field imaging behaviour from one quasicrystalline phase to another can be explained consistently if it is supposed that there are differences in the degree of chemical inhomogeneity from phase to phase. In particular, icosahedral phases which exhibit thickness fringes are necessarily more chemically uniform than those which exhibit speckle contrast. We account for these differences in behaviour in terms of the way in which atoms decorate the underlying 'tiling'. Furthermore, small differences in the chemical composition of the icosahedral phases formed in alloys with nominally the same composition examined by different investigators probably accounts for many of the apparent discrepancies found in the literature.

ACKNOWLEDGEMENTS

We would like to thank Dr. Marc Audier of the Laboratoire de Thermodynamique et Physico-Chimie Métallurgiques, E.N.S.E.E.G., St. Martin d'Heres for providing us with the $Al_{74}Mn_{20}Si_6$ melt-spun ribbon and we would like to thank Dr. R.F. Cochrane and Dr. P.V. Evans for experimental assistance with the differential scanning calorimetry measurements. We would also like to thank Prof. D. Hull, F.R.S. for the provision of laboratory facilities.

REFERENCES

Calvayrac, Y., Devaud-Rzepski, J., Bessiere, M., Lefebvre, S., Quivy, A. and Gratias, D., 1989, The nature of the topological disorder in the rapidly quenched $Al_{73}Mn_{21}Si_6$ icosahedral phase, *Phil. Mag. A*, 59:439.

Chandra, S. and Suryanarayana, C., 1988, Quasicrystalline-to-crystalline transformation in rapidly solidified $Mg_{32}(Al,Zn)_{49}$, *Phil. Mag. A*, 58:185.

Ebalard, S. and Spaepen, F., 1989, Long range chemical ordering in Al-Cu-Fe, Al-Cu-Mn and Al-Cu-Cr quasicrystals, submitted to *J. Mater. Res.*

Elser, V., 1985, Indexing problems in quasicrystal diffraction, *Phys. Rev. B*, 32:4892.

Gibbons, P.C., Kelton, K.F., Levine, L.E. and Phillips, R.B., 1989, Arcs of diffuse scattering from icosahedral Ti-Mn quasicrystals, Washington University preprint.

Guryan C.A., Goldman, A.I., Stephens, P.W., Hiraga, K., Tsai, A.P., Inoue, A. and Masumoto, T., 1989, Al-Cu-Ru: an icosahedral alloy without phason disorder, *Phys. Rev. Lett.*, 62:2409.

Henderson, D.W., 1979, Thermal analysis of non-isothermal crystallisation kinetics in glass froming liquids, *J. Non-Cryst. Solids*, 30:301.

Hiraga, K., Zhang, B-P., Hirabayashi, M., Inoue, A. and Masumoto, T., 1988, Highly ordered icosahedral of Al-Cu-Fe alloy studied by electron diffraction and high resolution electron microscopy, *Japanese Journal of Applied Physics*, 27:L951.

Ishimasa, T., Fukano, Y. and Tsuchimori, M., 1988, Quasicrystal structure in Al-Cu-Fe annealed alloy, *Phil. Mag. Lett.*, 58:157.

Kissinger, H.E., 1957, Reaction kinetics in differential thermal analysis, *Analytical Chemistry*, 29:1702.

Knowles K.M. and Stobbs W.M., 1987, The inhomogeneity of the icosahedral phase in Al-Mn and Al-V, *J. Microscopy (Oxford)*, 146:267.

Mukhopadhyay, N.K., Ranganathan, S. and Chattopadhyay, K., 1987a, On the short-range order in Al-Mn quasicrystals during low temperature ageing, *Phil. Mag. Lett.*, 56:121.

Mukhopadhyay, N.K., Thangaraj, N., Chattopadhyay, K. and Ranganathan, S., 1987b, A comparative electron microscopic study of Al-based and Mg-based quasicrystals, J. *Mater. Res.*, 2:299.

Socolar, J.E. and Steinhardt, P.J., 1986, Quasicrystals II: Unit cell configurations, *Phys. Rev. B*, 34:617.

ELECTRON MICROSCOPY OF MODULATED STRUCTURES

J. Van Landuyt, G. Van Tendeloo and S. Amelinckx

University of Antwerp, RUCA
Groenenborgerlaan 171
B-2020 Antwerp, Belgium

1. INTRODUCTION

The concept of "modulated structure" can be defined in a general way as
a new structure which consists of a periodic modulation of a parent
structure. The modulation can be "commensurate" with the basic structure or
"incommensurate", a characteristic which will be evident from diffraction
data. Two different kinds of modulated structures can be distinguished :

1. The modulation can consist of the periodic repetition of "modules"
   separated by interfaces such as antiphase boundaries, discommensuration
   walls (DCW), crystallographic shear planes (CSP), inversion boundaries,
   stacking faults and twins or orientation variants.
   The interface can be conservative (eg. stacking faults) or
   non-conservative (eg. CSP) enabling through incorporation of these
   defects deviations from stoichiometric composition to be accommodated.

2. A simple basic structure can be modulated by imposing a periodic deformation wave with a period which is larger than that of the basic structure.
   The average atom positions of the basic structure are then conserved.

Electron microscopy is particularly useful for the analysis of
modulated structures and enables crystallographic models to be proposed. It
has provided proof for the existence of a wide variety of these structures.
The strength of this technique mainly stems from the fact that the atomic
scattering factor of electrons is very high in comparison with X-rays and
neutrons (10^4 x that for X-rays). This yields signals of sufficient
intensity even from interaction of the electrons with very small crystal
quantities or areas; this interaction can be translated into direct images
and also into diffraction patterns. Both types of images can be interpreted
in terms of structures and structure defects.
The advantage of the use of an electron microscope is furthermore its
faculty for providing in <u>one</u> instrument electron diffraction patterns as
well as direct space images from the same small and intentionally selected
areas. Furthermore modern instruments now reach resolving powers down to
0.17nm thus providing atomic scale detail of the structures.

It is the purpose of this lecture to provide a short introduction on
the observation techniques and the interpretation of the resulting images
and diffraction patterns.
A further section will discuss a few case studies of modulated structures.

Geometry and Thermodynamics
Edited by J.-C. Tolédano
Plenum Press, New York, 1990

2. OBSERVABLES OF MODULATED STRUCTURES

2.1. Electron Diffraction
2.1.1. Periodic translation interfaces

If a crystal structure is modulated by a set of periodic interfaces along a unit normal $\bar{e}$ and with a period λ the structure can be considered as being modulated by a set of plane waves with wave vector $\bar{q} = (1/\lambda)\bar{e}$. If furthermore the interface is characterised by a displacement $\bar{R}_o$ between successive modules the displacement of the atoms in the nth block can be found as $\bar{a}(\bar{r}_n) + N\bar{R}_o$ N : integer, $\bar{a}$: a lattice vector of the undisplaced structure).

It is easily shown [1] that the scattered amplitude of the modulated crystal is then given by :

$$A(\bar{g}) = \sum_n \sum_m f_n(\bar{g}) \; A_m(\bar{g}) . \exp 2\pi i [\bar{g} + (\bar{g}.\bar{R}_o)\bar{q} + m\bar{q}].\bar{r}_m$$

where :
- $f_n(\bar{g})$ is the atomic scattering factor for the atom at $\bar{r}_n$
- $A_m(\bar{g})$ is the fourier coefficient of the expansion of the modulation wave
- m is the order of the expansion term and thus integer.

This amplitude will exhibit sharp peaks at :

$$\bar{H} = \bar{g} + (\bar{g}.\bar{R}_o)\bar{q} + m\bar{q} = \bar{g} + (m + \bar{g}.\bar{R}_o)\bar{q}$$

The diffraction spots at $\bar{H}$ of the interface modulated structure thus form linear arrays of spots with a spacing $|q|=1/\lambda$ and positions which are shifted with respect to the basic spots at $\bar{g}$ over a fraction $\bar{g}.\bar{R}_o$ of the interspot separation.
It is clear that measurement of this shift for three independent reflections $\bar{g}$ allows for the determination of the displacement vector $\bar{R}_o$ thus enabling crystallographic models to be proposed for the modulated structure.
The calculation procedure can be extended to account for more complex and also realistic interface descriptions which include [2] :
1° relaxation at the boundary plane
2° deviation from strict periodicity of the defects
The results of course confirm the above mentioned simple situation of periodic interfaces but also predict as in the case of the less periodic interface structures similar to Fujiwara structures [3], sharp satellites in incommensurate positions shifted over the fractional distance $\bar{g}.\bar{R}_o/d$ whereby the satellite intensity decreases with increasing order. A procedure outlined by Lu and Birman [4] for mistakes and long range correlation in one-dimensional quasi lattices could within this formalism be extended to two-dimensional (-planar) interfaces [2].

Also convergent beam electron diffraction techniques (CBED) can be used for studying incommensurate structures [5]. CBED is particularly effective for determining the symmetry characteristics of the phases. Proposals for using CBED for quasicrystal studies were reported by Bird & Withers [6] and applications e.g. to AlMnSi by Bendersky and Kaufman [7].

2.1.2. Periodically deformed crystal

Another type of modulated structure consists of a basic crystal which is periodically deformed. Simple Fourier transformation procedures allow for the amplitude $A(\bar{h})$ of the satellite reflections to be calculated as [8] :

$$A(\bar{h}) = \sum_{\bar{g},\bar{G}} V_{\bar{g}} . W_{\bar{g},\bar{G}} \; \delta (\bar{g}+\bar{G}-\bar{h})$$

$\bar{g}$: reciprocal lattice vector of the basic structure
$\bar{G}$: reciprocal lattice vector of the periodic deformation pattern
$V_{\bar{g}}$ and $W_{\bar{g},\bar{G}}$ are the fourier coefficients of the expansions of respectively the undeformed crystal and the crystal modulation.

For an infinite crystal the diffraction pattern thus consists of arrays of sharp peaks located at $\bar{h} = \bar{g}+\bar{G}$ where $\bar{g}$ are the basic reflections around which the satellites are positioned at $\bar{g}+\bar{G}$. Since $|G|$ is smaller than $|g|$ the "satellites" form an array associated with each Bragg spot.
It is furthermore easily shown that the ratio of the amplitudes of the satellite reflections to those of the corresponding basic reflections : A_{sat}/A_{basic} is proportional to $|g|$ and thus will increase with the order of reflection. This seems to be a characteristic feature of the diffraction patterns of deformation modulated structures and can thus be considered as some type of signature for their presence.

2.2. Imaging – Direct Space Evidence

Valuable information on modulated structures can also be obtained from transmission images in electron microscopy. Two imaging modes yield useful information on a different scale :

1. diffraction contrast images (at moderate magnification) of the defects in the modulations and the modulations themselves if sufficiently isolated.
2. high magnification high resolution images of the modulations and their repetitition into new structures.

2.2.1. Diffraction Contrast Images – Low Magnification

The elastic interaction of high energy electrons with crystalline material is very sensitive to local changes in periodicity and orientation. Images obtained using the transmitted or diffracted beam resulting from this interaction, rely for their interpretation mainly on the two-beam electron diffraction theory which enables the identification of structure defects by well understood and reliable procedures (e.g. [9]).

Structure factor changes and local orientation differences, will influence the diffracted intensities and consequently the transmission image which is an enlarged mapping of the beam intensity at the exit face of the crystal foil. As a consequence crystalline areas differentiated in structure factor, give rise to differences in intensity. It allows one to observe directly phase changes or transformation fronts between phases. Also the periodic alternation of phases or structure variants can be revealed by this contrast mechanism if the periods involved are sufficiently large. The evolution of phase fronts upon exterior influences such as heating, cooling, straining, etc. can also be studied in this way.

By using the dark field technique whereby a spot originating from only one of the phases is selected for imaging, only that area which belongs to this phase state will light up. This mode is also particularly useful to visualise defects associated with modulated structures; dark field images in satellite reflections (SDF) will reveal defects in that particular ordered structure which gave rise to the satellite reflections.

<u>2.2.2. Structure Imaging - High Resolution Mode</u>

In this mode many diffracted beams along a zone axis are allowed to interfere and give rise to an image after transfer through the microscope lens system. Modern electron microscopes in which lens defects and electronic and mechanical stability have been improved, allow for the higher angle beams to contribute in a non detrimental way to high magnification images. This multibeam-mode is called high resolution structure imaging; if all necessary precautions are taken concerning specimen thickness, precise alignment along a zone-axis controlled focusing conditions etc., resolution figures for structure imaging of 0.17nm can nowadays be reached.
This mode is very useful for the analysis of modulated structures since it enables to visualise the structure nearly down to the atomic scale; in particular it allows in some cases to determine the nature of the modulation. However it should be born in mind that the interpretation down to this scale should be <u>complemented by structure image calculations</u>. A review of these calculation methods was given by Spence [10], and recent advances with respect to multislice algorithms were reported by Van Dyck [11] and Self et al. [12].

The three basic observation modes : electron diffraction, diffraction contrast and high resolution imaging which are as mentioned before simultaneously available in electron microscopes, often complement each other to enable detailed and otherwhise unattainable information to be derived on modulated structures and e.g. also quasicrystals. These studies are often usefully complemented by optical diffraction analysis of the HR-negatives as will be illustrated in the case studies. Numerous studies on modulated structures made fruitful use of these techniques as it will appear clearly in this conference. A word of caution should be sounded for the interpretation of high resolution images of highly defective structures and also quasicrystals. The image is in these cases usually complicated by the fact that the atoms are no longer situated on a lattice which can cause unpredictable overlap along the beam direction and the consequent uncertainty in atom allocation.

3. CASE STUDIES

A few case studies will now be developed in some detail and are intended to illustrate the potentialities of electron microscopy for the study of modulated structures.

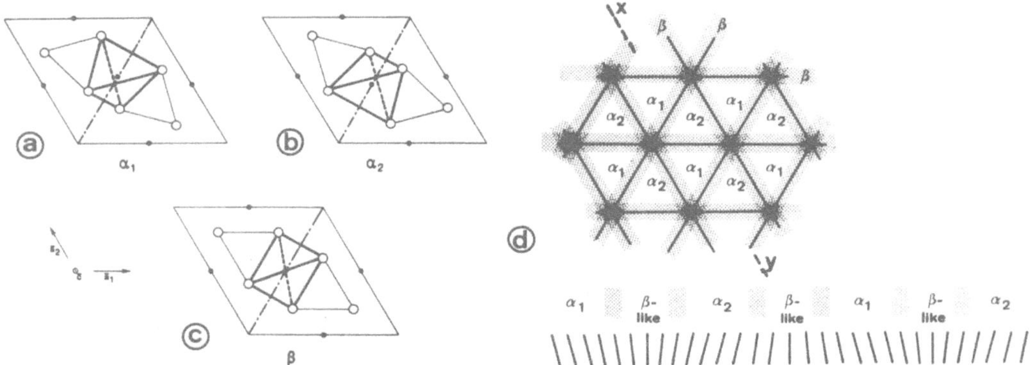

Fig. 1. Structure projection along the c-axis. (a) α-quartz, one variant e.g. α_1, (b) the other variant of α-quartz, α_2, (c) the projection of the β-phase (tetrahedron of oxygen atoms (open circles) projected as a rectangle). (d) Illustrates the model for the incommensurate intermediate phase, in terms of α_1-α_2 domains with β-like interfaces.

<u>3.1. The $\alpha \to \beta$ phase transition in quartz</u>

Although quartz is one of the oldest known and studied minerals, and its phase transition at 573°C was recognised for more than a hundred years it is only recently that the existence of a modulated structure as an intermediate phase between α and β has been established and later studied by various techniques [13]. Electron microscopy observations indeed revealed in 1975 [14] the presence of a complex microstructure in the vicinity of the transformation front between α and β.

α-quartz is rhombohedral and belongs to the pointgroup 32 whereas β-quartz is hexagonal with pointgroup 62. The loss in symmetry is evident on a projection along the c-axis in figs. 1a-c. Both the α and β structures are represented and for the α phase its two variants, α_1 and α_2. The structure consists of interlinked SiO_4 tetrahedra which are projected as trapezia in the α-phase and as rectangles in the β-phase. The relationship between the α_1 and α_2 variants is that of a 180° rotation twin with the c-axis as twin axis : the Dauphiné twin relationship.

The experimental conditions for the observation of the intermediate phase were the following :
Ion beam thinned specimens were heated to close to 573°C in the specimen heating holder and a temperature gradient was further produced by local heating with the electron beam. By choosing the gradient across the transition, a large number of defects is observed; they are very mobile at temperatures just below the transition [15]. In bright field images these defects are hardly visible, whereas a pronounced dark-light contrast is observed in the dark field images of particular reflections such as e.g. the $(30\bar{3}1)$ reflection in fig. 2. The contrast features together with the observation of only two variants allowed us to identify these defects as the Dauphiné twins (D.T.) described above, and the contrast between the domains as structure factor contrast. Indeed the extinction distances (structure factors) for some particular reflections which are simultaneously excited in the two crystal parts are drastically different e.g. for $30\bar{3}1$ $t_{\alpha 1}$ = 604nm, $t_{\alpha 2}$ = 144nm. As a consequence the diffracted intensity which is directly related to the structure factor is drastically different for the two α-variants.

In figure 2 three successive stages obtained upon heating towards the transition temperature are shown. The closer to the transition the smaller the mesh size of the regular network of triangular domains becomes until it reaches sizes below the resolution limit of the microscope which is about 10nm under the experimental conditions whereby the specimen is in a heating holder and the D.T. boundaries are constantly vibrating. A model for the intermediate phase can thus be proposed as represented in fig. 1d.

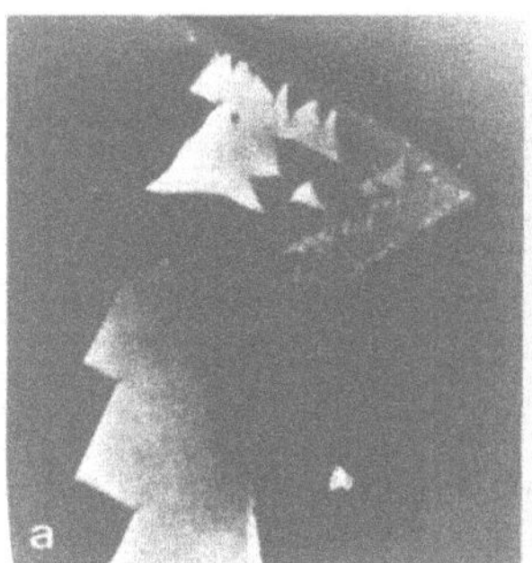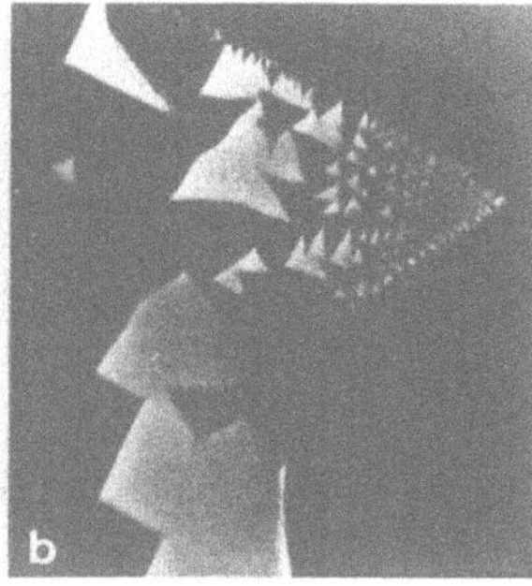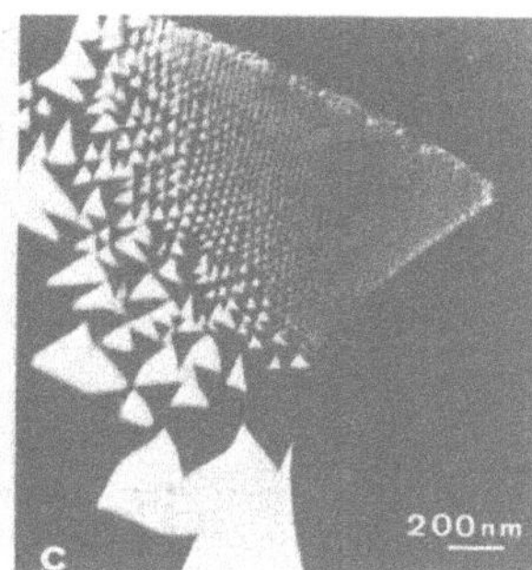

Fig. 2. Quartz crystal fragment heated in the furnace holder of the E.M. Three successive stages are observed. In (c) the crystal tip is at transition temperature.

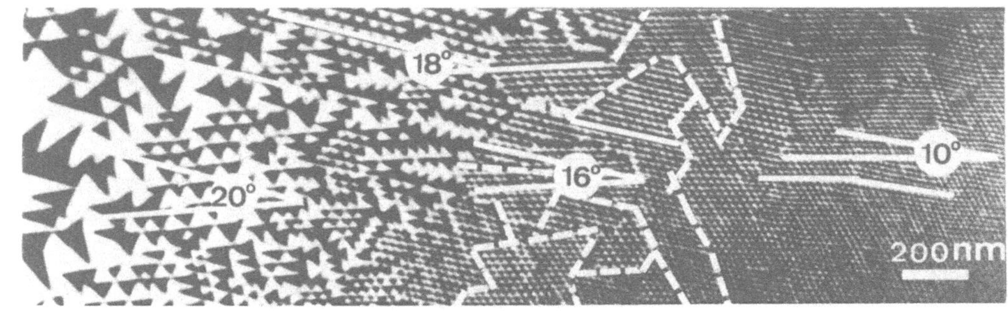

Fig. 3. Variation as a function temperature of the angle between q-vectors. The highest temperature in this gradient picture is reached to the right. A few "macrodomains" are delineated by white dashed lines.

These periodic arrays of Dauphiné twin columnar domains can be considered as an incommensurate superstructure of the quartz structure. The incommensurate nature is clearly established from the continuously variable size of the q-vector with temperature.

These observations inspired theoretical considerations of Aslanyan and Levanyuk [16]; based on Landau theory an incommensurate phase of dauphiné domains could be predicted. These authors conjectured the occurrence of two possible triple $\bar{q}$ vectors enclosing an angle of 10-20°. This angle depends on the order parameter i.e. on the temperature. Based on their theory Walker predicted the possible geometrical aspects for the domain wall configurations [17]. He also suggested the existence of macrodomains of periodic Dauphiné twin arrays of the two above mentioned orientations for the $\bar{q}$-vectors. Figure 3 clearly provides proof for the existence of these macrodomains and the angular difference as a function of ΔT from the transition temperature is also evident as marked [18].
The existence of this incommensurate phase has now also been confirmed by other techniques such as neutron diffraction [19], X-ray topography and micro Laue-techniques [20]. These studies confirm the idea of an incommensurate phase with a varying q-vector. The real space periodicity (1/q) is in the range of 10-15nm and can directly be read off from the electron microscopy images.
Recently Dolino et al. [21] have found diffraction evidence for the existence of a 1-q incommensurate state at the high temperature side of the above mentioned 3-q state. Some real space evidence for the existence of such a state can be found in electron micrographs.

3.2. Barium Sodium Niobate

Barium sodium niobate (BSN) is known to undergo various phase transitions, some of the phases being modulated structures. Table I summarizes the most important characteristics such as transition temperature, pointgroup of the structure, physical nature and unit cell parameters and interrelations. In particular the transition at 543K between the incommensurate (I) and the commensurate (C) phase has been studied extensively by Toledano et al. [22].
Detailed studies by electron microscopy were performed of the commensurate phase by Manolikas [23] and on the structure and defects of the incommensurate phase by Van Tendeloo et al. [24]. The present short report will mainly be based on the results of a recent complementary study by Verwerft et al. [25] on the modulated phases in BSN as further confirmed by low temperature observations. This system was also studied in detail by S. Barré [26].

124

Table 1. Summary of the phase transitions in barium sodium niobate.

	≈853K		≈573K		≈543K		≈105K
4/m 2/m 2/m (order 16)	—	4mm (order 8)	—	2mm (order 4)	—	2mm (order 4)	—	2mm (order 4)
paraelectric paraelastic		ferroelectric (2 variants) paraelastic		ferroelectric (2 variants) ferroelastic (2 variants) incommensurate		ferroelectric (2 variants) ferroelastic (2 variants) quasi commen- surate		ferroelectric (2 variants) ferroelastic (2 variants) modulated
a_t ≈ 1.24 nm c_t ≈ 0.4 nm				a_o ≈ $a_t\sqrt{2}$ ≈ 1 759 nm b_o ≈ a_o ≈ 1.762 nm c_o ≈ $2c_t$ ≈ 0.8 nm		a'_o ≈ $2\sqrt{2}\ a_t$ ≈ 3.54 nm		a^*_o ≈ $2\sqrt{2}\ a_t$ ≈ 3.54 nm b^*_o ≈ $2\sqrt{2}\ b_t$ ≈ 3.52 nm

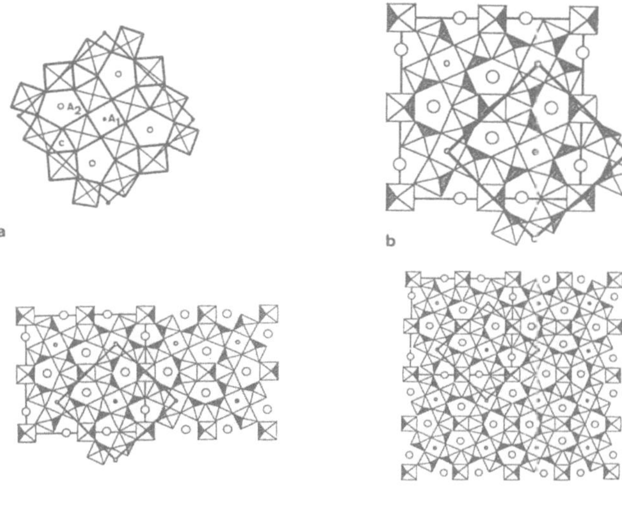

Fig. 4. Structure projections along c for BSN. a) T>573K tetragonal. b) 573K>R>543K orthorhombic incommensurate (shearing of octahedra is marked by shading one of the faces). c) 543K>T>105K orthorhombic commensurate a=2a$_O$; b=2b$_O$.

In figure 4 the unit cell projections of proposed structures for the different phases are represented. The interrelation between the tetragonal (a), the orthorhombic (b) the commensurate orthorhombic (c) and the low temperature orthorhombic structure (d) appears clearly from the drawings.

Special attention will be paid to the transition at 543K from incommensurate to commensurate and more in particular the nucleation stage of the incommensurate phase will be discussed. It was already observed that the degree of incommensurability was related with the number of discommensuration walls (DCW) present in the material [24]. From diffraction contrast experiments these structure defects can be characterised by a displacement vector 1/4[120] of the commensurate structure. This displacement vector was substantiated by the observation that often the DCW's merge in groups of four, thus restoring the perfect structure.

The density of DCW's increases markedly with temperature as observed in fig. 5a&b where two dark field diffraction contrast images are juxtaposed, of

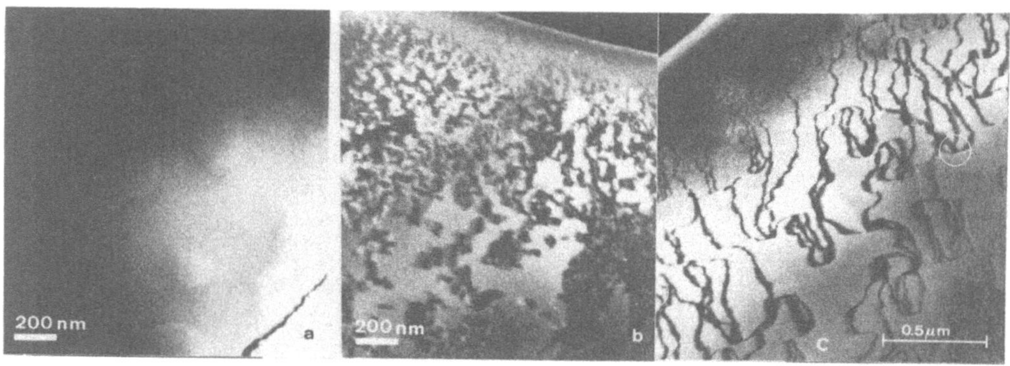

Fig. 5. Beam heating induced phase transition to the incommensurate phase :
increase of the density of discommensurations is observed in (b)
(same crystal area as in (a)). Fig. (c) shows another region where
the DC walls are clearly discerned.

the same crystal area before (a) and after heating (b) by combined furnace
holder and beam heating sources.
Some kind of at first sight featureless black lobes are observed in Fig. 5b,
which, if particular attention is paid to the nucleation stage of the inc.
phase as in Fig. 6 appear to be the nuclei for this phase and to consist of
four DCW's constricted at two ends.
This observation confirms one of the nucleation models proposed by Xiao-Qing
et al. [27] and is in agreement with the above mentioned observation of
residual configurations of 4 merging DCW's as seen in fig. 5c.

In situ observation of the formation and characteristics of the low
temperature phase were also performed. This phase which was by some authors
assumed to restore the tetragonal symmetry, was shown by Toledano et al. on
the basis of X-ray and neutron diffraction techniques to still present a
specimen dependent orthorhombicity but with enlarged unit cell (Table I)
[22].

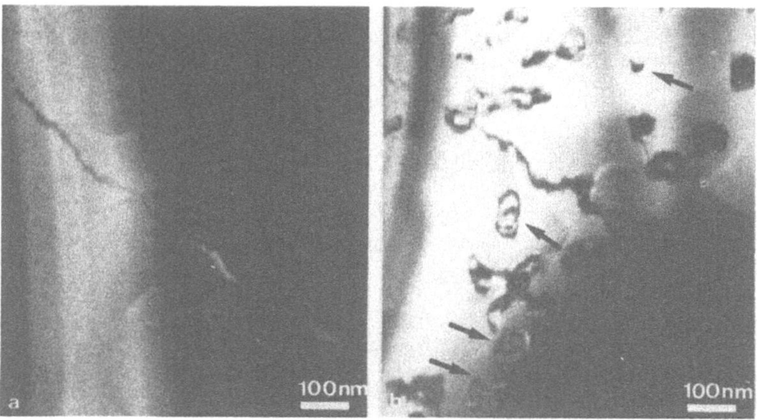

Fig. 6. After heating with the electron beam the same crystal area reveals
in (b) the nucleation stage of the incommensurate phase : closed
loops consisting of four discommensurations are formed (arrows).

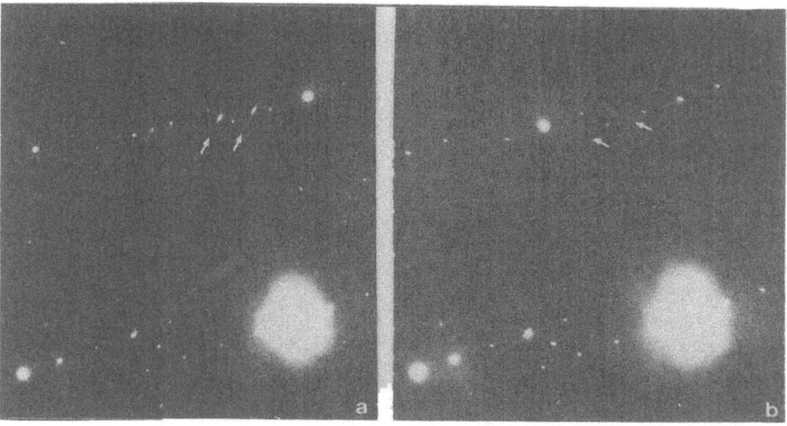

Fig. 7. Selected area diffraction in different ferroelastic domains after
cooling to liquid nitrogen temperature. The reflections appearing at
105K are marked by arrows. Notice the direction of the streaked
reflections.

Electron microscopy using a cooling stage enabled us to show evidence
in direct space as well as in reciprocal space of this apparent double $\bar{q}$-
modulation. Indeed Figs. 7 and 8 show respectively electron diffraction and
image evidence for this modulated phase.
The diffraction patterns in fig. 7 were taken from two different
ferroelastic domains; at the higher order Laue zone supplementary
reflections (marked by arrows) streak segments are observed which have in
the two domains mutually perpendicular directions.
The streaked reflections which appear at positions :

$$(h + \tfrac{1}{4}, \ k - \tfrac{1}{4}, \ \tfrac{2\ell+1}{2})^{*}$$

correspond with a modulation along b_0 whereby also the b-axis is doubled.
Dark field images taken in such a streaked reflection below 105K (fig. 8)
reveal a dense array of discommensurations approx. parallel with $(010)_0$.
They have an average spacing of 3.5nm i.e. approx. $2b_0$. Cooling down to
15K did not result in a further ordering of the discommensuration walls.

Fig. 8. Satellite dark field image of the low temperature phase. Dense
sequences of discommensurations are observed. The average distance
is about 3,6nm. The visibility and orientation of the CDW's is
correlated with the ferroelastic domains.

Recently a new family of high T_C superconductors with structures related to those of the Aurivillius phases was discovered in the Bi-Sr-Cu-O system [29] and in the Bi-Sr-Ca-Cu-O system [30]. Although precise structure determinations were still lacking, electron microscopy and electron diffraction have made it possible to propose approximate structural models for these materials and to reveal the microstructure.
Like the Aurivillius phases these structures consist of lamellae of two, three or four layers of perovskite like cubes, separated by bismuth oxide layers. The general composition formula is $Bi_2Sr_2Ca_nCu_{n+1}O_{2n+6}$. For $n=1$ the layer sequence $BiO-SrO-CuO_2-Ca-CuO_2-SrO-BiO$ adequately describes the structure along the c-axis. A structure model describing the average (unmodulated) structure for this compound is shown in fig. 9.

Diffraction patterns along different zones allow to reconstruct reciprocal space as shown in fig. 10 for $n=1$. The $[110]_p$ section (p=perovskite) is especially relevant since it contains linear sequences of relatively weak satellite spots. The pattern is incommensurate since satellite sequences associated with different rows of basic spots do not match where they meet.

Figure 11 shows actual diffraction patterns, along the $[110]_p$ zone, exhibiting the satellite sequences. The patterns are different for the compounds with $n=0$ and $n=1,2$. Whereas in the compounds with $n=1$ and 2 the (001) plane is a mirror plane for the satellite sequences (Fig. 11b) this is not the case in the compound with $n=0$ (Fig. 11a).

As a result the tetragonal symmetry of the basic perovskite-like structure represented in fig. 9 is only that of the average structure. The real structure has monoclinic symmetry in the compound $n=0$ and orthorhombic symmetry in the compounds with $n=1$ and $n=2$. A high resolution image along the zone exhibiting the satellite reflections is reproduced in fig. 12 for the compound $n=1$. The (001) lattice planes are visibly wavy and the pattern has a centered rectangular unit mesh in accordance with the orthorhombic symmetry.

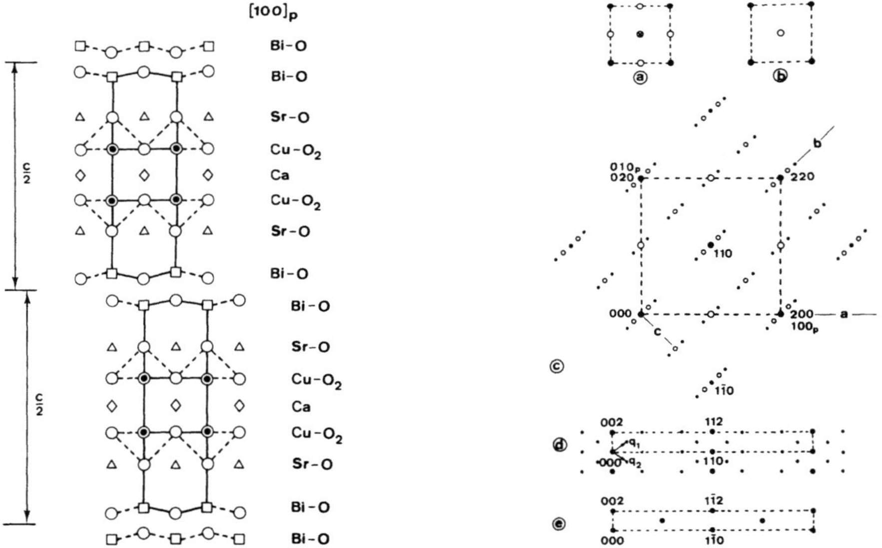

Fig. 9. Schematic representation of the n=1 compound in the homologous series of high T_C compounds $Bi_2Sr_2Ca_nCu_{n+1}O_{2n+6}$.

Fig. 10 Reciprocal space reconstructed from different diffraction patterns for n=1 in the Bi-Sr-Ca-Cu-O system.

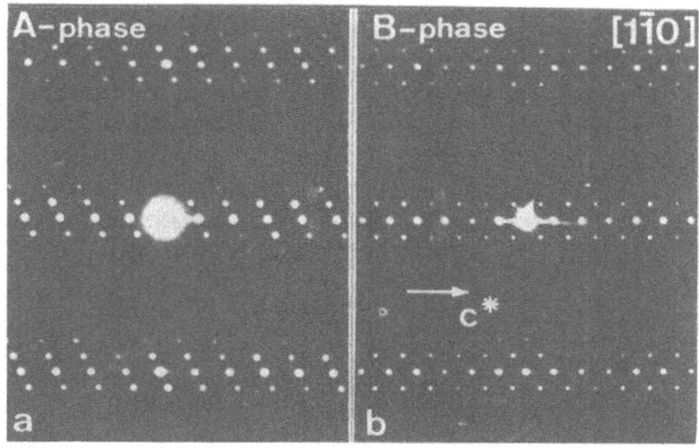

Fig. 11. Diffraction patterns along the $[110]_p$ zone in the Bi-Sr-Ca-Cu-O
sytem :   (a) for n=0   (A-phase)
              (b) for n=1, 2 (B-phase)
        exhibiting the satellite sequences due to the modulation of the
        structure.

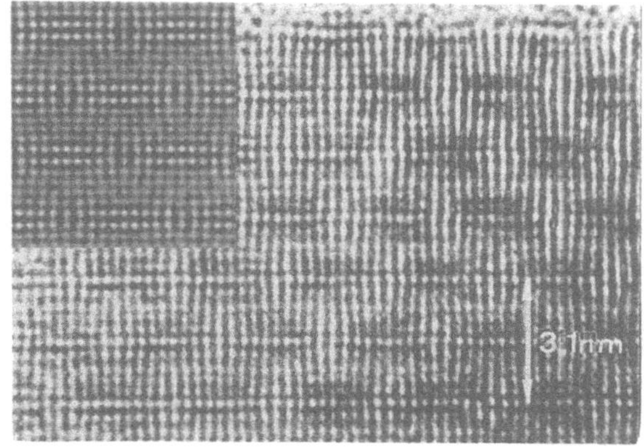

FIG. 12. High resolution image of the deformation pattern in the compound
        with n=1.   The inset shows a simulated image based on a model
    assuming excess oxygen and lattice relaxation in the BiO layers.

The origin of the modulation is still a matter of debate. It is probable
that it is due to excess oxygen in the BiO layers. A model whereby excess
oxygen was introduced in the BiO layers and the atom positions relaxed so as
to achieve reasonable interatomic distances was studied in [31]. The image
to be expected from such a model was simulated and compared with the
observed image in fig. 12; the fit is quite reasonable. The presence of
excess oxygen in the BiO layers can be understood by noting that the lattice
parameter of these layers does not fit very well with that of the underlying
perovskite block : the stresses resulting from this misfit can be relieved
by the introduction of excess oxygen, hereby changing also the effective
valence of Bi.

4. CONCLUSION

Various types of modulated structures can be studied in great detail with the use of electron microscopy electron diffraction sometimes complemented by optical diffraction.

An interface (twin) modulated structure in quartz a (translation) interface modulated structure in $Ba_2NaNb_5O_{15}$ and a composition deformation modulated phase in a high T_C-superconductor provided illustrations for the power of the method.

ACKNOWLEDGEMENTS

Thanks are due to M. Verwerft and H. Zandbergen for the use of material from co-authored papers.

REFERENCES

1. S. Amelinckx, G. Van Tendeloo, D. Van Dyck and J. Van Landuyt, Phase Transitions, 16–17, 3–40 (1989).
2. D. Van Dyck, D. Broddin, J. Mahy and S. Amelinckx, Phys. stat. sol.(a) 103, 357 (1987).
3. K. Fujiwara, J. Phys. Soc. Jap. 12, 7–13 (1957).
4. J.P. Lu and J.L. Birman, Journal de Physique, Colloque C3, supplément au n°7, Tome 47, juillet 1986, C3–251.
5. J.W. Steeds, R. Ayer, Y.P. Lin and R. Vincent, Journal de Physique, Colloque C3, supplément au N°7, Tome 47, juillet 1986, C3–437.
6. D.M. Bird and R.L. Withers, J. Phys. C, Solid State Physics 19, 3497 (1986).
7. L.A. Bendersky and M.J. Kaufman, Phil. Mag. B 53, n°3, L75 (1986).
8. J. Van Landuyt, G. Van Tendeloo and S. Amelinckx, Phys. stat. sol. (a) 26, K9 (1974).
9. S. Amelinckx and J. Van Landuyt, Diffraction and Imaging Techniques in Material Science, 1987, North Holland, Amsterdam, p. 105.
10. J. Ch. Spence, Experimental high resolution electron microscopy, 1981, Clarendon, Oxford.
11. D. Van Dyck, in Adv. Electronics & Electron Physics, Ed. P. Hawkes, 1985, 65, 295.
12. P.G. Self, M.A. O'Keefe, P.R. Buseck, A.E.C. Spargo, Ultramicroscopy 11, 35 (1983).
13. G. Dolino, in "Incommensurate phases in Dielectrics", Eds. R. Blinc & A.P. Levanyuk, Elseviers, North Holland Amsterdam, 1986, p. 205.
14. G. Van Tendeloo, J. Van Landuyt and S. Amelinckx, Phys. stat. sol. (a) 30, K11 (1975).
15. G. Van Tendeloo, J. Van Landuyt and S. Amelinckx, Phys. stat. sol. (a) 33, 723 (1976).
16. T.A. Aslanyan and A.P. Levanyuk, Sov. Phys. JETP Lett. 28, 70 (1978).
17. M.B Walker, Phys. Rev. B 28, 70 (1978).
18. J. Van Landuyt, G. Van Tendeloo, S. Amelinckx and M.B. Walker, Phys. Rev. B 31 (1985).
19. G. Dolino, J.P. Bachheimer, B. Berge, C.M.E. Zeyen, G. Van Tendeloo, J. Van Landuyt and S. Amelinckx, J. Phys. (Paris) 45, 901 (1984).
20. K. Gouhara and N. Kato, J. Phys. Soc. Jap. 53, 2177 (1984).
21. G. Dolino et al., Europhysics Letters 3, 601 (1987).
22. J.C. Toledano, J. Schneck and G. Errandonéa, in : Incommensurate phases in dielectrics, Eds. R. Blinck and A.P. Levanyuk. North-Holland, 1986, p.233–251.
23. C. Manolikas, Phys. stat. sol. (a) 68, 653 (1981).
24. G. Van Tendeloo, S. Amelinckx, C. Manolikas and Wen Shulin, Phys. stat. sol. (a) 91, 483 (1985).

25. M. Verwerft, G. Van Tendeloo, J. Van Landuyt and S. Amelinckx, Ferroelectrics 88, 27 (1988).
26. S. Barré, thesis Université de Toulouse, 1987.
27. Pan Xiao-Quing, Hu Mei-Shen, Yao Ming-Hui and Feng Duan, Phys. stat. sol. (a) 91, 57 (1985).
28. H.W. Zandbergen, P. Groen, G. Van Tendeloo, J. Van Landuyt and S. Amelinckx, Solid State Communications 66, n°4, 397-403 (1988).
29. C. Michel, M. Hervieu, M.M. Borel, A. Grandin, F. Deslandes, J. Provost and B. Raveau, Z. Phys. B68, 421 (1987).
30. H. Maeda, Y. Tanaka, M. Fukutomi and T. Asano, Jap. J. of Appl. Phys., 27, L209 (1988).
31. H. Zandbergen, W.A. Groen, M.C. Mijlhoff, G. Van Tendeloo and S. Amelinckx, Physica C 156, 325 (1988).

PHYSICAL MODELS OF PERFECT QUASICRYSTAL GROWTH

D. P. DiVincenzo

IBM Research Division
T. J. Watson Research Center
P.O. Box 218
Yorktown Heights, NY 10598 USA

In this summary I will briefly review our previous work, and the work of others, on the problem of growing quasicrystals, and in particular a Penrose tiling, according to "physical" rules, i.e., ones involving one-by-one addition of tiles to a growing cluster. I will then show some of our new work on the properties of defects in these growth algorithms.

Growing a perfect quasiperiodic tiling would appear to be a daunting task, since such a tiling never repeats itself exactly. An important clue that such growth rules might be possible is the existence of "matching rules", the arrow rules illustrated in Fig. 1. The precise statement of the meaning of matching rules is that

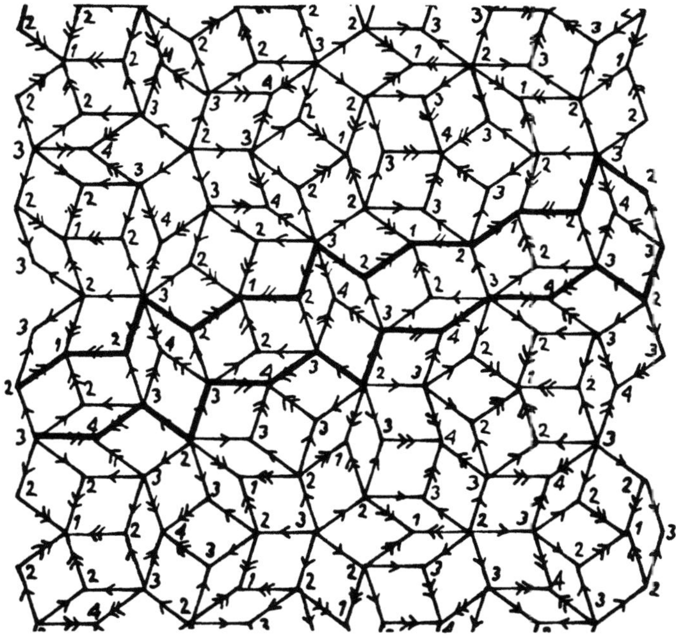

Fig. 1. The Penrose tiling showing the arrow matching rules (taken from Ref. 1). A "Conway worm" is highlighted.

Geometry and Thermodynamics
Edited by J.-C. Tolédano
Plenum Press, New York, 1990

"

"If a tiling which covers the plane is constructed which obeys the matching rules everywhere, then that tiling can only be a perfect quasicrystal." [1-3]

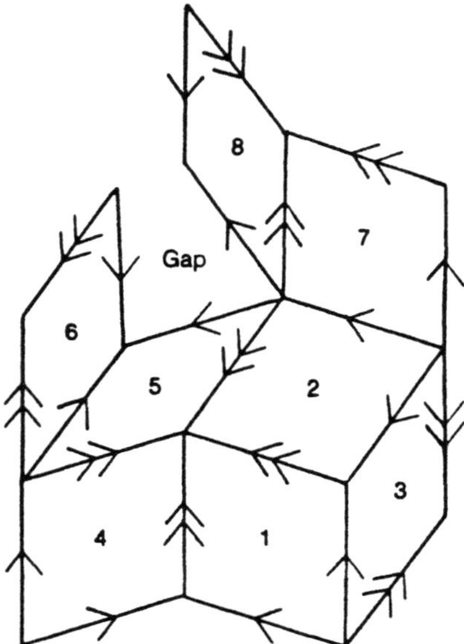

Fig. 2. Failure of the matching rules as a growth procedure.

However, matching rules do not immediately imply a growth rule, as Fig. 2 illustrates. This problem was recognized in the original work of Penrose (see Gardner);[4] it was discussed there that one important ingredient that was needed was the idea of "forcing", that is, in many circumstances during growth there is no choice of whether to add a fat or a skinny tile, even though the arrow rules imply that there is always such a choice. However, the original work did not pursue this idea to its logical conclusion.

Our recent work showed that a complete set of forcing rules can be constructed,[5] which is the most important ingredient in a growth rule. The forcing

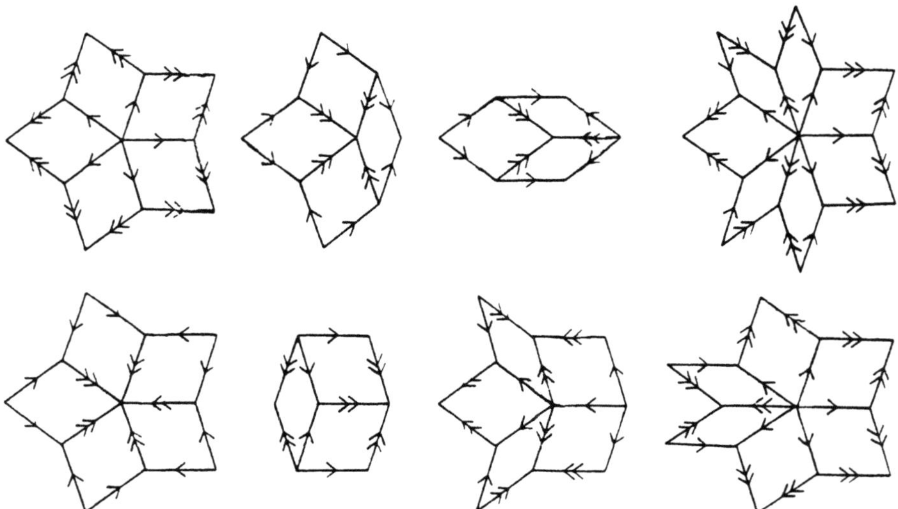

Fig. 3. Eight legal Penrose vertices, which are used in our forcing rules.

rules which we used (other variants of them are possible) work by comparing the site on the perimeter where growth is to occur with the set of legal arrangements of tiles around a vertex in the Penrose tiling (Fig. 3); if only one choice of a new tile to add is possible on the basis of comparison with these templates, then the tile is "forced". Unfortunately, adding of forced tiles is not a complete growth rule, because after some growth a tiling reaches a state where all the sites on the perimeter are "unforced", as in Fig. 4. We found that at this point, the growth can

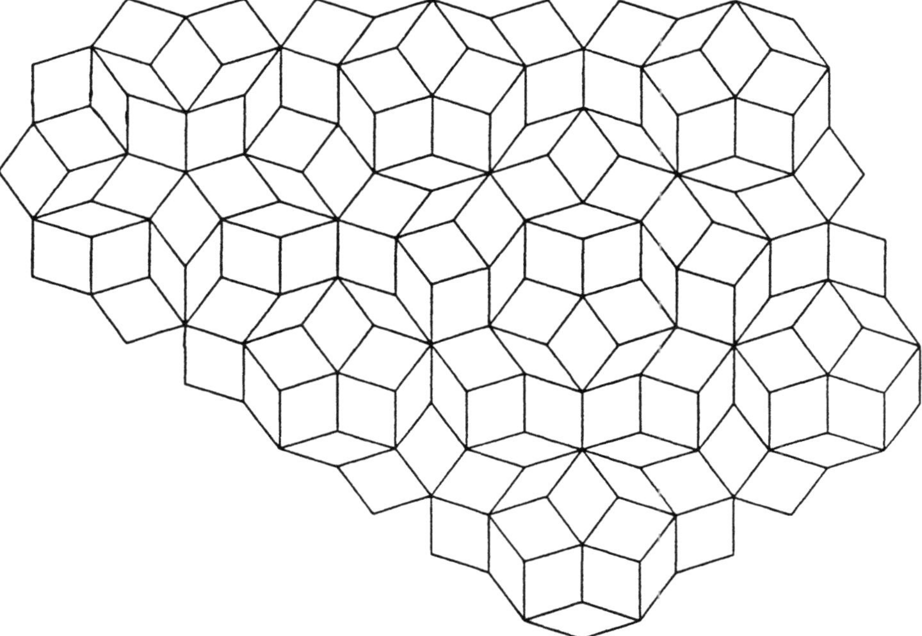

Fig. 4. A stage in perfect Penrose tiling growth where are the growth sites on the perimeter are unforced. Dots denote the tips of the 108° corners.

always be consistently continued with one of a number of possible simple rules; for example, "add a fat tile to the tip of the 108° corner" of the surface in Fig. 4. (See Ref. 5 and 6 for more details.) This one additional rule, applied whenever the perimeter is completely "unforced", nucleates new forced growth, which will lead to another unforced surface. We proved that this procedure will produce an infinite, perfect Penrose tiling.

It has been pointed out[6,7] that there are problems with interpreting the procedure that we developed as a physical growth process. It is "nonlocal" in the sense that the entire perimeter must be examined to assure that all the forced growth which is possible is actually performed. On the other hand, it is possible to construct a probabilistic model of growth where different kinds of sites (forced, unforced, corners) are assigned different "sticking probabilities", which is a perfectly realistic physical scenario. If the forced-site sticking probability is much greater than the corner-sticking probability, then the growth is very likely to proceed in exactly the way that the Onoda algorithm specifies, i.e. from one unforced cluster to another. Thus, the perfect-tiling growth algorithm represents a limiting case of a physical growth scenario, specifically in the limit in which the ratio of sticking probabilities approaches zero.

The new results which I want to discuss here concern the departures from perfect growth for models which are close to this limiting case. (Most of the ideas for how to deal with defective growth come from Joshua Socolar). If the ratios of sticking probabilities described above are small but non-zero, eventually (possibly after growth has proceeded for a long time) an incorrect tile will be attached. It is clear that after this happens, since we are considering models where annealing is not possible, the tiling <u>cannot</u> any longer be a perfect tiling. However, it does make sense to ask questions about the degree of subsequent perfection.

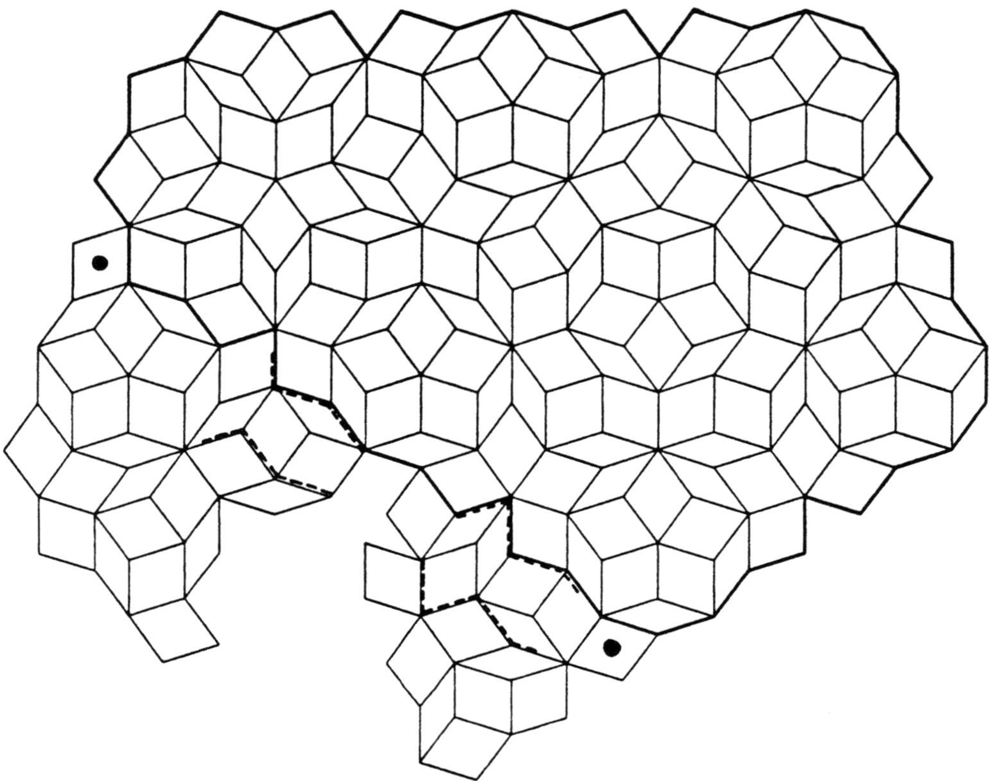

Fig. 5. Defective tiling growth. The filled circles indicate two tiles which have been added to the unforced surface of Fig. 2 (whose perimeter is indicated here by a dark line). These two choices are incompatible; compare the Conway worm here (dashed line) with the one in Fig. 1.

Fig. 5 illustrates the beginnings of a defective growth procedure; as indicated, two incompatible applications of the 108° corner occur simultaneously, which is a possible event in probabilistic growth. Growth has proceeded to the point where a violation of the Penrose matching rules is about to occur. As I described above, the rules which we use will cause mistakes to be made along unforced surfaces. Unforced surfaces are bounded by "Conway worms",[4] and the mistakes are virtually certain to occur in the interior of these worms, as Fig. 5 indicates.

Since the defects occur in certain well-defined environments in the tiling, there are local rules for dealing with the defect and continuing with perfect-tiling growth. The fix relies on the fact that the parts of the tiling immediately above and below the worm are related by a mirror symmetry (see Fig. 1). The essence of the fixing rule, then, is to ignore the matching rule violation inside the worm, add tiles

which are mirror-related to the ones on the other side of the worm until the tiling
perimeter has completely enclosed the defect. (This is illustrated in Fig. 6.) Then,
ordinary forced-rule growth can take over. The defect which is left behind (outlined in Fig. 6) falls into the class of objects called "decapods" by Gardner.[4] In
fact, only one of the 62 possible decapods can occur in our defect repair procedure; it is interesting that this decapod has already been identified as special, although for a different reason, by Conway.[4] (He calls it the "Asterix".)

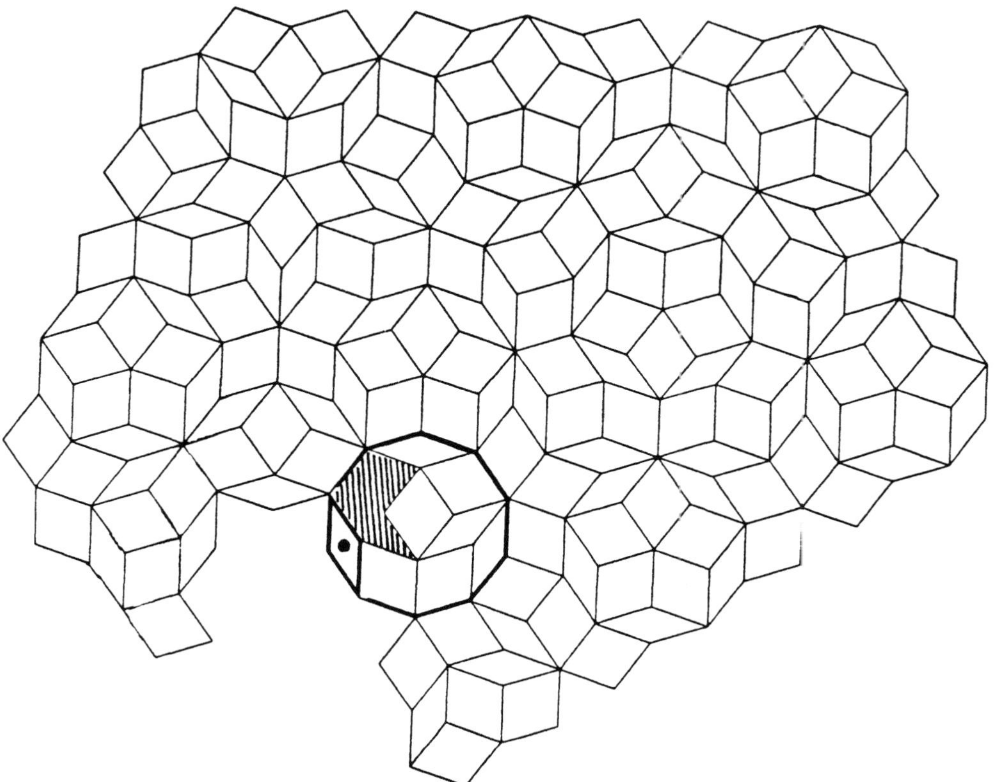

Fig. 6. The fix for the defect of Fig. 5. The highlighted skinny tile is added; a hole is left
behind (hatched), which obviously could be filled in with tiles (although at the cost of a
matching rule violation). This procedure creates the defective decagonal region indicated,
called a "decapod", as described in the text. The outer perimeter of the tiling now consists only of legal vertices (Fig. 3).

As we and others have discussed previously,[4-7] an isolated decapod defect
permits infinite perfect growth around it, thus, perfect growth will continue (see
Fig. 7) until the next probabilistically generated defect occurs. After several such
defects have accumulated, a strain field (of the "phason" type) develops in the
tiling, the presence of these strains will cause the appearance of further tiling
defects,[8] even in the absence of any further mistakes in the tiling algorithm.

The are still some important unanswered questions about the ultimate fate
of this probabilistic growth scheme which we have developed. It is clear from my
simulations (reaching about 20 000 tiles) that this algorithm has no problem with
repeatedly encountering defects, repairing and isolating them, and continuing

perfect-tiling growth. We know that if this continues indefinitely, the tiling will
have at least quasi-long range quasiperiodic order. (That is, the diffraction pattern
of the structure would display 5-fold symmetry and consist either of Bragg peaks
or power-law, zero-width peaks.)

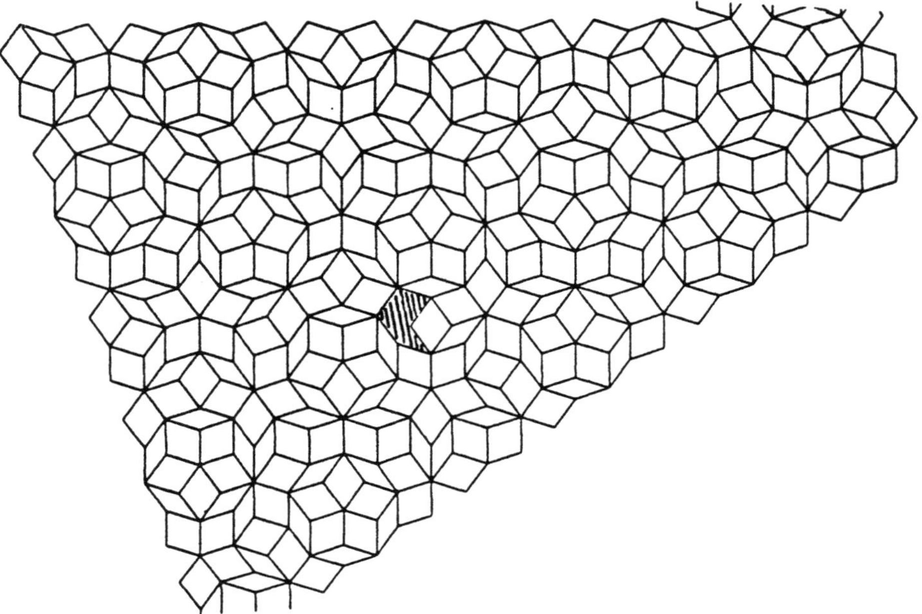

Fig. 7. Continuation of perfect growth around the defect of Fig. 6.

However, we cannot exclude the possibility that the tiling will eventually lose
long-range order. It could do this if the defect density gradually builds up during
growth; if the defects get too dense, the algorithm will fail, and perfect-tiling
growth will be unable to continue, leading to a glassy structure.[9] The same thing
might also happen if there are very rare defects which we have not yet encountered or understood which are not associated with Conway worms. More extensive
simulations should give us insights into these questions. Still, my opinion is that
despite these possible difficulties, probabilistic algorithms can easily be devised
which grow to macroscopic sizes before any unforseen problems develop.

If there prove to be probabilistic growth models which give structures with
truly long range quasiperiodic order, it will be also interesting to look at the transition into the disordered growth regime (as the sticking probabilities are made
more unfavorable for perfect growth). There is evidence from Ronchetti and
Jaric[7] that such a transition exists.

I wish to acknowledge discussions with S. Langer, G. Onoda, M. Ronchetti,
J. Socolar, and P. Steinhardt. Thanks also to S. Langer for the use of his growth
program.

REFERENCES

1. N. de Bruijn, Proc. K. Ned. Akad. Wet., Ser. A **84**, 39 (1981).
2. A. Katz, Commun. Math. Phys. **118**, 264 (1988).
3. L. S. Levitov, Commun. Math. Phys. **119**, 627 (1988); S. E. Burkov, Commun. Math. Phys. **119**, 667 (1988).
4. M. Gardner, Sci. Am. **236** (1), 110 (1977).
5. G. Y. Onoda, P. J. Steinhardt, D. P. DiVincenzo, and J. E. S. Socolar, Phys. Rev. Lett. **60**, 2653 (1988). See also R. Penrose, in <u>Introduction to the Mathematics of Quasicrystals,</u> ed. M. V. Jaric (Academic, New York, 1989), p. 53.
6. M. V. Jaric and M. Ronchetti, Phys. Rev. Lett. **62**, 1210 (1989).
7. M. Ronchetti and M. V. Jaric, Proceedings of the Trieste meeting on Quasicrystals, 1989, to be published.
8. J. E. S. Socolar, Phys. Rev. B **39**, 10 519 (1989).
9. A. I. Goldman and P. W. Stephens, Phys. Rev. B **37**, 2826 (1988).

GENERATION AND DYNAMICS OF DEFECTS IN TWO-DIMENSIONAL QUASICRYSTALS

Marco RONCHETTI, Mauro BERTAGNOLLI
Dipartimento di Fisica, Università di Trento, 38050 POVO (TN), ITALY

Marko V. JARIĆ
Center for Theoretical Physics, Texas A&M University
College Station, TX 77843, USA

1. ABSTRACT

We analyze defects in quasicrystals by simulating their generation during growth, and by studying their dynamics during relaxation at finite temperature. In a first simulation, based on a Eden-type growth algorithm of a Penrose Tiling, we observe defects similar to vacancies which do not destroy quasi–long–range order. In a second simulation, a molecular dynamics based on an atomistic model, we find that phason relaxation occurs only in presence of vacancies.

2. INTRODUCTION

The study of defects is very important for the understanding of solid matter and for the quest for possible applications of materials. Much work has been devoted to the generalization of well known crystal defects (e.g. dislocations and disclinations) to try to explain formation and stability in amorphous solids[1]. It was therefore very natural to apply the same concept to quasicrystals (QC) right after their discovery. Papers have appeared discussing existence and role of dislocations[2] and strain accumulation[3] in QC. Experimental evidence for dislocations has also been presented[4]. Other imperfections are known to exist in crystals: point defects, and in particular vacancies. No efforts, as far as we know, has until now been devoted to a theoretical study of stability and role of vacancies in QC. Part of the work we present here is dedicated to show that vacancies do exist in QC, and that they play an important role. In another part the formation of defects during growth is investigated.

Geometry and Thermodynamics
Edited by J.-C. Tolédano
Plenum Press, New York, 1990

Besides the extension of usual concepts to quasicrystals, the study of defects in QC has been focussed mainly on the typical defects of incommensurate structures: phasons[5]. In fact, most of the discussion about what QC really are can be put in terms of what kind of phasons are present in real QC. In fact the possible models of quasicrystals can be characterized in terms of their phason fluctuations.

Models postulating the existence of a quasiperiodic ground state, a prototype of which is the two–dimensional Penrose Tiling, can be described as the projection from a smooth strip in higher dimensional space (or equivalently as the cut operated by a smooth hypersurface trough higher dimensional space)[6]. Fluctuations in the strip (or cutting surface), i.e. phason fluctuations, are then needed to explain intrinsic peak width, peak shifts and shapes[5]. In this case phason fluctuations are such that long–range order is preserved through the QC.

The models known as icosahedral glasses[7,8] can also be described as projection from higher dimensional space due to the preserved orientation of their randomly stacked building blocks (e.g. decagons, icosahedra). These systems possess only short–range translational order. Phason fluctuations in this class of structures present a power–law behavior and lead to intrinsic peak broadening.

Another recently proposed model is known as random-tiling[9]. In this case the system does not possess a quasiperiodic ground state, and its structure is rather stabilized by entropy. It has quasi–long–range order, and its phason fluctuations are characterized by a logarithmic behavior. Its diffraction pattern (in the three-dimensional case) presents sharp Bragg peaks.

Even a description of QC as periodic approximants of quasiperiodic structures (i.e. as a periodic system with very large unit cell) can be given in the projection or cut language. In this case, starting from a quasiperiodic system, a uniform phason is needed, which is equivalent to a uniform tilt of the whole strip (or hyperplane).

The study of phasons is therefore of fundamental importance in the understanding of quasicrystals. In section 3 we study in terms of phason fluctuations the defect generation during growth. In section 4 we show that for a two–dimensional model QC vacancies play an important role in the phason relaxation and dynamics. Section 5 contains the summary.

3. DEFECT GENERATION DURING QUASICRYSTAL GROWTH.

The study of structure growth is a fascinating one, and has captured the interest of researchers in many areas[10]. In quasicrystals, the growth problem has been addressed mainly within the approach of icosahedral (decagonal) glasses, where the structure itself is defined via a growth process[8,9]. The elementary building block(s) possess the desired symmetry, and can be identified with clusters which are present in the unit cell decoration of related crystals (e.g. with the icosahedra sitting at vertices and body center in the BCC structure of Al-Mn-Si α-phase). Starting from a small seed (e.g. from a single unit), more blocks are added with the only restriction that orientational order is preserved when moving from one block to the next, and therefore throughout the whole sample. Diffraction patterns calculated from such structures present the same symmetry as the seed. However diffraction peaks show an intrinsic broadening, due to the fact that the growth-induced structure possesses only finite-range translational order. These structure therefore might be useful for describing some of the observed quasicrystals, but cannot be a model for the newly discovered quasicrystals[11] which exhibit peaks of sharpness comparable to that of crystals.

The task of finding an algorithm for growing perfect quasiperiodic structures has been hence regarded as a problem of central interest in the study of quasicrystals. Several papers have addressed the problem of existence of "matching rules"[12], i.e. of rules for packing the elementary tiles in a quasiperiodic way. Such rules were known to exist for the Penrose Tiling (PT) before the discovery of quasicrystals: the PT can in fact be built by assembling the two well known (*fat* and *skinny*) rhombi whose edges are decorated with colored arrows, with the condition that adjacent rhombi only share edges with arrows matching color and direction[13].

The existence of matching rules is not sufficient though for providing a growth process. In fact the matching rules force quasiperiodicity in the sense that if they are respected all over the tiling, then the resulting structure is quasiperiodic. However, some ambiguity is left on how to reach by accretion the desired state. In many cases in fact two different choices can be operated (e.g., one can choose to satisfy the rule by either adding a fat or a skinny rhombus). Sometimes the two choices just lead to different realizations of the perfect PT. In most cases though two such choices can influence each other (over arbitrarily long distances), so as to lead to the impossibility of continuing the growth without somewhere violating the rules. In this case, one has to undo a portion of the tiling and try to make different choices. The place where a violation is forced can be as far as the sample size: the required backtracking is therefore, in general, very relevant.

Until recently, no algorithm was known for growing such structures without requiring an extended amount of trial and error. Attention was focussed mainly on the Penrose Tiling (PT), as it provides a well studied prototype of quasiperiodic structures.

Recently Onoda et al.[14] found an algorithm for growing infinite perfect PT without making mistakes. They devised stronger matching rules, which not only request that the arrows be respected, but also require that tiles sharing a vertex form one of eight allowed vertex configurations. In fact only eight kinds of vertices occur in the perfect PT, although by respecting the arrow matching rules in a small environment of a vertex it is possible to build many other vertex-configurations. These other vertex-configurations necessarily lead to arrow mismatches somewhere else. Starting from a partial vertex there is sometimes only one possibility to obtain one of the eight legal vertex-configurations. In this case the partial vertex is named *forced*. *Partially forced vertices* are those which can be completed in different ways, but for which at least one tile can has only one allowed possibility. *Forced tiles* are those which can be added in a unique way to a (partially) forced vertex. Surfaces which do not present forced tiles are in average straight and are called *dead*.

The algorithm achieves the goal of building a PT without backtracking by asking that all forced tiles be placed first. Only when no forced tiles are left, a single tile is grown on a special corner site, producing more forced tiles (during a perfect growth process there are always at least two special corners when no forced tiles are left). This process avoids incompatible choices, and therefore it is guaranteed to produce an infinite perfect quasiperiodic tiling. It was though pointed out[15] that the step of deciding to grow on the special sites requires global knowledge of the sample (one has in fact to inspect the whole surface to verify that no forced tiles are left).

A variant of the algorithm requires that when no forced tiles are left, a single tile is grown on a random site on the surface. Although this variant does not guarantee that a perfect PT will be generated, it allows perfect growth with very high probability, and might look more appealing because it does not require the definition of an additional special site. Also in this case though global knowledge of the sample is required. A single tile should "know" if forced sites are present on any place on the surface before deciding to stick on a non-forced site. Infinite range interaction would therefore be required.

An argument given by Penrose[16] suggests that it is impossible to find rules which do not require a global knowledge of the system, and still allow to build an infinite quasiperiodic structure. The question therefore remains open: given finite range interactions, how can a quasicrystal grow?

Onoda et al. suggested that the problem could be solved by interpreting their algorithm in a probabilistic sense: different sticking probabilities can be introduced for forced tiles, unforced tiles, and special corners. They then argue that when the probabilities are very different the growth of perfect structures is recovered.

We used such a probabilistic approach to study the behavior of the system for various probability ratios, and found that perfect growth can occur only up to a critical size. We investigated the nature of such growth, and the defects which are created beyond the critical size. We found that the existence of defects drives the system away from quasiperiodicity, and that the resulting (defected) tilings behave either like random tilings, or like orientationally-ordered glasses, depending on the ratios between different sticking probabilities.

The algorithm we use mimics the growth of a quasicrystal from a liquid containing fat and skinny tiles in concentration τ:1. Using the definitions of forced sites proposed by Onoda et al., we investigated the growth based on different sticking rates for different sites, and in particular for forced vertices and generic sites. We also checked the influence of special corners. We found that although special corners play a crucial role in the algorithm producing perfect quasicrystals based on global knowledge of the surface, they do not add significant features to the behavior of the system when the probabilistic approach is chosen, as discusseded in more detail elsewhere[17]. We shall here restrict ourself to consider the special corners as any other generic site.

At each time step all sites on the surface are examined: tiles are stuck to the sites with probabilities which depend on the kind of site: P_f for forced sites, P_o for all other places. Kind and orientation of the new tile are obviously unique for forced sites. If the site is not forced there are two possibilities: a fat or a skinny tile can be added (the orientation is determined by the surface, since overlapping of tiles is not allowed). In this case the tile to be added is chosen to be skinny with probability $1/(1+\tau)$ and fat with probability $\tau/(1+\tau)$, according to the frequency ratios in the perfect PT. This procedure acts in parallel, simulating the growth of a solid from the corresponding liquid phase. Since the simulation was performed on computers with traditional Von Neumann architectures (VAXes and Cray), the intrinsic parallelism in the algorithm was not reflected by the hardware, but only emulated. To do so a list of all surface sites is built at each time step. All vertices in the list are then examined once in random sequence and an attempt of growing tiles on them is performed. When the list is exhausted we pass to the next time step. Conflicts which might arise between adjacent sites are here solved on the first come basis provided by the random order.

The process is iterated until the desired sample size is obtained. The control parameter is the ratio between sticking rates, i.e. $\eta = P_o/P_f$. To speed up the simulation all our results were obtained by maximizing one of the sticking rates: for various values of η we have either $P_f = 1, P_o = \eta$ or $P_f = 1/\eta$, $P_o = 1$. We found that for any value of η defects are necessarily generated, and studied their generation and their influence on the broadening of peaks.

One recovers the ideal case when η is so small that it is very unlikely to add tiles to unforced vertices before all forced vertices are completed. A patch of perfect PT is grown. However the growth has an intermittent character: the sample grows (with very high probability) only on forced sites until all surfaces are dead. One has then to wait for the unlikely event to happen (i.e. for the sticking on an unforced site). The waiting time is proportional to the inverse of P_o. When a critical size is reached it becomes very likely to stick a tile on a wrong place. Growth of perfect PT is therefore (in average) possible, but only up to a critical size, and only in an intermittent fashion. One can always increase the critical size by reducing the value of η, at the expenses of the time needed to grow (i.e. increasing the time spent on dead surfaces). However one has to notice that the limit $\eta = 0$ in which an infinite tiling could apparently be grown is singular: the sample cannot in fact grow beyond the first dead surface (i.e. just a few tiles) because the sticking probability on non-forced sites is zero.

Beyond the critical size the structure starts exhibiting defects, which in all cases manifest themselves (in our model) as holes in the structure. In fact we do not allow arrow mismatches to occur: whenever one is detected, the growth is locally inhibited. The sample

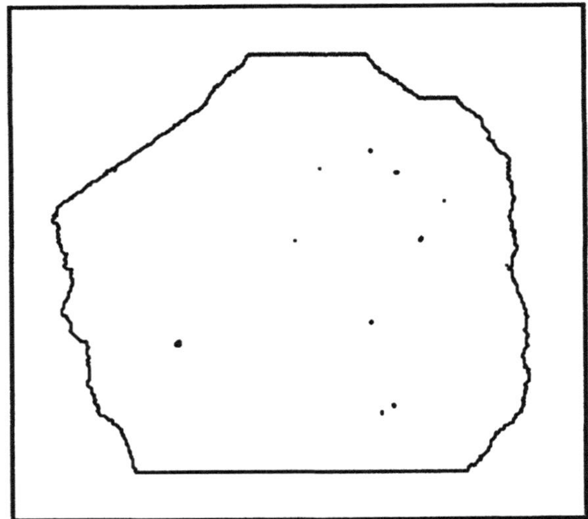

FIGURE 1. Sample grown with η = .0001. The few defects present are sufficient to destroy perfect quasiperiodicity, to grant a continous growth and to induce phason fluctuations which grow logarithmically with sample size.

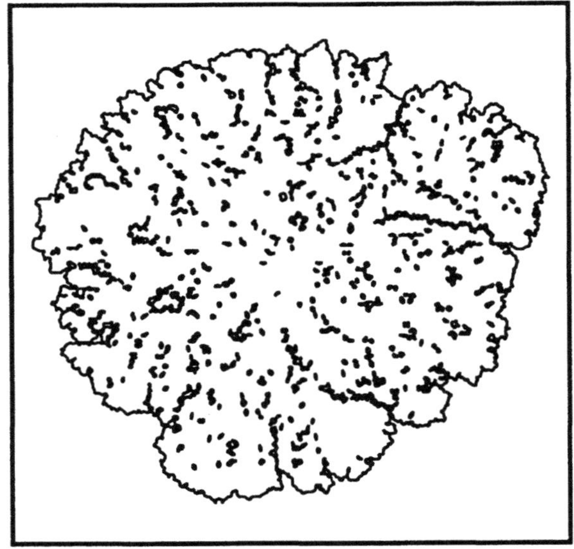

FIGURE 2. Sample grown with η = .01. A strong correlation between defects is induced by growth, so that sequences of defects equivalent to tears are developed.

can continue growing on the right and on the left of the defects, and the two sides can merge leaving behind a hole, which in most cases is just a single missing tile. The hole marks a point where translational order is broken. Sometimes it just marks a simple flip of one of the hexagons which form a Conway worm[18,19]. Two near flips can anneal each other. However when more worms cross, a flip is induced also in the crossing worms. In these cases defect propagation can start: the mismatch causes a new mismatch to occur, and so on. A series of point defects can grow on top of each other, giving rise to large tears, like the ones shown in figure 2. Each tear separates two regions of different translational order: the Amman bars[19] in the two regions follow different Fibonacci sequences. There are occasional coincidences between the two sequences, which lead to new merging of the two portion of the tiling (i.e. a tear heals). Since each region though follows its own sequence, new mismatches are bound to be created soon, so that a sequence of strongly correlated defects is produced.

We explored a rather large range of values of our parameter: $10^{-4}<\eta<10$. For each value of η we grew several samples consisting of 100,000 tiles each. Visual inspection of our samples seems to indicate that at least two regimes exist. In fact for the lowest values of our parameter ($\eta = 10^{-4,}10^{-3}$) only isolated defects are present (as shown in fig. 1) , while for larger values ($\eta = 10^{-2,}10^{-1}$) long tears appear. We also run smaller samples with even higher sticking rates for unforced sites, obtaining structures which appear to have a fractal surface (Fig.3).

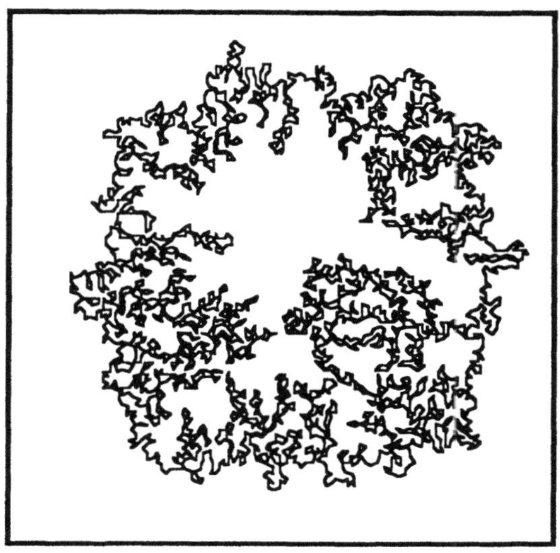

FIGURE 3 - Small sample obtained with $\eta=10$. Only the *external* surface is shown (internal defects are not represented). When η grows beyond 1 the surface becomes fractal-like.

A quantitative analysis, based on the evaluation of phason fluctuations, confirms the existence of (at least) two regimes. Phason fluctuations can be easily computed, since one can lift the structure from real space to its full description in 5 dimensional integer coordinates[8] and then project onto phason space. From the position $\mathbf{r}$ in 2-D (parallel) space one can determine[8] the five integer coefficients n_α and then get the phason coordinates:

$$\mathbf{r} = \sum_{\alpha=1}^{5} n_\alpha \mathbf{e}_\alpha^{\parallel} \quad ; \quad \mathbf{h}(\mathbf{r}) = \sum_{\alpha=1}^{5} n_\alpha \mathbf{e}_\alpha^{\perp} \quad ; \quad h_z(\mathbf{r}) = \sum_{\alpha=1}^{5} n_\alpha \quad .$$

where $(\mathbf{h}, h_z)$ is the position in phason space and $\mathbf{e}^{\parallel}$, $\mathbf{e}^{\perp}$ are the versors in parallel (real) and perpendicular (phason) space:

$$\mathbf{e}_\alpha^{\parallel} = \left(\cos \frac{2\pi}{5}\alpha, \ \sin \frac{2\pi}{5}\alpha \right) \ ; \ \mathbf{e}_\alpha^{\perp} = \left(\cos \frac{4\pi}{5}\alpha, \ \sin \frac{4\pi}{5}\alpha \right) \ ; \ \alpha = 1,...5$$

Phason fluctuations are given by:

$$\left\langle \, | \mathbf{h} - \langle \mathbf{h} \rangle |^2 \right\rangle \ = \ \frac{1}{N} \sum_{\mathbf{r}} \left| \mathbf{h}(\mathbf{r}) - \frac{1}{N} \sum_{\mathbf{r}'} \mathbf{h}(\mathbf{r}') \right|^2 \quad .$$

We examined the fluctuations for 4 values of η, averaging over 3/4 samples containing 10^5 tiles each. We find that for $\eta=10^{-2}$ and $\eta=10^{-1}$ the dependence of phason fluctuations on sample size exhibits a power law behavior. In the cases $\eta=10^{-3}$ and $\eta=10^{-4}$ instead $<|h-<h>|^2>$ scales instead as the logarithm of sample size. Growth of more heavily defected structures requires much more memory and computer time (the surface to be inspected grows in fact much more rapidly): we therefore were not able yet to examine samples of comparable size with $\eta \geq 1$.

The behavior of phason fluctuations is relevant because from it one can predict what kind of diffraction pattern would be observed from the structure. In fact, for a perfect PT, phason fluctuations do not depend on the sample size, and the diffraction peaks are sharp. A power-low dependence on sample size instead is a signature of the loss of long–range translational order, as in the case of decagonal glass model. The corresponding diffraction peaks exhibit an intrinsic broadening. Logarithmic dependence of the fluctuations instead means that quasi–long–range order extends over the sample. This kind of behavior is typical of the random tiling models. In three dimensions the diffraction peaks would be sharp, as in the case of perfect quasiperiodic systems.

Our results therefore indicate that depending on the value of the parameter η there is a transition from a random-tiling behavior to an orientationally ordered glass. It is not clear though whether the transition from one regime to the other only depends on the value of the parameter η, or if it also depends on sample size. It is in fact possible that samples which behave logarithmically up to the sizes we investigated start exhibiting defect accumulation at larger size. In alternative, it is possible that even at larger sizes the density of defects remains constant, so that a transition from logarithmic to linear behavior is only driven by η. To clear this point it will be necessary to examine larger samples.

It is possibile[20] to adopt a special algorithm for identifying defects during growth, allowing for some degree of matching rule violations. This would have the advantage to avoid a growth stop at defect sites, which seem to be a mechanism promoting defect accumulation. We believe that such an approach might in fact extend the range of η where logarithmic fluctuations are observed. However, we expect that systems which are already in this regime would not be driven toward a more perfect stage (for such systems the only difference would be that defects would not be identified by isolated holes but by isolated mismatches).

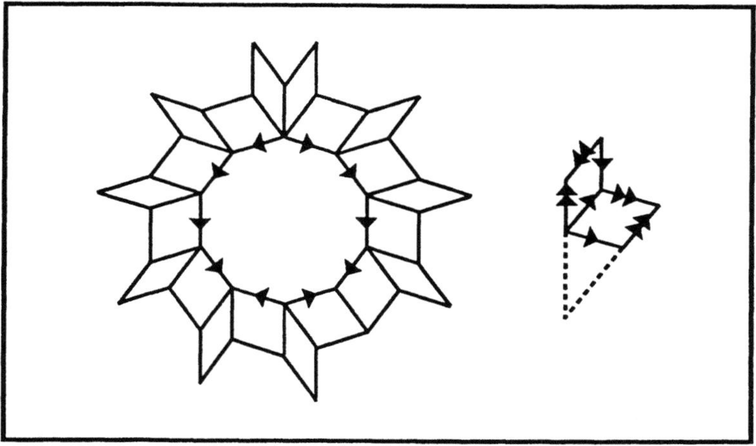

FIGURE 4 - Left: a decapod. The interior of the decagon cannot be filled without violating matching rules. All possible decapods can be built starting from a decagon decorated with single arrows, or equivalently by putting together 10 triangles like the one shown on the right and its mirror image.

It is also worth mentioning that special defects exist which force the entire space around them: decapods [18]. If the initial seed contains one of these defects and if growth occurr only on forced sites, then the presence of more forced sites is guaranteed at any stage of growth (except for two out of the 62 possible decapods): growth could therefore continue forever without creating additional defects. Although the final structure would not be the ideal PT, it would contain a single defect: the decapod. Decapods could therefore play in PT a role analogous to that of screw dislocations in crystals. A decapod is shown in Fig. 4.

4. VACANCY MEDIATED PHASON DYNAMICS

In a second simulation, we studied stability and diffusion of vacancies in a two-dimensional quasicrystalline model, consisting of a decorated Penrose Tiling. The decoration is the one suggested by Lançon et. al.[21] and by Widom et al.[22], and is based on disks of two kinds: small (S) and large (L). They are such that 5 L-disks fit around one S-disk and 10 S-disks of type one L-disk (See Fig. 5). The geometry fixes the ratios $d_{ss}/d_{sL} = 2 \sin \pi/10$ and $d_{LL}/d_{ss} = \sin \pi/5 / \sin \pi/10 = 2 \cos \pi/10$, requiring therefore the radii to be non-additive: in fact $d_{sL} > (d_{ss} + d_{LL}) / 2$.

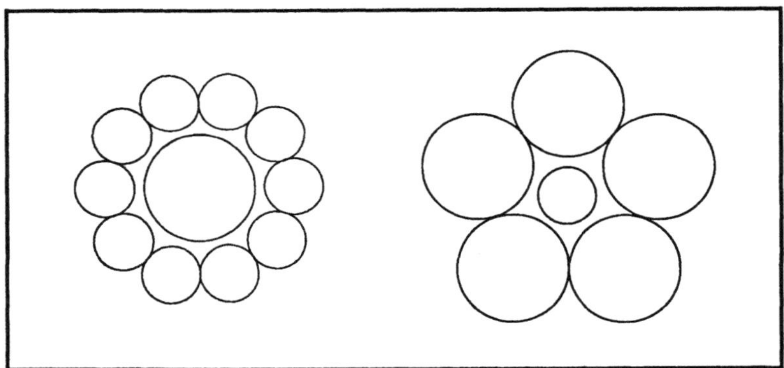

FIGURE 5 - The disks used for the decoration of the Penrose Tiling: 10 S-disks fit around one L-disk and 5 L-disks fit around one S-disk. Unlike disks do not touch each-other, showing the non-additivity of the radii.

Each fat tile is decorated with two large and two small disks, and to each skinny tile are associated one large and two small disks. A PT can in this way be completely decorated with S and L disks. Figure 6-A shows a portion of PT covered with balls.

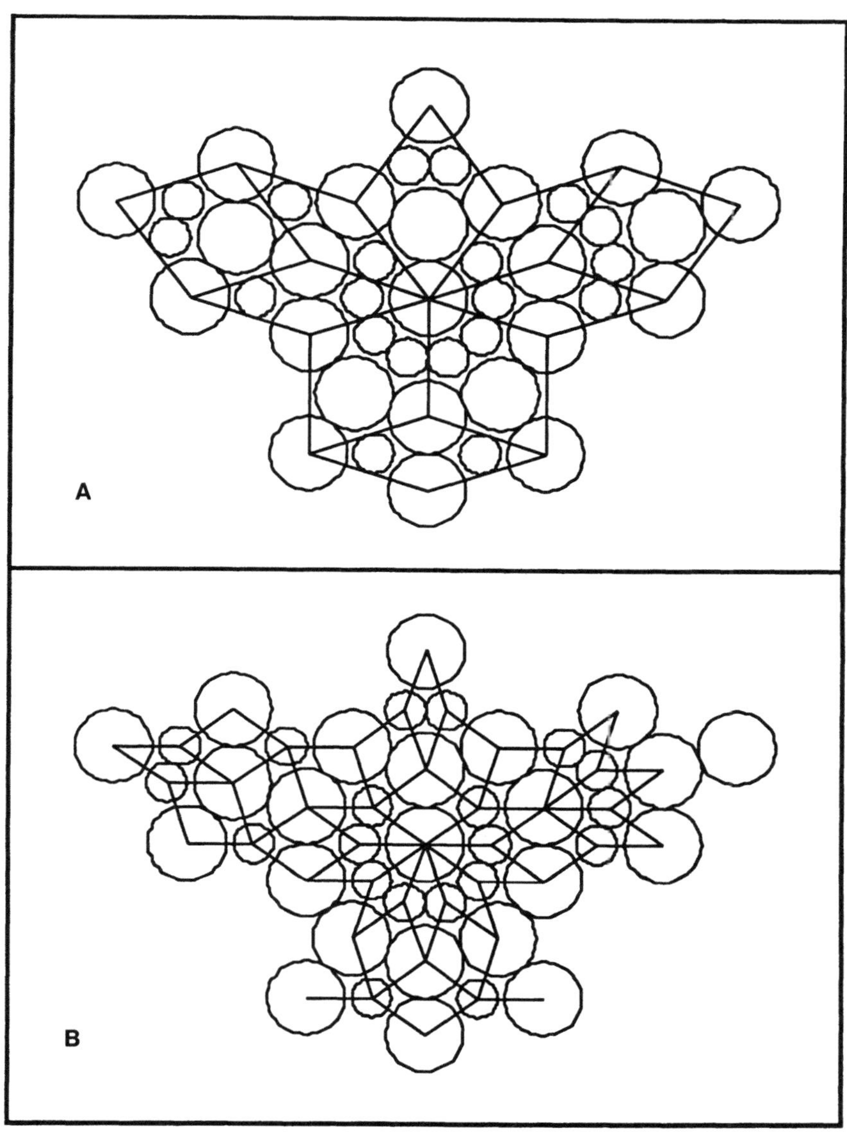

FIGURE 6. A: A set of Penrose Tiles decorated with L- and S-disks.
B Dual Tiling obtained from the same set of disks, by joining unlike nearest
neighbours.

Once the set of disks is defined, it is useful to associate it with another tiling, which following Lançon et al. we shall call "dual". The dual tiling is obtained by joining all S disks with their nearest neighbors of type L. In this way one gets a generalized tiling made of Penrose rhombi scaled by a factor $1/(2\cos\pi/20)^2$, which belongs to a different local isomorphism class. The usefulness of introducing the dual tiling comes from the possibility of describing the set of disks as a tiling even after the initial ideal configuration has been scrambled by atomic diffusion. Figure 6-B shows the dual tiling associated with the disk configuration of figure 6-A. The tiling obtained in this way is a realization of the class of random tilings studied by Tang et al.[24] which can be described by a matching rule allowing on each corners either only even or only odd angles to be present. In fact each such tiling can be decorated L and S disks, although not from every such tiling one can then go back to the PT.

Computer simulation studies performed by Lançon et al.[22] (Molecular Dynamics) and by Widom et al.[23] (Monte Carlo) have shown that such system has a quasicrystalline state, and that it undergoes a first order transition to the liquid state.

Molecular Dynamics simulation are usually performed in periodic boundary conditions, so as to eliminate boundary effects. No portion of the Penrose Tiling can be enclosed in a periodic system, due to its non-periodic nature. However, quasiperiodic structures obtained by the cut or projection methods (as the PT) can also be seen as the limit of an infinite sequence of periodic structures, in the same sense as irrational numbers can be seen as the limit of a sequence of rational numbers. It is therefore possible to find a periodic structure which approximate the PT. We chose such an approximant, and decorated it with disks.

The potential we used was a truncated and shifted Lennard-Jones:

$$U_{\alpha\beta}(r) = \left(V_{\alpha\beta}(r) - V_{\alpha\beta}(r_{cut})\right)\,\theta\!\left(r_{cut} - |r|\right)$$

where r_{cut} is the cutoff distance, α and β can assume the values L and S, θ is the usual step function defined as $\theta(x)=0$ if $x\leq0$ and $\theta(x)=1$ if $x>0$. $V_{\alpha\beta}(r)$ is the unshifted Lennard-Jones :

$$V_{\alpha\beta}(r) = \varepsilon_{\alpha\beta}\left[\left(\frac{\sigma_{\alpha\beta}}{r}\right)^{12} - 2\left(\frac{\sigma_{\alpha\beta}}{r}\right)^{6}\right]$$

where $\varepsilon_{\alpha\beta}$ is the depth of the potential well and $\sigma_{\alpha\beta}$ is the position of the potential minimum. Truncation allows to save computer time, and shifting is then needed to ensure a smooth behavior of the energies.

We normalized $\sigma_{SL} = 1$, and consequently the values for the positions of the other two minima were $\sigma_{SS} = 2 \sin \pi/10 \cong 0.6180$ and $\sigma_{LL} = 2 \sin \pi/5 \cong 1.1756$. Unlike nearest neaghbors were favoured by the choice of the $\varepsilon_{\alpha\beta}$: we used $\varepsilon_{SL} = 2 \varepsilon_{SS} = 2 \varepsilon_{LL}$. The cutoff was chosen to be $r_{cut}=1.3\ \sigma_{\alpha\beta}$.

We used molecular dynamics at constant pressure[25]. The dynamics of the system can be derived from the following lagrangian

$$L(s,\dot{s},\Omega,\dot{\Omega}) = \frac{m}{2}\sum_i \left[\frac{\partial}{\partial t}\left(B_x s_{ix} + B_y s_{iy}\right)\right]^2 - U + \frac{1}{2}W\dot{\Omega}^2 - p\Omega \ ;$$

where B_x and B_y are lenght and heigth of the simulation box, $\Omega=B_x B_y$ is the box volume and W is a parameter which controls the inertia of the walls motion. The reduced coordinates of the disks are $s_{ix}=r_{ix}/B_x$, $s_{iy}=r_{iy}/B_y$ (with (r_{ix}, r_{iy}) giving the position of disk i), and m is the mass (unlike disks are considered to have equal masses). The corresponding equations of motion are

$$\ddot{s}_{i\alpha} = \frac{1}{mB_\alpha}F_{i\alpha} - 2\frac{\dot{B}_\alpha}{B_\alpha}\dot{s}_{i\alpha}\ ;$$

$$\ddot{\Omega} = \frac{1}{W}\left(\frac{B_x}{B_y}m\sum_i s_{ix} + \frac{1}{2B_y}\sum_{i \neq j}F_{ijx}\left(s_{ix}-s_{jx}\right) - p\right)$$

where F_{ijx} is the force on particle i due to particle j, p is the pressure and volume expansions and contractions are assumed to be isotropic.

Derivatives were substituted by finite differences with a time step h, using the Verlet algorithm:

$$\dot{\Omega}(t) = \frac{\Omega(t+h) - \Omega(t-h)}{2h} \quad ; \quad \dot{s}_{i\alpha}(t) = \frac{s_{i\alpha}(t+h) - s_{i\alpha}(t-h)}{2h} \quad ;$$

$$\ddot{s}_{i\alpha}(t) = \frac{s_{i\alpha}(t+h) - 2s_{i\alpha}(t) + s_{i\alpha}(t-h)}{h^2} \ .$$

The system behaved as expected: the quasicrystalline state was stable, and a melting transition from quasicrystalline to liquid was observed. Upon rapid cooling from the liquid, the system did not go back in the original quasicrystalline state, but was trapped in a metastable amorphous state (more details are given in ref. 26).

A vacancy can be introduced in the system at any stage, by removing a particle. We studied four cases, by removing a small or a large disk in the quasicrystal and in the amorphous. We then checked stability and dynamics of the vacancy. To do so, we superimposed to the system a fine grid composed of 275 x 105 square cell. The cells were much smaller than the dimension of the disks, so that the surface of a S-disk measures approximately 36 cells and one L-disks contains 130 cells. An atomic configuration can be examined by setting equal to 1 all cells whose distance from a particle of type α is less than $\sigma_{\alpha\alpha}$, and equal to 0 all other cells (periodic boundary conditions are taken into account). The clustering of cells with a value equal to zero is then studied using a coarse-graining procedure wich allows to identify the large, non elongated clusters [26]. When a vacancy is introduced, one such cluster is obviouly formed. Stability and dynamics of the vacancy can then be studied by examining the time evolution of such clusters. The configurations examined are actually the average of atomic positions over a few atomic vibrations, so as to mask the noise due to thermal motion.

A first result concerns the stability of vacancy in the quasicrystal. A vacancy introduced by removing a S-disk is stable, in the sense that after dynamic relaxation a cluster of empty cells can still be found in the system, and that the volume represented by the cluster coincides with the original free volume of the vacancy.

In the amorphous state instead vacancies are not stable. A similar result for a three-dimensional one-component glass was found some years ago by Bennett et al.[27] The free volume generated by removing either an S-disk or an L-disk does not in fact remain localized, and after a short relaxation the vacancy disappears. During dynamic relaxation we also observe local free volume fluctuations which are not correlated with the original vacancy: empty sites are occasionally created and destroyed throughout the sample.

The case of vacancies created by removing a L-disk from the quasicrystalline state is sligtly more complicated than the case of S-vacancies and needs more investigations. We observed that after dynamic relaxation two situations can be reached. A case similar to the one of S-disks can be observed, in that a single large cluster of volume close to the original free volume of the vacancy can be found. In other cases the vacancy can split in two smaller volumes, and the dual tiling can be somewhat distorted.

We also observe vacancy diffusion. In the case of the S-vacancy in particular the vacancy remains always well defined, and therefore its history can be easily followed. One can always describe each state by giving the corresponding dual tiling, where the presence of the vacancy is manifested by a few missing tiles. In fig. 7 we show two dual tilings, corresponding to two stages of the simulation: initial, after and after 750 τ (τ is the natural time unit for the simulation and corresponds to approximately to 4 atomic oscillations, i.e.

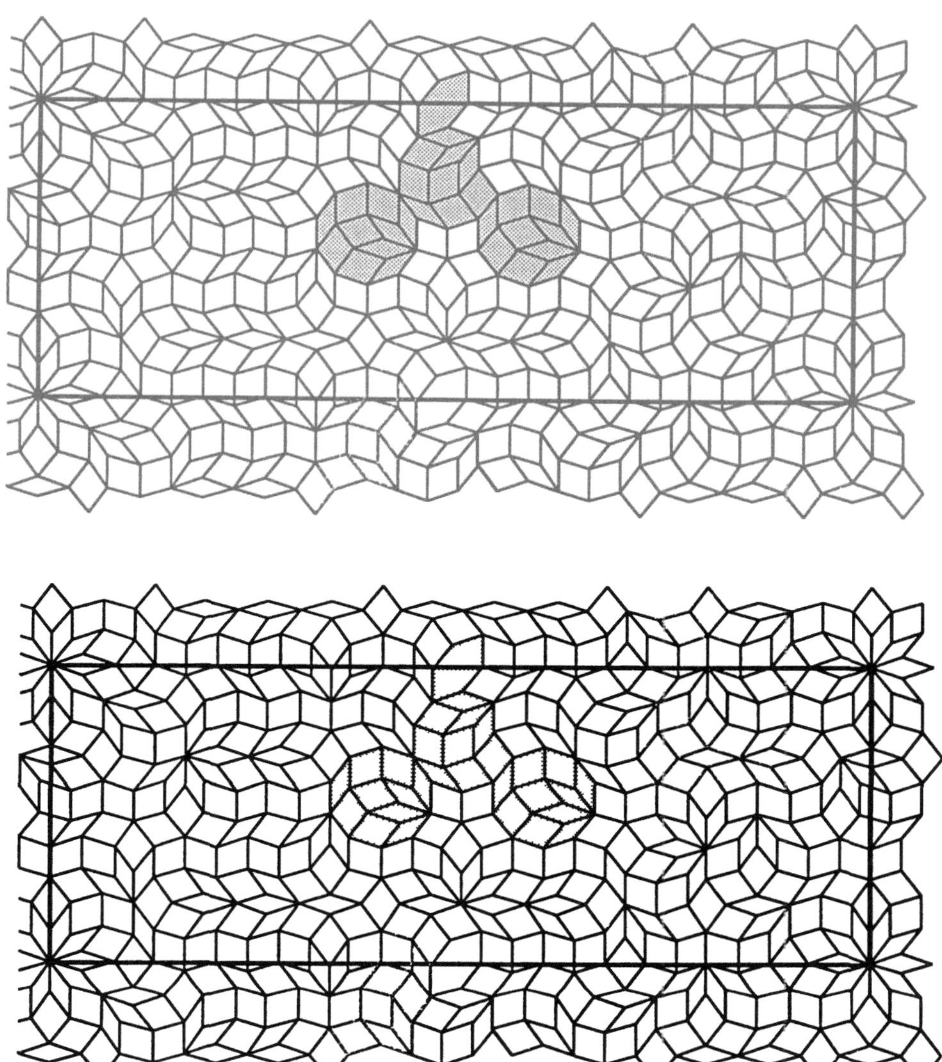

FIGURE 7. Top: dual tiling at the beginning of the simulation. The thick rectangle shows the cell used for the simulation. The vacancy is approximately at the cell center.

Bottom: dual tiling at the end of the Molecular Dynamics. The vacancy has moved to the upper edge of the cell. The only tile rearrangements are observed in the shaded region, i.e .in the region visited by the vacancy.

200 integration steps each). An analysis of the dual tilings shows that tile rearrangement (i.e. phason relaxation) only has taken place in the small region which is shaded in figure 7. Checking the diffusion path of the vacancy, we observed that the trajectory goes exactly through that region. We therefore find that the presence of vacancies strongly enhances phason relaxation, and in fact makes it possible: we do not observe any tile flip not connected with vacancy migration. Our result is obtained in two dimensions, where atomic diffusion is extremely difficult: in three dimensions the picture is certainly different, in that atomic rearrangements can take place more easily. However we suggest that also in three dimensions the presence of vacancies strongly enhance phason relaxation, which is important in all models of QC, and in particular is a necessary ingredient for the random tiling model.

5. SUMMARY

All models based on local growth presented in literature produce structures which exhibit a glassy behavior, leading to finite width of the peaks in their diffraction patterns. On the other hand, models granting the growth of perfect, infinite quasiperiodic structures need global inspection of the sample. In these models, when only local knowledge is used, an ideal system can only be grown up to a size which is determined by the sticking rates, and its growth presents a strongly intermittent behavior.

By using a local algorithm based on different sticking rates for different sites, we have found instead the existence of different regimes, according to a parameter describing the ratio between sticking rates. In all regimes defects are created, different regimes being characterized by the impact of defects on translational order. For high values of our parameter very strong disorder is produced, leading to a seemingly fractal surface. In an intermediate regime the system behaves like an orientationally ordered glass. The most interesting behavior is observed for low values of our parameter, when quasi-long-range translational order is preserved in the resulting structure, so that its three-dimensional analog would present Bragg peaks.

Allowing phason relaxation during the growth process might anneal some of the defects, moving the boundaries between regions of different behavior and favoring the regime in which quasi-long-range translational order is preserved. Phason relaxation is supposed to be a very slow process, since it involves rather complex atomic rearrangements. Atomic diffusion benefits very much from the presence of vacancies, and in fact in the second part of this paper we have demonstrated via molecular dinamics simulation that vacancy do exist is quasicrystals, and that their presence allows phason relaxation.

6. ACKNOWLEDGMENTS

This work has been supported in part by the Italian CNR "Progetto Finalizzato Sistemi Informatici e Calcolo Parallelo" (U.O.Reatto) and by the Center for Theoretical Physics through the Board of Regents Advanced Technology Materials Program at Texas A&M University.

7. REFERENCES

1) For a review see Chaudhari P., Spaepen F., Steinhardt P.J. in *Glassy Metals II (Topics Appl. Phys. Vol 53)* , H.J. Guntherodt, H. Beck eds., (Springer Berlin, 1983) pag. 127, and Ronchetti, M. in *Computer Simulation in Physical Metallurgy*, G.Jacucci ed., (Reidel, Dordrecht 1986) pag. 129

2) Socolar, J.E.S., J. Phys. (Paris), Coll.C3, 47 (1986) 217
Kléman, M. and Pavlovitch, A., J. Phys. (Paris), Coll.C3, 47 (1986) 229
Kléman, M., in *Quasicrystalline Materials*, Ch.Janot and J.M.Dubois eds., World Scientific (Singapore, 1988) pag.318

3) Nori, F., Ronchetti, M. and Elser, V., Phys.Rev.Lett. 61, 24 (1988)

4) Wang, D.N., Ishimasa,T., Niessen ,H.U. and Hovmöller, S., Material Sci. Forum, 22-24 (1987) 381
Wang, D.N., Ishimasa,T., Niessen, H.U. Hovmöller, S. and Rhyner, J., Phil.Mag.A 58 (1988) 737
Zhang, Z. and Urban, K., in *"Quasicrystals. Proceedings of the Adriatico Anniversary Research Conference"*, M.V.Jarić and S.Ostlund eds, (World Scientific, Singapore 1990), pag. 269

5) Bak, P., Phys.Rev.Lett. 54 (1985) 1517
Lubensky,T.C., Socolar,J.E.S., Steinhardt,P.J., Bancel,P.A., Heiney,P.A. Phys.Rev.Lett. 57, 1440 (1986)
Socolar,J.E.S., Lubensky,T.C., Steinhardt,P.J., Phys.Rev. B34, 3345 (1986)

6) For a review see the book series *Aperiodicity and Order,* M.V. Jarić and D.Gratias eds. (Academic Press, Boston 1988,1989)

7) Shechtman, D. and Blech, I., Metall.Trans. 16A,1005 (1985)
Stephens, P.W., Goldman, A.I., Phys.Rev.Lett. 56, 1168 (1986); 57,2331 (1986); 57, 2770 (1986); Steinhardt, P.J., ibid. 57, 2769 (1986)
Elser,V. in *Fractals, Quasicrystals, Chaos, Knots and Algebraic Quantum Mechanichs* eds. A.Amman, L.S. Cederbaum and W.Gans (Kluwer Academic Publishing, Dordrecht 1988)

8) Elser,V. in *Proceedings of the XVth International Colloquium on Group Theoretical Methods in Physics*, Vol 1, eds R.Gilmore and D.H.Feng (World Scientific Press, Singapore, 1987)

Elser,V. , in *Icosahedral Structures*, M.V. Jarić ed. (Academic Press, Boston 1989)

9) Henley, C.L., J.Phys.A: Math.Gen. <u>21</u>, 1649 (1988)

    Widom, M. , Deng, D.P. and Henley, C.L., Phys.Rev.Lett. <u>63</u>, 310 (1989)

    Strandburg, K.J, Tang, L.-H. and Jarić, M.V., Phys.Rev.Lett. <u>63</u>, 314 (1989)

10) *Kinetics of Aggregation and Gelation*, F.Family and D.P.Landau eds, (North Holland, New York, 1984),

    On Growth and Form, H.E.Stanley and N.Ostrowsky eds. (Martinus Nijhoff, Dordrecht 1986),

    Fractal in Physics, L.Pietronero and E.Tosatti eds. (North Holland, N.Y., 1986)

11) Guryan, C.A., Goldman, A.I., Stephens, P.W., Hiraga, K., Tsai, A.P., Inoue,A., and Masumoto, T., Phys.Rev.Lett. <u>62</u>, 2409 (1989)

    Hiraga,K. Zhang,B.P., Hirabayashi,M. Inoue,A. and Masumoto, T., Jpn.J.Appl. Phys. Part 2 <u>27</u>, L951 (1988)

12) Levitov, L.S., Comm.Math.Phys. 119,627 (1988)

    Socolar, J.E.S., in *"Quasicrystals. Proceedings of the Adriatico Anniversary Research Conference"*, M.V.Jarić and S.Ostlund eds, (World Scientific, Singapore 1990) pag. 182

13) de Bruijn, N.G., Nederl.Akad.Wetensch.Proc.,Ser. <u>A44</u>, 39,53 (1981)

14) Onoda, G.Y., Steinhardt, P.J., DiVincenzo, D.P. and Socolar, J.E.S., Phys.Rev.Lett. <u>60</u>, 2653 (1988); <u>62</u>, 1210 (1989)

15) Jarić, M.V. and Ronchetti, M., Phys.Rev.Lett. <u>62</u>, 1209 (1989)

16) Penrose, R., in *Introduction to the Mathematics of Quasicrystals*, M.V. Jarić ed. (Academic Press, Boston 1989)

17) Ronchetti, M., Jarić, M.V., in *"Quasicrystals. Proceedings of the Adriatico Anniversary Research Conference"*, M.V.Jarić and S.Ostlund eds, (World Scientific, Singapore 1990) pag 227

18) Gardner, M., Sci.Am. <u>236</u>, 2653 (1988).

19) For a discussion of the Amman bars and Conway worms see chapter 10 of *Tilings and Patterns*, B.Grünbaum and G.C.Shephard (W.H.Freeman, New York 1987)

20) Socolar, J. personal communication, and DiVincenzo, D., these proceedings.

22) Lançon, F., Billard, L. and Chaudhari, P. Europhys.Lett. 2, 625 (1986)

    Lançon, F., Billard J. de Phys., <u>49</u>, 249 (1988)

23) Widom, M., Strandburg, K.J. and Swendsen R.H., Phys.Rev.Lett. <u>58</u>, 706 (1987)

24) Tang , L.H. and Jarić, M.V., to appear in Phys.Rev.B

25) Andersen, H.C. J.Chem.Phys. <u>72</u>, 2384 (1980)

26) Mauro Bertagnolli, Tesi di Laurea, Facoltà di Scienze, Università degli Studi di Trento (1988) (in Italian)

27) Bennett, C.H., Chaudhari, P., Moruzzi, V. and Steinhardt, P.J., Phil.Mag.A <u>40</u>, 485 (1979)

INTRINSIC STABILITY OF QUASICRYSTALS AND BEHAVIOUR UNDER A LOAD OF FRENKEL DEFECTS

Johannes Roth

Institut für Theoretische und Angewandte Physik
der Universität Stuttgart
Pfaffenwaldring 57, D-7000 Stuttgart 80, Federal Republic of Germany

1 INTRODUCTION

Until the discovery of quasicrystals in 1984 solid matter was assumed to occur in three different types of structures: periodic, incommensurate and aperiodic. But then it was found that noncrystalline symmetry is also possible combined with sharp diffraction peaks.

As the existence of noncrystalline translational symmetry was not recognized before, the new materials required a new description of structure and stability. The first approaches were phenomenological using Landau-type theories (see for example [1,2,3,4]). It was found that icosahedral structures may be stable. The next steps were computer simulations of quasicrystalline models in two dimensions by Lançon et $al.$ [6,7], Sasajima et $al.$ [8] and Widom et $al.$ [9]. Sasajima showed that monatomic decorated Penrose patterns relaxed to a hexagonal state, whereas Lançon and Widom could demonstrate the stability of diatomic quasicrystals using different potentials and different methods. Recently Leung et $al.$ [10] could also prove the stability of a similar diatomic dodecagonal quasicrystal.

Concerning three-dimensional quasicrystals there are a number of analytical results now proving the stability of quasicrystals by Burkov and Levitov [13] and by Narasimhan and Jarić [14], but no microscopic models have been studied in detail (but notice [15]). The only approach to an analysis of stability was made by Janssen using a spring model to calculate the density of phonon states [11].

We started in 1987 to study threedimensional quasicrystals, but because of the complexity of the system we used numerical minimization $i.e.$ the $steepest$ $descent$ $method$ instead of computer simulation methods. Our first results were published in Ref. [16]. A detailed description of all results in two and three dimensions for mono- and diatomic quasicrystals both at 0 K and at finite temperature can be found in Ref. [17]. Very recently we became aware of similar results by Lançon and Billard [18]. Although they used the conjugate gradient method and other potential types they found results very similar to ours.

In this article we will deal only with monatomic quasicrystals. After presenting the model for a Lennard-Jones system in section 2 and the method of relaxation and structural analysis in section 3, there are two major parts: the first concerns the stability of "perfect" quasicrystals with different decorations (Sec. 4) and the second quasicrystals with Frenkel defects (Sec. 5).

Geometry and Thermodynamics
Edited by J.-C. Tolédano
Plenum Press, New York, 1990

In three dimensions the primitively decorated Penrose pattern (an atom on each vertex) relaxes to a glassy state. If, however, atoms are removed according to the unit sphere packing discussed by Henley [19], even the *monatomic* quasicrystals remain stable, or more precisely metastable. Thus, our results demonstrate that it is *not* necessary for stability of three-dimensional quasicrystals to have at least two sorts of atoms.

The degree of metastablity of quasicrystals was also investigated. This point addresses the question: how much may the atoms fluctuate from their positions in the Penrose pattern such that the system still relaxes to a quasicrystalline structure. Using a Gaussian distribution for these fluctuations ("noise") we have found that the unit sphere packing is stable if the mean fluctuation does not exceed a critical value which is about 7% of the edge length of the rhomboedra. For larger values the system turns into an amorphous structure.

The second part of the article deals with the problem of radiation damages in quasicrystals. Urban *et al.* [20] have studied quasicrystals with electron microscopy and found that AlMn and AlV can be transformed into an amorphous state, the transition being reversible.

According to Kronmüller [21] radiation defects are Frenkel defects. We have generalized the concept of vacancy-interstitial pairs for quasicrystals and studied interstitials with octahedral and pyramidal neighbourhood.

A small density of Frenkel pairs does not alter the stability of the quasicrystal, but when the number increases amorphous domains occur, and at about 25% defects per lattice site the whole structure becomes amorphous. Contrary to the study of noise there is no clear distinction between the various states after relaxation, but three regions of different behaviour are found. An intermediate range with disordered domains clearly shows an icosahedral bond order while the translational order decays.

2 THE MODEL

The perfect quasicrystalline rhombohedral tiling is constructed by the grid method. Initially, all vertices of the rhombs or rhombohedra are covered with atoms. Modifications are made by removing subsets of the vertices.

The centers of the atoms are the origins for the pair interaction potential. We use a truncated Lennard-Jones potential of the following form:

$$V_{LJ}(r) = \begin{cases} E\left[\left(\frac{\sigma}{r}\right)^{12} - 2\left(\frac{\sigma}{r}\right)^{6}\right]\cos^2\left(\frac{\pi}{2}\frac{1}{\beta}\frac{r}{\sigma}\right) & \text{if} \quad 0 < \frac{r}{\sigma} \leq \beta \\ 0 & \text{if} \qquad \frac{r}{\sigma} \geq \beta \end{cases} \tag{1}$$

Here r denotes the distance $|\mathbf{r}_i - \mathbf{r}_j|$ between the atoms i and j, σ the radius of the potential minimum, β a truncation factor (always taken as 2 here) to make the potential range finite, and E sets the energy scale. The truncation of the potential defines a sufficiently small shell of interaction for each atom which is necessary for the simulation. Using a finite range, the time for calculation grows only linearly with the number of atoms, otherwise it would grow with the square.

The radius σ of the potential minimum was optimized before relaxation by fixing the atoms to their initial positions and minimizing the total interaction energy. The optimal value of σ is always close to the shortest separation of atoms in the decoration. This minimum separation R_{min}, however, frequently is smaller than the radius R of the first *pronounced* shell of atoms (having appreciable occupation number). As a rule stability was registered if both $\sigma \approx R_{min}$ and $R_{min} \approx R$. But the system was unstable if $\sigma \approx R_{min}$ and $R_{min} < R$, and even when testing several values of σ in the whole range between R_{min} and R we could not stabilize it, except in trivial cases (see Sec. 4).

To avoid surface effects we had to use periodic boundary conditions, which we implemented by rational approximants in the following way: The sixdimensional points in the strip which are to be projected into three dimensions can be found easily by dualising the cells of the grid. We replaced the usually rationally independent grid vectors by rationally dependent ones but left the tiling vectors and therefore the shapes of the tiles unchanged. The possible range of the approximants p/q of τ for our calculations was 2/1, 3/2 and 5/3, yielding periodicity lengths $2(\tau p + q)(\tau + 2)^{-1/2}$.

3 THE METHOD

We have used a numerical method to study the stability of the quasicrystals and analytical tools to examine structure and temporal changes. The basic features of the relaxation procedure and analysis are described in the following section.

3.1 Relaxation

The method of *steepest descent* is static, however, it can be interpreted dynamically as describing an overdamped atomic motion at zero temperature. In a first step we calculate the force

$$\mathbf{F}_i = -\frac{\partial V}{\partial \mathbf{x}_i} = -\sum_{j \neq i}^{N} \frac{\partial V_{LJ}(\mathbf{r}_{ij})}{\partial \mathbf{r}_{ij}} \; , \; \mathbf{r}_{ij} = |\mathbf{x}_i - \mathbf{x}_j| \tag{2}$$

which atom i feels from all atoms j in its own potential range. Then each atom is displaced by a vector $\boldsymbol{\delta}_i = \lambda \mathbf{F}_i$. Through normalizing $\boldsymbol{\delta}_i$ by the maximal force $\max|\mathbf{F}_k|$ we insure that the largest displacement is λ. The procedure is iterated until equilibrium (minimal energy or $\mathbf{F}_i \approx 0$) is obtained. We used the relative change $\Delta E/E$ of energy per atom to decide whether the atoms have reached the final state. The number of iterations ranges from less than 100 to 1000. In most cases we stopped relaxation, when the relative change of energy $\Delta E/E$ per iteration was less than the numeric precision. Relaxation for unstable open clusters was slowest. Here we stopped at $\Delta E/E = 10^{-5}$.

3.2 Structural analysis

The straightforward way to examine a structure is by depicting it in real space. To display the vertices of a *threedimensional* quasicrystal, we projected them onto a plane perpendicular to a fivefold axis. Then a point in the plane represents an entire row of vertices which can smear out during relaxation.

A more quantitative way for structural characterization is the radial pair distribution function $G(r)$ or its Fourier transform, the powder structure factor $S(q)$:

$$S(q) = 1 + \int_0^{\infty} 4\pi r^2 \; [\rho(r) - \rho_0] \frac{\sin qr}{qr} dr + \int_0^{\infty} 4\pi r^2 \; \rho_0 \frac{\sin qr}{qr} dr \tag{3}$$

ρ_0 denotes the average density of atoms around a central atom. The second term of equation 3 yields the characteristic scattering intensity of the quasicystal whereas the third term describes the homogenous part which for clarity is taken out further on. The information contained in $G(r)$ allows to locate nearest, next nearest and eventually higher order neigbour separations as well as to determine coordination numbers. But it does not measure the orientational bond order which is crucial for quasicrystals. In the amorphous case most of the peaks of $G(r)$ vanish, but if $G(r)$ displays no peaks other than for nearest neighours the structure need not be amorphous. It can have an anisotropic bond orientational ordering. We therefore calculated, according to Steinhardt *et al.* [23], the modulus

$$Q_l = \left[\frac{4\pi}{2l+1} \sum_{m=-l}^{l} |\bar{Q}_{lm}|^2 \right]^{1/2} \tag{4}$$

of the l-th bond orientation multipole moment

$$\bar{Q}_{lm} = \frac{1}{N_B} \sum_{bond_i} Y_{lm}(\theta_i, \phi_i) \tag{5}$$

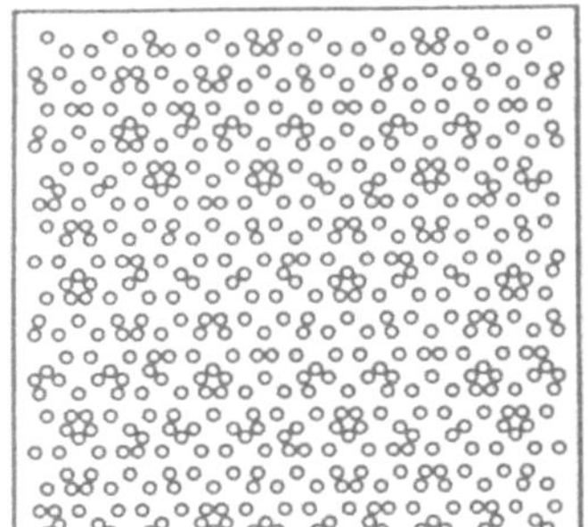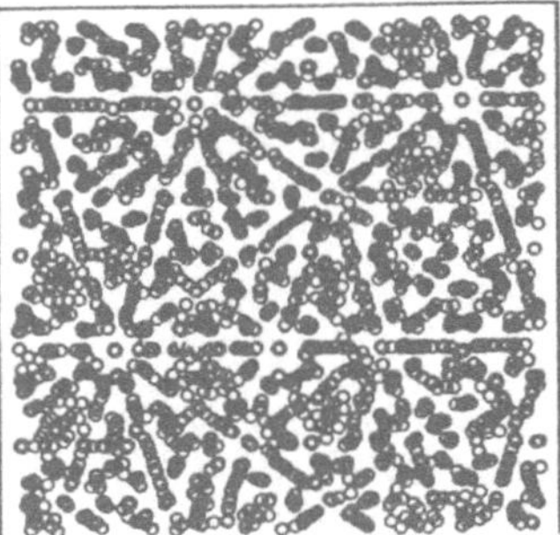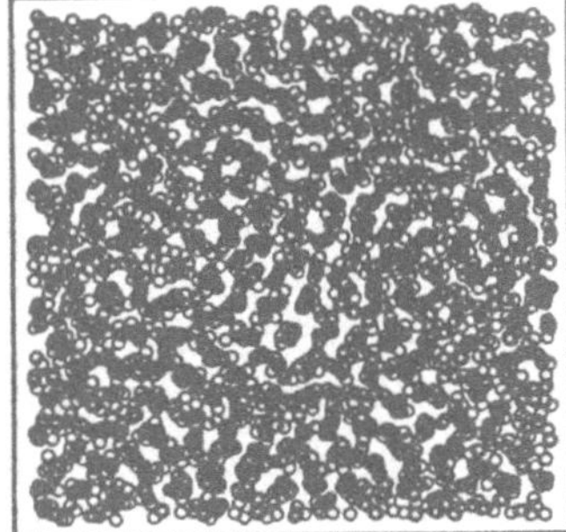

Figure 1. Projection of a threedimensional quasicrystal resulting from a primitive monatomic decoration. Two representative rhombohedra have been drawn.
Left side: initial state. Middle part: intermediate state. Right side: final state.

Here θ_i and ϕ_i denote polar and azimuthal angle of the i-th nearest neighbour bond relative to a fixed laboratory coordinate system, and N_b is the number of bonds. As nearest neighbours we declared all atoms separated a distance less than 1.2 times the length of the first coordination shell like in Ref. [23]. In an isotropic medium all Q_l vanish for $l > 0$, whereas in an icosahedral medium of long-range bond orientational order Q_6 dominates, and Q_4 and Q_8 are zero.

4 THREE-DIMENSIONAL MONATOMIC SYSTEMS

When we started our calculations there were no quantitative microscopic models of the atomic structure of icosahedral materials. Therefore we examined several decorations of Penrose patterns, starting with an occupation of all vertices.

4.1 Systems with all vertices occupied

The smallest separations of atoms in a three-dimensional Penrose tiling is the short diagonal of the oblate rhombohedron, having length $d = 0.563$ in units of the edge-size. If *all* vertices are occupied the width of the optimized potential radius σ is about 0.58. The atoms of the next larger separation, $d = 1$ and $d = 1.052$, feel a very small force only, which is rather unphysically. Increasing σ in the range between 0.58 and 1.03 drives the system into a glassy state. We have examined several values for σ but we will only talk about $\sigma = 1.0$, because the other samples do not yield significantly different results

In projection along a fivefold axis (Fig. 1) the atoms are lined up in rows. During relaxation, the alignment is lost, and the projected atoms cover the projection plane homogenously (Fig. 1).

It is interesting to note, that in the course of the decay the quasicrystalline phase does not pass directly into the amorphous state but *via* a sequence of intermediate phases, of which we do not know yet whether they are an icosahedral glass or a new type of icosahedral structure. In the first relaxation steps the atoms arrange in planes which in the projection along a fivefold axis constitute a generalized pentagonal grid with two spacings of ratio 1:τ ("Amman-grid", Fig. 1). The polygons formed by the intersecting planes are golden triangles with edge lengths 1,1,τ or τ,τ,1. In the following steps new planes of reduced density form in between the old ones, leading to a deflated Amman grid which is scaled down by τ. The golden triangles are successively subdivided into smaller ones of similar shape. Amorphization begins when the edge length of the triangles reaches the average separation of the atoms.

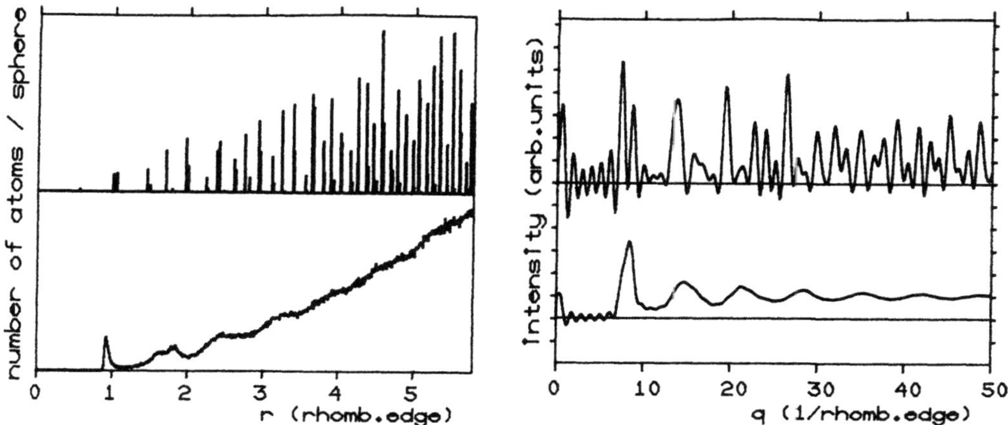

Figure 2 Left side radial distribution functions of the initial quasicrystalline state with the primitive monatomic decoration (upper part) and of the final amorphous state (lower part)
Right side structure factor of the initial state (upper part) and the final state (lower part)

This behaviour reveals the hierarchy of local atomic arrangements Those atoms which have the most anisotropic environments in the initial pattern move first, because they feel the largest force On the other hand there are many vertices which are surrounded by twenty prolate rhombohedra, forming a neighbourhood of *perfect* icosahedral symmetry These vertices become the intersections of five Amman planes in the course of the relaxation The central atom and its first neighbours feel no force and thus remain fixed until the last stage of amorphization

Whether this new phase has a physical significance is unclear to us It is stabilized by a repulsive potential Further investigations are in progress

Perfect quasicrystals have discrete peaks in the radial distribution function The coordination numbers are noninteger due to averaging over all possible vertex environments (Fig 2) Depending on the value of σ, the *relaxed* crystal shows three or more coordination shells, but the RDF smoothes off at larger separations as for a homogenous, amorphous medium

In the initial state there is a small coordination shell at 0 563 with frequency of 0 76, but the most pronounced coordination shells are at 1 0 and 1 051 with frequencies of 6 00 and 6 47 The value 1 051 belongs to the short diagonal of a rhombic surface element Thus a double shell exists with center of gravity at about 1 03 having intensity of 12 47 like in dense packings After relaxation, the initial radius of the first coordination shell is always equal to the optimal σ During relaxation, the first peak increases up to a maximum coordination number of about 9 8 Independent of the choice of σ, there are further peaks with maxima at about 1 7σ and 2σ This splitted second nearest neighbour peak is characteristic for metallic glasses When σ is about 1 0 further peaks exist with decreasing intensity due to the dense packing of atoms

The powder structure factor of the perfect system (Fig 2) displays peaks which can be identified with and indexed as those obtained by projection from sixdimensional reciprocal space The width of the peaks, however, is finite, because in the integral (Eq 3) we have cut off the radial distribution function at the radius of that sphere, which just fits into the periodicity cell of the approximant of the Penrose pattern The cutoff is also responsible for the negative values of the structure factor and the remaining oscillations at low wavelengths

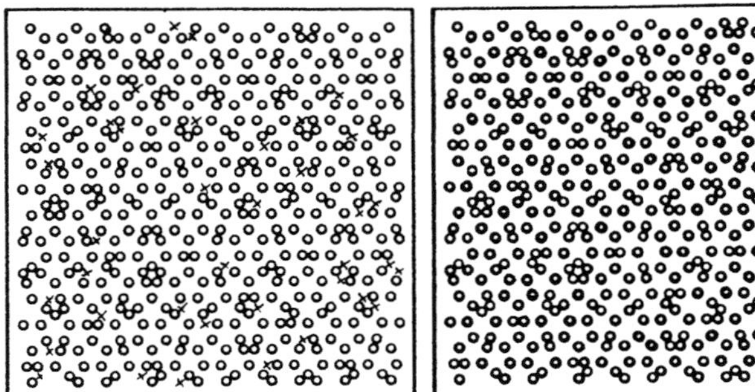

Figure 3. Threedimensional quasicrystal with the unit sphere packing. The places where atoms have been taken out of the primitively decorated monatomic quasicrystal are marked with a cross.
Left side: initial state. Right side: final state.

After relaxation, the quasicrystalline dominant peaks disappear. Only the typical structure factor of an amorphous material remains with short-wavelength modulation due to the finite range of integration.

In the perfect state Q_2, Q_4 and Q_8 vanish as expected, but Q_6 takes the value 0.153 indicating long-range icosahedral bond orientation. The reduction of Q_6 to $9 \cdot 10^{-3}$ in the course of relaxation indicates that the long range bond order vanish. But it does not mean that the icosahedral ordering of the nearest neighbours, which is typical in dense packed amorphous materials, is reduced, too. To examine the short range order one would have to calculate correlation functions and not only multipole moments.

4.2 Monatomic unit sphere packing

The smallest length of 0.563 turns out to be the reason for the instability of the fully covered primitive Penrose lattice. The atoms separated by this distance form rings of ten members in the shape of a crown or chains being fractions thereof. If a chain has an even number of links, then we remove every second atom in order to leave the greatest possible number of atoms in the cluster. If a chain has an odd number of links and in case of rings we also removed every second atom then beginning at an arbitrary site. With this construction none of the remaining atoms is separated by less than 1.0 from a neighbour. There are also other constructions providing a packing with the same properties by Henley (see [19]) and by Olami and Alexander (see [24]).

The number of atoms removed is only about 5% (Fig. 3). The structure of the projected pattern remains pentagonal. In the course of relaxation the modified structure changes by minor local fluctuations only and remains stable (Fig. 3).

The structure of the RDF is unchanged when the subset of atoms is removed, only the peak at 0.563 has vanished. During relaxation some very close peaks in the RDF (Fig. 4) merge to one; altogether the peaks stays distinct. The coordination number remains nearly decreases and is now 10.67.

The peaks of the structure factor (Fig. 4) retain their position, only the intensity is altered slightly. Local distortions cannot be identified since the peak width is finite due to the periodic boundary conditions. Since the width scales like l^{-1} where l is the periodicity length, we expect that an infinite system shows Bragg peaks and that the long-range positional order is not disturbed in the relaxation process.

The bond order parameter Q_6 increases slightly from 0.180 to 0.183 indicating only minor changes of the icosahedral bond angles, while Q_2, Q_4, Q_8 are approximately zero before and after relaxation.

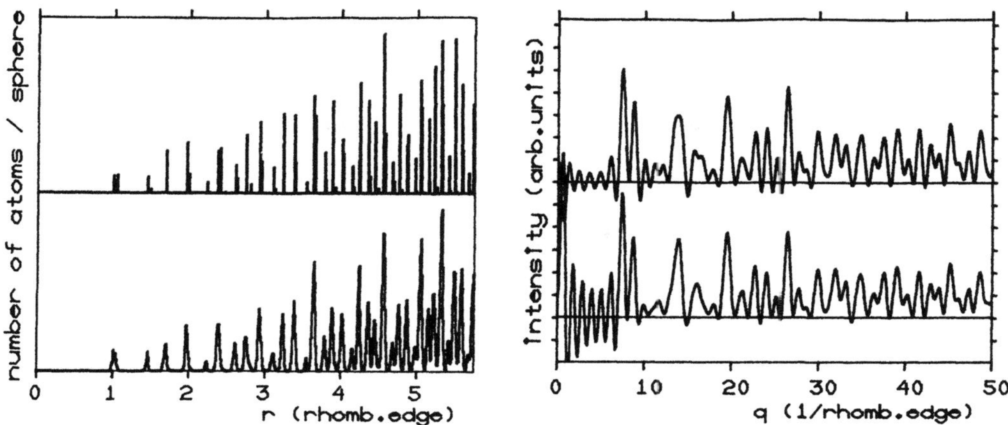

Figure 4 Left side radial distribution function of the unit sphere packing Upper part initial state
Lower part relaxed state
Right side structure factor Upper part initial state Lower part relaxed state

To investigate the degree of metastability we compared the energy per atom of
the quasicrystalline unit sphere packing with that of a bcc and fcc packing using the
same potential type, potential radius and potential range We found an energy of 2 8
for bcc and 3 2 for fcc (both without relaxation) whereas the energy of the relaxed
quasicrystalline state with periodic boundary conditions is about 2 8 and was decreased
only by 1% during relaxation This fact tells us that our quasicrystalline packing can be
as stable as crystalline packings

The Landau theory up to the third order in $\rho(\mathbf{r})$ used in Ref [2] leads to a similar
result, $i\ e$ the energy of the quasicrystal is below that of the bcc lattice However, this
agreement may be accidental, because Biham *et al* [5] have shown that this is no longer
true if also the fourth order term is taken into account (see footnote in Ref [5])

Very recently, Olami and Alexander [24] have investigated the density of several
packings in more detail From their results one cannot conclude that Henley's unit sphere
packing is the densest one But it is at least the densest known up today

Control calculations have been performed also with finite samples of the usp
monatomic quasicrystal and open boundary condition For 1 888, 6 070 and 13 189 atom
clusters the energy per atom was 2 52, 2 61 and 2 65 respectively, the last two configu
rations not having been relaxed fully These values converge towards the energy per atom
 2 86 of the sample with 1 902 atoms and periodic boundary conditions (being the ratio
nal approximant 5/3) The higher energy of the open clusters is due to surface effects
For all cases the structure in the interior of the samples was identified Thus we may
conclude that periodic boundary conditions actually simulate the behaviour of samples
of infinite size

4 3   The stable decoration with noise

Having discovered a stable monatomic quasicrystal, it appeared to us of considerable
interest to find a measure for the width of the metastable valley of the monatomic unit
sphere decoration Therefore we displaced the atoms of the unit sphere packing randomly
before relaxation The orientational displacement was chosen isotropically, whereas for
the radial displacement we used a Gaussian distribution If the average displacement is
increased to about 6 % of the unit edge length we note the following properties of the
initial state There is so much noise that the quasicrystalline structure has nearly vanished
in the projection The RDF has lost its isolated peaks although there remain certain

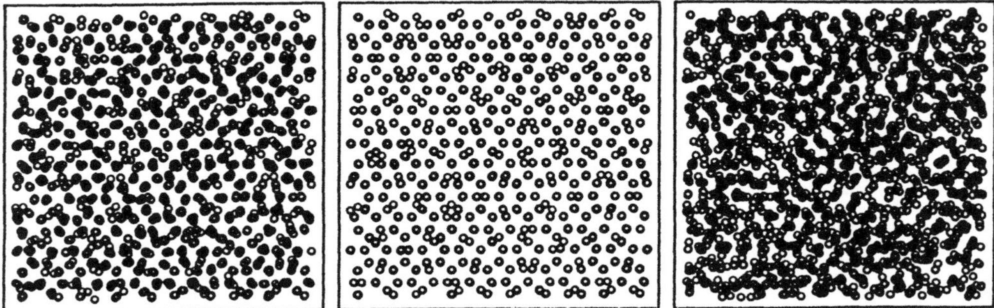

Figure 5. Unit sphere packing at nonzero temperature. Atoms which are lined up in the perfect quasicrystal, are now displaced from their rows.
Left side: initial state for 6% average displacenment. No difference is seen to the state of 8% average displacement. Middle part: final state, small noise (6%). Right side: final state, large noise (8%).

distinct shells and the structure appears not completely amorphous. In the structure factor the quasicrystalline peaks are small and peaks greater than 30 reciprocal edge lengths cannot be seen due to the Debye-Waller-factor produced by the noise.

If the average displacement is less than 6 %, then in course of the relaxation pentagonal symmetry returns with remarkable clarity (Fig. 5). The RDF now has small discrete peaks whereas the coordination number remains constant. The quasicystalline peaks in the structure factor have acquired the height of the perfect quasicrystallinic ones, the peaks greater than 30 wavelengths are seen again. During relaxation, the icosahedral order *increases* slightly and the energy per atom is lowered again from -2.5 to -2.85. The original pattern is restored completely.

But if the average displacement is chosen to be greater than 7 % of the unit length, the final state is amorphous, although the initial state does not look much different (Fig. 5). If we consider the maximal displacement of atoms in the initial state we find some which are displaced by 40 % of the unit edge length and hence penetrate deeply into the repulsive range of other atoms. The energy per atom is only about -1.4, and the icosahedral parameter Q_6 is changed from 0.17 to 0.11, thus the icosahedral order has not disappeared completely. The amorphous character of the final state is revealed by the RDF as well as by the structure factor. The energy per atom decreases during relaxation to -2.3 but remains higher than that of the stable phase.

These results show that the "basin of attraction" of the metastable icosahedral configurations has a *finite* measure. Therefore it is not unlikely that a liquid-like initial configuration relaxes to a quasicystalline structure which would not be possible if the final state is in a "golf hole".

The results do not prove that real monatomic quasicrystals must exist; but for their existence these properties are necessary.

5 FRENKEL DEFECTS IN QUASICRYSTALS

Urban *et al.* [20] have studied quasicrystals with high voltage electron microscopy and they that quasicrystals of the AlMn and AlV type can be transformed into an amorphous state when they are exposed to electron beams. The transition is reversible: the perfect quasicrystalline state returns after annealing of the irradiated sample.

Table 1. description of the vertices and holes in the usp decoration

type of solid	abbreav.	type of interstitial	radius	number / solid	number
vertices	-	-	1.000		1902
prolate rhomb.	O_{PR}	octahedral	0.726	1	720
"	T_{PR}	tetrahedral	0.629	2	1440
oblate rhomb.	T_{OR}	tetrahedral	0.596	6	0
rh. dodecahedron	T_{PR}	tetrahedral	0.629	4	1576
"	Pr	prismatic	0.786	2	788
"	Py	pyramidal	0.726	4	1576

The prolate octahedral holes of the perfect decoration are partially transformed into pyramidal holes in the usp decoration. There total number is 2296. There have been 5592 T_o holes in the primitive decoration.

A possible mechanism for such a transformation was studied by Kronmüller *et al.* [21,22]. They showed that the radiation defects can be understood as pairs of vacancy and interstitial atoms called *Frenkel pair*. This kind of defects is well known for crystals and also for non-crystalline materials. Our aim will be to define the concept of Frenkel defects for quasilattices and to study their behaviour.

5.1 Definition of Frenkel Defects for Quasicrystals

A Frenkel defect consists of an empty atomic site, and a nearby interstitial. But what are the interstitials? In an ordinary crystal lattice *e.g.* an fcc lattice, there are two types: octahedral and tetrahedral interstitials. They can be found using the Voronoi construction: to each atom (of the perfect lattice) is assigned a domain containing all points of space which are closer to that atom than to all others. For the fcc lattice the Voronoi cells are rhombic dodecahedra. As can be seen easily by geometric consideration, the points most distant from the atoms are the corners of the Voronoi cells. Therefore they are called "deepest holes". The radius of a deepest hole is equal to its distance from the center of the cell and is the same for all atoms adjacent to that hole. Thus the deepest holes have a unique circumradius, which can be used to classify the holes (in addition to the number of atoms around the hole). We call the deepest holes "interstitial sites".

The concept of Voronoi decomposition is by no means limited to the case of lattices, but is often used to describe liquid and amorphous structures. There a large number of different Voronoi cells occurs. They are handled with statistic tools: the interstitials are now called vacancies. In quasicrystals we have a smaller number of Voronoi cells (depending on the neighbourhood of the vertices) and therefore we can define the interstitials in the same way as those of an ordinary lattice: the interstitials of a quasicrystal are the corners of the Voronoi cell decomposition and they are distinguished by their circumradius, number of neighbouring atoms and shape.

5.2 Types of Frenkel Defects for the primitive Decoration and the usp Packing

The interstitial sites of a perfect 3d Penrose pattern decorated in primitive way with an atom at each vertex can be found by decomposing the prolate and oblate rhombohedra into tetrahedra and octahedra. The prolate rhombohedron can be divided into two tetrahedra at the acute corners and an octahedron between them. The oblate rhombohedron is composed of six tetrahedra of equal shape. The properties of the interstitals are listed in Tab. 1.

The description of the interstitials in the usp-decoration is more complicated. According to Henley [19] the neigbourhood of a chain of short bonds is a rhombohedral dodecahedron containing two oblate and two prolate rhombohedra. When the usp decoration is formed the atom in the dodecahedron is removed. To find the interstitials we

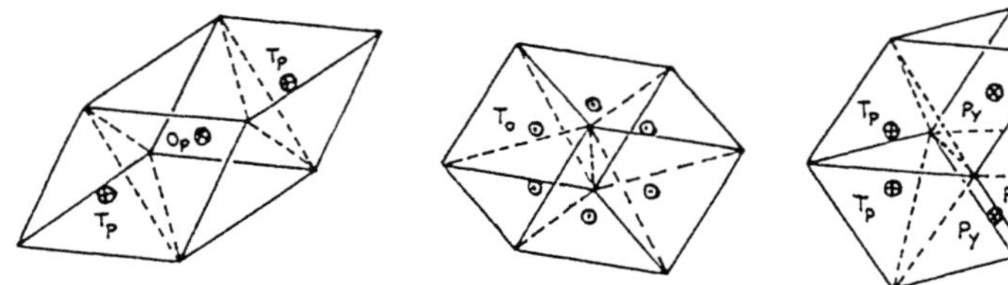

Figure 6. The interstitials in the quasicrystal cells.
Left side: prolate rhombohedron. Middle part: oblate rhombohedron. Right side: Rhombic Dodecahedron. The abbreaviations are explained in the table above

have to subdivide the solid into Voronoi cells. We get four tetrahedral cells of the prolate type, four pyramidal cells and two prismatic cells as shown in Fig. 6. There are further holes connected with the single, not fully decorated oblate rhombohedra, but they have low density and have not been determined yet.

5.3 Formation of the Frenkel Defects in the usp Decoration

The quasicrystal with Frenkel defects was constructed in two steps. First the vertex coordinates of the perfect pattern and the hole coordinates were calculated. Then atoms of the primitively decorated pattern where chosen at random and removed. Also holes were selected at random and occupied with atoms. Thus we obtained Frenkel pairs which were *not* correlated, contrary to what is found experimentally, where most of the Frenkel pairs consist of an empty atomic site and a *nearby* interstitial [21].

Until now, we only studied octahedral interstitials of the prolate type and the pyramidal ones of the same size, which are smaller than the prismatic holes but are much more in number (2296 interstitials for 1902 atoms). Our results cover the whole range up to 400 defects (25% defects per lattices site).

5.4 Results

The radius of the interstitial is about 73% of the edge length in the quasilattice (see Tab. 1). Therefore the bond distance is smaller than the optimum (1.046 of the edge length) and the bond energy per atom must decrease when defects are built in. We find that the activation energy per defect at the start of the relaxation increases linearly over the whole range which was under study.

There is no clear distinction between different types of behaviour as the number of defects grows although three different regions can be distinguished in a projection of the relaxed quasilattice: up to 30 defects are compensated completely. Only minor displacements can be seen similar to the case of the relaxed usp decorated quasicrystal. From 30 to 100 defects cause disordered domains due to the fact that the density of defects is not homogenous enough. The sample is covered by quasicrystalline and amorphous regions. When the number of defects is higher than 150 the whole sample appears disordered and the quasicrystalline symmetry has vanished.

The activation energy per defect (Fig. 7) is 0.085 in units of the potential minimum ($E_{min} = 0.5$ for the cut off potential) and independent of the number of defects. After relaxation there exist three ranges for the activation energy: up to 30 defects the activation energy decays until it reaches a plateau between 30 and 100 at $9.4 \cdot 10^{-4}$. If the number of defects is larger than 110, the activation energy also decreases until it reaches $4.5 \cdot 10^{-4}$ at 400 defects.

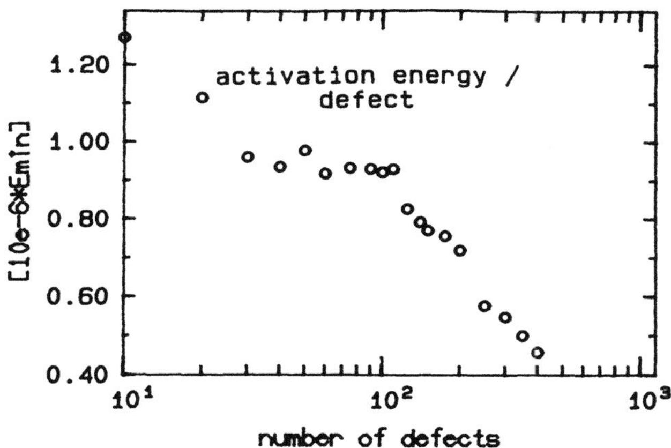

Figure 7. Activation energy after relaxation

Closely related to the activation energy is the bond energy per atcm (Fig. 8) which becomes worse when defects are filled in. Before relaxation it reaches zero at about 30 defects and is negative for a higher number. But after relaxation the bond energy per atom is positive in the whole range up to 400 defects. Displaying the bond energy versus the number of defects linearly we find that it decays rapidly and seems to tend to the value of an amorphous state (about 2.55 for the primitive decoration).

The bond orientation order parameter Q_6 for the initial state is reduced linearly with the number of Frenkel pairs. After relaxation it is constant up to 150 defects and decreases rapidly for a higher number. This means that the bond order is *growing* between 0 and 150 Frenkel pairs during relaxation because it reaches the value of the perfect quasicrystalline state. When there are more than 150 defects the bond order parameter shows that the orientational order is lost.

To check this somewhat surprising result we plotted the bond correlation: the atomic distances were projected onto a plane which was chosen perpendicular to a fivefold axis. The plots clearly show that the fivefold (and icosahedral) symmetry is preserved up to 150 defects. Contrary to that fact the peaks in the RDF are smeared out and it looks like an RDF obtained for truely an amorphous systems. A similar behaviour is found for the SF, which also presents an average over all angles.

We conclude that the RDF and SF do not tell us much about the structure of a sample, and therefore do not admit a characterization of the relaxed state. Translational ordering can exist even if RDF does not contain sharp peaks because of the averaging over all angles.

The peaks in the bond correlation plot are distinctly resolved for a perfect quasilattice and clearly show a fivefold symmetry. Introducing defects causes the peaks to be broadened radially and smeared out after relaxation. Until the number of defects is very large the major peaks remain distinct and do not merge, but are clearly visible over a homogenous background. The nearest neighbour peaks can be found up to 300 defects, but then they begin to form a closed ring as would be expected for a truely amorphous material.

5.5 Discussion of the Results

There is a mechanism which helps to stabilize the quasicrystal with few defects: the pyramidal holes in the dodecahedron are connected via their basis to the empty prismatic cells, and the interstitial sites in a pyramidal cell are lying on its basis. Therefore an atom will drift from the pyramidal to the prismatic hole which is 5% larger than the original site, and the bond energy will increase.

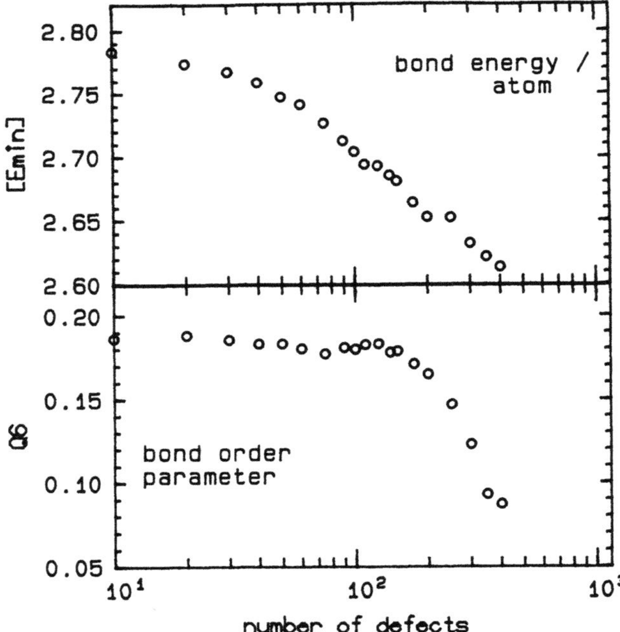

Figure 8. Bond energy and bond ordering after relaxation

We found that a small number of Frenkel defects formed in a quasicrystal by irradiation can be compensated although the bond energy decreases. This is the range where reversible phase transitions are possible. To find these transitions it would probably be necessary to study the quasicrystal at nonzero temperature. In the range between 1.5% up to 5% defects per lattice site we have an inhomogenous sample: There are disordered and quasicrystalline domains, but the disordered regions already have an icosahedral orientational order. The disordered domains grow with the number of defects. More than 6% defects destroy the quasicrystalline structure completely. But there is still some translational and orientational ordering found in the plot of the bond correlation. It would be very interesting to determine the correlation lengths of this structure, but we think that our samples are too small for such calculations. The quasicrystal becomes completely amorphous when the number of defects is very large (20%). No order is left, and the characteristica of an amorphous phase are clearly visible.

We considered only monatomic quasicrystals whereas all real quasicrystals found up to now are at least diatomic. Our results can therefore only give us preliminary hints to the radiation defects of quasicrystals, and it is not clear whether the behaviour of Frenkel defects is the same for real diatomic quasicrystals as for our monatomic prototypes.

6 SUMMARY AND DISCUSSION

We have simulated the relaxation of quasicrystalline structures using the method of steepest descent. In three dimensions we find that a monatomic quasicrystal with all vertices occupied is unstable because certain pairs of atoms with very short separations disturb the unit sphere packing. If one partner of these pairs is removed a stable decoration is obtained, not only at 0 K, but also at higher temperatures. Thus within the limitations of our model *monatomic* quasicrystals can really exist. Similar results

170

were found for diatomic quasicrystals [17]. We think that our model provides a basis for
further investigations of stability using computer simulation methods.

During our calculations we found more and more that the crucial criterium of stability of a quasicrystalline decoration is the density of the packing of one or two types
of spheres. This fact can be understood by recalling the properties of the potential we
have used: the Lennard-Jones potential is isotropic. Therefore it does not prefer any
bond orientational symmetry. It favours dense packings of spheres, because the denser a
packing is the more bonds do exist per atom exist and the larger does the binding energy
per atom grow.

Although there are no monatomic quasicrystals found up to now, our calculations
for monatomic (and diatomic) quasicrystals yield helpful hints how to model realistic
decorations or atomic hypersurfaces. They also represent a basis for dynamical calculations.

Concerning the Frenkel defects we do not find a transition from quasicrystalline
state to an amorphous state, but we have seen that a small number of defects does
not alter the stability of the quasicrystal. When the number of defects is large, the
icosahedral structure disappears, as would be expected from ordinary materials, and
the quasicrystalline order cannot be restored. The question, if the relaxed state with a
moderate number of defects showing orientational ordering together with translational
disorder is an "icosahedratic" phase remains open, because our samples are too small to
calculate correlation lengths with reliable statistics.

The author wants to thank Prof. H.-R. Trebin and R. Schilling for there participation in this project and for many useful discussions.

References

[1] M. V. Jarić, Phys. Rev. Lett. **55**, 607 (1985)

[2] P. A. Kalugin, A. Y. Kitaev, L. S. Levitov, J. Phys. Lett. **46**, L601 (1985) JETP
Lett. **41**, 145 (1985)

[3] N. D. Mermin, S. M. Troian, Phys. Rev. Lett. **54**, 1524 (1985)

[4] P. Bak, Phys. Rev. Lett. **54**, 1517 (1985)

[5] O. Biham, D. Mukamel and S. Striktman, Phys. Rev. Lett. **56**, 2191 (1986)

[6] F. Lançon, L. Billard, P.Chaudhari, Europhys. Lett. **2**, 625 (1986)

[7] F. Lançon and L. Billard, J. Phys. France **49**, 249 (1988)

[8] Y. Sasajima, T. Miura, M. Ichimura, M. Imabayashi and R. Yamamoto, J. Phys.
F17, L53 (1987)

[9] M. Widom, K. J. Strandburg, R. H. Swendson, Phys. Rev. Lett. **58**, 706 (1987)

[10] P. W. Leung, C. L. Henley, G. V. Chester, Phys. Rev. **B39** 446 (1989)

[11] T. Janssen, in **Quasicrystalline Materials**, Proc. of the I.L.L./CODEST Workshop
in Grenoble, ed. Ch. Janot, J. M. Dubois, World Scientific, Singapore, p. 327
(1988)

[12] K. Y. Szeto and J. Villain, Phys. Rev. **B36**, 4715 (1987); notice also the comment
by D. P. Shoemaker and C. B. Shoemaker in Phys. Rev. **B38**, 6319 (1988)

[13] S. E. Burkov and L. S. Levitov, Europhys. Lett. **6**, 233 (1988)

[14] S. Narasimhan and M. V. Jarić, Phys. Rev. Lett. **62** 454 (1989)

[15] D. Levine and P. J. Steinhardt, Phys. Rev. **B34**, 596 (1986)

[16] J. Roth, J. Bohsung, R. Schilling and H.-R. Trebin, in **Quasicrystalline Materials**, Proc. of the I.L.L./CODEST Workshop in Grenoble, ed. Ch. Janot, J. M. Dubois, World Scientific, Singapore, p. 65 (1988)

[17] J. Roth, R. Schilling, H.-R. Trebin, Phys. Rev. **B**, Febr. 1990

[18] F. Lançon and L. Billard, preprint, to be published in J. Non-Cryst. Sol.

[19] C. L. Henley, Phys. Rev. **B34**, 797 (1986)

[20] K. Urban, J. Mayer, M. Rapp, M. Wilkens, A. Csanady and J. Fidler, J. de Phys. **47**, C3-465 (1986)

[21] H. Kronmüller, Cryst. Latt. Def. & Amorph. Mat. **14** 137 (1987)

[22] H. Kronmüller, A. Hoffmann and M. Bauerin in **Quasicrystalline Materials**, Proc. of the I.L.L./CODEST Workshop in Grenoble, ed. Ch. Janot, J. M. Dubois, World Scientific, Singapore, p. 29 (1988)

[23] P. J. Steinhardt, D. R. Nelson, M. Ronchetti, Phys. Rev. **B28**, 784 (1983)

[24] Z. Olami and S. Alexander, Phys. Rev. **B37**, 3973 (1988)

RECONSTRUCTIVE PHASE TRANSITION TO THE ICOSAHEDRAL PHASE

P. Tolédano and C. Katarci

Laboratoire des Transitions de Phases
Université d'Amiens
80000 Amiens, France

V.P. Dmitriev, Yu.M.Gufan, S.B. Rochal

Institute of Physics
Rostov State University
344006 Rostov on Don, USSR

INTRODUCTION

The various models which have been proposed to explain the stability of the icosahedral phases[1-6] start from the isotropic symmetry of the liquid phase, assumed to be the parent phase. This assumption stems from the fact that initially, the icosahedral structure could only be obtained in certain cooling conditions, below the range of stability of the liquid phase. Further experiments revealed however, that one could get the icosahedral phase under as equilibrium conditions[7], and that it could coexist[8] with, and recrystallize into[9], other crystalline structures.

Following these considerations, the model presented in this paper, starts from a solid crystalline phase as the parent structure for the icosahedral phase. More precisely, the icosahedral phase is shown to result from a reconstructive transition of the displacive type[10] from a cubic phase. Two distinct order-parameters are associated with the transition mechanism : a displacement field which transforms some of the cubic sublattices into icosahedral clusters, and an incommensurate propagation of the icosahedral symmetry,which breaks the initial cubic lattice. The essential features of the theory are introduced through the example of AlMnSi.

FORMATION OF ICOSAHEDRAL CLUSTERS

Let us start from the structure of the α-AlMnSi phase as described by Cooper[11]. The 138 atoms contained in one unit-cell decompose into 11 sublattices, which according to the labelling of ref. 11 can be shared into three groups : Al (1), Al (2) and Al (7) are characterized by a 6-fold position ; Al(3), Al(4), Al(8), Al(9),Mn(1)

Mn(1) and Mn(2) possess a 12-fold degeneracy ; Al(5) and Al(6) display a 24-fold degeneracy. We first focus on the atoms of the Al(7) sublattice. From Fig. 1 one can see that antiparallel displacements of each pair of atoms, denoted (1,4), (2,5), and (3,6) in Fig. 1, along the respective cubic directions $[100]$, $[010]$, and $[001]$, bring the twelve atoms forming the Al(7) sublattice, to form an icosahedron, for the specific displacements of magnitude 0.244 Å = 0.0194 a where a = 12.56 Å is the lattice parameter of the cubic unit-cell. The same displacement field applied to the atoms of the sublattices Mn(1), Mn(2), Al(3), and Al(4) also give icosahedral clusters. Thus, Al(3), Mn(1) and Al(7) form three embedded icosahedra of increasing volume, centered at the origin of the unit-cell, whereas Al(4) and Mn(2) form two embedded icosahedra centered at ($\frac{1}{2}$, $\frac{1}{2}$,$\frac{1}{2}$) (Fig. 2). Besides, one can show that under the same displacement field, the other sublattices form regular polyhedra which are not icosahedra.

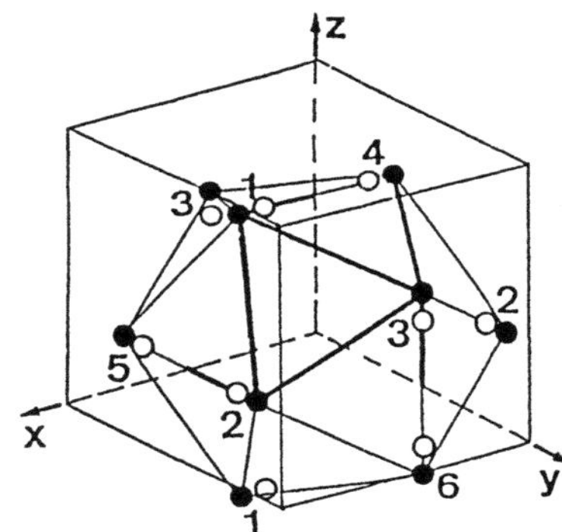

Fig. 1. Icosahedron formed by the atoms belonging to the Al(7) sublattice. White and black dots represent respectively the cubic and icosahedral positions.

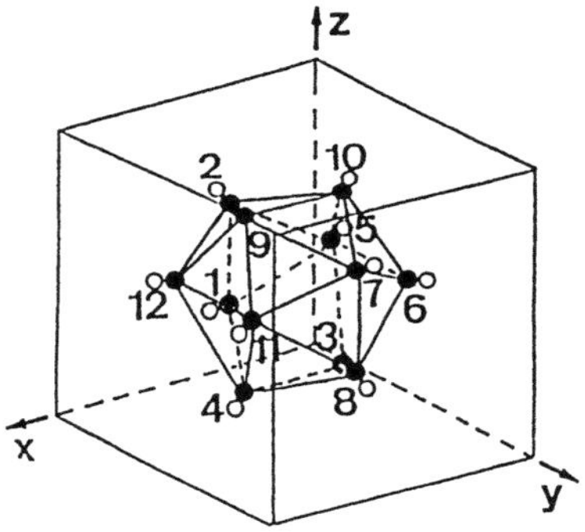

Fig.2 Icosahedron formed by the atoms belonging to the Mn(2) sublattice

Let us now show that the preceding cubic structure with icosahedral clusters is thermodynamically stable. The basis function spanning the antiparallel displacements of the Al(7) atoms, can be written, using the notation of Fig. 1 :

$$\psi = x_1 - x_4 + y_2 - y_5 + z_3 - z_6 \qquad (1)$$

It transforms as the identity representation of the Pm3 space group, associated with the α-structure, at the center (k = 0) of the primitive cubic Brillouin zone. One can easily verify that although the atomic displacement associated with the other sublattices are different (i.e. they are spanned by different basis functions), they transform as the same identity representation. Accordingly, the whole displacement field is associated with a one-component order-parameter η which transforms as the identity representation. The corresponding thermodynamic potential can thus be written[12] :

$$F(\eta) = F_o + \frac{a}{2}\eta^2 + \frac{b}{3}\eta^3 + \frac{c}{4}\eta^4 \qquad (2)$$

Because of the magnitude of the atomic displacements which lead to
the icosahedral clusters, it should be unphysical to put forward a
linear relationship beyween the order-parameter and the displacements
as it is usual in the standard Landau theory of structural transitions[13]
Following the generalisation of the Landau theory which has been
recently proposed for reconstructive transitions of the displacive
type[10], we will assume a transcendental dependence of in functions
of the displacements[10]. One finds here :

$$n = n_o \left\{ \sin \left[\frac{2\pi}{a} (x_i - x_j) - \frac{\pi}{2} \right] + \cos \frac{2\pi}{\phi} \right\} \qquad (3)$$

where $\Delta x = x_i - x_j$ represents the distance between two antiparallel
atoms pertaining to the same sublattice. ϕ is the golden mean. The
form of the function (3) which holds for the whole set of displacements, is represented in Fig. 3. Thus for the periodic displacements
of magnitudes

$$\Delta x_1 = a\phi^{-1} + an , \qquad \text{and } x_2 = a(\phi - 1) \phi^{-1} + an \qquad (4)$$

Where n is an integer, the Al(7) atoms form two sets of icosahedral
clusters respectively, which can be deduced one from another by a
shifting of $(\frac{a}{2}, \frac{a}{2}, \frac{a}{2})$ and a rotation of 90° around the $[001]$ cubic
axis. For the specific displacements corresponding to such icosahedral
domains, the point-group symmetry of the Al(7) sublattice increases
from m3 to m$\overline{3}$5.

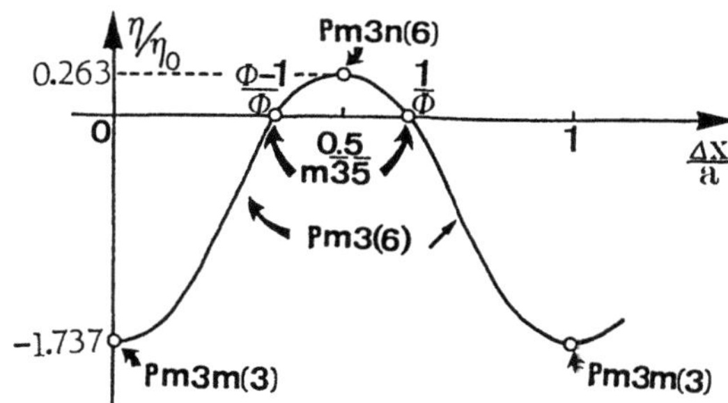

Fig. 3 . Periodic dependence of the order-parameter for the
Pm3 (cubic clusters) to Pm3 (icosahedral clusters)
transition.

The thermodynamic stability of the icosahedral clusters can be
verified by the introduction of (3) in the potential (2), and by
minimization of F with respect to the displacements[10]. One gets the
equation of state :

$$n(a + bn + cn^2) \frac{\partial n}{\partial \Delta x} = 0 \qquad (5)$$

The stable solution $\eta = 0$ is obtained for $\Delta x_1 = a \phi^{-1} + an$ or
$\Delta x_2 = a(\phi - 1)\phi^{-1} + an$, i.e. for the icosahedral clusters.

The preceding displacive field mechanism, preserves the periodicity of the initial cubic structure. In order to loose the cubic translational order, one should find, in the primitive cubic Brillouin-zone, a non-zero k-vector which propagates the icosahedral symmetry of the clusters, producing an infinite quasi-crystalline diffraction pattern in the three-dimensional reciprocal space. The suitable k-vectors should fulfil the sufficient condition to possess a twelve-arms star k* pointing towards the vertices of a regular icosahedron. A systematic investigation of the primitive cubic Brillouin-zone, using Kovalev tables[14], leads to the conclusion that only two incommensurate k-vectors fulfil the preceding condition :

$$\vec{k}_1 = (2\mu_1 \frac{\pi}{a}, 0, 2\mu_2 \frac{\pi}{a}) \text{ with } \frac{\mu_1}{\mu_2} = \phi \text{ , and } \vec{k}_2 = (\phi \frac{\pi}{a}, 0, \frac{\pi}{a}) \qquad (6)$$

$\vec{k}_1$ and $\vec{k}_2$ are located along the same direction in the Brillouin-zone, $\vec{k}_1$ lying inside, and k_2 at the boundary of the zone.

The order-parameter associated with k_1^* and k_2^* possess respectively twelve and six components, denoted ζ_i. Observation of equal diffraction intensities in the icosahedral phase require the equilibrium condition : $\zeta_i = \zeta$ for all i. One can thus write the (twelve or six) order-parameter expansions under the effective form :

$$F'(\rho) = F'_o + \frac{a'}{2} \zeta^2 + \frac{c'}{4} \zeta^4 \qquad (7)$$

Accordingly the whole transition mechanism from the cubic to the quasicrystal phase, i.e. the formation of icosahedral clusters plus their incommensurate propagation can be accounted by the two-order parameter potential

$$F(\eta,\rho) = F(\eta) + F(\zeta) + d\eta\zeta^2 \qquad (8)$$

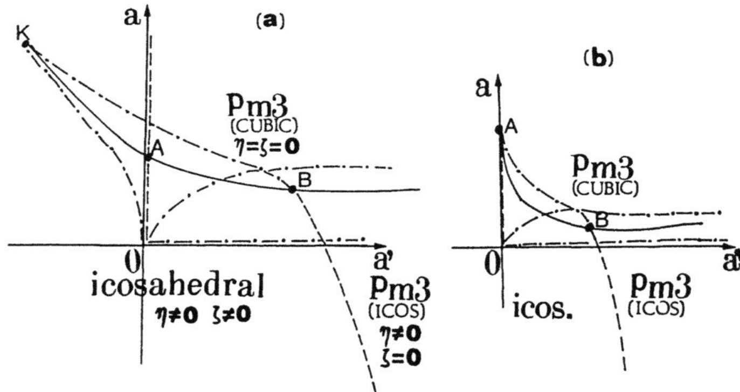

Fig. 4 . Phase diagram representing the potential (8), respectively for bd < 0 (Fig. A) and bd > 0 (Fig. B). Solid, dashed and mixed lines are first-order, second-ORDER transition and limit of stability lines.

which corresponds to the theoretical phase diagrams represented
in Fig. 4. Among the interesting features of these diagrams, let
us note the second-order transition line which separate the icosahedral
phase from a cubic phase with icosahedral clusters.

CONCLUSION

In summary, we have presented in this paper a theoretical
model which explains the formation of icosahedral structures by
a coupling of two independent mechanisms : 1) a displacement field
which transforms groups of atoms into icosahedral clusters whithin
a cubic unit-cell,   2) a propagation of the local icosahedral
symmetry via an incommensurate wave-vector. The fact that the two
mechanisms are phenomenologically independent suggests that their
experimental realization in Nature is not necessarily correlated.

REFERENCES

1. P. Bak, Phys. Rev. Lett. $\underline{54}$, 1517 (1985)
2. N.D. Mermin and S.M. Troian, Phys. Rev. Lett. $\underline{54}$, 1524 (1985)
3. P.A. Kalugin, A.Yu.Kitaev and L.S. Levitov, JETP Letters $\underline{41}$, 145 (1985)
4. M.V. Jaric, Phys. Rev. Lett. $\underline{55}$, 607 (1985)
5. V. Dvorak and V. Holakovsky, J. Physics, $\underline{C19}$, 5289 (1986)
6. D.R. Nelson and S.Sachdev, Phys. Rev. $\underline{B32}$, 689 (1985)
7. B. Dubost, J.M. Land, M. Tanaka, P. Sainfort and M. Audier, Nature, $\underline{324}$, 48 (1986)
8. A. Loiseau and G. Lapasset, J. de Physique (Paris) $\underline{47}$, C3-331 (1986)
9. M.K. Sanyal, V.C. Sahni and G.K. Dey, Nature $\underline{328}$, 704 (1987)
10. V.P. Dmitriev, S.B. Rochal, Yu.M.Gufan and P.Tolédano, Phys. Rev. Lett. $\underline{60}$, 1958 (1988)
11. M.Cooper, Acta Cryst. $\underline{23}$, 1106 (1967)
12. The order-parameter expansion associated with the isostructural Pm3-Pm3 transition induced by the identity representation, admits invariants of all degrees, however, the linear invariant can be cancelled out providing a suitable change in the variables.
13. L.D. Landau, Phys. Z. Soviet $\underline{11}$, 26, 545 (1937)
14. O.V. Kovalev, Irreducible representations of the Space-Groups (Gordon and Breach, New York, 1965)

STRUCTURE AND GROWTH OF TWO- AND THREE-DIMENSIONAL HEXATIC LIQUID CRYSTALS

J. D. Litster

Department of Physics
Massachusetts Institute of Technology
Cambridge, MA 02139, USA

Introduction

We usually consider two extremes when thinking about condensed states of matter. One extreme is the perfect crystalline solid, in which atoms form a perfectly periodic array that extends to infinity in all three directions. The other extreme is a fluid, in which the atoms or molecules are completely disordered and the material is both orientationally and positionally isotropic – that is, it looks the same when viewed from any direction.

I am going to talk today about an intermediate state which some physicists have known for many decades was possible, but which would have seemed quite strange to most solid state physicists until recently. This is a fluid state of matter in which the molecules are distributed at random, as in a liquid or a gas, but which is orientationally anisotropic on a macroscopic scale, as in a crystal. Some of the properties of this fluid are different in different directions. One way to describe the order in this fluid is to say that if you sit on any molecule in the fluid, you know what directions to go in order to reach your nearest neighbors, *and* these directions are the same for all molecules in the system. This type of order is commonly called "bond-orientational" order, which I shall abbreviate as BO. While Landau and Peierls knew in the 1930's that such a state of matter was theoretically possible, the concept was unknown to most solid state physicists and it is only within the last decade that it has been observed and studied. The best physical example is to be found among the materials known as smectic liquid crystals, and the concept of BO order appears to be the missing concept which was required to be able to classify the structure of the known liquid crystal phases.

Since this workshop involves solid state physicists with a variety of interests and backgrounds, I shall make my discussion a tutorial one. Some of the more esoteric details may be found in the references. As I mentioned, Landau and Peierls knew about the possibility of a state with BO order, but no positional order, in the 1930's; in fact it is interesting to quote the relevant paragraph from Landau's 1937 paper:[1]

Geometry and Thermodynamics
Edited by J.-C. Tolédano
Plenum Press, New York, 1990

If the body is isotropic, then ρ = const; however, from ρ = const it does not follow that the body should necessarily be isotropic. If ρ = const, then this means that all positions of an atom, more precisely its center of mass, in the body are equally probable. Nevertheless in this case different orientations in the body can be non-equivalent. Namely, when the position of any particular atom No. 1 is given, then the probability of different positions of a neighboring atom No. 2 is a function of their relative positions (*i.e.* of the vector $\mathbf{r}_{12}$ connecting atom No. 1 and No. 2). This probability ρ_{12} can depend on the direction of $\mathbf{r}_{12}$...

These ideas reappeared about 40 years later in the context of new theories of melting in two dimensions. I should like to begin by reviewing the important concepts in these theories. The idea of BO order was introduced by Halperin, Nelson,[2] and Young[3] in their extension of the Kosterlitz-Thouless model[4] for dislocation-mediated melting in two dimensions.

Figure 1A illustrates a dislocation in a two-dimensional triangular lattice. Such dislocations play an important role in the the mechanism proposed by Kosterlitz and Thouless for the melting of a two-dimensional solid. If we follow a closed loop in a perfect crystal lattice and count the number of lattice parameters we have moved in each direction, the net result will be zero. If we follow a closed loop which contains the dislocation shown above, the net number of lattice parameters traversed will be one because an extra row of atoms has been inserted to the right of the dislocation core (the atoms shown as open circles). This net displacement is called the Burgers vector. If the extra row of atoms were to the left of the core, the dislocation would have a Burgers vector of the opposite sign. In Figure 1B two dislocations of equal and opposite Burgers vectors are drawn close together. It is apparent the the strain in the lattice away from the cores has been greatly reduced by this arrangement. Thus, the elastic strain energy of the crystal causes an attractive interaction between two dislocations of opposite Burgers vectors. The quantitative theory of this interaction is identical to the electrostatic interaction between charged particles. The lowest energy thermally excited defects in the two-dimensional triangular lattice consist of bound pairs of dislocations, as in Figure 1B. While they cause a local disturbance, they do not interfere with the longer range translational order in the lattice.

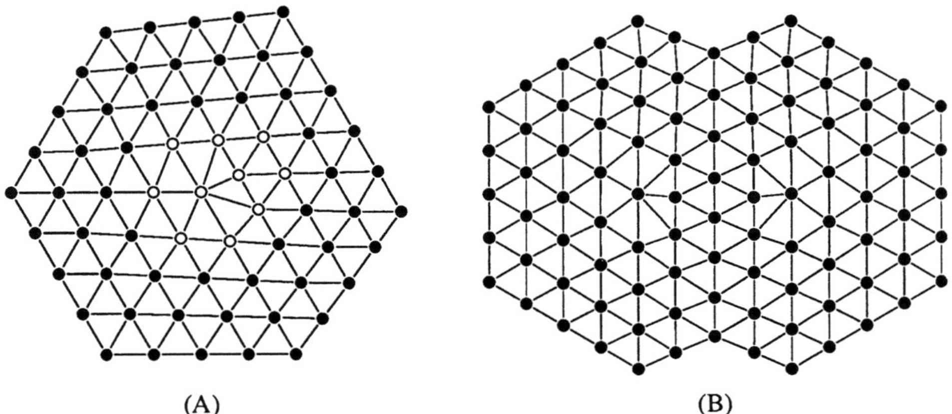

(A)   (B)

Figure 1. At left (A), an isolated dislocation in a triangular lattice; at right (B), a bound pair of dislocations in a triangular lattice.

It is necessary for us to discuss the translational order, and to do so it is convenient to define a correlation function

$$\mathcal{G}(\mathbf{r}) = \langle e^{i\mathbf{q}\cdot[\mathbf{u}(0)-\mathbf{u}(\mathbf{r})]} \rangle \tag{1}$$

in which $\mathbf{q}$ is the X-ray scattering momentum transfer relative to a Bragg peak position (reciprocal lattice vector), and $\mathbf{u}(\mathbf{r})$ is the displacement of the atom or molecule at $\mathbf{r}$ from its equilibrium position. The displacements $\mathbf{u}$ are due to thermally excited lattice vibrations (phonons). The Fourier transform of $\mathcal{G}$ gives the X-ray scattering intensity near the Bragg peak. In a three-dimensional crystal $\mathcal{G}(0) = 1$, and it falls to a constant finite value, the Debye-Waller factor e^{-2W}, as r becomes very large. The finite value of $\mathcal{G}(\infty)$ indicates there is true long-range order, and the Fourier transform of the constant gives an infinitely sharp Bragg peak in the scattering. The mathematical function which describes how $\mathcal{G}(\mathbf{r})$ decreases from unity to the Debye-Waller factor can be obtained by calculating the thermal average of Eq. (1) over all the phonon modes. When the calculation is done in the harmonic approximation, the Fourier transform is proportional to $|\mathbf{q}|^{-1}$; this is the classical thermal diffuse X-ray scattering from lattice vibrations.

The correlation function $\mathcal{G}(\mathbf{r})$ in a liquid decays exponentially to zero, $viz.$ $\sim e^{-r/\xi}$, where ξ is the pair correlation length between molecules in the liquid. The two-dimensional solid phase is peculiar, because it does not have the long-range translational periodic order of a three-dimensional crystal. The thermally excited phonons destroy the long-range positional order, and a simple calculation[5] shows that $\mathcal{G}(\mathbf{r})$ decays algebraically with distance

$$\mathcal{G}(\mathbf{r}) \sim r^{-\eta}. \tag{2}$$

Because $\mathcal{G}(\mathbf{r})$ does not have a finite constant value at large distance, the X-ray scattering from a two-dimensional lattice does not have Bragg peaks. However, a two-dimensional crystal would still have most of the other properties, for example a shear modulus, that we associate with solids. The power-law decay of the correlation function has other interesting consequences, which I shall discuss presently.

Now we can return to the Kosterlitz-Thouless model for melting in two dimensions. In two dimensions, the elastic contribution to the free energy of a pair of dislocations with opposite Burgers vectors is proportional to the logarithm of their separation, analogous to the electrostatic potential of two line charges. This contribution gives the energy resulting from the attraction between the dislocations. There is, however, an increase in the entropy of the system as the dislocations are separated; it too is proportional to the log of their separation. At high enough temperature one expects the entropy increase to overcome the elastic attraction and the two dislocations of Figure 1B will unbind. That is the mechanism for melting in the Kosterlitz-Thouless model. In 1978 Halperin, Nelson, and Young pointed out that the melting process could occur in two stages in two dimensions. From their calculations they found that unbinding of the pairs of dislocations did not necessarily destroy the BO order. That might be expected from the free dislocation shown in Figure 1A; the extra row of atoms upsets the translational periodicity, but the orientation of the local triangles outside of the dislocation core remains unchanged. What the calculations actually showed was that the BO order decayed algebraically with distance. To state this more quantitatively, we may define the BO order parameter as

$$\Psi_6(\mathbf{r}) = \langle e^{-i6\theta(\mathbf{r})} \rangle \tag{3}$$

where $\theta(\mathbf{r})$ is the angle of the imaginary "bonds" between molecules or atoms with, say, the x axis. Analogous to Equation (2), we may also define a correlation function for the

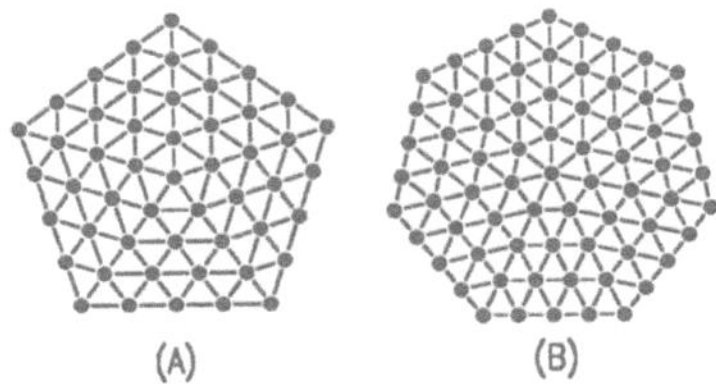

Figure 2. Five-fold (A), and seven-fold (B)
disclinations in a triangular lattice.

BO order

$$O_6(\mathbf{r}) = \langle e^{i6[\theta(0)-\theta(\mathbf{r})]} \rangle \tag{4}$$

which decays algebraically with distance

$$O_6(\mathbf{r}) \sim r^{-\eta_6}. \tag{5}$$

This intermediate phase with short-range positional order and power-law decay of the six-fold BO correlations was called "hexatic" by Halperin and Nelson. It has the continuous translational symmetry of a liquid, but the continuous rotational symmetry is still broken. The hexatic phase can have defects in the BO ordering which are analogous to the dislocations in the positional order. They are called *disclinations*, and two of them are shown in Figure 2. When a five-fold and seven-fold disclination bind closely together, the long range distortion in the BO field is reduced; hence they will be attracted in a way analogous to disclinations of opposite Burgers vector. In fact, if you look closely at the core of the isolated dislocation in Figure 1A, you can see that it is made of a pair of five-fold and seven-fold disclinations bound together. Halperin, Nelson, and Young showed that the free energy arguments of Kosterlitz and Thouless also applied to the bound disclinations. As the temperature was increased, the free dislocations of the hexatic phase could come apart to produce free disclinations. When this happens, the hexatic "melts" into a two-dimensional liquid with continuous rotational and translational symmetry.

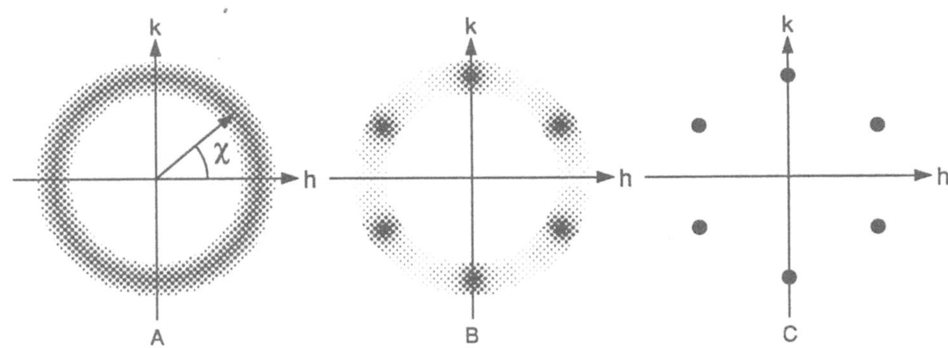

Figure 3. In-plane X-ray scattering for the liquid phase (A), hexatic phase (B), and solid phase (C) of a two-dimensional triangular lattice.

Application to Liquid Crystals

In 1978, Bob Birgeneau and I were trying to understand what the structure might be for the alphabet soup of the known smectic phases of liquid crystals. When we saw the Halperin-Nelson paper, we asked ourselves what would be the consequences if these ideas were applied to three-dimensional liquid crystals[6]. The smectic phases are layered, and so we thought of them as layers of two-dimensional phases stacked on one another. We knew from investigations of critical phenomena that power-law decay of correlation functions was associated with a corresponding susceptibility being infinite. Thus, if two-dimensional "solids" were stacked up, infinitesimal interactions would lead to long-range positional order and a three-dimensional crystal. If two-dimensional liquids were stacked, one would get the conventional smectic A phase with a one-dimensional density modulation and liquid-like order in the plane of the layers. Stacking up two-dimensional hexatics could result in a three-dimensional liquid that had a density modulation in one direction and long-range three-dimensional bond orientational order. The question was, then, to see if such a phase could be found in nature. If it existed, it could be detected by X-ray scattering. In Figure 3A, I show the expected in-plane X-ray scattering from a two-dimensional liquid. The scattering is a diffuse ring independent of the azimuthal angle χ; a radial scan through the ring would give the usual broad liquid structure factor peak. Figure 3C shows the six first order Bragg spots expected from the solid phase with a triangular lattice. Strictly speaking, because of the power-law decay of the positional correlations, the Bragg peaks are cusps rather than the resolution-limited peaks one would see for a three-dimensional crystal. It requires, however, a very careful experiment[7] to distinguish the difference.

In the hexatic phase, the diffuse scattering of the liquid phase acquires a six-fold modulation in the angle χ, as illustrated in Figure 3B. The scattering remains diffuse, and the sharpness of the scattering is determined by the positional correlation lengths. This modulated diffuse ring of scattering is the signature of the hexatic phase and the existence of long-range bond orientational order. It was first observed[8] by Pindak *et al.* in freely suspended thin films of materials with a hexatic smectic B (B_H) phase. While freely suspended thin smectic films provide well oriented smectic layers, the scattering is weak, often requiring storage ring sources. The chief experimental difficulty is to obtain single domain samples of the hexatic phase. This is essential; a powder average of Figure 3B would be indistinguishable from the liquid scattering, and the presence of several domains would make quantitative studies of the hexatic order impossible.

Thermotropic liquid crystal molecules are anisotropic, being approximately rod-like in shape. For example, most of the results presented in this article were obtained on studies of a racemic mixture of the compound 4-(2-methylbutyl) phenyl 4'-(octyloxy)-(1,1')-biphenyl-4-carboxylate (8OSI) whose structure is shown below.

$$C_8H_{17}O-\langle\!\!\langle\bigcirc\rangle\!\!\rangle-\langle\!\!\langle\bigcirc\rangle\!\!\rangle-\overset{O}{\underset{\|}{C}}-O-\langle\!\!\langle\bigcirc\rangle\!\!\rangle-CH_2-CH\overset{CH_3}{\underset{C_2H_5}{<}}$$

In the smectic A and B phases the molecules are oriented, on average, normal to the smectic layers. The smectic A phase, as mentioned above, has a one-dimensional density wave oriented along the average orientation direction (called the *director*) of the molecules. Most smectic B phases are actually three-dimensional crystals, and a few, the B_H phases, have hexatic order. However, there are also smectic phases in which the molecules are tilted with respect to the plane of the smectic layers. The tilted analogue of the smectic A is called the smectic C, and there are two common tilted analogues of the

B_H phase with the director tilted either towards the nearest neighbors, smectic I, or between them, smectic F. The molecules have a sufficiently anisotropic diamagnetic susceptibility that the director is readily oriented by a magnetic field of a few kOe. The direction of the molecular tilt in the plane of a freely suspended film of these materials can thus be determined by small permanent magnets, which, through coupling between the director and the hexatic order, produces a single domain bond orientationally ordered phase for study. This coupling means that the hexatic phase transition occurs in the presence of an ordering field, and the coupling will induce some hexatic order in a smectic C phase. Strictly speaking, then, the smectic C is not a thermodynamically distinct phase. In practice the coupling is weak, and enables one to measure a number of interesting things about the establishment of BO. It does mean, however, that quantitative interpretation of certain measurements, such as the specific heat, require knowledge of the equation of state. The tilt also breaks the six-fold rotational symmetry in the plane of the smectic layers, but that does not alter fundamentally the interpretation of the measurement.

We have carried out X-ray scattering measurements[9] for the compound 8OSI which has the following sequence of phases (temperatures are given in °C):

$$Isotropic \xrightarrow{174.5°} N \xrightarrow{170.0°} S_A \xrightarrow{133.4°} S_C \xrightarrow{79.9°} S_I \xrightarrow{75.1°} S_J \xrightarrow{61.9°} S_K.$$

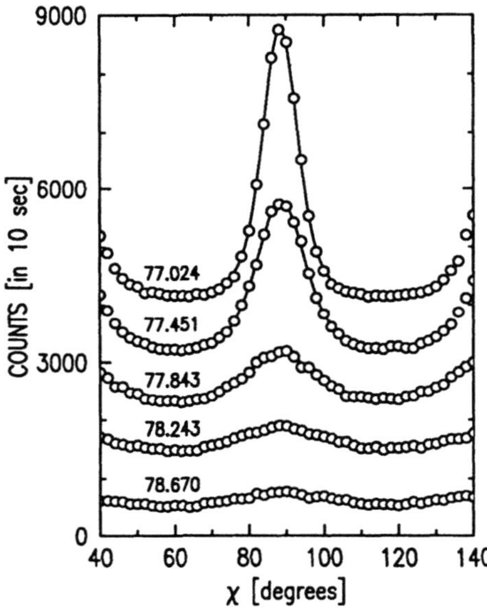

Figure 4. X-ray scattering intensity as a function of the azimuthal angle χ for several temperatures in the S_I phase of a thick film of 8OSI. The numbers above each scan show the temperature in °C. The curves for 78.243, 77.843, 77.451, and 77.024 °C have been displaced upwards by 1000, 2000, 3000, and 4000 counts, respectively. The solid lines show the results of fitting to Eq. (6) with the coefficients C_{6n} determined by Eq. (7).

It is the S_I phase which has hexatic order. There is one experimental procedure to observe with tilted phases; because of the molecular form factor, the scattering is strongest, not in the plane of the layers, but in a plane tilted with the molecules. Figure 4 shows the X-ray scattering from several temperatures in the S_I phase of 8OSI, measured in the plane of maximum intensity. The sort of scattering pattern shown in Figure 3B can be seen to emerge as the temperature is lowered. The magnetic field direction is at $\chi = 0$, so the peaks come at the angles expected for the S_I, not the S_F, phase. Radial scans through the diffuse peaks, not shown here, show that the positional correlations are always short-ranged with correlation lengths ranging from 2.5 nm to about 20 nm.

It is possible to analyze the data of Figure 4 to obtain quantitative information about the BO order. One may perform a non-linear least-squares fit of the χ scans between 60° and 120° to the Fourier cosine series

$$S(\chi) = I_0 \left[\frac{1}{2} + \sum_{n=1}^{\infty} C_{6n} \cos 6n (90° - \chi) \right] + I_{BG} \tag{6}$$

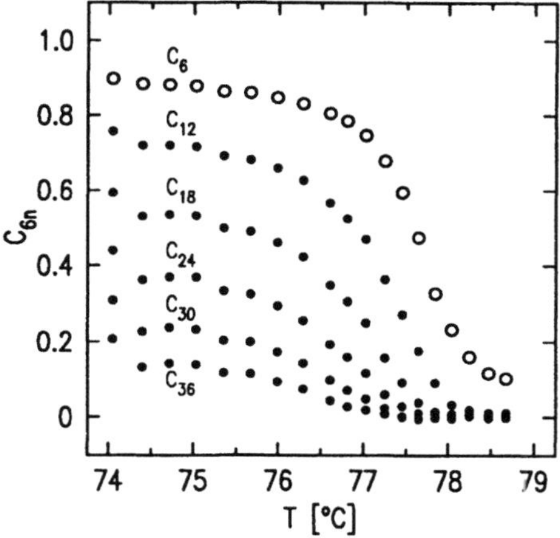

Figure 5. The temperature dependence of the Fourier coefficients obtained by fitting χ scans to Eq. (6) in the smectic I phase of 8OSI.

where χ is the angle between the in-plane component of the scattering vector $\mathbf{q}$ and the magnetic field. The coefficients C_{6n} measure the amount of $6n$-fold ordering in the sample. This approximation neglects the residual effects of the two-fold symmetry arising from the molecular tilt; these effects are important when one studies the normal modes of the molecular director, but do not affect our principal conclusions from the X-ray scattering measurements. The results of these fits are shown in Figure 5. You can easily see there is not a sharp boundary between the S_I and S_C phases; this is a consequence of the BO ordering field caused by the tilting of the molecules. The data for C_6 are similar to the order parameter for a ferromagnet in an applied field.

In the language commonly used in phase transitions, C_6 is the primary order parameter and each of the higher harmonics is a secondary order parameter. Since $C_6 \sim e^{i6\theta}$, $C_{12} \sim e^{i12\theta}$, and so on, we might expect $C_{6n} \sim C_6^n$. That the result of a mean-field theory. Experiments showed that this simple scaling was not correct, hence the mean-field approximation is not valid; thermally induced fluctuations of the order parameter are important. In that case, we know from studies of critical phenomena that the behavior near the phase transition depends on, among other things, the symmetry of the order parameter. The fundamental order parameter Ψ_6 has two degrees of freedom (magnitude and phase) and so it has XY symmetry; C_6 may be written as the real part of $x + iy$. However, C_{12} scales like the real part of $(x + iy)^2$, or $(x^2 - y^2)$, C_{18} scales like $(x^3 - 3xy^2)$, C_{24} as $4(x^4 + y^4) - 3|\Psi_6|^4$, and so on. Thus each of the secondary order parameters has different symmetry in the xy plane, and one might expect different behavior near the transition temperature. It turns out that one may obtain a complete theoretical understanding of the scaling of the harmonics of the BO order parameter using theories of multicritical phenomena developed to explain phase diagrams in magnets. Using these theories, in a way that I shall explain presently, one obtains[10] the result

$$C_{6n} \sim C_6^{\sigma_n} \quad \text{where} \quad \sigma_n = n + 0.3n(n-1). \tag{7}$$

The agreement between the calculation and the measurements seen in Fig. 6 is surprisingly good, as the calculation was expected to hold only asymptotically close to the XY critical point, and for 8OSI works up to C_6 as large as 0.9.

The theory of multicritical crossover behavior in magnets is complicated and rather esoteric. Without going into much detail, I would like to say a bit about how the concepts arose and the general applicability of the results, for they *are* relevant to phase transitions in a number of different kinds of materials. In an XY magnet, the spins order parallel to one another in a plane; however there is isotropy in the plane, so no particular direction is taken unless some applied force breaks the orientational symmetry. Suppose we have an isotropic XY magnet which orders at a temperature T_C; then, thermodynamic properties of the magnet near the transition will scale as a power of the reduced temperature $t \equiv (T - T_C)/T_C$. In particular, the specific heat diverges as $|t|^{-\alpha}$. If a field be applied to break the orientational symmetry in the xy plane, the behavior will "cross

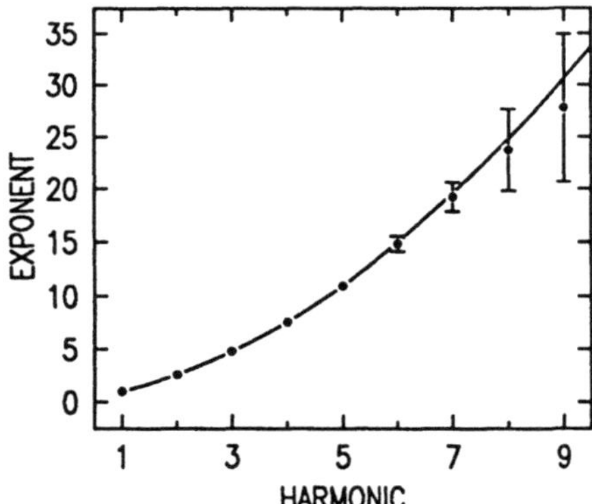

Figure 6. The harmonic scaling exponents σ_n obtained from fitting the scattering data for 8OSI to Eq. (6). The solid line is the calculation of Eq. (7).

over" to that of a transition with a different symmetry. The field could be written as

$$H_n = g_n \int d\mathbf{r}\, \mathrm{Re}\,(x+iy)^n \tag{8}$$

where g_n is the amplitude of the field. Consider the example of uniaxial symmetry; a field

$$H_2 = g_2 \int d\mathbf{r}\, (x^2-y^2)$$

will cause a higher ordering temperature T_M for either the x component, if $g_2 < 0$, or the y component, if $g_2 > 0$. Only one of the x or y components will order[11] and so the behavior crosses over to that of the Ising model. One will still observe XY behavior at the *bicritical point*, which occurs when $g_2 = 0$. Details of the crossover are determined by the *crossover exponent*, ϕ_2. In particular, the two Ising phase transition lines for positive and negative g_2 approach one another as $|T - T_M|^{\phi_2}$. Generally, for a field given by Eq. (8), the free energy asymptotically close to the XY critical point scales as:

$$F(t,g_n) \sim |t|^{2-\alpha} f\left[\frac{g_n}{|t|^{\phi_n}}\right],$$

where $f(0)$ is a constant. A standard thermodynamical calculation gives

$$C_{6n} = \left[\frac{\partial F}{\partial g_n}\right]_{g_n=0} \sim |t|^{2-\alpha-\phi_n} \sim C_6^{\sigma_n}. \tag{9}$$

From Eq. (9) we obtain

$$\sigma_n = \frac{2-\alpha-\phi_n}{2-\alpha-\phi_1} = \frac{2(d-\lambda_n)}{d-2+\eta} \tag{10}$$

where $\lambda_n = \phi_n/\nu$, d is the space dimension, and the correlation length diverges as $|t|^{-\nu}$. The first equality in Eq. (10) follows directly from Eq. (9). Since the XY order parameter vanishes as $|t|^\beta$, we identify $2-\alpha-\phi_1$ with β; using the standard scaling laws, $d\nu = 2-\alpha = \gamma + 2\beta$, and $\gamma = (2-\eta)\nu$, we arrive at the second equality of Eq. (10). The solution to the problem requires calculation of the crossover exponents ϕ_n. Numerical renormalization group calculations are needed, and the form of the result in the XY critical region is $\sigma_n = n + x_n n(n-1)/(d-2+\eta)$. The best result uses diagrammatic integrals obtained by Jug[12] for $d = 3$, rather than an ε expansion, and is given in Eq. (7). Details may be found in the reference of Aharony *et al.*[10]

Previously, determination of the critical behavior for systems of these differing symmetries had required different experiments, and most of them had not been measured[13]. It is possible to generate crossover to uniaxial anisotropy by applying a uniform magnetic field along the axis of a uniaxial antiferromagnet[14]. The field tends to orient the spins normal to itself, thus competing with the uniaxial anisotropy; at a field H_M a *spin flop* transition occurs. Then $g_2 \sim (H^2 - H_M^2)$, and the crossover exponent can be determined from the shape of the phase boundaries near the spin flop transition. In this way Rohrer and Gerber[14] measured $\phi_2 = 1.17 \pm 0.02$ in GdAlO$_3$. We find $\phi_2 = 1.14 \pm 0.04$ in 8OSI.

The third harmonic of the BO order parameter has three-fold rotational symmetry, and ϕ_3 will describe crossover to the behavior of the three-state Potts model[15]. (The Ising model is a two-state Potts model.) Experimentally, this crossover can be produced in a cubic three-component system. We must begin with a system which prefers to order along one of the three cube axes, and then introduce sufficient anisotropy to cause ordering along the cube body diagonal well above the cube axis ordering temperature. The transverse component of the order parameter will have three preferred directions in a plane normal to the body diagonal; its ordering on further cooling is described by a three-state Potts model. In three dimensions, this model has a first-order transition, and the discontinuity in the order parameter is proportional to $|g_3|^{\beta/\phi_3}$. This system was realized[16] by applying uniaxial stress to the structural phase transition in $SrTiO_3$ and a value $\phi_3 \approx 0.5 \pm 0.1$ was obtained. In 8OSI, we find $\phi_3 \approx 0.4 \pm 0.1$.

Rohrer and Gerber[14] were able to estimate $\phi_4 = 0 \pm 0.03$ by applying the field transverse to the axis of the uniaxial antiferromagnet. Symmetry-breaking fields for $n = 5$ are not physically realizable in periodic lattices. At one time I thought that quasicrystals were glasses with quenched icosahedral BO order, however recent samples which show resolution-limited X-ray scattering peaks seem to rule out that possibility. Thus the liquid crystal BO ordering seems to be the only system in which, via measurement of C_{30}, one may determine ϕ_5. One could, in principle, determine ϕ_6 from the transverse susceptibility of hexagonal layered antiferromagnets[17], such as $CrCl_3$ or $CoCl_2$, but I am not aware of any such measurements.

It seems that the behavior of BO order in the tilted smectic hexatic phase of 8OSI is well described by a three-dimensional XY model in an orienting field. The strongest test of the theory is at temperatures somewhat below the phase transition. Very close to the transition, the higher harmonics are small and the scaling cannot be determined with sufficient accuracy to test the asymptotic behavior. The finite ordering field has turned out not to hide interesting behavior, but rather has made possible a direct measurement of the C_{6n}. Thus a single experimental measurement of the X-ray scattering as a function of the azimuthal angle χ yields a simultaneous measurement of many exponents σ_n, or alternatively many crossover exponents ϕ_n.

The harmonic-scaling theory for the liquid crystal hexatic transition should also apply to other systems where the real part of a complex number is measured as the order parameter. An example is the spiral spin density wave found in erbium. Using neutron scattering, Habenschuss et al[18] were able to measure the temperature dependences of the harmonics of the C-axis moment up to the 17th harmonic. Their data are qualitatively similar to those of Fig. 5, but a quantitative determination of the harmonic scaling is not possible. Future experiments on erbium, or similar systems, should provide a further test of the theory. Variants of the theory which include explicitly the quasi-two-dimensional nature of the transition describe rather well the thermodynamic properties of the fluid-hexatic transition in thin films, as will be discussed presently, and should describe equally well features of the magnetic phase transition in CuO_2 lamellar superconductors.

Heat capacity measurements[19] show that most of the entropy comes out of the system at the fluid-hexatic transition, rather than at the hexatic-crystal phase transition. In the tilted hexatics[20] the heat capacity varies as $|t|^{-\alpha}$ over a limited temperature range with $\alpha \approx 0.5$, but the divergence is rounded by the field from the tilt of the director. An equation of state is therefore needed to analyze the data. A phenomenological parametric equation provides an excellent description of the 8OSI data[20], but the parameters required do not have values that seem physically reasonable. That may be because the transition is weakly first-order, and the discontinuity is masked by the field. The value $\alpha \approx 0.5$ suggests that the behavior very near the transition may be influenced by a nearby tricritical point. The scaling exponents very close to the transition cannot be determined accurately enough to check this. Studies of the S_C to S_F transition in the TBnA homologous series[21] show a change from a discontinuous to a continuous transition. However, nearly all liquid crystal fluid-hexatic transitions show a value $\alpha \approx 0.5$ for some

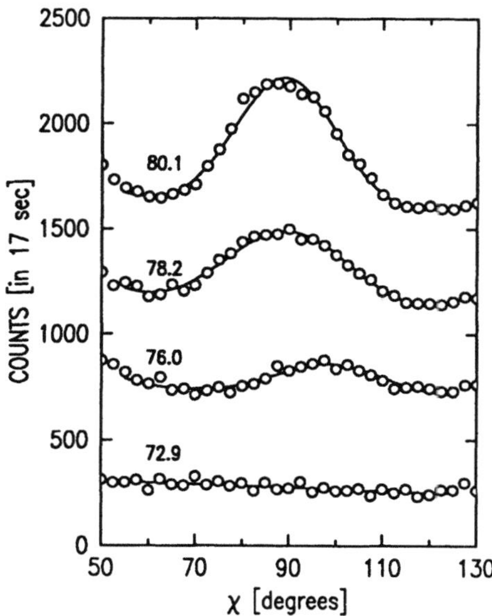

Figure 7. X-ray scattering intensity as a function of the azimuthal angle χ for several temperatures in the S_I phase of a 23 molecule thick film of 8OSI. The numbers above each set of data show the temperature in Celsius. For display, the curves for 76.0, 78.2, and 80.1 °C have been displaced upwards by 500, 1000, and 1500 counts, respectively. The solid lines show the results of fitting to Eq. (6).

temperature range; it seems unlikely that all would happen to be just the right distance in some parameter space from a tricritical point. A more promising approach is to treat the hexatic phase as weakly coupled two-dimensional layers. The two-dimensional fluctuations are treated precisely[4] and they are coupled in a mean-field approximation. The approach describes the behavior of C_6 rather well, but fails the more stringent test of predicting the heat capacity; the agreement is only qualitative[9]. It seems that one will have to go beyond the mean-field approximation.

Approach to Two Dimensions

I would like to close by describing the results of some experiments on the fluid-hexatic transition in thin films of 8OSI. These are of interest to see if it is possible to obtain information on the fluid-hexatic transition in two dimensions, which motivated the original search for liquid crystal hexatics. Thermal fluctuations become much more important in two-dimensions and the harmonic scaling will be quite different. In an infinite system there will be no long-range order, only an algebraic decay of correlation functions as in Eq. (5). In a finite system the fluctuations are reduced, but even there one expects[22] $\sigma_n \sim n^2$. In an applied field, the algebraic decay of correlations implies a susceptibility to an applied field of the form

$$\chi_{6n} \sim L^{2-\eta_6 n^2},$$

where L is the sample size and $\eta_6 \approx 0.25$ in the Kosterlitz-Thouless theory[4]. This implies that for $n \geq 3$, $\chi_{6n} \to 0$ and the tilt field does not induce not induce long-range $6n$-fold BO order. Therefore, as the system crosses over from three to two dimensions one expects to see the higher harmonics of the BO order parameter "turn off". The results of an X-ray scattering experiment for an 8OSI film 23 molecules thick are shown in Fig. 7. When they are fit to Eq. (6), one finds $C_6 < 0.5$, C_{12} detectable at the few percent level only in the lowest temperature data, and no detectable harmonics for $n > 2$. This behavior is consistent with our expectations. The experiments are not easy, but thin films of smectic liquid crystals offer interesting potential[23] to study dimensional crossover and the interplay between two-dimensional and three-dimensional physics. They also offer a theoretical challenge, but one which does not seem to be insurmountable.

The ideas that have come out of a study of stacked two-dimensional hexatic phases seem to have provided the missing element required for a complete classification of the large variety of thermotropic liquid crystal phases. I had thought, at one time, that they would also provide a simpler way to understand the icosahedral order in quasicrystals, but that now seems unlikely to be so. I have tried to give a somewhat tutorial overview to our current understanding of bond-orientational order in smectic liquid crystals, and have included a large number of references for readers who wish to know more detail.

Acknowledgements

My original introduction to storage ring sources for X-ray scattering as a tool to study the structure of liquid crystals was supported by a NATO grant, and I am grateful to NATO for support to attend this workshop. Much of the research described in this article formed the Ph.D thesis of Joel D. Brock; both his and my participation in the research was supported by NSF grant DMR-8619234. Other collaborators, as well as the construction of the scattering spectrometer at the National Synchrotron Light Source were supported by the NSF MRL program under grant DMR-8418718 and by the IBM corporation.

References

1. L.D. Landau, *Phys. Z. Sowjun.* **11**, 545 (1937). An English translation appears in the book: *Collected Works of L. D. Landau*, edited by D. ter Haar, Gordon and Breach, New York (1965), p. 209.

2. B.I. Halperin, and D.R. Nelson, Phys. Rev. Lett. **41**, 121 (1978); D.R. Nelson, and B.I. Halperin, Phys. Rev. B **19**, 2457 (1979).

3. A.P. Young, Phys. Rev. B **19**, 1855 (1979).

4. J.M. Kosterlitz, and D.J. Thouless, J. Phys C **6**, 1181 (1973); J.M. Kosterlitz, J. Phys C **7**, 1046 (1974).

5. L.D. Landau, and E.M. Lifshitz, *Statistical Physics*, 3rd ed., Pergamon, New York (1980), Chapter XIII.

6. R.J. Birgeneau and J.D. Litster, J. Phys. (Paris), Lett. **38**, 1399 (1978).

7. J. Als-Nielsen, J.D. Litster, R.J. Birgeneau, M. Kaplan, C.R. Safinya, A. Lindegaard-Andersen, and S. Mathiesen, Phys. Rev. B **22**, 312 (1980).

8. R. Pindak, D.E. Moncton, S.C. Davey, and J.W. Goodby, Phys. Rev. Lett. **46**, 1135 (1981); J. Budai, R. Pindak, S.C. Davey, and J.W. Goodby, J. Phys. (Paris), Lett. **45**, 1053 (1984).

9. J.D. Brock, A. Aharony, R.J. Birgeneau, K.W. Evans-Lutterodt, J.D. Litster, P.M. Horn, G.B. Stephenson, and A.R. Tajbakhsh, Phys. Rev. Lett. **57**, 98 (1986); J.D. Brock, D.Y. Noh, B.R. McClain, J.D. Litster, R.J. Birgeneau, A. Aharony, P.M. Horn, and Jason C. Lang, Z. Phys. B **74**, 197 (1989).

10. A. Aharony, R.J. Birgeneau, J.D. Brock, and J.D. Litster, Phys. Rev. Lett. **57**, 1012 (1986).

11. M.E. Fisher, and P. Pfeuty, Phys. Rev. B **6**, 1889 (1972); F.J. Wegner, Phys. Rev. B **6**, 1891 (1972).

12. G. Jug, Phys. Rev. B **27**, 609 (1983).

13. A. Aharony, in *Phase Transitions and Critical Phenomena*, C. Domb, and M.S. Green, eds., Academic, New York (1976), p. 357.

14. H. Rohrer, and Ch. Gerber, Phys. Rev. Lett. **38**, 909 (1877).

15. R.B. Potts, Proc. Camb. Phil. Soc. Math. Phys. Sci. **48**, 106 (1952).

16. A. Aharony, K.A. Müller, and W. Berliner, Phys. Rev. Lett. **61**, 2855 (1977).

17. D.R. Nelson, Phys. Rev. B **13**, 2222 (1976).

18. M. Habenschuss, C. Stassis, S.K. Sinha, H.W. Deckman, and F.W. Spedding, Phys. Rev. B **10**, 1020 (1974).

19. T. Pitchford, G. Nounesis, S. Dumrongrattana, J.M. Viner, C.C. Huang, and J.W. Goodby, Phys. Rev. A **32**, 1938 (1985).

20. C.W. Garland, J.D. Litster, and K.J. Stine, Mol. Cryst. Liq. Cryst. **170**, 71 (1989).

21. D.Y. Noh, J.D. Brock, J.D. Litster, R.J. Birgeneau, and J.W. Goodby, Phys. Rev. B **40**, 4920 (1989).

22. M. Paczuski and M. Kardar, Phys. Rev. Lett. **60**, 861 (1988).

23. M. Cheng, J.T. Ho, S.W. Hui, and R. Pindak, Phys. Rev. Lett. **59**, 1112 (1987).

THE TILING STRUCTURE OF SIMPLE LIQUIDS SQUARES AND TRIANGLES
IN TWO DIMENSIONS

Matthew A Glaser and Noel A Clark

Condensed Matter Laboratory, Department of Physics
University of Colorado, Boulder, Colorado 80309, USA

1 INTRODUCTION

It has long been appreciated that dense liquids exhibit some degree of transient local structural
order Local triangular lattice-like order has been observed in dense two-dimensional (2D) liquids by
a number of workers [1-6] In this paper, we present results of a molecular dynamics (MD) simulation
study that explores the geometrical structure of 2D liquids in more detail In particular, we show that
solid-like fluctuations in the dense 2D liquid are a manifestation of a more general local structural
organization resembling plane tilings composed of squares and equilateral triangles (ST tilings), and
that the 2D liquid can be characterized as an ST tiling containing numerous low-strength tiling faults
A similar geometrical structure is seen in 2D colloidal liquids [5] and in analog simulation studies of
2D liquids [6] Our results indicate that square-triangle fluctuations should be a feature of realistic
models of dense liquid structure and the melting transition in 2D

2 STATISTICAL GEOMETRY OF TWO-DIMENSIONAL LIQUIDS

We performed microcanonical ensemble MD simulations using periodic boundary conditions and
the WCA (truncated Lennard-Jones (LJ)) pair potential [7] The WCA potential is short-ranged
and purely repulsive — as a result, we expect its thermodynamic properties to be similar to those
of the hard-disk system We simulated 896- and 3584-particle systems at a temperature of $T = 0\,6$
[8] for decreasing number densities ρ in the range $0\,70 \leq \rho \leq 0\,91$ At each density the system was
thermalized for $50,000$ timesteps, starting from a well-thermalized configuration from a slightly higher
density We observed discontinuities in pressure (Fig 1) and potential energy (Fig 2) indicative of a
first-order melting transition, with two-phase coexistence for $0\,85 \lesssim \rho \lesssim 0\,88$

Our techniques for probing local geometrical and topological structure in the liquid are based on
the Voronoi construction [9] The Voronoi construction assigns a "cell" to each particle, namely that
region of the plane which is nearer to the given particle than to any other particle We consider any
pair of particles whose Voronoi cells share an edge to be nearest neighbors The number of nearest
neighbors (coordination number) for each particle characterizes the local topology, or connectivity, of
the nearest-neighbor network, and makes it possible to identify topological defects (dislocations and
disclinations) in 2D systems by finding particles with a coordination number different than six (the
triangular-lattice value)

We identified local solid-like regions, or "triangular ordered clusters" (TOCs) in the liquid by
evaluating the local sixfold bond-orientational order parameter

$$\psi \equiv \psi_i = (1/n_i) \sum_{j=1}^{n_i} e^{i6\theta_{i,j}} \tag{1}$$

Geometry and Thermodynamics
Edited by J -C Toledano
Plenum Press, New York, 1990

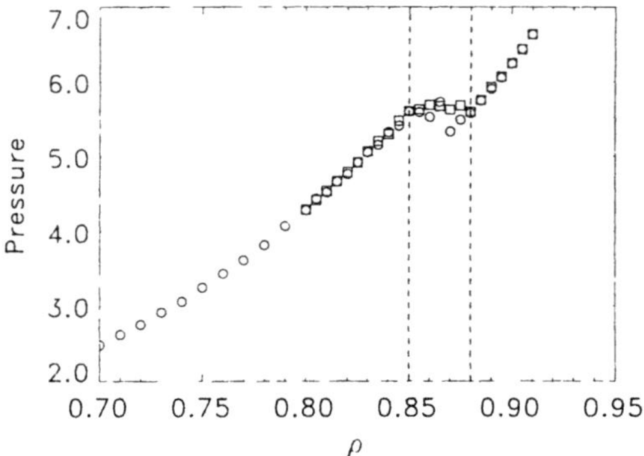

Figure 1. Equation of state for the 2D WCA system at $T = 0.6$. Circles, 896-particle system; squares, 3584-particle system. The vertical dashed lines indicate the approximate boundaries of the coexistence region.

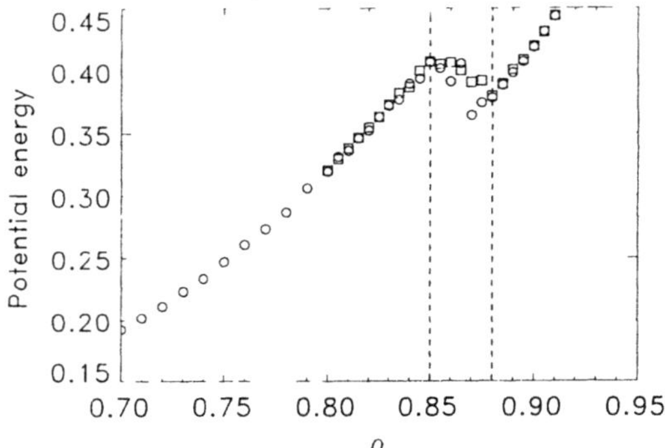

Figure 2. Potential energy for the 2D WCA system at $T = 0.6$. Circles: 896-particle system; squares: 3584-particle system. The vertical dashed lines indicate the approximate boundaries of the coexistence region.

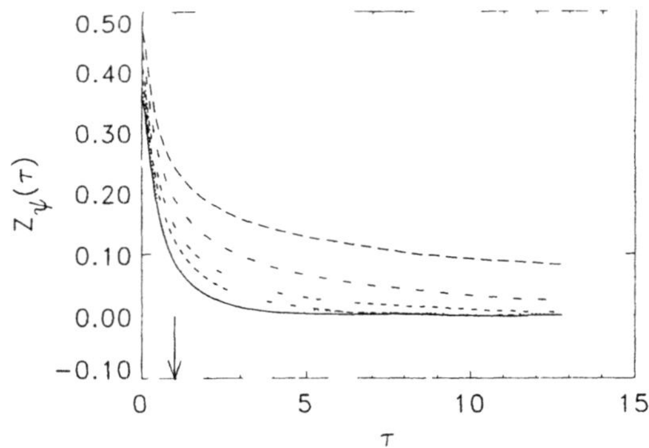

Figure 3. $Z_\psi(\tau)$ for several liquid densities near coexistence. Solid line, $\rho = 0.80$; dotted line, $\rho = 0.81$; dashed line, $\rho = 0.82$; dot-dash line, $\rho = 0.83$; three dots-dash line, $\rho = 0.84$; long dashes, $\rho = 0.85$. The averaging time discussed in the text is indicated by an arrow.

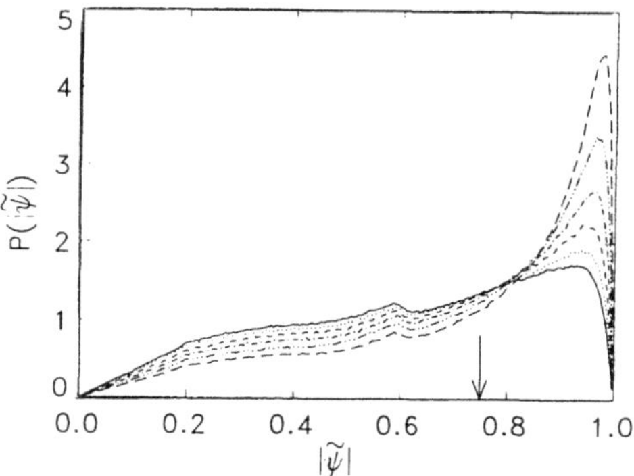

Figure 4. Probability distributions of $|\tilde{\psi}|$ for several liquid densities near coexistence. The lines have the same meaning as in Fig. 3. The arrow indicates the cutoff employed in assigning particles to triangular order clusters.

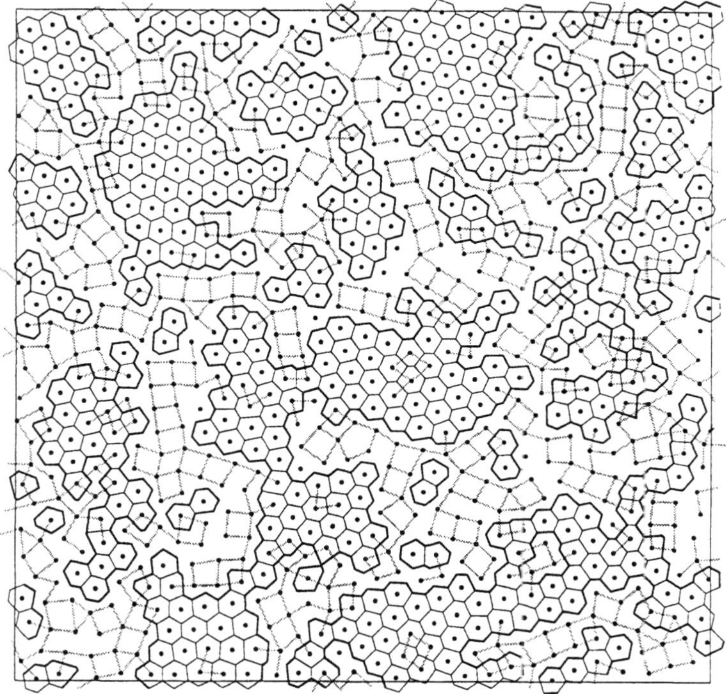

Figure 5. Triangular-ordered clusters (TOCs) for a typical liquid configuration at $\rho = 0.83$. The Voronoi cells for particles in TOCs have been drawn with interior TOC boundaries indicated by light solid lines and exterior boundaries by heavy solid lines. Nearest-neighbor bond angles exceeding 75° are also shown (patterned lines), indicating square lattice-like coordination in non-TOC regions.

Here n_i is the number of nearest neighbors for particle i, and $\theta_{i,j}$ is the orientation of the nearest-neighbor bond between particle i and its jth neighbor relative to an arbitrary axis The dynamics of ψ are characterized by the correlation function

$$Z_\psi(\tau) = \langle \psi(t)\psi^*(t+\tau)\rangle - |\langle \psi(t)\rangle|^2 \qquad (2)$$

In the dense 2D liquid, $Z_\psi(\tau)$ exhibits relaxation on two distinct time scales, as is evident from Fig 3, which shows $Z_\psi(\tau)$ for several densities near freezing The fast relaxation is due to rapid single-particle vibrational motions, and the slow relaxation is due to the collective relaxation of solid-like regions (TOCs) The characteristic time associated with the slow relaxation increases steeply with increasing density near freezing due to a corresponding increase in the typical size of TOCs To visualize the underlying structure of the liquid, we average configurations over a time window of 1 0 LJ units, indicated by the arrow in Fig 3, which is long enough to suppress vibrational fluctuations yet short compared to the mean lifetime of TOCs Quantities evaluated from such averaged configurations will be denoted by a $\tilde{\ }$, e g $\tilde{\psi}$ These time-averaged configurations were quite similar to those we obtained from steepest-descents "quenches" [2] of the corresponding "raw" configurations, suggesting that the time-averaged configurations correspond to local potential energy minima

Particles were assigned to TOCs by evaluating the magnitude of the local sixfold order parameter, $|\tilde{\psi}|$, for each particle Fig 4 shows the probability distribution of $|\tilde{\psi}|$ for several liquid densities near freezing A striking feature of these distributions is the growth of a well-defined peak in the region $0\,75 \lesssim |\tilde{\psi}| \lesssim 1\,0$ as the density approaches the freezing density, due to an increase in the number of particles in solid-like regions (the origin of the small feature near $|\tilde{\psi}| = 0\,6$ is discussed below) Particles having $|\tilde{\psi}|$ in the peak ($|\tilde{\psi}| \geq 0\,75$) are therefore assigned to TOCs — nearest-neighbor pairs of particles that both satisfy this condition are assigned to the same TOC Fig 5 shows the resulting geometry in the liquid phase at $\rho = 0\,83$, where the Voronoi cells of particles in TOCs are plotted TOCs are typically globular, with a characteristic size distribution that is well described by the Fisher droplet model [10]

An interesting feature of Fig 4 is the presence of an isosbestic point, i e a point where all of the (normalized) distributions cross, near $|\tilde{\psi}| = 0\,8$ This observation strongly suggests that the overall distribution is really the sum of two underlying distributions — one distribution, at large values of $|\tilde{\psi}|$, associated with geometrically ordered regions (TOCs), and the other associated with disordered regions (non-TOC regions) Varying the density of the liquid changes the relative proportions of particles in the two types of region (and therefore changes the relative contributions of the two underlying distributions) but does not change the local geometry of either type of region (so the form of each un derlying distribution does not change) This observation (together with Fig 5) illustrates the essential inhomogeneity of the 2D liquid, and validates the procedure we use for identifying TOCs

Once we have identified particles in TOCs, we can analyze the geometrical properties of triangular ordered and disordered regions separately A key finding is shown in Figs 6 and 7 Fig 6 shows probability distributions of nearest-neighbor bond lengths $\tilde{r} \equiv \tilde{r}_{i,j}$ for several liquid densities near freezing, and Fig 7 shows the corresponding probability distributions of nearest-neighbor bond angles $\tilde{\varphi} \equiv \tilde{\varphi}_{i,j} = \tilde{\theta}_{i,j+1} - \tilde{\theta}_{i,j}$ In each case we show distributions for particles in TOCs (dotted line), for particles not in TOCs (dashed line) and for all particles (solid line) For particles in TOCs, the distributions have single peaks at $\tilde{r} \simeq 1\,11$ and $\tilde{\varphi} \simeq 58°$ (close to the expected value of $60°$) By constrast, the distributions for particles not in TOCs are distinctly multi-peaked, with peaks at $\tilde{\varphi} \simeq 48°, 56°$, and $83°$, and $\tilde{r} \simeq 1\,10$ and $1\,53$ The extra peaks appear close to positions expected for a square lattice having the same lattice parameter as the TOCs [11] $\tilde{\varphi} = 45°$ and $90°$, and $\tilde{r} = 1\,11\sqrt{2} = 1\,57$ Clearly, non-TOC regions are distinguished from TOCs by the presence of bonds longer than $\sim 1\,4$ and bond angles larger than $\sim 75°$ In Fig 5 we have plotted nearest-neighbor bonds adjacent to bond angles larger than $75°$ (patterned lines), showing that the non-TOC regions are largely occupied by local groups of particles having a nearly square arrangement These "squares" have nearly the same lattice parameter as TOCs, and typically form ladder-like clusters

Another noteworthy feature of Figs 6 and 7 is that the separate distributions for TOC and non-TOC particles have a form that is only weakly dependent on density, although the relative contributions of TOC and non-TOC regions to the overall distribution varies steeply with density This suggests that the primary effect of varying the density is to alter the relative proportions of TOC and non-TOC regions, with the local geometry of either type of region relatively insensitive to varying density This is consistent with our observations regarding the $|\tilde{\psi}|$ distributions

196

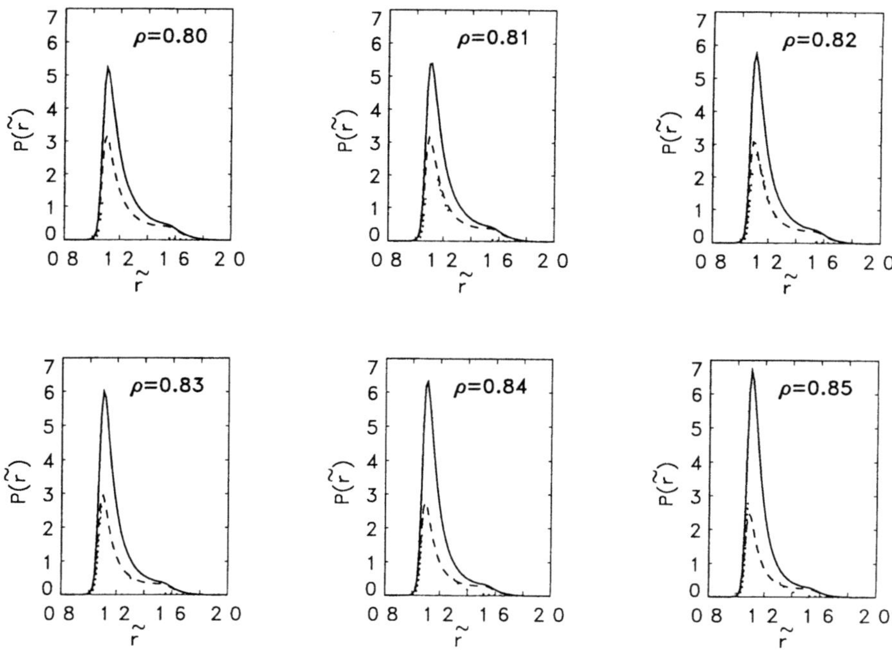

Figure 6 Probability distributions of nearest neighbor bond length $\tilde{r}$ for several liquid densities near coexistence Dotted lines, distributions for particles in TOCs, dashed lines, distributions for particles not in TOCs, solid lines, all particles

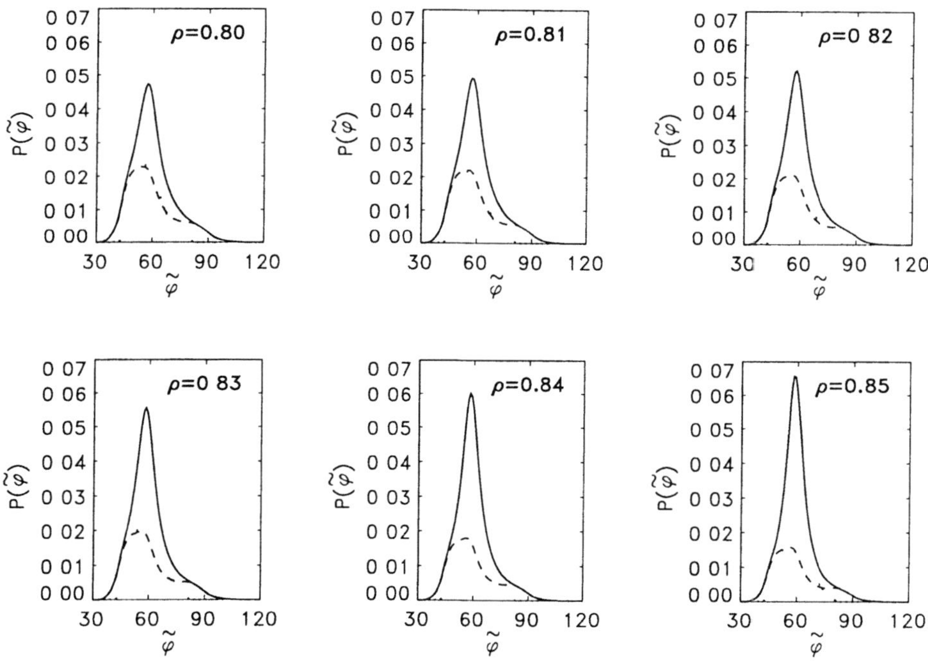

Figure 7 Probability distributions of nearest-neighbor bond angle $\tilde{\varphi}$ for several liquid densities near coexistence Dotted lines, distributions for particles in TOCs, dashed lines, distributions for particles not in TOCs, solid lines, all particles

The nearest-neighbor bond network obtained from the Voronoi construction partitions the plane into triangular areas. We can probe the local geometry of the liquid in more detail by diluting this bond network, removing nearest-neighbor bonds that contribute to the secondary peaks in the probability distributions of $\tilde{r}$ or $\tilde{\varphi}$. The specific dilution procedure used to obtain the results described here involves the removal of bonds opposite nearest-neighbor bond angles larger than 75° — alternatively, we could remove bonds significantly longer than the most probable nearest-neighbor distance. The resulting "polygon construction" is shown in Fig. 8, for the same particle configuration as Fig. 5. The polygon contruction results in a partitioning of the plane into polygonal areas having three or more sides. Because it is obtained from the Voronoi construction, the polygon construction contains topological information, but it also contains information about the local geometry of the 2D liquid that the Voronoi construction lacks. This procedure removes, for example, the diagonal bond from each group of four neighboring particles having a nearly square arrangement, leaving a four-sided polygon in place of two triangles [11]. Fig. 8 exhibits large patches of triangles, corresponding to TOCs, and ladder-like clusters of "squares". Polygons with 5 and 6 sides are also evident in Fig. 8. Because polygons having more than three sides represent a less efficient packing than the triangular regions, these polygons represent voids in the 2D liquid. Four-sided polygons are the most common variety of void, while five- and six-sided polygons identify larger voids.

The local arrangements of "squares" and triangles in Fig. 8 resemble those found in tilings composed of squares and equilateral triangles of equal edge length (ST tilings), which have been studied at least since the time of Kepler [12]. ST tilings are the basis for idealized models of the 2D melting transition [13-15] and for a recently discovered class of 2D dodecagonal quasicrystals [16]. A typical "random" ST tiling is shown in Fig. 9. Comparison of Figs. 8 and 9 shows that the 2D WCA liquid has a tendency to form local ST arrangements of "squares" and triangles, with many particles occupying one of the four types of vertex allowed under the ST tiling rules (inset, Fig. 9). To quantitatively compare the topology of the WCA liquid to that of an ideal ST tiling, polygons having five or more sides are decomposed by successively replacing their shortest internal bonds until they are divided into three- and four-sided polygons (triangles and "squares") (Fig. 10). We can then classify vertices in the resulting "reduced" polygon construction as either ST-allowed (one of the four types shown in the inset of Fig. 9) or as disclinations (tiling faults) [17]. The resulting vertex classification is shown in Fig. 11 — ST-allowed vertices are identified with the corresponding symbol from the inset of Fig. 9, and disclinations are marked by plus or minus signs. The fraction of ST-allowed vertices varies from 63% at $\rho = 0.85$ to 41% at $\rho = 0.70$, and the fraction of strength 1 disclinations varies from 35% to 47% over the same range of densities. There is clearly a tendency for particles to adopt local arrangements characteristic of ST tilings for a wide range of densities near freezing, with the overwhelming majority ($\gtrsim 90\%$) of the vertices being either ST-allowed vertices or strength 1 disclinations.

Recognition of this local tiling structure allows us to explain some of the interesting features of the $|\tilde{\psi}|$ distributions shown in Fig. 4. If we calculate the value of $|\tilde{\psi}|$ corresponding to each of the four types of ideal ST-allowed vertex shown in the inset of Fig. 9, we obtain values of 1.0, 0.6, 0.2, and 0.0. The small feature in the $|\tilde{\psi}|$ distributions near $|\tilde{\psi}| = 0.6$ thus appears to be associated with type 2 vertices (the second type of vertex shown in the inset of Fig. 9), where the value 0.6 represents an upper cutoff on the magnitude of the sixfold order parameter for such vertices. Similarly, the break in the curves seen near $|\tilde{\psi}| = 0.2$ is associated with type 3 vertices (the third type of vertex shown in the inset of Fig. 9).

Because geometrical packing constraints are known to be responsible for many aspects of dense liquid structure [7], it seems likely that the tendency to form local tiling structures is a result of packing constraints. To test this hypothesis we generated dense random packings (DRPs) of hard disks [18] and examined the polygon structure of the resulting configurations. The reduced polygon construction for a representative DRP at a packing fraction of 0.827 is shown in Fig. 12. This DRP configuration shows a local ST-tiling structure even more pronounced than that of the dense, time-averaged WCA liquid. Thus it appears that packing constraints are the primary factor producing local tiling structures. Local ST-tiling structures are probably favored because they provide relatively efficient packing while giving a finite entropy contribution due to the many possible arrangements of tiles.

As mentioned before, the polygons having more than three sides in the polygon construction can be thought of as voids. The average area of triangles in the polygon construction for the liquid at coexistence ($\rho = 0.85$) is $a_{tl} = 0.572$, nearly the same as the area per triangle in the solid at coexistence ($a_{ts} = 0.568$ at $\rho = 0.88$), indicating that most of the area expansion upon melting ($\sim 3.5\%$) is associated with the appearance of polygons having four or more sides. At $\rho = 0.85$, 14%

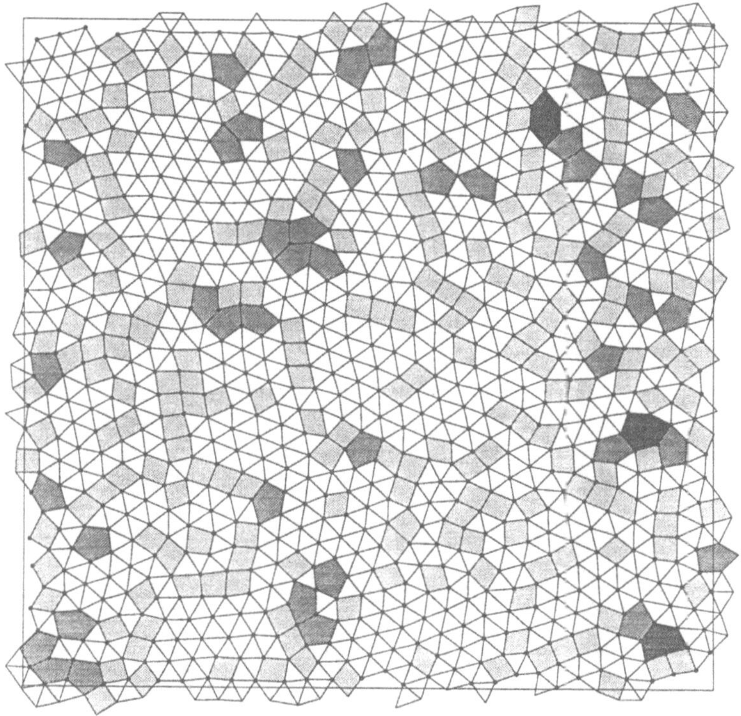

Figure 8. Polygon construction for the configuration shown in Fig. 5, with polygons shaded according to number of sides.

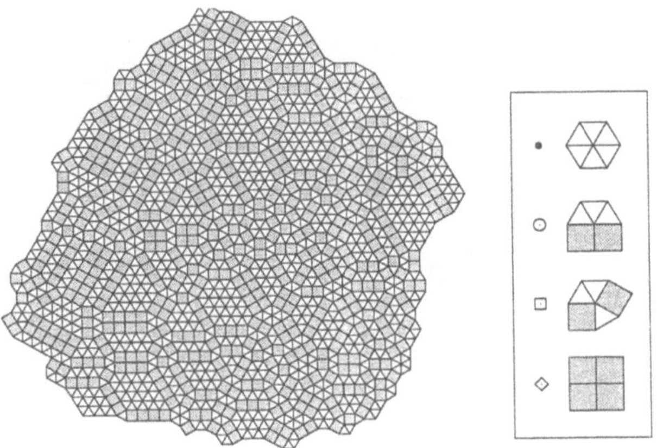

Figure 9. A representative random square-triangle (ST) tiling. Inset: the four types of vertices allowed by the ST tiling rules.

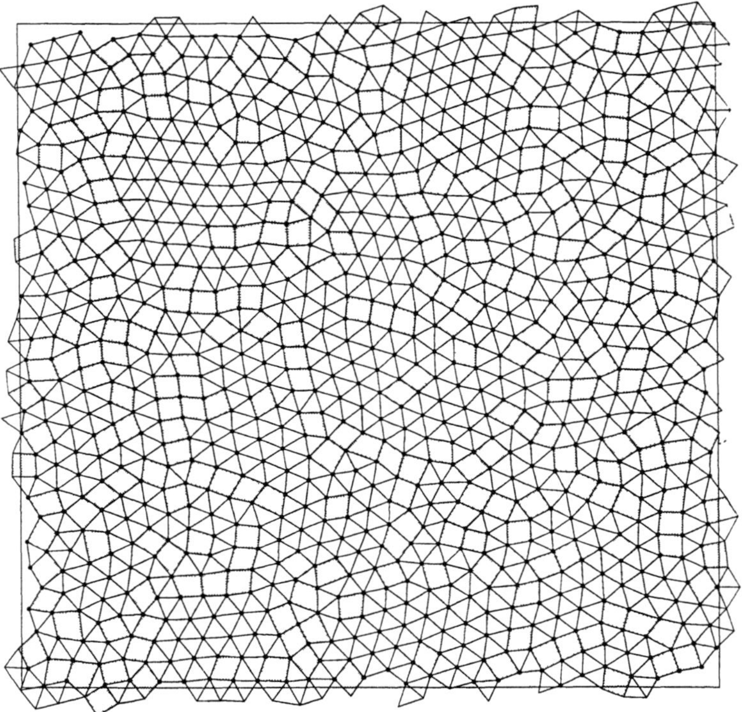

Figure 10. Reduced polygon construction for the configuration of Fig. 5.

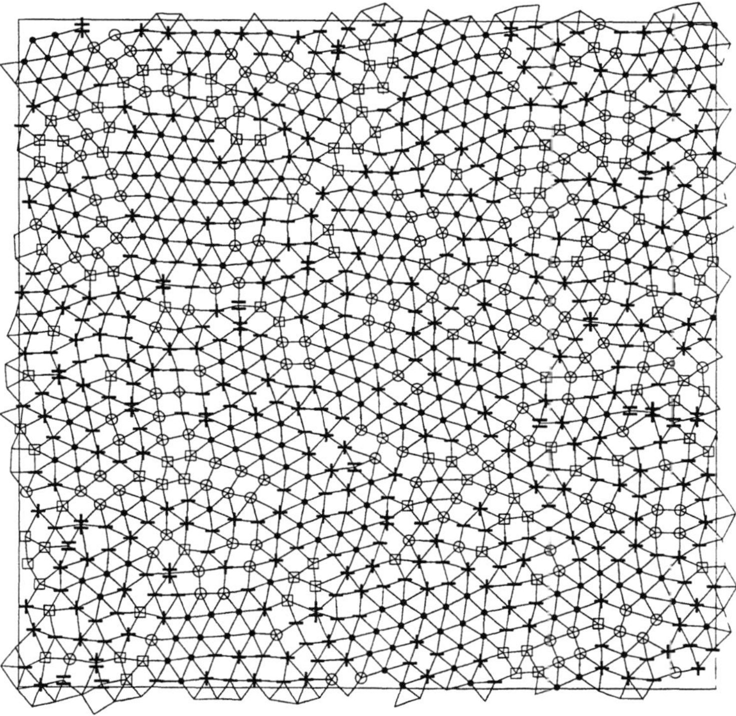

Figure 11. Vertex classification for the configuration of Fig. 5. ST-allowed vertices are marked with the corresponding symbol from the inset of Fig. 9, disclinations of strength 1 are marked with plus or minus signs, indicating their sign, and disclinations of strength 2 are marked with double plusses or double minuses.

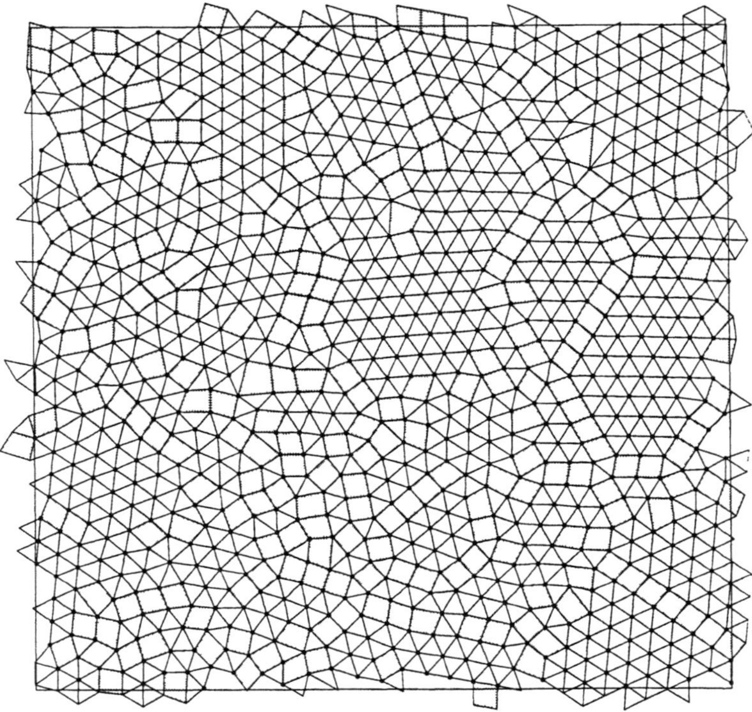

Figure 12. Reduced polygon construction for a dense random packing of hard disks at a packing fraction of 0.827.

of the "free area" [19] is in triangles, 56% is in "squares", 25% is in five-sided polygons, and 5% is in larger polygons. Thus the area increase associated with melting is primarily associated with the creation of geometrical defects (polygons with more than four sides). This observation suggests a novel geometrical melting mechanism based on the creation of geometrical defects which aggregate into characteristic tiling structures. Work on a melting model based on this observation is in progress [20]. It is interesting that the free volume tends to be distributed in small voids (four- and five-sided polygons) — the distribution of free volume in small packets may be entropically favorable because there are more possible spatial arrangements for a large number of small voids than for a small number of large voids.

3. CONCLUSION

The most striking aspect of 2D liquid structure is its inhomogeneity — the 2D liquid can be naturally partitioned into geometrically ordered (solid-like) and disordered regions. Ordered and disordered regions each have a characteristic geometrical signature, which remains nearly invariant over a broad range of densities near freezing. The local geometrical structure of disordered regions shows a surprizing degree of local "square" coordination, and such "squares" are grouped into characteristic local arrangements that strongly resemble ST tilings. The overall goemetrical structure of the 2D liquid can be characterized as an ST tiling containing numerous low-strength tiling faults. The local tiling structure observed in the 2D liquid is evidently the result of packing ccnstraints, as dense random packings of hard disks exhibit the same characteristic local geometrical structure. The area increase upon melting can be directly traced to the formation of "tiles" having more than three sides, suggesting a novel geometrical mechanism for 2D melting [20].

We would like to thank Allen Armstrong, Rainer Malzbender, Paul Beale, Yeke Tang, and David Cohen for many helpful comments and suggestions. This work was supported by NSF Low-Temperature Physics Grant DMR-8807443.

REFERENCES

[1] Observations of local triangular order in the dense 2D liquid are reviewed by K. J. Strandburg, Rev. Mod. Phys. **60**, 161 (1988).

[2] F. H. Stillinger and T. A. Weber, Phys. Rev. A **25**, 978 (1982).

[3] H. J. M. Hanley, T. R. Welberry, D. J. Evans, and G. P. Morriss, Phys. Rev. A **38**, 1628 (1988).

[4] N. A. Clark, B. J. Ackerson, and A. J. Hurd, Phys. Rev. Lett. **50**, 1459 (1983); C. A. Murray and D. H. Van Winkle, Phys. Rev. Lett. **58**, 1200 (1987);

[5] Y. Tang, A. J. Armstrong, R. C. Mockler, and W. J. O'Sullivan, Phys. Rev. Lett. **62**, 2401(1989).

[6] B. Pouligny, R. Maltzbender, P. Ryan, and N. A. Clark, to appear in Phys. Rev. A, Rapid Comm.

[7] J. D. Weeks, D. Chandler, and H. C. Andersen, J. Chem. Phys. **54**, 5237 (1971).

[8] Unless otherwise noted, all quantities are expressed in LJ units [7], in which the LJ ϵ is the unit of energy, the LJ σ is the unit of length, and the particle mass m is the unit of mass.

[9] G. F. Voronoi, J. Reine. Agnew. Math. **134**, 198 (1908).

[10] M. A. Glaser and N. A. Clark, to be published.

[11] In particle groups having a nearly square arrangement, the shortest diagonal of the "square" is a nearest-neighbor bond.

[12] B. Grünbaum and G. C. Shephard, *Tilings and Patterns*, (Freeman, New York, 1987), frontispiece.

[13] R. Collins, Proc. Phys. Soc. **83**, 553 (1964).

[14] H. Kawamura, Prog. Theor. Phys. **70**, 352 (1983).

[15] Y. M. Yi and Z. C. Guo, J. Phys.: Condens. Matter **1**, 1731 (1981).

[16] H. Chen, D. X. Li, and K. H. Kuo, Phys. Rev. Lett. **60**, 1645 (1988); P. W. Leung, C. L. Henley, and G. V. Chester, Phys. Rev. B **39**, 446 (1989); Q. B. Yang and W. D. Wei, Phys. Rev. Lett. **58** 1020 (1987); K. H. Kuo, Y. C. Feng, and H. Chen, Phys. Rev. Lett. **61**, 1740 (1988).

[17] We first compute the sum Σ of "nominal" bond angles around a vertex (a triangle contributes $60°$ and a "square" $90°$); the strength of a disclination is then given by $n = (\Sigma - 360°)/30°$. ST-allowed vertices are those having $n = 0$.

[18] J. L. Finney, Mater. Sci. Engin. **23**, 207 (1976).

[19] The free area was calculated by comparing the average area per triangle at $\rho = 0.88$ with that in each type of polygon at $\rho = 0.85$ (where a "square" is counted as two triangles, a five-sided polygon is counted as three triangles, etc.).

[20] M. A. Glaser, N. A. Clark, A. J. Armstrong, and P. D. Beale, to be published.

DOES CHOLESTERIC BLUE PHASE III HAVE AN ICOSAHEDRAL STRUCTURE?

R.M. Hornreich and S. Shtrikman*

Department of Electronics
Weizmann Institute of Science
76100 Rehovot, Israel

INTRODUCTION

Recently, attempts have been made to apply the quasicrystal concept to the physics of liquid crystals (Hornreich and Shtrikman, 1986a; Rokhsar and Sethna, 1986; Filev, 1986). Although it is still unclear whether this concept is actually relevant to understanding the puzzling phase found in certain cholesteric liquid crystals, the possibility is intriguing and the combination of liquid crystal and quasicrystal behavior is, in any case, interesting in itself. Our objective here is to introduce the essential elements of quasicrystal structures in liquid crystals. A more detailed exposition may be found elsewhere (Hornreich, 1989).

The Landau theory of cholesteric liquid crystals (Grebel *et al.*, 1983, 1984; Belyakov and Dmitrienko, 1985; Hornreich and Shtrikman, 1988; Wright and Mermin, 1989) provides a straightforward explanation for the existence of complex phases between the disordered (isotropic) phase and the usual cholesteric (helicoidal) one when the pitch of the cholesteric structure is sufficiently short. We shall be interested in the so-called *blue phases* (BP), showing in particular that one of them (BP III) may have icosahedral (quasicrystalline) order. After developing the theoretical concepts, we shall discuss briefly the experimental evidence for and against quasicrystalline ordering, noting where further studies might assist in resolving whether such ordering indeed exists.

LANDAU THEORY OF CHOLESTERICS

In Landau theory, one first selects an appropriate *order parameter*. For the case of thermotropic liquid crystals, neither a scalar or vector order parameter suffices (de Gennes, 1975). These systems exhibit orientational (*not* translational) order below a critical temperature; this appears as an anisotropy in their second-order tensor properties

---

* Also with the Department of Physics, University of California, San Diego, La Jolla, CA 92093

(e.g., the dielectric or diamagnetic tensor). A *tensor* order parameter is thus appropriate and a convenient choice is

$$\epsilon_{ij}(\vec{r}) \equiv \epsilon_{ij}^{d}(\vec{r}) - \frac{1}{3}Tr(\epsilon^{d})\delta_{ij}. \tag{1}$$

This quantity vanishes in the disordered phase where the dielectric tensor ϵ^{d} is isotropic and becomes non-zero when there is orientational order.

For *cholesteric* liquid crystals, the free energy is (de Gennes, 1975)

$$\begin{aligned}
F =V^{-1} \int d\vec{r}[\frac{1}{2}(a\epsilon_{ij}^{2} - 2de_{ijl}\epsilon_{in}\epsilon_{jn,l} + c_{1}\epsilon_{ij,l}^{2} + c_{2}\epsilon_{ij,i}^{2}) \\
- \beta\epsilon_{ij}\epsilon_{jl}\epsilon_{li} + \gamma(\epsilon_{ij}^{2})^{2}].
\end{aligned} \tag{2}$$

Here a is proportional to a reduced temperature, d, β, γ, c_{1}, and c_{2} are temperature-independent parameters, V is volume, $\epsilon_{ij,l} \equiv \partial\epsilon_{ij}/\partial r_{l}$, e_{ijl} is the usual antisymmetric third-order tensor, and we sum on repeated indices. For thermodynamic stability, γ and c_{1} must be positive and c_{2} must be greater than $-\frac{3}{2}c_{1}$ (Grebel *et al.*, 1983). The sign of d is fixed by the handedness of the chiral molecules; we take it to be positive.

Using the scaled quantities (Grebel *et al.*, 1983):

$$\begin{aligned}
\mu_{ij} = \epsilon_{ij}/(\beta/\sqrt{6}\gamma), \quad f = F/(\beta^{4}/36\gamma^{3}), \quad \frac{1}{4}t = (3\gamma/\beta^{2})a, \\
\frac{1}{4}\xi_{R}^{2} = (3\gamma/\beta^{2})c_{1}, \quad c_{2}/c_{1} = \rho, \quad \kappa = (d/c_{1})\xi_{R}, \\
\vec{x} = \vec{r}/\xi_{R}, \quad \mu_{ij,l} = \mu_{ij}/\partial x_{l} \quad v = V/\xi_{R}^{3},
\end{aligned} \tag{3a}$$

Eq. (2) becomes

$$\begin{aligned}
f =v^{-1} \int d\vec{x}[\frac{1}{4}(t\mu_{ij}^{2} - 2\kappa e_{ijl}\mu_{in}\mu_{jn,l} + \mu_{ij,l}^{2} + \rho\mu_{ij,i}^{2}) \\
- \sqrt{6}\mu_{ij}\mu_{jl}\mu_{li} + (\mu_{ij}^{2})^{2}].
\end{aligned} \tag{3b}$$

For periodic and quasiperiodic structures, $\mu_{ij}(\vec{x})$ is expanded in Fourier components (Grebel *et al.*, 1983, 1984)

$$\mu_{ij}(\vec{x}) = \sum_{\sigma,n} N^{-\frac{1}{2}}(\sigma)\mu_{ij}(\sigma,n)\exp(i\vec{q}_{\sigma,n} \cdot \vec{x}), \tag{4a}$$

$$= \sum_{h,k,l} N^{-\frac{1}{2}}(\sigma)\mu_{ij}(h,k,l)\exp[iq(hx_{1} + kx_{2} + lx_{3})]. \tag{4b}$$

Here n is over the set of $N(\sigma)$ wave vectors of magnitude $q_{\sigma} =\mid \vec{q}_{\sigma,n} \mid = \sigma^{\frac{1}{2}}q$ with $\sigma_{0} \equiv 1, \vec{q}_{\sigma,-n} = -\vec{q}_{\sigma,n}$; σ (not necessarily integer) indexes wave vectors of different

206

magnitudes. For cubic structures, (h, k, l) are Miller indices, $\sigma = h^2 + k^2 + l^2$, and $N(\sigma) = (3!)2^{3-n_o}/n_1!$ where $n_0(n_1)$ is the number of vanishing (equal) $\mid h \mid, \mid k \mid, \mid l \mid$ values for each distinct set $\{h, k, l\}$.

The amplitudes are next written in terms of five basis tensors. For $[\mu_{ij}(\sigma, n)]$, we have

$$
\begin{aligned}
[\mu_{ij}(\sigma, n)] &= \sum_{m=-2}^{2} \mu_m(\sigma) e^{i\psi_m(\sigma, n)}[M_m(\sigma, n)] \\
&= \frac{1}{2}\Bigg[\mu_2 e^{i\psi_2} \begin{pmatrix} 1 & i & 0 \\ i & -1 & 0 \\ 0 & 0 & 0 \end{pmatrix} + \mu_1 e^{i\psi_1} \begin{pmatrix} 0 & 0 & 1 \\ 0 & 0 & i \\ 1 & i & 0 \end{pmatrix} + \sqrt{\frac{2}{3}}\mu_0 e^{i\psi_0} \begin{pmatrix} -1 & 0 & 0 \\ 0 & -1 & 0 \\ 0 & 0 & 2 \end{pmatrix} \\
&\quad + \mu_{-1} e^{i\psi_{-1}} \begin{pmatrix} 0 & 0 & -1 \\ 0 & 0 & i \\ -1 & i & 0 \end{pmatrix} + \mu_{-2} e^{i\psi_{-2}} \begin{pmatrix} 1 & -i & 0 \\ -i & -1 & 0 \\ 0 & 0 & 0 \end{pmatrix} \Bigg].
\end{aligned}
$$

$$(5)$$

Here $\mu_m(\sigma) \geq 0$ and the five basis matrices for each (σ, n) are in a local right-handed coordinate system $(\hat{\xi}, \hat{\eta}, \hat{\zeta})$ with the polar axis $\hat{\zeta} \equiv \hat{q}_{\sigma,n}$. The directions of the other two axes are connected to the choice of phases $\psi_m(\sigma, n)$. Note that q in Eq. (4) is dimensionless; it is fixed by minimizing the free energy.

Summarizing, the scaled average free energy density of cholesterics is

$$
\begin{aligned}
f = \frac{1}{4}\sum_{\sigma,m}\{t - \kappa m q \sigma^{1/2} + [1 + \frac{1}{6}\rho(4 - m^2)]q^2\sigma\}\mu_m^2(\sigma) \\
+ v^{-1}\int d\vec{x}[-\sqrt{6}\mu_{ij}\mu_{jl}\mu_{li} + (\mu_{ij}^2)^2],
\end{aligned}
$$

$$(6a)$$

with q determined via $\partial f/\partial q = \partial f_2/\partial q = 0$. Our objective is to minimize $f([\mu])$ for arbitrary (κ, t). Note that the phase diagram obtained by minimizing f can depend only upon one parameter, ρ.

To accentuate the essential elements of the free energy calculation, we shall consider here only a *modified* free energy f', obtained by replacing the quartic term in Eq. (6a) by

$$
f_4 = v^{-1}\int d\vec{x}(\mu_{ij}^2)^2 \to f_4' = [v^{-1}\int d\vec{x}\mu_{ij}^2]^2.
$$

$$(6b)$$

ICOSAHEDRAL STRUCTURES

The basic nature of ordered phases in cholesterics can be understood by considering f_2. For chiral systems, the minimum of f_2 is at $q \neq 0$ and the f_2 ground state is always on the $m = 2$ branch of the spectrum. Such Fourier components are thus an *essential ingredient* in the structure of all ordered cholesteric phases.

A second key factor is that the phase transition from the I phase cannot be to a helicoidal cholesteric (CH) one for all values of κ. By considering the *exact* form of the CH-phase order parameter and minimizing the resultant free energy, Grebel *et al.* (1983) showed that $q = \kappa$ and that, for $\kappa \geq 3$, any $I - CH$ phase transition *must be of second-order* and take place at $t = t_{I-CH} = \kappa^2$.

Given this, is a *first-order* transition from I to some ordered phase is possible for $\kappa \geq 3$? The answer is affirmative and it follows that such a transition *necessarily* occurs at some $t > t_{I-CH}$ and that the resulting ordered state's free energy must be *lower* than that of either I or CH in some region of the (κ, t) phase diagram. A central objective of liquid crystal physics has been the theoretical description of these new ordered structures and a comparison of their properties with those found experimentally for the cholesteric BP.

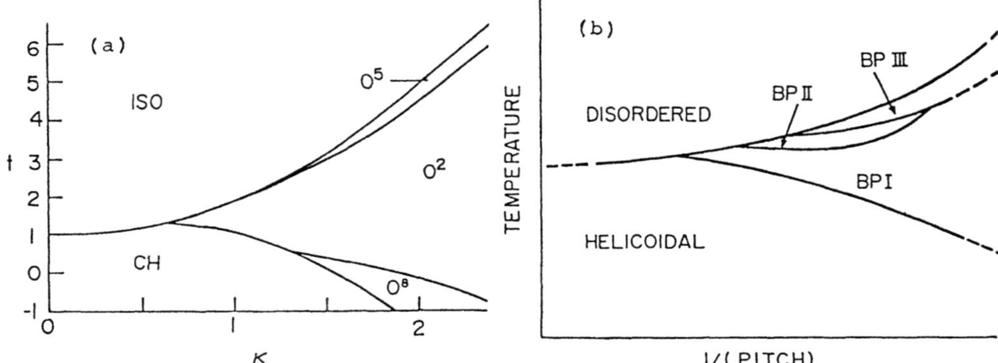

Figure 1.   (a) Theoretical phase diagram of cholesteric liquid crystals when the possible allowed phases include the disordered (I), cubic (O^2, O^5, O^8) and helicoidal cholesteric (CH) phases (from Grebel *et al.*, 1984).   (b) Schematic phase diagram, showing the three experimentally observed blue phases (BP).

As a global minimization of Eq. (6) is not possible, the approach taken has been to consider, individually, possible candidates for non-helicoidal ordered phases (Grebel *et al.*, 1983, 1984). Since we require first-order transition from I, a key element in selecting candidates has been the third-order contribution to f. This naturally leads to the consideration of structures with order parameters whose Fourier components have $m = 2$ tensor amplitudes and associated reduced wave vectors of magnitude κ. Wave vector directions are chosen such that equilateral triangles are formed, thereby giving the required third-order contribution to f.

In the simplest such three-dimensional structure, the wave vectors are the edges of a regular *tetrahedron*. All the cubic phases analyzed in the literature ($O^5 - I432$, $O^8 - I4_132$, and $O^2 - P4_232$) take this tetrahedral "skeleton" as their starting point; the results obtained (Grebel *el al.*, 1984) are summarized in Fig. 1a. For comparison, a schematic diagram showing the experimentally observed BP is given in Fig. 1b. Note particularly that the experimentally observed BP III appears in the region of the phase diagram in which, among cubic structures, O^5 has the lowest free energy. We must therefore compare the free energy of any new model for BP III with that of O^5. Of particular interest will be structures based upon a skeleton formed by a regular *icosahedron* of wave vectors (Kleinert and Maki, 1981).

A simple quasicrystalline structure is obtained by taking a set of wave vectors which form 15 linearly independent edges of a regular icosahedron. (Their negatives are the remaining 15 edges.) To minimize f_2, we set the magnitude of these wave vectors equal to κ and associate with each a Fourier amplitude composed of an $m = 2$ basis tensor, a phase factor, and a common scalar magnitude (labeled $\mu_2(\sigma_0)$). The latter is a consequence of the icosahedral (532) point group symmetry.

Even before considering the phases $\psi_2(\sigma_0, n)$, we can calculate the quadratic and quartic contributions to f' as they are phase-independent. For $m = 2$ Fourier components, the ρ-dependent term in Eq. (6) vanishes and we obtain $f_2 = \frac{1}{4}(t - \kappa^2)\mu_2^2(\sigma_0)$. and $f_4' = \mu_2^4(\sigma_0)$.

We now turn to f_3, where the choice of phase angles is crucial. The optimum would clearly be to fix the phases such that *each* of the twenty triangles forming the faces of the icosahedron contributes maximally to f_3. This, however, is *impossible* (Hornreich, 1989) as the twenty triangles form a closed surface. The naive minimum value of f_3 is thus *unattainable* for *any* choice of the $\psi_2(\sigma_0, n)$.

For cubic structures, Grebel *et al.* (1983, 1984) showed how to determine phase factors *a priori* from the space group symmetry. For icosahedral structures, one might think that only $\psi_2(\sigma_0, n) = 0$ is possible, since any other choice would seemingly break (532) point symmetry. This, however, is not so, due to the *phason degrees of freedom* existing in incommensurate structures. Such a phason mode may combine with a rotation to give a hybrid rotation-phason element in the space group, analogous to the rotation-phonon or rotation-translation (screw axis) elements found in non-symmorphic space groups. We therefore allow such hybrid symmetry elements in the icosahedral space group.

Consider a rotation of $2\pi/5$ about an axis joining a vertex to the icosahedron center. As five such permutations return us to the initial state, the phase angles between adjacent edges at this vertex can differ only by $2\pi p_1/5$, with p_1 integer and mod 5. However, the quantities relevant to f_3 are not the phase angles of individual edges but rather the phases ϕ_Δ associated with oriented triangular loops. Each such phase is the sum of two contributions, an "explicit" phase ψ_Δ (i.e., the algebraic sum of the $\psi_2(\sigma_0, n)$ associated with the triangle's legs) and an "implicit" phase due to the angles between the local vectors $\hat{\xi}$ associated with each edge of a given triangle and the normal to that triangle.

The structure having minimizing f_3 is that for which the coefficient of $\mu_2^3(\sigma_0)$ in $| f_3 |$ is maximum. From Eq. (8), this occurs when $p_1 = 2$; $f_3(\sigma_0; 2) = -0.720\mu_2^3(\sigma_0)$. Note, however, that this value differs by less than 1% from $f_3(\sigma_0; 1) = -0.715\mu_2^3(\sigma_0)$.

The corresponding value for the transition to the *bcc* O^5 skeleton structure is (Grebel *et al.*, 1983) $f_3(\sigma = 2) = -(23\sqrt{2}/32)\mu_2^3(\sigma = 2) = 1.016\mu_2^3(\sigma = 2)$. One might therefore

When these are evaluated, we obtain (Hornreich, 1989)

$$
\begin{aligned}
p_1 &= 0; \phi_\Delta = \phi = \arccos(-7\sqrt{5}/27) = 54.57^\circ, \\
p_1 &= 1; \phi_\Delta = \phi - \pi/5 = 18.57^\circ, \\
p_1 &= 2; \phi_\Delta = \phi - 2\pi/5 = -17.43^\circ, \\
p_1 &= 3; \phi_\Delta = \phi - 3\pi/5 = -53.43^\circ, \\
p_1 &= 4; \phi_\Delta = \phi - 4\pi/5 = -89.43^\circ.
\end{aligned}
\tag{7}
$$

There are thus *five different structures* compatible with (532) point symmetry. Their respective f_3 contributions are (Hornreich, 1989)

$$
f_3(\sigma_0; p_1) = -(27\sqrt{5}/80) \cos(\phi - \pi p_1/5)\mu_2^3(\sigma_0).
\tag{8}
$$

conclude that a cubic structure is energetically preferred to an icosahedral one. We now show that such a conclusion is premature.

In applying Landau theory to the cholesteric BP, it was quickly realized (Hornreich and Shtrikman, 1981) that a model restricting the order parameter to a set of Fourier wave vectors having a *single* magnitude could not possibly describe both of the experimentally observed cubic structures, BPI and BPII. *Harmonics* of this magnitude play a key role (Grebel *et al.*, 1983, 1984).

What then is the effect of harmonics on icosahedral structures? In analogy with the cubic case, there are two fundamentally different types of icosahedral structures. In the first, a second set of wave vectors is introduced such that new triangles are formed by the combination of *two* vectors from the new set and *one* from the original set. For cubic structures, this was done by supplementing the $< 110 >$ wave vectors forming the tetrahedral skeleton by a set of $< 100 >$ wave vectors. The original *eight* triangles are then augmented by an additional *twelve*, composed of two $< 100 >$ and one $< 110 >$ wave vectors. As a result, a new O^2 *sc* structure (see Fig. 1a) supplants the original O^5 *bcc* one in part of the the (κ, t) phase diagram (Grebel *et al.*, 1983, 1984).

For icosahedral structures, the same procedure can be followed. Supplement the original 30 wave vectors forming the edges of the icosahedron structure by a set of 12 wave vectors from the center to the vertices. This yields *thirty* new triangles, formed by combinations of two vertex and one edge vector. We designate the vertex vector set by $\sigma = \sigma_{-1}$ and again restrict ourselves to $m = 2$ states (Hornreich and Shtrikman, 1986a; Rokhsar and Sethna, 1986; Filev, 1986). This model is attractive for two reasons: One, the magnitude $q_{\sigma_{-1}}$ of the vertex vectors is almost equal to that of the edge vectors, thus f_2 remains very close to its minimal value. Two, the phases $\psi_2(\sigma_{-1}, n)$, $\psi_2(\sigma_0, n)$ can be chosen so that *each* of the thirty additional triangles contributes maximally to the third-order term in the free energy.

To establish these points, note that the ratio $q_{\sigma_{-1}}/q_{\sigma_0} = \sigma_{-1} = \sin(2\pi/5) = 0.951$. As the quadratic part f_2 of the free energy is minimized when the reduced wave vector magnitudes equal κ, let $q_{\sigma_{-1}} = (1-\delta_{-1})\kappa$, $q_{\sigma_0} = (1+\delta_{-1})\kappa$, with $\delta_{-1} = (1-\sigma_{-1})/(1+\sigma_{-1}) = 0.0251$. It follows that f_2 remains essentially at its minimum value when the set $\pm\vec{q}_{\sigma_{-1},n}$ is added to $\pm\vec{q}_{\sigma_0,n}$; i.e., the δ_{-1} correction is *negligible* in the region of physical interest (Hornreich, 1989).

Turning to the second point, note that the $\pm\vec{q}_{\sigma_0,n}$ side of each of the triangles formed by two $\pm\vec{q}_{\sigma_{-1},n}$ and one $\pm\vec{q}_{\sigma_0,n}$ wave vector is *not shared* with any other congruent triangle. Therefore, the phase associated with this wave vector may by fixed so as to maximize this triangle's f_3 contribution. However, in order for the contribution from *each* of the thirty vertex-vertex-edge triangles to f_3 to be simultaneously maximum in magnitude and negative, we must set $p_1 = 3$ and add π to ϕ_Δ in Eq. (7). Then (Hornreich, 1989) minimizing f_3 with respect to $\mu_2(\sigma_{-1})$ and $\mu_2(\sigma_0)$ with $\mu^2 = \mu_2^2(\sigma_{-1}) + \mu_2^2(\sigma_0) = $ constant gives $[f_3(\sigma_1,\sigma_0)]_{min} = 1.010\mu^3$. The corresponding value for the $I - O^5$ phase transition is (Hornreich and Shtrikman, 1986b) $f_3(\sigma = 2,4) = 1.016$. Thus these two structures are essentially *degenerate* when a modified free energy model is used.

As noted at the beginning of this section, there is an alternate way of augmenting the icosahedral skeleton and thereby reducing f'. The idea is to introduce a second set of wave vectors such that the new triangles are composed of *one* vector from the second set and *two* from the original skeleton. For the cubic case, this was done by supplementing the $< 110 >$ tetrahedral skeleton by a set of $< 200 >$ wave vectors. This resulted in an additional *twelve* triangles and, with suitably chosen phases, to a new O^8 bcc structure (see Fig. 1a).

We now apply this modification to the icosahedral skeleton (Hornreich and Shtrikman 1987b, Hornreich, 1989). We supplement the set of 30 $\pm\vec{q}_{\sigma_0}$ edge wave vectors with a set of 60 wave vectors of magnitude $q_{\sigma_1}/q_{\sigma_0} = \sigma_1 = 2\sin(\pi/5) = 1.176$. This set, $\pm\vec{q}_{\sigma_1,n}$, is formed by the vector sum $(\vec{q}_{\sigma_0,1} + \vec{q}_{\sigma_0,3})$ and all those equivalent to it under icosahedral (532) point symmetry. Proceeding as before, we again find that the correction to f_2 is negligible.

Consider now f_3. Detailed calculations (Hornreich, 1989) shows that the relevant icosahedral structure is that with $p_1 = 2$ and that $[f_3(\sigma_0,\sigma_1)]_{min} = -1.031\mu^3$. Comparing this with the result given earlier, we see that the *modified* Landau model has a *lower* free energy for a $p_1 = 2$ structure than for a $p_1 = 3$ one when two harmonics are included in the order parameter. In fact, this free energy is also *less* than that of the bcc O^5 structure.

DISCUSSSION

In the previous section we considered a simplified model of ordering in cholesteric liquid crystals. A modified Landau theory framework was used and the free energy was calculated for several icosahedral structures as well as a cubic one. It was found that when harmonics are included in the icosahedral phase order parameter, the resulting modified free energy is slightly *lower* than that of a cubic structure. *This is no longer the case when the unmodified model is used* (Hornreich, 1989). However, Landau free energy calculations, even when several harmonics are included in the order parameter, should only be regarded as indicative and *not* as an absolute determination of the structures of experimentally observed phases. The energy differences between the various cubic, icosahedral, and helicoidal phases are small (less than 2%) over substantial regions of the (κ, t)-plane and the theoretical phase boundaries are very sensitive to small shifts in the free energies (Grebel *et al.*, 1984). Thus even though results obtained using the unmodified Landau free energy with order parameters composed of $m = 2$ tensors only are, in principle, universal, such factors as higher-order terms in $\mu_{ij}(\vec{x})$ and its spatial derivatives, $m \neq 2$ Fourier components of the order parameter, and fluctuations, none of which has been considered, could significantly effect the (κ, t) phase diagram even if their free energy contributions small. The possibility that cholesteric BP III is characterized by an icosahedral structure thus deserves further consideration.

An icosahedral BP III stucture is consistent with many (but not all) of the experimental observations on this phase (Crooker, 1989). These include:

(a) Nonbirefringence - Due to their (532) point symmetry, icosahedral structures are nonbirefringent.

(b) Optical activity - All cholesteric icosahedral structures lack inversion centers and are therefore strongly optically active.

(c) Latent heat (Kleiman *el al.*, 1984; Thoen, 1988) - The latent heat associated with an I-icosahedral phase transition will be essentially the same as that of an I to cubic one (Grebel *et al.*, 1984). This is in qualitative agreement with experiment.

(d) Shear elasticity (Kleiman *et al.*, 1984) - Phason modes are theoretically allowed in icosahedral structures. This could result in their having a lower shear modulus than cubic BP, in agreement with experiment.

(e) Bragg scattering - One would expect any icosahedral phase to exhibit a characteristic Bragg scattering pattern, similar to that seen for metallic quasicrystals. Experimental BP III spectra, on the other hand, show only a single broad peak (Demikhov *et al.*, 1985). This is one of the strongest arguments *against* an icosahedral BP III structure. However, the experimental spectrum is sensitive to the ratio of Bragg peak widths to their spacing. The latter is particularly small ($\sim 5\%$) for σ_{-1} and σ_0 reflections. Also, quasicrystallite size is crucial, typical quasicrystallite dimensions of at least several microns are necessary in order for closely spaced peaks to be resolvable. While such dimensions have been obtained for BP I and BP II, attempts to obtain similar crystallites by "annealing" BP III have been unsuccessful (Demikhov *et al.*, 1985).

A possible method of experimentally testing for an icosahedral BP III structure is by NMR measurements. Theoretical spectra have been calculated for both cubic and icosahedral structures (Hornreich and Shtrikman, 1987a; Filev, 1987). It was found that the theoretical spectrum for an icosahedral structure differs significantly from those of cubic BP. Also, the former differs (Filev, 1987) from that expected from a collection of randomly oriented "double-twist cylinders," another possible model for BP III (Hornreich *et al.*, 1982). NMR studies could thus provide a way of differentiating between different models for BP III.

Other possible approaches to verification of an icosahedral structure for BP III are (a) to produce a surface-aligned BP III structure and probe for additional Bragg peaks, and (b) to study the electric field dependence of the reflected light. Both approaches have been used to study cubic BP (Heppke *et al.*, 1983, 1989; Porsch and Stegemeyer, 1986, 1989; Pieranski *et al.*, 1986) and have recently been applied to BP III. For (a), Yang *et al.* (1988) found no evidence supporting either a periodic or quasiperiodic structure for BP III while for (b), experiments on *positive* dielectric anisotropy materials were inconclusive (Yang and Crooker, 1988) while those on *negative* ones (Yang and Crooker, 1989) were more supportive of non-quasicrystalline models. However, definite conclusions are not yet possible. Finally, inelastic light scattering studies on BP III may be useful. There should be marked differences in scattering mode lifetimes for the different cholesteric phases.

Summerizing, icosahedral structures may possibly characterize BP III. However, further studies are needed in order to decide whether this structure indeed exists.

REFERENCES

Belyakov, V.A., and Dmitrienko, V.E., 1985, The blue phase of liquid crystals, Usp. Fiz. Nauk. 146:369. [trans.: Sov. Phys. Usp. 28:535.]

Collings, P.J., 1984, Optical rotary dispersion measurements in the third cholesteric blue phase, Phys. Rev. A 30:1990.

Crooker, P.P., 1989, The blue phases: a review of experiments, Liq. Cryst. 4:in press.

de Gennes, P.G., 1975, "The physics of liquid crystals." Clarendon, Oxford.

Demikhov, E.I., Dolganov, V.K., and Krylova, S.P., 1985, Selective optical reflection in the fog phase, Pis'ma Zh. Eksp. Fiz. 42:15. [trans.: JETP Lett. 42:16.]

Filev, V.M., 1986, Icosahedral ordering in cholesteric liquid crystals, Pis'ma Zh. Eksp. Teor. Fiz. 43:523. [trans.: JETP Lett. 43:677.]

Filev, V.M., 1987, NMR spectrum of an icosahedral fog phase, Pis'ma Zh. Eksp. Teor. Fiz. 45:190. [trans.: JETP Lett. 45:235.]

Grebel, H., Hornreich, R.M., and Shtrikman, S., 1983, Landau theory of cholosteric blue phases, Phys. Rev. A 28:1114.

Grebel, H., Hornreich, R.M., and Shtrikman, S., 1984, Landau theory of cholesteric blue phases: The role of higher harmonics, Phys. Rev. A 30, 3264-3278.

Heppke, G., Krumrey, M., and Oestreicher, F., 1983, Observation of electro-optical effects in blue phase systems, Mol. Cryst. Liq. Cryst. 99:99.

Heppke, G., Jérôme, B., Kitzerow, H.-S., and Pieranski, P., 1989, Electrostriction of BP I and BP II for blue phase systems with negative dielectric anisotropy, J. Phys. (Paris) 50:549.

Hornreich, R.M., and Shtrikman, S., 1981, Theory of structure and properties of cholesteric blue phases, Phys. Rev. A 24:635.

Hornreich, R.M., Kugler, M., and Shtrikman, S., 1982, Localized instabilities and the order-disorder transitions in cholesteric liquid crystal, Phys. Rev. Lett. 48:1404.

Hornreich, R.M., and Shtrikman, S., 1986a, Broken icosahedral symmetry: a quasicrystalline structure for cholesteric blue phase III, Phys. Rev. Lett. 56:1723.

Hornreich, R.M., and Shtrikman, S., 1986b, A quasicrystalline structure for the cholesteric fog phase, Phys Lett. A 115:451.

Hornreich, R.M., and Shtrikman, S., 1987a, Theoretical nuclear magnetic resonance spectrum for quasicrystalline ordering in cholesteric blue phase III, Phys. Rev. Lett. 59:68.

Hornreich, R.M., and Shtrikman, S., 1987b, Unpublished work.

Hornreich, R.M., and Shtrikman, S., 1988, Landau theory of blue phases, Mol. Cryst. Liq. Cryst. 165:183.

Hornreich, R.M., 1989, Theory of Icosahedral Liquid Crystal Structures, in "Aperiodicity and Order; Vol. 3," M. Jarić and D. Gratias, eds., Academic Press, New York.

Kleiman, R.N., Bishop, D.J., Pindak, R., and Taborek, P., 1984, Shear modulus and specific heat of the liquid-crystal blue phases, Phys. Rev. Lett. 53:2137.

Kleinert, H., and Maki, K., 1981, Lattice textures in cholesteric liquid crystals, Fortschr. Phys. 29:219.

Pieranski, P., Cladis, P.E., Garel, T., and Barbet-Massin, R., 1986, Orientation of crystals of blue phases by electric fields, J. Phys. (Paris) 47:139.

Porsch, F., and Stegemeyer, H., 1986, The effect of an electric field on the selective reflection bands of liquid crystalline blue phases, Chem. Phys. Lett. 125:319.

Porsch, F., and Stegemeyer, H., 1989, Electrostriction of liquid crystalline blue phases, Chem. Phys. Lett. 155:620.

Rokhsar, D.S., and Sethna, J.P., 1986, Quasicrystalline textures of cholesteric liquid crystals: the blue fog? Phys. Rev. Lett. 56:1727.

Thoen, J., 1988, Adiabatic scanning calorimetric results for the blue phases of cholesteryl nonanoate, Phys. Rev. A 37:1754.

Wright, D.C., and Mermin, N.D., 1989, Crystalline solids: the blue phases, Rev. Mod. Phys. 61:385.

Yang, D.K., and Crooker, P.P., 1988, BP III phase of chiral liquid liquid crystals in an electric field, Phys. Rev. A 37:4001.

Yang, D.K., and Crooker, P.P., and Tanimoto, K., 1988, Reflectivity of surface-aligned blue phase III of chiral liquid crystals, Phys. Rev. Lett., 61:2685.

Yang, D.K., and Crooker, P.P., 1989, The effect of electric fields on blue phase III with negative dielectric anisotropy Phys. Rev. A, submitted.

INTRINSIC FRUSTRATION AND SPACE CURVATURE IN SMECTIC A LIQUID CRYSTALS

J.B. Fournier, J.F. Sadoc, and G. Durand

Laboratoire de Physique des Solides, associé au CNRS, LA2
Bât. 510, Université Paris-Sud, 91405 ORSAY Cedex France

ABSTRACT

Smectic A connex droplets in Isotropic melt are frustrated systems. Frustration comes from two incompatible constraints : layer equidistance in the bulk and layer orientation perpendicular to the Smectic A-Isotropic interface. In a 2D model, this frustration is released in the S_2 curved space. Uncurving S_2 to R_2, one finds back the bâtonnets and "zig-zag" shapes observed in thin samples.

INTRODUCTION

Smectic A (SA) liquid crystals are lamellar systems[1]. Each lamella is a 2D liquid layer of molecular thickness. In each layer, rod-like molecules present a nematic ordering, i.e. the average molecular axis (director $\mathbf{n}$) is perpendicular to the lamella plane. These phases show usually a large number of macroscopic defects of the "focal conic" (F.C.) type[2]. In these defects, the SA layers are curved at constant thickness. These arrangements require topologically the creation of two disclination lines: an ellipse and its confocal hyperbola, in the most general case.

It has been recently shown[3], that SA "single crystals", in contact with air on one side and isotropic (I) phase on the other, with layers initially parallel to both interfaces, are unstable, over a SA thickness threshold, with respect to the creation of a dense periodic network of F.C. This instability was quantitatively explained by a frustration related to the anisotropy of interfacial energies : the SA layers want to be parallel to the air interface and perpendicular to the I interface. This can only be achieved in thick samples by the onset of F.C. bulk defects network.

In this paper, we show that surface frustration is an intrinsic feature of SA-I liquid crystals : a SA drop nucleating in the I melt is also frustrated, even in the absence of any air interface. For 2D models, we show that these surface frustrations are released in the S_2 curved space. We obtain straightforward predictions on the possible 2D physical configurations in ordinary flat space.

Geometry and Thermodynamics
Edited by J.-C. Tolédano
Plenum Press, New York, 1990

The method of releasing frustrations by the introduction of space curvature is in fact well-known for amorphous solids[4][5]. In the case of liquid crystals, it was later applied to mesomorphic lyotropic systems[6][7], and to the blue phases of cholesterics[8].

THE GEOMETRICAL FRUSTRATION

In a real physical system, the SA-I and SA-air interfaces, are characterized by finite and antagonistic anisotropies of surface tensions, which lead to the observed angular frustration. To go from physics to geometry, we assume these anisotropies to be infinitely large, and consider them as geometrical constraints : the smectic layers must be *parallel* to the air interface and *perpendicular* to the Isotropic interface.

In the same spirit, following the first de Gennes model[9], the layer curvature energy, which is "weak", must be now taken to be zero. The layer compression, which implies a "large" elastic energy, is accordingly taken as infinite, resulting in a bulk geometrical constraint : the equidistance of (more or less curved) layers. Finally, all these physical properties lead to three geometrical constraints :
- smectic layers parallel to the air interface,
- smectic layers perpendicular to the I interface,
- equidistance of smectic layers in the bulk.

DEFRUSTRATION IN CURVED SPACE

We restrict our discussion to a 2D model, where the layers become equidistant lines. We start in the flat euclidian space R_2 and consider the SA "single crystal" island at the air-I interface, shown on the left of fig. 1. The (a) and (b) side satisfy the angular constraint at the SA-I interface, but not the (c) side. If we simply supress the (c) side at constant air-SA "area", we must curve the SA-I interface (fig. 1 right). This obviously violates the SA equidistance of layers.

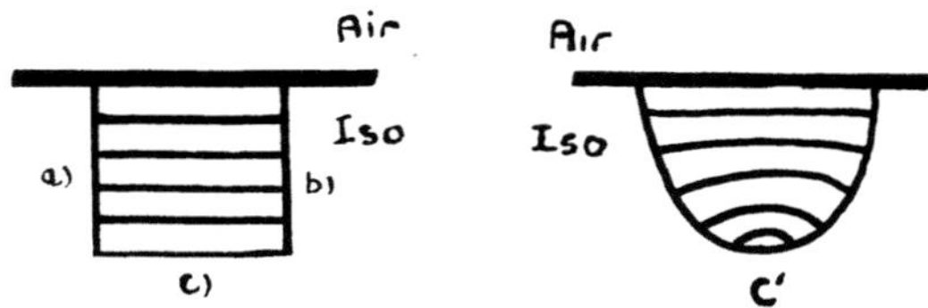

Fig.1.   *2D Smectic island at air-Isotropic interface.*
In the R_2 euclidian space, the system is frustrated :
*- **left** : layer equidistance is respected but the angular condition is violated on the (c) side.*
*- **right** : angular conditions are respected but layer dilation is introduced.*

In this 2D R_2 flat space, the system is frustrated. The frustration comes from the incompatibility of the three geometrical constraints : the two surface orientations and the equidistance of layers conditions.

Another way to supress the (c) side is to cut the SA island as shown on the left of fig. 2, and then stick back the lips together. This procedure naturally leads us to the curved space S_2, in which the three geometrical constraints are fully satisfied. It remains only one singular point c'

In fact, among the three geometrical constraints, only two are intrinsic to the liquid crystal material : the layer equidistance and the perpendicular orientation of layers at the SA-I interface. The parallelism of layers to the air interface, probably caused by the density drop of the material at the contact with air, can be considered as an external, and not intrinsic constraint. Let us show that these two intrinsic constraints are sufficient to induce a frustration.

We consider now (fig. 3 left) a SA island fully surounded by the Isotropic melt. In fact, this island is simply obtained from the previous one by symmetrizing with respect to the previous (but now non existant) air interface. We see now two "unhappy" sides : (c) and (d), where layer perpendicularity to the I-SA interface is not respected.

We now cut the smectic island as previously, following the same symmetric procedure. We see that the frustration is also released on the S_2 curved space (fig. 3 right). The smectic layers are now equidistant and everywhere perpendicular to the I interface. Note that it now remains two singular points c' and d'.

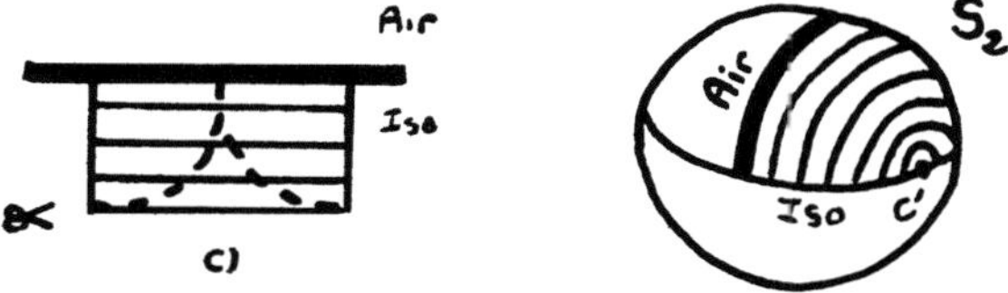

Fig.2.   *Defrustration on the curved space S_2.*
*Cutting along the dotted line to suppress the (c) side (**left**) leads to the S_2 sphere (**right**).*
On S_2, the system is no more frustrated : layer equidistance and angular conditions (layers perpendicular to the Smectic--Isotropic interface and parallel to the air interface) are both respected.

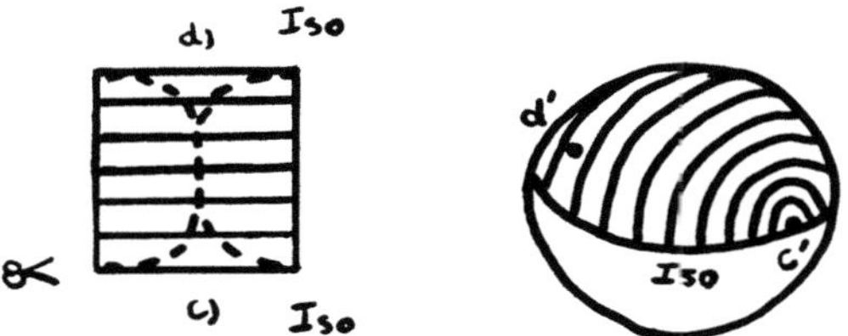

Fig.3.   *A 2D Smectic island inside isotropic melt (**left**) is still frustrated. Frustration is also released on the S_2 sphere (**right**). On S_2 the layers are equidistant and everywhere perpendicular to the Isotropic boundary.*

BACK TO EUCLIDIAN SPACE

To predict the equilibrium shapes of SA "droplets" in our flat space R_2 (with or without air interface), we must start from the non frustrated texture on S_2 (right of fig. 3 or right of fig. 2), and uncurve S_2. To uncurve S_2, we must choose a particular procedure. During this procedure, the geometrical constraints must be somehow relaxed, since in the flat R_2 space, these constraints are antagonistic. Obviously, uncurving must introduce as less stresses as possible.

We consider the half sphere S_2 shown on the left of fig.4. Layer dilation will necessarily be introduced during the uncurving process. We shall assume that the half sphere S_2 has a "rubber-like" behavior. If we simply flatten the "rubber-like" half sphere, we introduce high stresses concerning layer equidistance and interface angular requirements.

Therefore, we first make meridian cuts of small angle θ, as the one indicated in the left of fig.4. As this meridian cut is very "close " to the cylinder C_2 tangent to S_2, flattening it onto the cylinder requires very low stresses. Since the cylinder is a developpable surface, of zero gaussian curvature, the uncurving process is achieved. The elementary objects obtained by this process are the flattened meridian cuts shown on the right of fig. 4, in which the constraints are now slightly relaxed by small layer dilatation and angular mismatch at the SA-I interface.

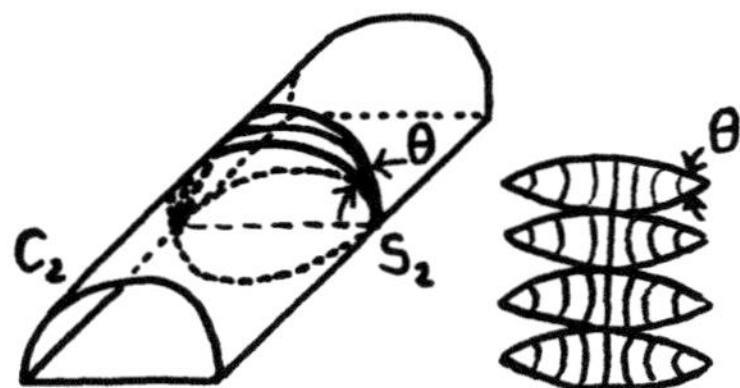

Fig.4.   *Uncurving procedure : A meridian cut of small angle θ is flatten onto the cylinder C_2 tangent to S_2 (**left**). The flat elementary objects obtained (**right**) are relaxed by small layer dilation and angular mismatch at the Smectic-Isotropic interface.*

We now compare the uncurved objects with the experimentally observed structures. The elongated form of the individual meridian cuts fit the shape of the 2D real "bâtonnets"[2] observed in the case of SA nucleation in I phase inside thin samples, i.e. the long axis is parallel to the layering direction (fig. 5 left).

Let us now reintroduce the air interface as previously and let the twin meridian cuts relax elastically, we can predict a 2D instability of SA sandwiches between air and I phases (fig. 5 right). This 2D "zig-zag" instability has indeed been observed in thin samples[10].

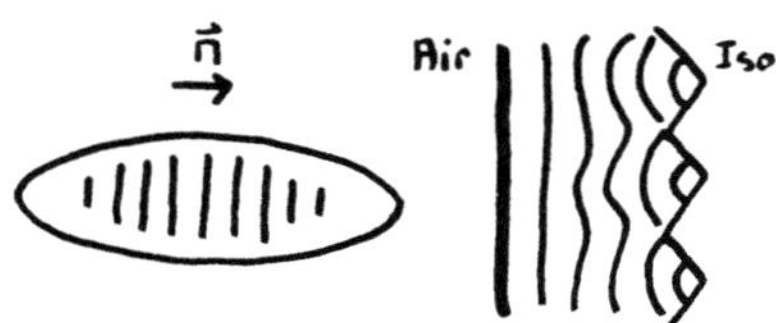

Fig.5.   *Predictions experimentally observed. 2D "bâtonnets" (**left**) and "zig-zag" 2D instability in thin samples (**right**).*

CONCLUSION

In summary, we have shown that smectic A materials are intrinsically frustrated systems in presence of their isotropic phase. The intrinsic frustration comes from the impossibility, for a connex volume of Smectic A in the Isotropic melt, to satisfy both perpendicularity of the layers to the interface everywhere along the boundary and equidistance of the layering into the smectic bulk. In a 2D model, this frustration can be released on S_2 and leads in R_2 to the elongated shape of "bâtonnets" and to the "zig-zag" instability observed in thin samples.

In the same spirit , we expect 3D models to explain the onset of 3D focal conics. The focal conics could then be considered as intrinsic defects of smectic materials.A more complete 3D description is now in progress.

REFERENCES

(1)   P.G. De Gennes, "The Physics of Liquid Crystals", Academic Press, New York (1974).

(2)   G. Friedel and F. Grandjean, Bull. Soc. Fr. Miner. , 33, 409 (1910).

(3)   J.B. Fournier, I. Dozov, G. Durand. Oral communication at VI European Liquid Crystal Winter Conference, Schladming (1989). To be published.

(4)   G. Venkataraman and Debendranath Sahoo, Contemp. Phys., 26 n°6, 579 (1985).

(5)   G. Venkataraman and Debendranath Sahoo, Contemp. Phys., 27 n°1, 3 (1986).

(6)   J.F. Sadoc and J. Charvolin, J. Phys. 47 , 683 (1986).

(7)   J. Charvolin and J.F. Sadoc, J. Phys. 48, 1559 (1987).

(8)   B. Pansu and E. Dubois-Violette, J. Phys. 48, 1861 (1987).

(9)   P. G. De Gennes, J. Phys. 30, C4-65 (1969).

(10) J.B. Fournier, G. Durand, to be published.

CRITICAL BEHAVIOR OF POLYMORPHIC SMECTIC-A LIQUID CRYSTALS

C. W. Garland

Department of Chemistry and Center for Materials
Science and Engineering
Massachusetts Institute of Technology
Cambridge, Massachusetts 02139 USA

I. INTRODUCTION

Liquid crystal molecules with long aromatic cores and
polar head groups are characterized by competition between two
order parameters with incommensurate natural wave vectors. As
a result, a rich variety of polymorphic smectic-A (and also
smectic-C) phases are observed. Such polymorphism in
"frustrated" smectic liquid crystals was first discovered
experimentally [1] and discussed theoretically [2] in 1979 by
the Bordeaux liquid crystal group. By 1983, this group, in
collaboration with A. M. Levelut, had characterized the
structures, determined the phase diagrams, and explored some
of the physical properties of a large variety of these new
thermotropic liquid crystal compounds [3-5]. The generic
molecular formula for a frustrated smectic compound is

$$R{-}\langle\bigcirc\rangle{-}X{-}\langle\bigcirc\rangle{-}Y{-}\langle\bigcirc\rangle{-}CN\,(or\ NO_2) \tag{1}$$

where R is an alkyl or alkoxy group and X, Y are small groups
such as $CH{=}CH$, $C{\equiv}C$, $CH{=}N$, COO. Polymorphism is sensitive to
the length of the nonpolar tail R and to the relative
orientations of the dipole in the strongly polar head group
and those in the linking groups X and Y of the three-ring
aromatic core.

Experimental results and comparison with appropriate
theory will be reviewed for phase diagrams, phase transition
behavior, and critical fluctuation effects in pure compounds
and binary mixtures at 1 atm. The emphasis will be on high-
resolution x-ray scattering and ac calorimetry studies. The
former provide information about structure and about positional fluctuations (i.e., the smectic mass density order-
parameter susceptibility σ and correlation lengths $\xi_{\parallel}$ and $\xi_{\perp}$).

Geometry and Thermodynamics
Edited by J.-C. Tolédano
Plenum Press, New York, 1990

The latter characterize the thermal fluctuations by providing values of the heat capacity C_p. These two types of fluctuation response are closely related via hyperscaling and two-scale-factor universality [6]. The focus will be on smectic-A (SmA) systems, where uniaxial commensurate phases and incommensurate phases can occur as well as laterally modulated (biaxial) smectic and reentrant nematic phases in which frustration is resolved without incommensurability. Interesting smectic-C (SmC) analogous phases are mentioned briefly but not dealt with in detail.

The flavor of this review will be rather personal in that much of the work described has been done at MIT in my laboratory (ac calorimetry) and that of Birgeneau (x-ray scattering). There have also been close interactions with Shashidhar, formerly at the Raman Institute in Bangalore, and Sigaud and Hardouin at Bordeaux. Important synthetic work has been carried out by Tinh [5], Goodby, and Heppke. Earlier reviews concerning one or another aspect of SmA polymorphism are available in Refs. [3,4,7,8].

II. THEORY

In its simplest terms, the SmA liquid crystal phase results from the development of a 1-D density wave in an orientationally ordered nematic (N) phase. The smectic wave-vector q_A is parallel to the nematic director (z axis), and the SmA order parameter $\psi = |\psi|e^{i\varphi}$ can be defined by

$$\rho(z) = \rho_o[1 + \text{Re } \psi\, e^{iq_Az}] \tag{2}$$

Thus the SmA order parameter has two components — a magnitude and phase angle — like superfluid helium and belongs to the XY (n=2 vector model) universality class [9]. For nonpolar materials, there is only one type of smectic-A (let us denote it by SmA_m for *monomeric*), for which the layer spacing $d=2\pi/q_A$ equals the molecular length L. For polar materials with a molecular structure like that shown in formula (1), several smectic-A phases can exist. The most common examples are the *monolayer* SmA_1 with $d \cong L$, the *partial bilayer* SmA_d for which $L < d < 2L$, and the *bilayer* SmA_2 with $d \cong 2L$. Structures with in-plane (xy) modulation are the *fluid antiphase* $Sm\tilde{A}$ and the *crenelated* SmA_{cren} phase. Analogous tilted smectic phases, in which the director (average molecular axis direction) is tilted with respect to the layer normal, are SmC_m, SmC_d, SmC_2, $Sm\tilde{C}$. Schematic smectic-A structures are shown in Fig. 1, and the associated x-ray scattering patterns are given in Fig. 2.

For fluctuation-dominated phase transitions, one expects power law singularities in the response functions C_p and σ and in the correlation lengths $\xi_{\parallel}$ and $\xi_{\perp}$:

$$\sigma = \sigma_o|t|^{-\gamma}, \quad \xi_{\parallel} = \xi_{\parallel 0}|t|^{-\nu_{\parallel}}, \quad \xi_{\perp} = \xi_{\perp 0}|t|^{-\nu_{\perp}} \tag{3a}$$

222

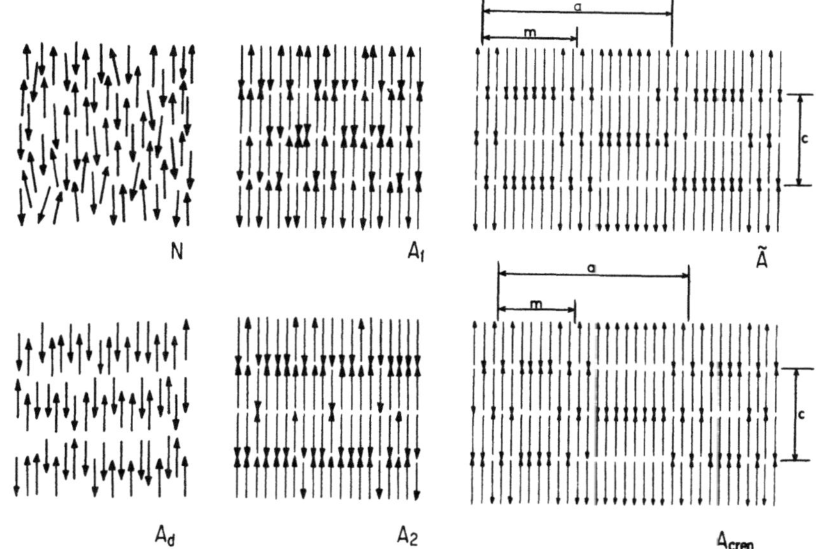

Fig. 1. Schematic representations of various smectic-A
structures. For SmÃ, m = a/2; whereas for
SmA$_{cren}$, m < a/2 and m approaches zero at the
SmA$_{cren}$-SmA$_2$ transition.

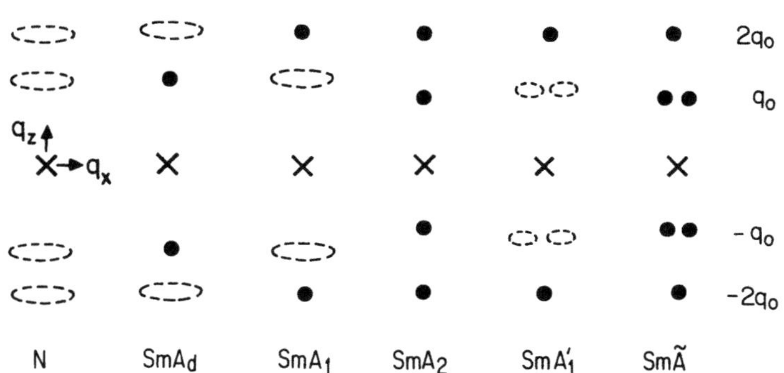

Fig. 2. Sketches of the x-ray scattering
patterns for polar smectic-A phases,
showing two wavevectors q_1 and q_2
where $q_0 \leq q_1 < 2q_0$ and $q_2 = 2q_0 =$
$2\pi/L$. The pattern for SmA$_{cren}$, similar to that for SmÃ, is not shown
(see[48]). The dark dots represent
condensed quasi-Bragg peaks, and the
dashed elliptical areas are diffuse
spots associated with short-range
fluctuations.

$$\Delta C_p^{\pm} = A^{\pm}|t|^{-\alpha}(1 + D^{\pm}|t|^{0.5}) + B_c \qquad\qquad (3b)$$

where the reduced temperature $t \equiv (T - T_c)/T_c$, $\xi_{\parallel}$ is parallel to the director (along the long axis of the rodlike molecule and normal to the layers), $\xi_{\perp}$ lies in the smectic plane, and $\Delta C_p = C_p - C_p$(background) is the excess heat capacity above the noncritical C_p(background) variation expected in the absence of the transition. Each universality class (Ising, XY, etc.) has its own characteristic set of exponent values [10]. Furthermore, there are general relationships among certain of these exponents and associated amplitudes: $\alpha + v_{\parallel} + 2v_{\perp} = 2$ (anisotropic hyperscaling) and $A\xi_{\parallel o}\xi_{\perp o}^2 = $ universal constant (two-scale-factor universality) [6].

If fluctuations are unimportant, the phase transition can often be described by the Landau model with a free energy in the form

$$G = G_o + at\psi^2 + b\psi^4 + c\psi^6 \qquad\qquad (4)$$

In this case, $\gamma = 1$, $v_{\parallel} = v_{\perp} = 0.5$, and ΔC_p is given by

$$\Delta C_p = A^*(T_k - T)^{-1/2} \qquad\qquad T < T_c$$
$$= 0 \qquad\qquad T > T_c \qquad\qquad (5)$$

where $T_k \equiv T_c + (b^2 T_c/3ac)$ and $A^* \equiv (a^3/12cT_c)^{1/2}$ [10].

Three types of theoretical effort have been undertaken on polymorphic SmA phases: phenomenological mean-field theories of frustrated smectics described by two order parameters with incommensurate wave vectors k_1 and k_2, developed by Prost and Barois [2,4,8,11-21]; spin-gas microscopic models in which both antiferro dipole frustration and tail-tail interactions play key roles, developed by Berker and Indekeu [22-31]; renormalization-group (RG) and elastic defect models in which fluctuation effects are taken into account, developed by Lubensky and coworkers [9,32-36]. In many cases, excellent agreement between theory and experiment is achieved, but several puzzling and challenging problems remain.

A. Frustrated Smectic Model

Prost and coworkers [4,8] have developed a mean-field treatment of frustrated smectics in terms of two complex long-range order parameters

$$\Psi_1(r) = \psi_1(r)\exp(iq_1 \cdot r) + c.c.$$
$$\qquad\qquad (6)$$
$$\Psi_2(r) = \psi_2(r)\exp(iq_2 \cdot r) + c.c.$$

where $\Psi_1 = |\Psi_1(r)|e^{i\alpha}$ and $\Psi_2 = |\Psi_2(r)|e^{i\beta}$. These order parameters have incommensurate natural wavevectors k_1 and k_2, which should be distinguished from the observed wavevectors q_1 and q_2 that result from coupling and elastic interactions. These order parameters can be considered as a polarization vector $P_z(r) \sim \Psi_1(r)$ associated with an antiparallel pair of dipolar molecules (thus $k_1 = 2\pi/L'$ is the "dimer" length $L < L' \leq 2L$) and a mass-density vector $\rho(r) \sim \Psi_2(r)$ (for which $k_2 = 2\pi/L$), but such an identification is not necessary. The Landau-Ginzburg free energy is given by

$$F - F_N = a_1(T-T_{1c})|\Psi_1|^2 + D_1|(\Delta+k_1^2)\Psi_1|^2 + C_1|\nabla_\perp\Psi_1|^2$$

$$+a_2(T-T_{2c})|\Psi_2|^2 + D_2|(\Delta+k_2^2)\Psi_2|^2 + C_2|\nabla_\perp\Psi_2|^2 \qquad (7)$$

$$+u_1|\Psi_1|^4+u_2|\Psi_2|^4+2u_{12}|\Psi_1|^2|\Psi_2|^2-w\mathrm{Re}\Psi_1^2\Psi_2^*-v\mathrm{Re}\Psi_1\Psi_2^*$$

The elastic terms $D_i|(\Delta + k_i^2)\Psi_i|^2$ favor $q_1^2 = k_1^2$ and $q_2^2 = k_2^2$, while the gradient terms $C_i|\nabla_\perp\Psi_i|^2$ favor q_1 and q_2 parallel to the director n. The third-order coupling term with coefficient w favors lock-in with $q_2 = 2q_1$, and the bilinear term with coefficient v favors a lock-in at $q_2 = q_1$. In general, these two coupling terms are not of comparable importance. For compounds with $k_2/k_1 \cong 1.7$ (e.g., DB_nCN), the third-order coupling term dominates; for different compounds with $k_2/k_1 \cong 1.3$ (e.g., T8), the bilinear coupling term governs the behavior [8].

The presence of two types of order that want to condense at incommensurate length scales leads to the rich polymorphism of these compounds. The nematic phase corresponds to $\Psi_1 = 0$ and $\Psi_2 = 0$ and yields only diffuse scattering. The SmA_1 phase is identified with $\Psi_1 = 0$ and $\Psi_2 \neq 0$ and yields a quasi-Bragg spot at $q_2 = k_2 = 2q_0$ plus diffuse scattering at q_1. Both the SmA_d and SmA_2 phases have $\Psi_1 \neq 0$ and $\Psi_2 \neq 0$. In the case of SmA_2, $|\Psi_1| \cong |\Psi_2|$ and Bragg spots are observed at $q_1 = q_0$ and $q_2 = 2q_0$. For SmA_d, $|\Psi_1|$ is large and $|\Psi_2|$ very small, which yields Bragg scattering at $q_1 \neq q_0$ and only diffuse scattering at $q_2 = 2q_0$. This theory also predicts under special circumstances an incommensurate smectic phase SmA_{ic} in which two collinear incommensurate density modulations coexist with q_2/q_1 an irrational fraction [4,8,18]. The incommensurability parameter

$$z^2 \equiv \frac{2D_1u_{12}}{w^2}(k_1^2 - \tfrac{1}{4}k_2^2)^2 \qquad (8)$$

governs the stability of such SmA_{ic} phases. However, frustration can be relieved in a simpler way by the two-dimensional lock-in associated with the fluid antiphase $Sm\tilde{A}$ or the ribbon phase $Sm\tilde{C}$ if the gradient term coefficient C_1 is sufficiently small [8,12,14]. Competition between incommensurate order

parameters could also lead to reentrant nematic behavior if
fluctuation effects are taken into account [4,11,13]. Finally,
the frustrated smectic model can yield a considerable variety
of critical and multicritical points — SmA_d–SmA_2 critical, N–SmA_1
tricritical, N–SmA_2 tricritical, SmA_1–SmA_2 tricritical, and
SmA_d–N_r–SmA_1 bicritical.

B. Frustrated Spin-Gas Model

Berker and Indekeu [28] have developed a microscopic
model that provides a detailed description of the SmA_d, SmA_1,
and reentrant nematic N_r phases of polar liquid crystals.
Indeed, this model is capable of predicting complex reentrant
sequences on cooling such as N–$SmA_{d'}$–N_r–SmA_d–N_r–SmA_1. The polar
molecule is modelled by a rigid linear core, a longitudinal
dipolar head, and a semi-flexible corrugated (all-trans)
aliphatic tail. There is a competition between short-range
positional order laterally in the smectic layer and antiferroelectric dipole order. The close packing that can occur
in smectic layers gives rise to dipole-dipole frustration.
Furthermore, tail-tail van der Waals attraction and steric
repulsion play a role by providing "substrate fields" for the
positional motion of neighboring molecules normal to the
smectic layer. The calculation is carried out on a nearest-
neighbor triplet of molecules and uses a prefacing transformation that effectively reduces the problem to one of long-
range ferro- or antiferroelectric order in a planar Ising
model. However, the molecules, which are located at the
corners of an equilateral triangle of side a in the xy plane,
can permeate in the z direction (translate normal to the
smectic layer). This permeation has a strong influence on the
stability of SmA_d phases.

The spin-gas model predicts phase stability by examining
the statistical weights of various distinct microscopic configurations, such as those shown in Fig. 3. Atomic permeation
tion (AP) corresponds to vertical shifts of the dipolar heads
by $\Gamma \equiv L/n$, where L is the length of the molecule and n is the
number of "notches" in the aliphatic chain. Librational permeation (LP) involves vertical shifts on a length scale δ that
is much smaller than Γ. Librational permeation can relieve
the frustration for configurations (b) and (c) in Fig. 3 and
allow local antiferroelectric order. Such species stabilize
the low-temperature SmA_d phase, while atomic-permeated configuration (d) occurs in the high-temperature $SmA_{d'}$ phase.
Configuration (e) is associated with SmA_1.

Phase stability depends quite sensitively on the temperature, on the ratio a/L of the lateral spacing to the effective
molecular length, on the values of a/Γ, and on the relative
strength of dipole-dipole interaction compared to tail-tail
interaction. Happily, spin-gas theoretical phase diagrams
reproduce the essential topology of experimental ones for
mixtures [29] and pure compounds [30] using physically
plausible parameter values. Since the calculation ignores
tilted configurations, smectic-C phases do not occur. Also

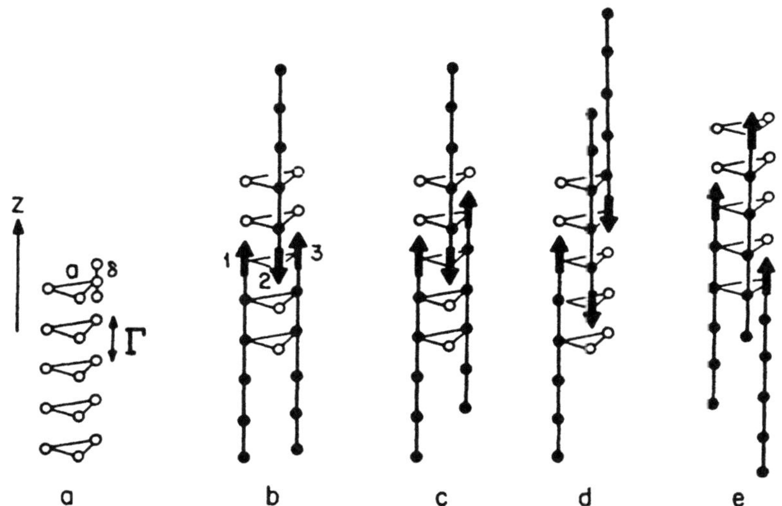

Fig. 3. Configurations of triplets of
molecules in the spin-gas model
[28]. (a) Atomic permeation positions $\Gamma = L/n$ and librational permeation positions δ; (b) and (c)
frustrated configurations; frustration relieved by atomic permeation
allowing local antiferroelectric
order(d) or ferroelectric order(e).

the SmA_2 phase is missing due to the two-dimensional character
of the present frustration model; and, of course, $Sm\tilde{A}$ and
SmA_{cren} phases do not occur since they are lateral antiphase
modulations of the SmA_2 structure.

C. Renormalization-Group Calculations

Lubensky has carried out very extensive RG calculations
on the N-SmA transition without regard to the detailed type of
SmA phase; see Ref. [9] for a review. More recently, he and
various coworkers have applied RG techniques to describe the
fluctuation effects involved in second-order SmA_1-SmA_2 transitions [32] and to deal with a new type of critical point
associated with SmA_d-SmA_1 and SmA_d-SmA_2 transitions [35,36].
Since the macroscopic symmetries of these SmA phases are
identical, there is no symmetry-breaking transition. Thus
there can be a first-order transition line (across which q_A
jumps discontinuously from q_A^- to q_A^+) terminating in a
critical point, beyond which there is a supercritical region
with a continuous non-singular evolution of q_A and no thermodynamic transition at all. Alternatively, the SmA-SmA' transition line can terminate in a closed re-entrant nematic "bubble".

Prost and Toner [19] have used a dislocation loop theory to show that either a nematic bubble in a smectic sea or a SmA-SmA$'$ critical point can exist. Park et $al.$ [35,36] have studied a nonlinear elastic model of the SmA-SmA$'$ critical point and found that it belongs to a new universality class with upper critical dimension $d_c = 6$ (as opposed to $d_c = 4$ for the liquid-gas Ising critical point). However, RG analysis predicts that small symmetry-breaking fields (electric, magnetic, or strain) would drop d_c to 2.5 and make the transition mean-field-like. The role of elasticity, defects and hydrodynamic modes in SmA$_{ic}$ liquid crystals has also been investigated by Lubensky et $al.$ [34], and it is predicted that second-order SmA-SmA$_{ic}$ transitions should belong to the 3-D XY universality class.

A summary of theoretical predictions and presently available experimental results for the character of various N-SmA and SmA-SmA$'$ transitions is given in Table I.

Table I. Theoretical predictions and current experimental results concerning the character of various polymorphic SmA transitions. 1st = first-order, 2nd = second-order, CP = critical point, CEP = critical end point, BP = bicritical point, TCP = tricritical point; Ising and XY are n=1 and n=2 vector model second-order transitions.

Transition	Theory [Ref.]	Experiment
N-SmA$_1$	XY [9,15]	XY 2nd for C_p
N-SmA$_2$	XY$\rightarrow$TCP$\rightarrow$1st [4]	1st
SmA$_1$-SmA$_2$	Ising [2,32]	Fisher renormal. Ising
SmA$_d$-SmA$_2$	1st$\rightarrow$CP [35]	1st$\rightarrow$CP (unusual)
SmA$_d$-SmA$_1$	1st$\rightarrow$CP [35]	−
	1st$\rightarrow$BP [17]	1st up to N-SmA$_1$ CEP
	1st$\rightarrow$N bubble [19]	1st$\rightarrow$N bubble
SmA$_1$-Sm$\widetilde{\text{A}}$	1st [4]	weak 1st, large pretrans.
SmA$_1$-Sm$\widetilde{\text{C}}$	1st [4,8,15]	weak 1st, some pretrans.
Sm$\widetilde{\text{A}}$-SmA$_{cren}$-SmA$_2$	Sm$\widetilde{\text{A}}$-SmA$_2$ 1st [8]; no SmA$_{cren}$	truncated Sm$\widetilde{\text{A}}$-SmA$_2$ C_p peak
Sm$\widetilde{\text{C}}$-SmA$_2$	1st [4,14]	−
Sm$\widetilde{\text{C}}$-SmC$_2$	1st [4,15]	weak 1st, inverted Landau
SmA$_2$-SmC$_2$	2nd [−]	odd 2nd order C_p
SmA$_d$-SmA$_{i1}$-SmA$_1$	1st [8];XY [8,34]	uncertain character
SmA$_d$-SmA$_{i2}$-SmA$_2$	1st [8]	uncertain character
SmA$_d$-N$_r$-SmA$_1$	XY [9,23,28]	non-XY 2nd SmA$_d$-N$_r$, near-XY N$_r$-SmA$_1$

III. N-SmA$_1 \rightarrow$ SmA$_2$ SYSTEMS

In this and subsequent sections presenting experimental results, it is convenient to use the common short names for liquid crystals instead of the full chemical nomenclature. The conventional names of all liquid-crystal materials mentioned in this review are defined in terms of their proper chemical names and formulae in Table II.

Table II. Identification of various liquid crystal molecules exhibiting polymorphic SmA phases. ϕ denotes $-\bigcirc\!\!-$.

Common name	Chemical name and formula
C$_n$stilbene	cyanobenzoyloxy-alkyl stilbene
	C$_n$H$_{2n+1}$-ϕ-CH=CH-ϕ-OOC-ϕ-CN
DB$_n$CN	alkylphenyl cyanobenzoyloxy benzoate
	C$_n$H$_{2n+1}$-ϕ-OOC-ϕ-OOC-ϕ-CN
DB$_n$OCN	alkoxyphenyl cyanobenzoyloxy benzoate
	C$_n$H$_{2n+1}$-O-ϕ-OOC-ϕ-OOC-ϕ-CN
DB$_n$ONO$_2$	alkoxyphenyl nitrobenzoyloxy benzoate
	C$_n$H$_{2n+1}$-O-ϕ-OOC-ϕ-OOC-ϕ-NO$_2$
TBBA	terephthal-bis-butylaniline
	C$_4$H$_9$-ϕ-N=CH-ϕ-CH=N-ϕ-C$_4$H$_9$
Tn	alkoxybenzoyloxy cyanostilbene
	C$_n$H$_{2n+1}$-O-ϕ-COO-ϕ-CH=CH-ϕ-CN
nOBCB	alkoxybiphenyl cyanobenzoate
	C$_n$H$_{2n+1}$-O-ϕ-ϕ-OOC-ϕ-CN
nOBCAB	alkoxybenzoyloxy cyanoazobenzene
	C$_n$H$_{2n+1}$-O-ϕ-COO-ϕ-N=N-ϕ-CN
nOPCBOB	alkoxyphenyl cyanobenzyloxy benzoate
	C$_n$H$_{2n+1}$-O-ϕ-OOC-ϕ-O-CH$_2$-ϕ-CN
nOCB	alkoxycyanobiphenyl
	C$_n$H$_{2n+1}$-O-ϕ-ϕ-CN

All of the systems considered in this section have a second-order N-SmA$_1$ transition followed at lower temperatures by direct or indirect transitions from SmA$_1$ to SmA$_2$. SmA polymorphism was first discovered and analyzed in DB$_n$CN + TBBA mixtures [1,2], where a direct SmA$_1$-SmA$_2$ transition occurs.

A recent experimental phase diagram for DB$_6$CN + TBBA [37] is shown in Fig. 4a, and the corresponding frustrated smectic theory diagram is given in Fig. 4b. In DB$_n$CN + C$_5$stilbene mixtures, modulated $\widetilde{\text{SmA}}$ and SmA$_{cren}$ phases occur between SmA$_1$ and SmA$_2$; see Figs. 5a and 5b. In the homologous series of compounds nOPCBOB, there is a Sm$\widetilde{\text{C}}$ ribbon phase between SmA$_1$ and SmA$_2$ as shown in Fig. 6. A general theoretical topology like Fig. 6 can also be obtained with the frustrated smectic mean-field model [4,8] but will not be shown here.

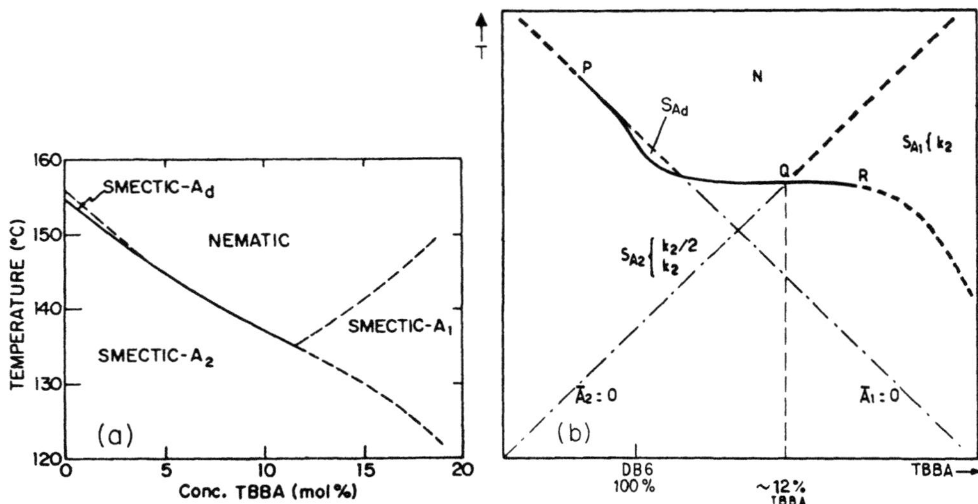

Fig. 4. Partial phase diagrams exhibiting SmA$_1$-SmA$_2$ transition. The solid lines are first-order phase boundaries, and the dashed lines are second-order transition lines. (a) Experimental DB$_6$CN + TBBA diagram [37]. (b) Corresponding theoretical diagram (adapted from [4]). The incommensurability parameter is $z^2 = 0.25$. The dot-dash lines indicate where the conditions $\bar{A}_1 \equiv a_1(T-T_{1c}) = 0$ and $\bar{A}_2 \equiv a_2(T-T_{2c}) = 0$ are satisfied.

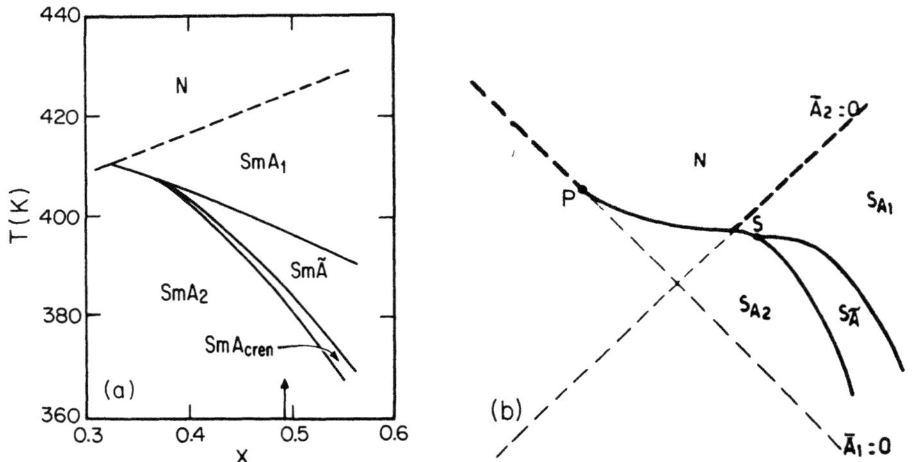

Fig. 5. Partial phase diagrams involving biaxial SmA phases. (a) Experimental diagram for DB$_5$CN + C$_5$stilbene mixtures [38]. X denotes the mole fraction of C$_5$stilbene; the arrow marks the composition investigated with ac calorimetry. (b) Theoretical phase diagram corresponding to 5a (adapted from [4]). The possibility of a SmA$_{cren}$ phase is not included in the theoretical development.

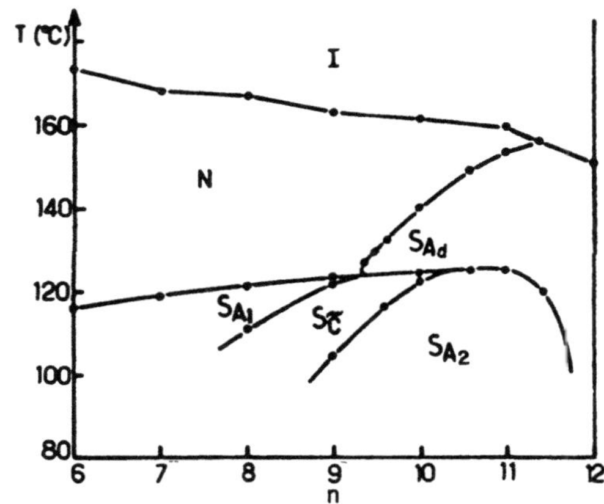

Fig. 6. Generalized phase diagram
for homologous nOPCBOB
compounds and their binary
mixtures. [39]

A. N-SmA₁ Transition

Second-order N-SmA$_1$ transitions are theoretically predicted
to belong to the 3-D XY universality class like all other N-SmA
transitions [4,9]. Previous experimental work on N-SmA$_m$ and
N-SmA$_d$ transitions showed system-dependent nonuniversal critical
behavior with $\nu_\| \neq \nu_\perp \neq \nu_{XY}$ and $\alpha \neq \alpha_{XY}$ [6,9,40]. However, recent
calorimetric studies of several N-SmA$_1$ transitions give excellent
agreement with the XY theoretical C_p behavior [41-44]. Figure 7
shows a typical example of N-SmA$_1$ experimental C_p data and an XY
fit. All of the universal aspects expected for the XY universality class are well satisfied by these 8OPCBOB data [41]:
$\alpha = -0.008$ (XY -0.007), $A^-/A^+ = 0.987$ (XY 0.971), $D^-/D^+ = 1.03$
(XY ~ 1.0), and $A^+|D^+|^{2\alpha}/B_c = -1.074$ (XY -1.057). Comparable
parameters are obtained for four other N-SmA$_1$ systems.

An earlier x-ray and calorimetric investigation of T7 and
T8 gave N-SmA$_1$ critical exponents $\gamma = 1.22 \pm 0.06$, $\nu_\| = 0.69 \pm 0.03$,
$\nu_\perp = 0.63 \pm 0.03$, and $\alpha = 0.06 \pm 0.06$ [45]; see Fig. 8 for the
behavior of σ, $\xi_\|$, and $\xi_\perp$. These exponent values are close, but
not identical, to the ideal XY values $\gamma = 1.32$, $\nu_\| = \nu_\perp = 0.67$, and
$\alpha = -0.007$. Unfortunately, there is some chemical instability and
T_c drift for these Tn compounds, and a high-resolution x-ray study
of the more stable 8OPCBOB and DB$_n$CN + C$_5$stilbene is in progress.
If this study shows that XY critical behavior is valid for all the
properties and not just for the thermal fluctuations alone, the
N-SmA$_1$ transition will satisfy the expected universality for N-SmA
transitions. A possible reason for realizing this XY behavior,

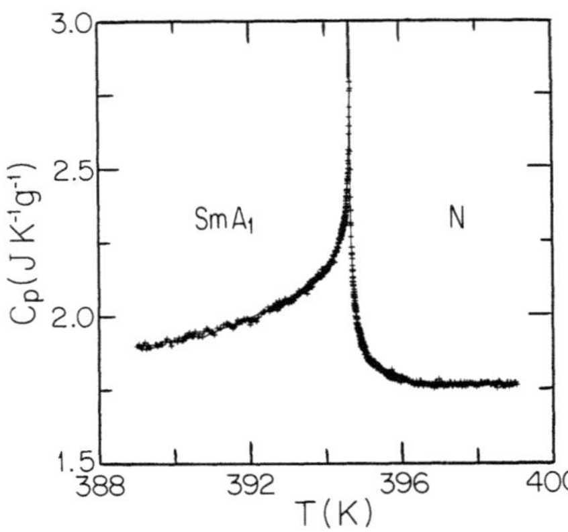

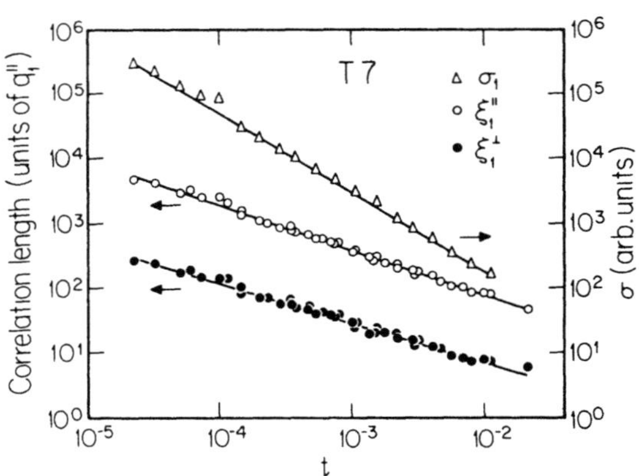

Fig.7.Heat capacity of
80PCBOB associated
with the N-SmA₁ tran-
sition [42]. Smooth
curve is an XY fit.

Fig.8. Smectic susceptibility σ,
longitudinal and transverse
correlation lengths ξ∥, ξ⊥
just above the N-SmA₁
transition in T7 [45].

which is not seen for N-SmA$_m$ and N-SmA$_d$ transitions, is the wide
N range and the resulting saturation of the orientational
nematic order parameter S prior to smectic layer formation.

B. SmA$_1$-SmA$_2$ Transition

Second-order SmA$_1$-SmA$_2$ transitions are theoretically
predicted to belong to the 3-D Ising universality class [2,32]
since they involve an antiferro ordering of the dipolar heads
(see Fig. 1). An overview of the C_p variation associated with
the N-SmA$_1$ and SmA$_1$-SmA$_2$ transitions in DB$_6$CN + TBBA [46] is
shown in Fig. 9. This system has also been studied by x-ray
techniques [37]. The resulting SmA$_1$-SmA$_2$ critical exponents are
$\gamma = 1.46 \pm 0.05$, *isotropic* correlation exponents $\nu_\parallel = \nu_\perp = 0.74 \pm$
0.03, and $\alpha = -0.15 \pm 0.06$ [37,46]. These values obey
hyperscaling ($\alpha + 3\nu = 2.08 \pm 0.15$) but do not agree with the
expected Ising exponents $\gamma = 1.24$, $\nu = 0.63$, $\alpha = +0.11$. However,
they can be understood as Fisher renormalized values

$$\gamma_R = \frac{\gamma}{1-\alpha} \; , \quad \nu_R = \frac{\nu}{1-\alpha} , \quad 2-\alpha_R = \frac{2-\alpha}{1-\alpha} \tag{9}$$

which for the Ising case are $\gamma_R = 1.39$, $\nu_R = 0.71$ and $\alpha_R = -0.124$.
Such renormalized exponents are expected [47] in mixtures along a
path of constant mole fraction if the unrenormalized amplitude of
the C_p peak is large and the phase boundary is steep (dT_c/dX
large), as it is in the present case. A comparable Fisher
renormalization of the SmA$_1$-SmA$_2$ transition is also seen in DB$_5$CN
+ TBBA mixtures [43].

232

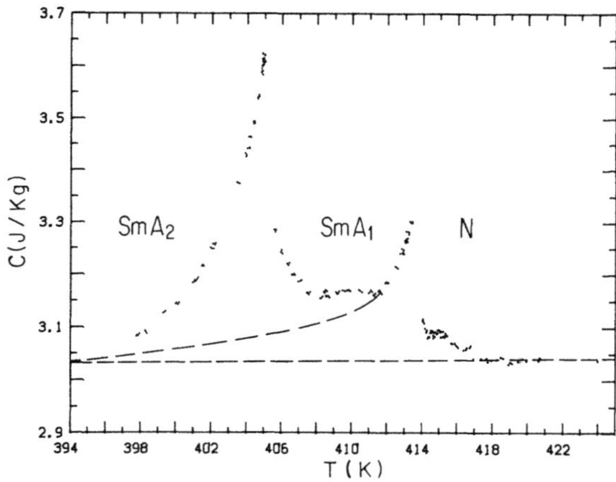

Fig. 9. Heat capacity of DB_6CN + 15% TBBA [46]. The dot–dash line represents an estimate of the tail of the N–SmA_1 peak that serves as baseline for the large $\Delta C_p(SmA_1$–$SmA_2)$ peak.

In summary, the straight-forward N–SmA_1 and SmA_1–SmA_2 transitions seem to agree quite well with theoretical expectations. Let us now consider the more complex modulated phases.

C. SmA_1–$Sm\tilde{A}$–SmA_{cren}–SmA_2 Sequence

The phase diagram for DB_5CN + C_5stilbene given in Fig. 5a shows two biaxial modulated phases occurring between the SmA_1 and SmA_2 phases. The overview of the C_p variation shown in Fig. 10 provides a striking contrast to that for the direct SmA_1–SmA_2 transition shown in Fig. 9. The N–SmA_1 transition is one of those discussed previously and having XY critical behavior [41]. The excess heat capacity ΔC_p associated with the SmA_1–$Sm\tilde{A}$ transition is shown in Fig. 11 for two similar mixtures. This transition is predicted to be first order due to a fluctuation-induced Brazovskii instability [4]; indeed, two-phase coexistence is observed over a very narrow temperature range. However, there are large and unusual pretransitional fluctuation effects, especially in the SmA_1 phase. Strong $Sm\tilde{A}$ fluctuations are observed in x-ray studies [48] and confirmed by the anomalous shear viscosity [38,49]. Analogous but smaller fluctuation effects have also been observed near the SmA_1–$Sm\tilde{C}$ transition [50]. Additional experimental and theoretical work is needed to clarify the nature of these $Sm\tilde{A}$ and $Sm\tilde{C}$ fluctuations.

The C_p variation in the $Sm\tilde{A}$–SmA_{cren}–SmA_2 region is shown in Fig. 12. This C_p peak is a truncated version of the analogous $Sm\tilde{C}$–SmC_2 peak discussed in Sec. V (see Figs. 23 and 25). The pretransitional C_p increase on cooling the $Sm\tilde{A}$ phase is

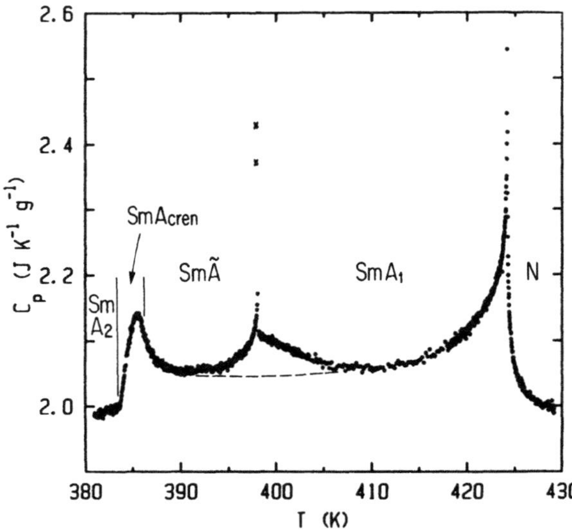

Fig. 10. Heat capacity of DB₅CN + 49.2% C₅stilbene [38]. The two data points marked with crosses indicate coexistence of two phases at the weakly first-order SmA₁-SmÃ transition.

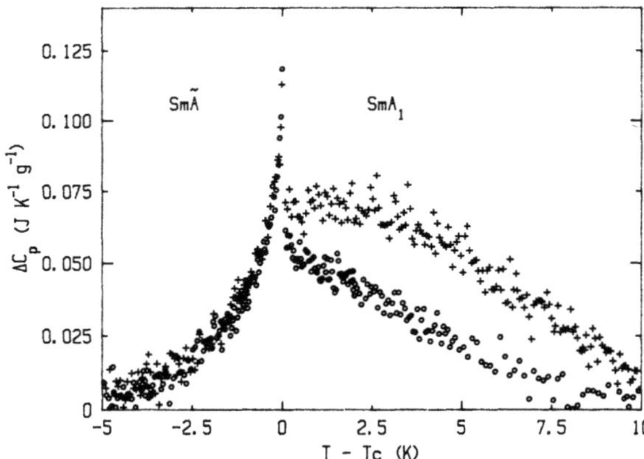

Fig.11. $\Delta C_p = C_p - C_p$(background), where C_p(background) is given by dashed curves like that in Fig. 10. Circles denote DB₅CN+C₅stilbene, and plus signs denote DB₆CN + C₅stilbene [38].

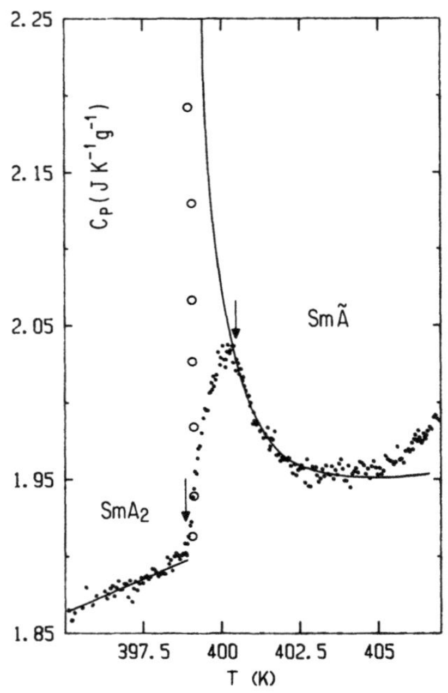

Fig.12. Heat capacity associated with the SmÃ-SmA$_{\mathrm{cren}}$-SmA₂ transition sequence in DB₆CN+C₅stilbene [38]. The solid curve and open circles show the analogous SmC̃-SmC₂ heat capacity in a DB$_n$ONO₂ mixture; the open circles denote data in the two-phase coexistence region (see Fig. 23).

associated with the evolution of the $\widetilde{\text{SmA}}$ structure, in which the lateral repeat distance a grows from ~120Å to ~330Å and the SmA_2-like ordering inside each antiphase region becomes more completely developed [48]. The SmA_{cren} phase is quite different: the value of a is independent of T and m/a approaches zero on cooling [48]. Thus the SmA_{cren} phase is rather like a special (periodic) structure of coexisting $\widetilde{\text{SmA}}$ and SmA_2 slabs, which would be consistent with the truncation of the $\widetilde{\text{SmA}}$–SmA_2 peak between the two arrows in Fig. 12; see Ref. [38] for more details.

IV. INCOMMENSURATE SYSTEMS

When the gradient term $C_1|\nabla_\perp\psi_1|^2$ in Eq. (7) is large, tilting the wavevectors costs more energy and biaxial modulated phases become unstable. If this is the case and the incommensurability parameter z^2 defined in Eq. (8) is large, SmA_{ic} phases can occur in the frustrated smectic model [4,8,14,16,18]. Such a phase will exhibit two collinear incommensurate SmA mass density modulations. Schematic pictures of SmA_{ic} structures are shown in Fig. 13. In the weak-coupling limit (close to the SmA_d–SmA_{ic} phase boundary), two independent extended incommensurate periodic modulations interpenetrate each other, $i.e.$, percolate through each other on a microscopic scale, as shown in Fig. 13a. In the strong-coupling limit (close to the SmA_{ic}–SmA_2 phase boundary), a soliton regime occurs with walls of thickness ξ_k containing interpenetrating independent periodicities k_1 and k_2 separating thicker regions of SmA_2 type, as shown in Fig.13b. The periodicity Z of the discommensurations diverges as the SmA_{ic}–SmA_2 transition is approached [8]. The picture shown in

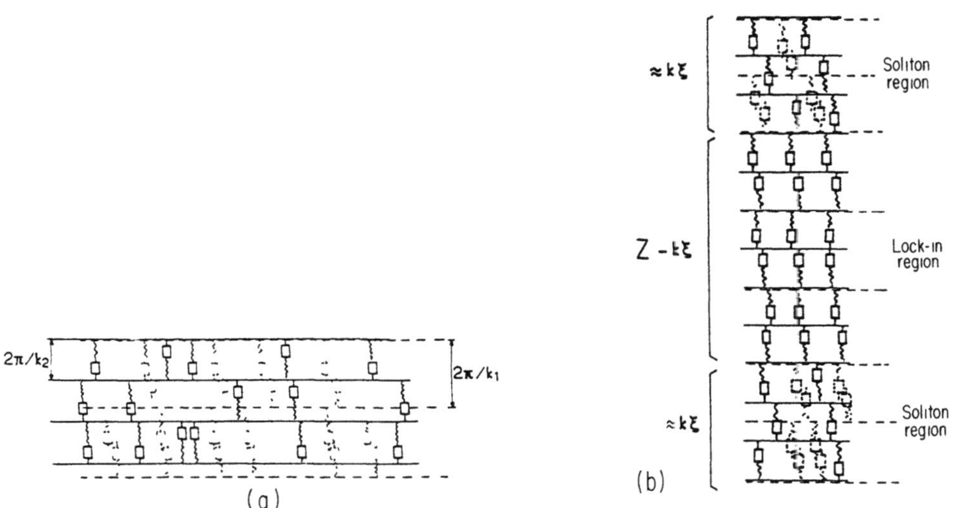

Fig. 13. Schematic SmA_{ic} structures in (a) the weak-coupling limit and (b) the strong-coupling limit with solitons [8]. Most of the molecules fit into one of the two modulation patterns.

Fig. 13a involves the interpenetration of SmA$_d$ and SmA$_1$ modulations and will be denoted as SmA$_{i1}$. When there is a three-dimensional percolation of SmA$_d$ and bilayer SmA$_2$ modulations, the phase will be denoted as SmA$_{i2}$ (not shown). The soliton incommensurate structure in Fig. 13b will be denoted as SmA$_{is}$.

The first observation of an incommensurate smectic-A phase was by Ratna *et al.* [51] in binary mixtures of DB$_7$OCN + 8OCB. The experimental and frustrated-smectic theoretical phase diagrams are given in Figs. 14a and 14b. There is excellent agreement between the transition temperatures obtained from microscopic and x-ray studies [51] and those from an ac calorimetric investigation [52]. The modulation periodicities obtained from sharp Bragg spots are shown in Fig. 15. In the SmA$_{i2}$ phase, there is simultaneous scattering at q_0, q_0' and $2q_0$. A detailed study of the temperature dependence of the layer spacing for a series of 8OCB concentrations rules out the possibility that SmA$_{i2}$ is merely a region of two-phase coexistence [15].

An overview of the C_p variation for two DB$_7$OCN + 8OCB mixtures is given in Fig. 16. The 12.5% sample goes directly from

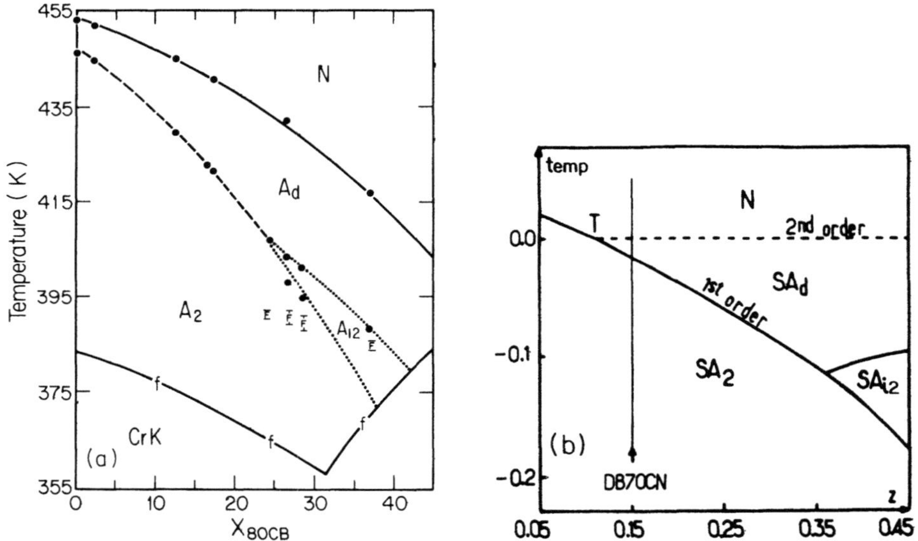

Fig. 14. Partial phase diagrams exhibiting SmA$_{i2}$ phase. (a) Experimental DB$_7$OCN + 8OCB system [52]. The lines are from [51] and the points are from [52]. The dashed SmA$_d$–SmA$_2$ line represents the locus of maxima in C_p and the layer thermal expansion in a supercritical region where no thermodynamic transition occurs. (b) Theoretical phase diagram corresponding to 14a [18]. Note that the SmA$_d$–SmA$_2$ transition is predicted to be first order in this calculation.

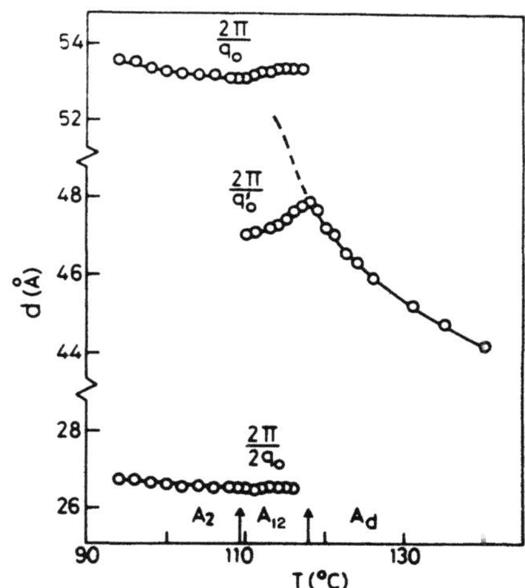

Fig. 15. Temperature variation of the layer
spacing d=2π/q for
a DB₇OCN+34.8% 8OCB
mixture [51]. The
dashed line shows
the expected trend
in $2\pi/q_o'$ if the
SmA_{i2} region were
merely two-phase
coexistence.

SmA_d to SmA_2, while the 27% sample exhibits a SmA_{i2} phase. These
ac calorimetric data contrast interestingly with DSC results
[51] on this system, for which DB₇OCN and low X_{8OCB} mixtures show
sharp peaks like those seen at first-order transitions and
higher X_{8OCB} mixtures show no DSC peaks in the SmA_d-SmA_{i2}-SmA_2
region. Differential scanning calorimetry (DSC) is a generally
sensitive qualitative technique but may fail when the system
response time is sluggish since fast scan rates are required.
Note the distinct changes in dC_p/dT at the transitions involving
the SmA_{i2} phase, marked by the arrows in Fig. 16. The available
information leaves the detailed character of the SmA_d-SmA_{i2} and
SmA_{i2}-SmA_2 transitions uncertain [8,34,51,52]. They do not
behave like first-order transitions but also do not conform to
the usual second-order behavior associated with singular fluctuations. In a formal way, they resemble Ehrenfest third-order
transitions in that there is a discontinuity in a third derivative of the free energy. Note also the rounded C_p peak for the
direct SmA_d-SmA_2 transformation; such rounding with distinct
points of inflection is noted also in pure DB₇OCN. This means
that the system is supercritical with respect to the SmA_d-SmA_2
transition, as allowed by theory [35]. No thermodynamic transition occurs and the dashed "phase line" in Fig.14a represents
the locus of finite maxima in the response functions such as C_p.

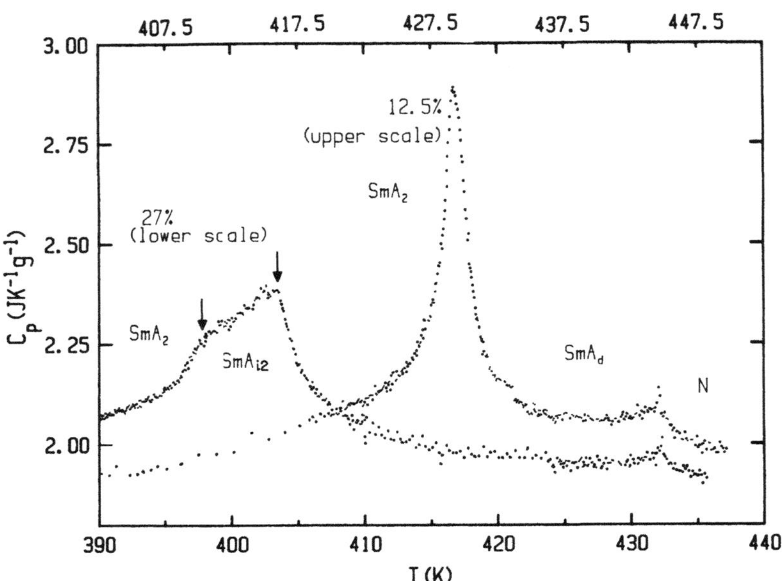

Fig. 16. Heat capacity of two DB$_7$OCN+8OCB mixtures [52]. The mole percent of 8OCB is given. Note that the second-order N-SmA$_d$ transition peak is very small compared to the transformation peaks involving bilayer smectic ordering.

This conclusion is confirmed by x-ray measurements of the smectic layer thickness, which shows a continuous *nonsingular* evolution from SmA$_d$-like to SmA$_2$-like scattering on cooling [52].

Recently, Shashidhar and Ratna [7] have reported evidence of a strongly coupled incommensurate SmA$_{is}$ phase in mixtures of 8OBCAB + DB$_8$OCN. The experimental phase diagram is given in Fig. 17a, and the corresponding theoretical diagram is presented in Fig. 17b. In the latter diagram there is a N-SmA$_d$-SmA$_{ic}$-SmA$_1$ tetracritical point, but it is known that the fourth-order coefficients u_1 and u_2 in Eq. (7) have an important effect on the topology [8,18]. Lower values of u_2 will yield a diagram with a N-SmA$_d$-SmA$_1$ bicritical point and a SmA$_{ic}$ domain disconnected from the N phase as in the experimental situation.

X-ray diffraction data [53] on a 8OBCAB + DB$_8$OCN sample with X = 81.8 mol % DB$_8$OCN show a SmA$_{i2}$ structure between 144°C and 147.5°C with collinear q_o' (SmA$_d$-like), q_o and $2q_o$ (SmA$_2$-like) Bragg spots, just like the behavior shown in Fig. 15. It should be noted that $I_{2q_o} \cong 3I_{q_o}$ for the intensities in the SmA$_2$ phase, in agreement with theory [7,33] and not usually seen. X-ray diffraction data [53] on a sample with X = 74.8 show a SmA$_{i1}$ region between 136°C and 137.6°C with collinear q_o' and $2q_o$ Bragg

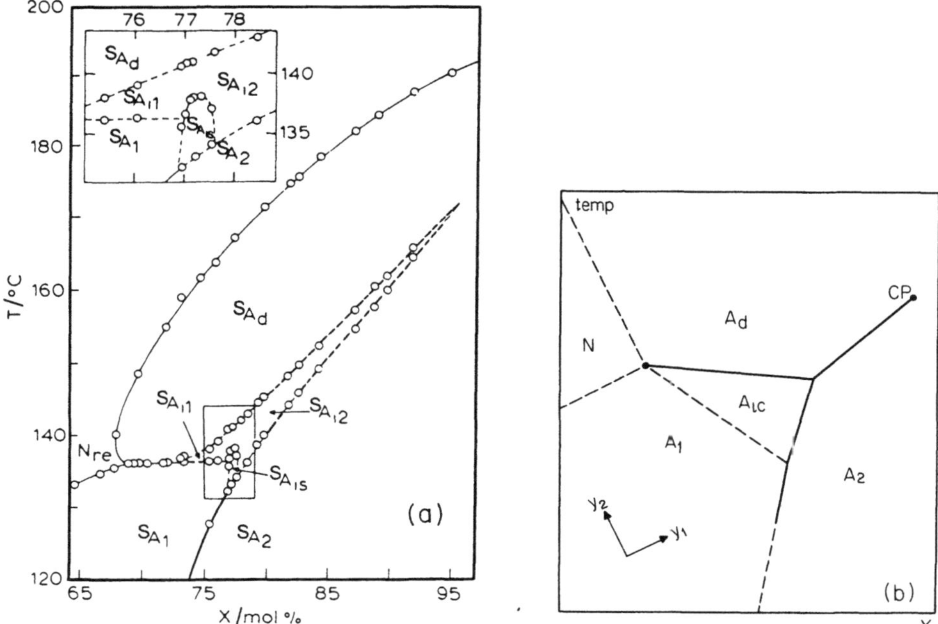

Fig. 17. Partial phase diagrams exhibiting SmA$_{ic}$ phases.
(a) Experimental diagrams for 8OBCAB+DB$_8$OCN mixtures, where X is the mol % of DB$_8$OCN [7]. For X greater than 96, the SmA$_{i2}$ phase disappears and a supercritical (continuous nonsingular) direct transformation from SmA$_d$-like to SmA$_2$-like structures is observed.
(b) Theoretical phase diagram related to that shown in 17a (adapted from [8,18]). The incommensurability parameter, $z^2 = 0.325$ to 0.5, is quite large. The theoretical axes y_1 and y_2 are functions of T and X, and the diagram has been rotated to make clear its correspondence to 17a.

spots. In the SmA$_1$ phase, diffuse SmA$_d$-like scattering is observed at $q_o'' > q_o$. This wavevector shifts rapidly on cooling and goes to q_o when the SmA$_1$–SmA$_2$ transition occurs. The temperature dependence of the wavevectors observed in a sample with X = 77.2 is shown in Fig. 18. Note the SmA$_{is}$ region where three Bragg reflections are observed: two incommensurate wavevectors $2q_o$ and q_o'' and a sharp combination reflection at $q_s = 2q_o - q_o''$ indicating strong coupling between the basic wavevectors. This is the soliton regime for an incommensurate phase that is denoted SmA$_{i1}$ in the plane-wave regime. As the SmA$_{is}$ phase is cooled, the soliton periodicity $Z = 2\pi/(q_o'' - q_s)$ appears to diverge at the SmA$_2$ lock-in transition, as predicted theoretically [8]. No high-resolution x-ray or calorimetric work has yet been done on this remarkably rich system.

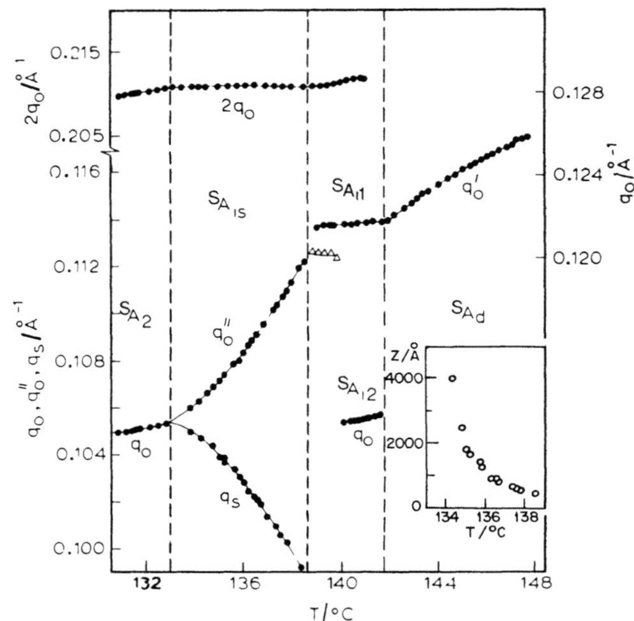

Fig. 18. Temperature dependence of the wavevectors characterizing a 8OBCAB+DB$_8$OCN mixture with 77.2 mol % DB$_8$OCN [7]. The wavevector q_s is equal to $2q_o - q_o''$, and the open triangles represent diffuse scattering. The soliton periodicity $Z = 2\pi/(q_o'' - q_s)$.

An x-ray and calorimetric study has also been carried out on the DB$_5$CN + T8 system [54,55], which exhibits a complex polymorphism. The SmA$_1$ phase of this system shows two collinear *diffuse* x-ray peaks at incommensurate wavevectors q_o'' and $q_s = (2q_o-q_o'')$, indicating a tendency to form the SmA$_{is}$ phase. However, the correlation lengths are smaller than the soliton periodicity of 500-1000Å. Thus this phase can be described as a SmA$_1$ phase containing an imperfect melted array of domain walls [54]. The observation of simultaneous sharp SmA$_2$ and SmA$_d$ peaks in samples with lower T8 concentrations is most likely due to bulk two-phase coexistence, but the presence of a SmA$_{i2}$ phase cannot be ruled out [55].

Considerably more work on incommensurate liquid crystals is required to clarify the detailed character of these phases and their phase transitions. An experimental review is given in [7] and theoretical considerations are described in [4,8,14,18,34].

240

V. REENTRANT SYSTEMS

Instead of incommensurate phases or biaxial modulated phases, liquid crystal systems can respond to frustration by destroying the smectic layers to yield a reentrant nematic N_r phase. The mean-field frustrated smectics model does not yield such reentrant behavior, but it should be achieved by considering fluctuation effects [4,13]. The frustrated spin-gas model, on the other hand, leads naturally to a rich variety of reentrant phenomena [27-30]. Indeed, these two quite different models are almost complementary — the phenomenological frustrated smectics model leads easily to bilayers, modulated biaxial phases and incommensurate phases, whereas the spin-gas model yields multiple N, SmA_d, SmA_1 reentrance (a direct SmA_d–SmA_1 transition has only recently been achieved [31]).

An early example of simple nematic reentrance at 1 atm is the Tn system [56], the phase diagram for which is shown in Fig. 19a. The spin-gas theory phase diagram is given for comparison in Fig. 19b. Topologically identical P-T phase diagrams have also been observed in pure 9OBCAB [57]. The Tn system has already been mentioned in connection with the almost XY character of the N–SmA_1 transitions (see Sec. IIIA). The upper N–SmA_d and lower SmA_d–N_r transitions have the same character

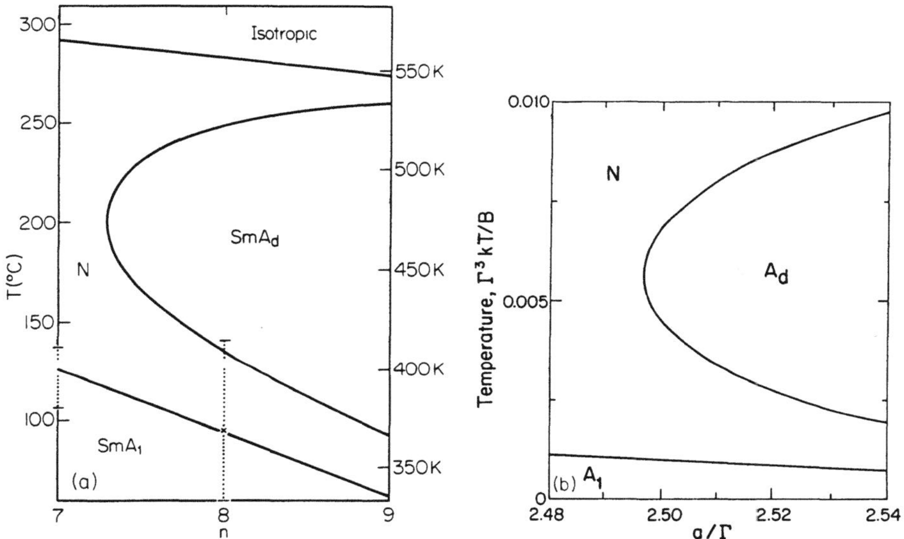

Fig. 19. Phase diagrams exhibiting N–SmA_d–N_r–SmA_1 topology. (a) Experimental diagram for Tn compounds and their binary mixtures [45]. X-ray and C_p data were obtained over the ranges shown by dotted vertical lines. (b) Spin-gas theoretical phase diagram [29]. The vertical axis is a scaled temperature. The quantity a/Γ can be related to pressure changes at constant composition or composition changes at constant pressure.

[25], which is non-XY-like in agreement with all previously
investigated N-SmA$_d$ transitions [45]. The smectic fluctuations
in the N$_r$ phase of T8 exhibit a quite complex behavior as shown
in Fig. 20. The SmA$_d$ and SmA$_1$ fluctuations are essentially
independent. As the SmA$_d$ fluctuations weaken on cooling, they
change over from SmA-like to SmC-like. This unusual fluctuation
behavior may perhaps help to explain the slight deviations from
XY critical behavior at this N$_r$-SmA$_1$ transition.

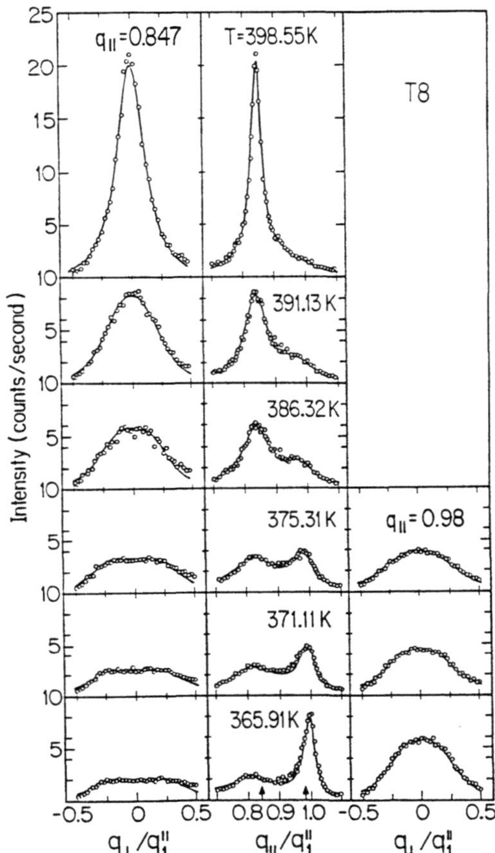

Fig. 20. Transverse and longitudinal
x-ray scans in the reentrant
nematic phase of T8 [45].
$T_{A_dN_r}$ = 401.41K and $T_{N_rA_1}$ =
361.17K. The left-hand
column shows transverse
scans through the diffuse
SmA$_d$ peak, and the right-hand
column shows such scans
through the diffuse SmA$_1$ peak.
Both peaks are visible in the
logitudinal scans.

The complex smectic polymorphism and multiple reentrance of frustrated liquid crystals is well illustrated by DB9ONO2 [58], DB10ONO2 [30], and homologous mixtures of DB8ONO2 + DB10ONO2 [50,59]. The experimental phase diagram for the mixture is given in Fig. 21a, and a closely related spin-gas theoretical phase diagram is shown in Fig. 21b. As pointed out in Sec. II, atomic permeations (Fig. 3d) stabilize the high-temperature SmA$_d'$ phase while librational permeations (Fig. 3b,c with librations) stabilize the low-temperature SmA$_d$ phase, and ferro-ordered triplets (Fig. 3e) dominate in the SmA$_1$ phase.

Figure 22 shows a detailed view of the SmA$_d$–N$_r$–SmA$_1$ "bicritical" region in DB8ONO2 + DB10ONO2. Instead of the expected bicritical point [17], one actually finds a SmA$_d$–N$_r$ critical end point (CEP) and a tricritical point (TCP) along the N$_r$–SmA$_1$ line [50,59]. This transformation of the bicritical topology into a TCP and a CEP was not originally predicted by the mean-field frustrated smectics model, but a recent renormalization calculation shows that such splitting of a bicritical point can occur [60].

The heat capacity has been studied for six different compositions [50], and an overview of the data on the sample with

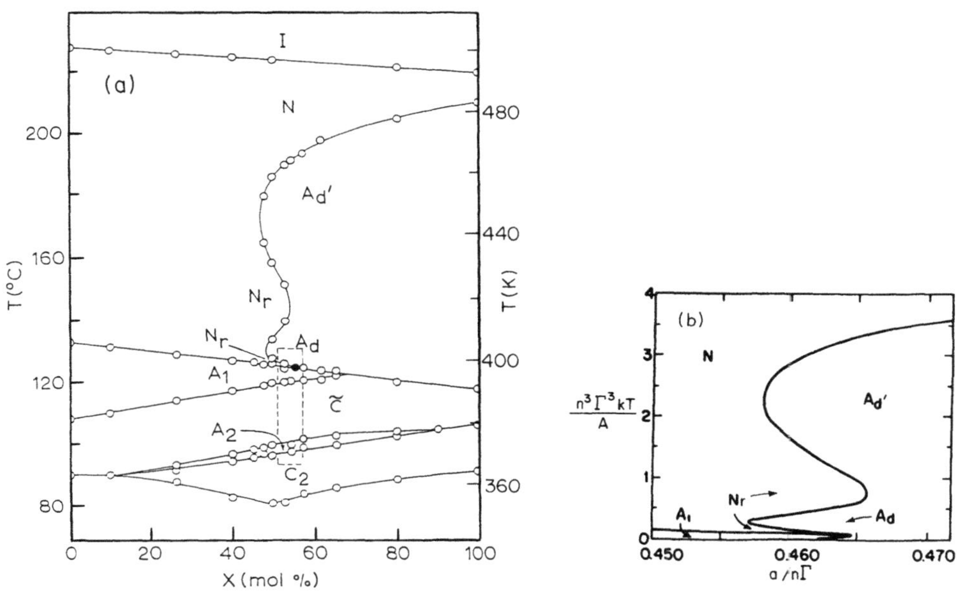

Fig. 21. Phase diagrams exhibiting quadruple N–SmA$_d$ reentrance.
(a) Experimental diagram for DB8ONO2+DB10ONO2 mixtures [59]. X is the mole % DB10ONO2. The region investigated in [50,59] is indicated by the dashed rectangle. (b) Spin-gas theoretical phase diagram exhibiting quadruple reentrance [23].

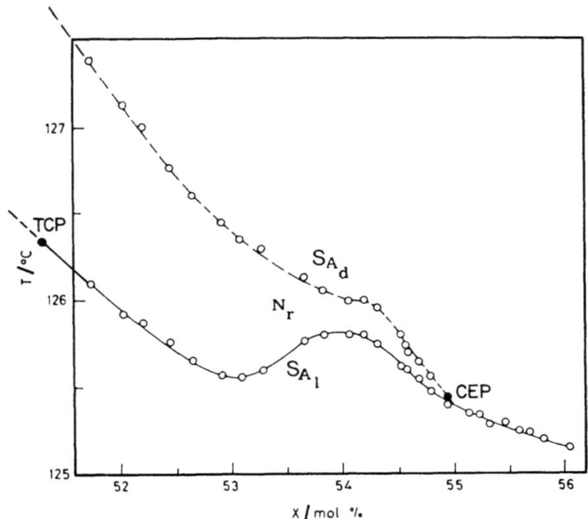

Fig. 22. Detail of the "bicritical"
region for $DB_8ONO_2+DB_{10}ONO_2$,
[59] augmented by the results
of [50]. Dashed lines are
second order and solid are
first order; see text for
details.

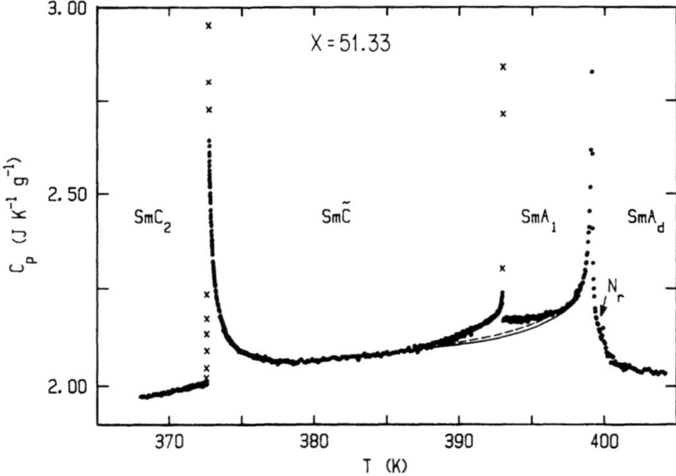

Fig. 23. Heat capacity of a $DB_8ONO_2+DB_{10}ONO_2$
mixture containing X = 51.33 mol %
$DB_{10}ONO_2$ [50]. The points × denote
two-phase coexistence at the first-
order SmA_1–$Sm\tilde{C}$ and $Sm\tilde{C}$–SmC_2 transitions.

51.33 mol % $DB_{10}ONO_2$ is shown in Fig. 23 as an example. This
sample corresponds to the tricritical composition, and a
detailed view of the C_p variation in the SmA_d-N_r-SmA_1 region is
shown in Fig. 24a. The relative magnitudes of the two peaks
are very different, and this feature is in excellent agreement
with semi-quantitative spin-gas model predictions, as indicated
by Fig. 24b. This qualitative difference between the integrated
enthalpies of N-SmA_1 and N-SmA_d transitions is not sensitive to
variations in the input model parameters and can be understood
in terms of the smectic ordering mechanisms for the spin-gas
model. The SmA_d phase (interdigitated partial bilayer) has
dominant antiferroelectric nearest-neighbor couplings, and frustration is lifted by the librational modification of local AFE
couplings. However, only a subset of molecules participates
in the smectic order defining the layer, and many dimers and
other n-mers are present that maintain a nematic background.
Thus, the ordering is quite imperfect and the transition
enthalpy is small. In contrast to this, the SmA_1 phase has
dominant ferroelectric couplings which do not compete with
each other. On ordering, all such couplings are satisfied (no
frustration) and the enthalpy is large. A power-law analysis
of the N_r-SmA_1 heat capacity data for X = 51.33 mol % yields a
good tricritical fit with α = 0.5 and $A^-/A^+ \cong 1.6$ [50]. For
smaller X values, XY-type second order N-SmA_1 transitions have
been observed [44] and crossover between XY and tricritical
behavior is being studied.

Brief mention will be made of the $Sm\widetilde{C}$ ribbon phase and SmC_2
tilted bilayer phase that occur in DB_8ONO_2 + $DB_{10}ONO_2$ at lower
temperatures. The SmA_1-$Sm\widetilde{C}$ transition is qualitatively like the
SmA_1-$Sm\widetilde{A}$ transition discussed in Sec. IIIC except that the pretransitional heat capacity above the transition is much smaller.

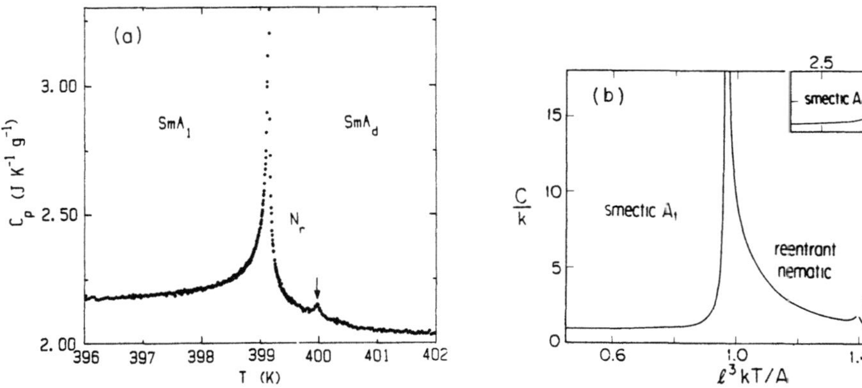

Fig. 24. (a) C_p variation in the SmA_d-N_r-SmA_1 transition region
for a DB_8ONO_2+$DB_{10}ONO_2$ mixture with X = 51.33 [50].
(b) Spin-gas model prediction of the heat capacity
variation for SmA_d-N_r and N_r-SmA_1 transitions [25].
The inset shows the predicted high-temperature N-SmA_d
peak.

The $Sm\tilde{C}$–SmC_2 transition is obviously related to the $Sm\tilde{A}$–SmA_2
transition (which has an intermediate SmA_{cren} phase that leads to
a truncation of the C_p peak). As shown in Fig. 25, the $Sm\tilde{C}$–SmC_2
data are well described by an *inverted* Landau model, for which
the amplitude A* is nonzero above the transition and zero below.
Note the small heat-capacity step at the transition (dashed line),
which is closely related to the observed C_p behavior at a SmA_2–
SmC_2 transition [61]. The details of the $Sm\tilde{C}$–SmC_2 fitting
procedure are described elsewhere [50].

VI. SmA–SmA′ CRITICAL POINTS AND THE N BUBBLE

Since the macroscopic symmetries of the SmA_d and SmA_2 phases
are the same, there cannot be a second-order symmetry-breaking
transition between them. Thus, the SmA_d–SmA_2 transition is
first order with a discontinuous jump in the wavevector associated with the layer thickness. This coexistence line terminates
in a critical point (CP), beyond which there is only a supercritical evolution of q with temperature and no thermodynamic
transition. According to mean-field theory, the SmA_d–SmA_2
transition is equivalent to the liquid-gas transition [17]. This
suggests that the SmA_d–SmA_2 critical point might belong to the
Ising universality class. However, a renormalization-group

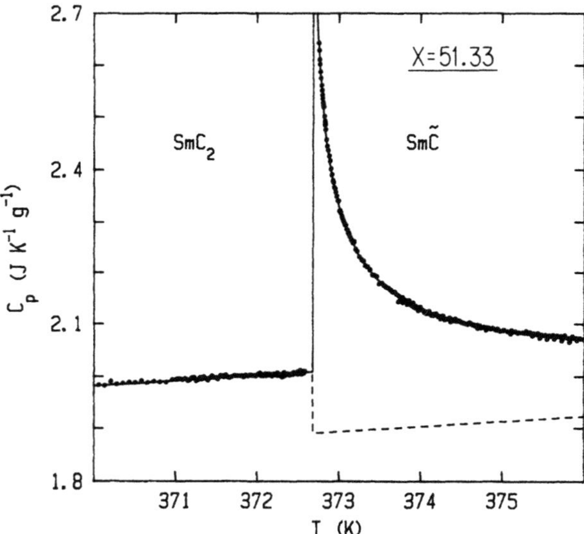

Fig. 25. Inverted Landau fit to the
first-order $Sm\tilde{C}$–SmC_2 transition in a 51.33 mol % mixture
of DB_8ONO_2+$DB_{10}ONO_2$ [50].
Data points obtained in the
two-phase coexistence region
are not shown and were not
used in the fit.

analysis of a nonlinear elastic model that takes fluctuations
into account predicts that this critical point belongs to a new
universality class with $d_c = 6$ for the upper critical dimension
[35,36].

DB7OCN is an example of a material with a SmA_d-SmA_2 transformation that has been shown to represent a continuous supercritical evolution from SmA_d structure to SmA_2 structure rather
than a real phase transition [62]. The first realization of the
SmA_d-SmA_2 critical point was in 9OBCB + 11OPCBOB mixtures [63],
and the phase diagram for this system is shown in Fig. 26.
Detailed x-ray data as a function of composition and temperature
are shown in Fig. 27. For $X < X_c \cong 0.64$, there is two-phase
coexistence and a discontinuous jump in the magnitude of q (i.e.,
the layer spacing) at the transition. As $X \to X_c$, the jump and
coexistence range decrease and go to zero at X_c. For $X > X_c$, q
evolves in a continuous nonsingular manner.

An analysis [64] of the $X = 0.642$ x-ray data shows that the
smectic layer thickness $d(T) = 2\pi/q$ along a constant near-critical composition path is well described by

$$d = d_c + a^{\pm}|t|^{1/\delta} + e(T-T_c) \tag{10}$$

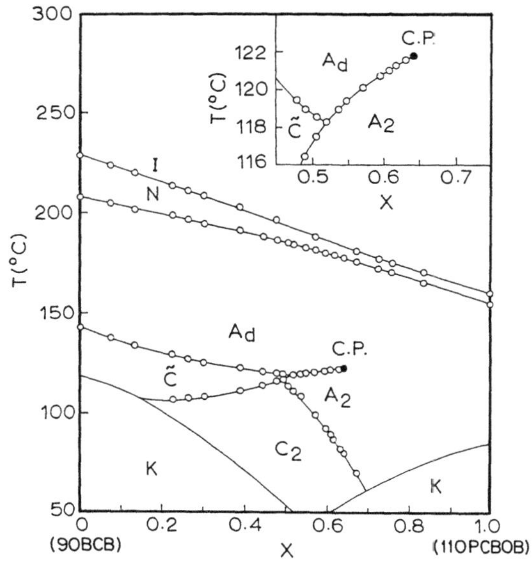

Fig. 26. Phase diagram for 9OBCB+
11OPCBOB mixtures [63].
X is the mole fraction of
11OPCBOB. The inset shows
an expanded view of the
vicinity of the SmA_d-SmA_2
critical point.

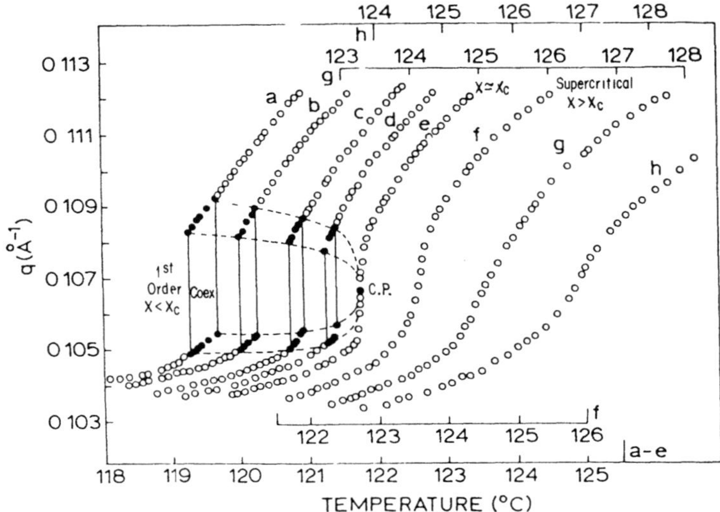

Fig. 27. The condensed smectic wavevector q as
a function of T for eight 11OPCBOB
concentrations [63]. A few X values
are (d) first-order 0.619, (e) $X_c \cong$
0.642, (f) supercritical 0.715.

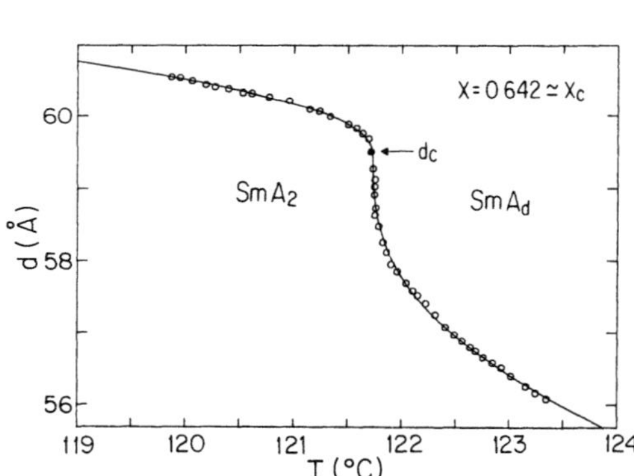

Fig. 28. Least-squares fit to
d(T) with Eq. (10) in
a near-critical
9OBCB+11OPCBOB mixture
with X=0.642 [64].

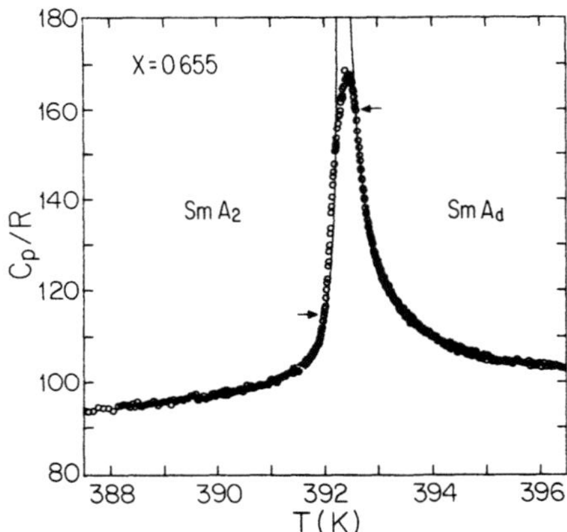

Fig. 29. Least-squares fit
to C_p(T) with the
critical exponent
γ/Δ=0.6 for a near-
critical sample
with X = 0.655 [64].
Points in the range
392.0–392.7K (see
arrows) were omit-
ted from the fit.

as shown in Fig. 28. By scaling, the exponent $1/\delta = 1 - (\gamma/\Delta)$. Thus the best fit value $1/\delta = 0.38 \pm 0.03$ yields $\gamma/\Delta = 0.62 \pm 0.03$.

High-resolution calorimetric measurements have also been made on five different compositions of 9OBCB + 11OPCBOB [64]. In these experiments near-critical behavior was observed for both $X = 0.641$ and 0.655, and the C_p variation of the latter sample is shown in Fig. 29. For such SmA_d-SmA_2 systems, C_p along a path of constant composition $X \cong X_c$ has a theoretical form like Eq. (3) but with the exponent α replaced by γ/Δ [35]. Least-squares fits with γ/Δ as a free parameter yielded $\gamma/\Delta = 0.71 \pm 0.11$ for $X = 0.655$ and $\gamma/\Delta = 0.50 \pm 0.22$ for $X = 0.641$; but fits of comparable quality can also be achieved with γ/Δ held fixed at 0.60, in agreement with the $d(T)$ fitting result. A fit of the latter type is shown in Fig. 29. The substantial rounding of the C_p peak (points of inflection are marked by arrows in Fig. 29) is a puzzling but systematic feature of all the C_p data. Possibly this is related to the unusual temperature dependence observed in the x-ray intensity for the quasi-Bragg diffraction spot [64].

The critical exponent value $\gamma/\Delta \cong 0.6$ can be compared with 3-D Ising ($\gamma/\Delta = 0.79$) and mean-field ($\gamma/\Delta = 0.67$) values as well as the first-order-in-ε estimate for the new universality class ($\gamma/\Delta \cong 0.31$) [60]. Since $d_c = 6$, $\varepsilon = 6-d = 3$ for three-dimensions, and the latter exponent value may not be very reliable. Clearly more theoretical and experimental work is needed to clarify the detailed nature of this new type of SmA_d-SmA_2 critical point.

The possibility of SmA_d-SmA_1 critical points has also been predicted [35] and the conditions for their existence explored with a dislocation-loop model [19]. A wide variety of theoretical phase diagrams with SmA_d-SmA_1 critical points can be generated, but no such experimental system has yet been studied with high-resolution techniques. Another phase topology that can preempt the SmA_d-SmA_1 critical point is a first-order SmA_d-SmA_1 line that terminates in a closed nematic domain. According to a fluctuation corrected mean-field theory of dislocation loops [19], this "nematic bubble in a smectic sea", shown in Fig. 30, has several unusual features: two disjoint N–SmA critical end points A and B, a first-order N–SmA transition line A–B, a first-order N–N transition line B–C across which the nematic order parameter S shows a discontinuous jump, and a N–N critical point C. The two types of N phase in the bubble are N_1 with strong SmA_1 fluctuations and N_d with strong SmA_d fluctuations. Such a phase topology has been seen experimentally [65], as shown in Fig. 31, but the resolution is insufficient to distinguish the detailed features predicted by the theory. This type of system represents an interesting and challenging problem in the study of frustrated smectic liquid crystals.

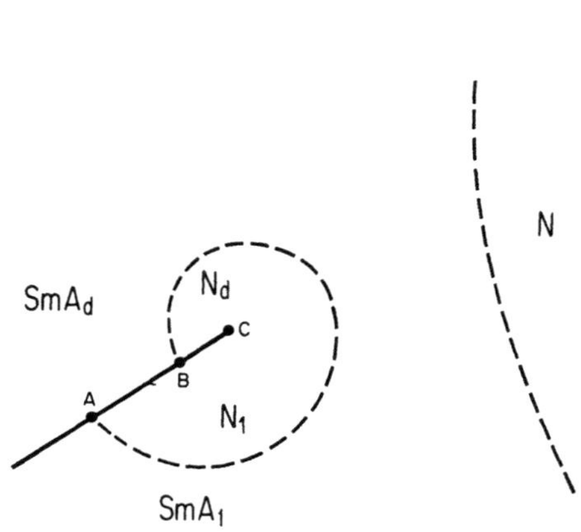

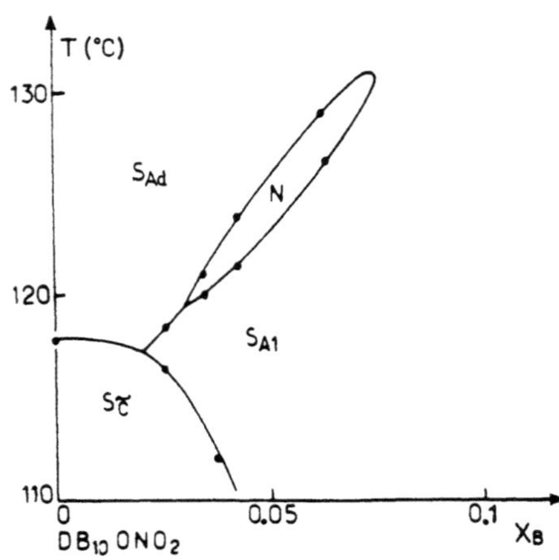

Fig. 30. Theoretical phase diagram for a frustrated
liquid crystal system
exhibiting a nematic
bubble at the end of
the SmA_1-SmA_d line [19].
Solid lines are first
order, and dashed lines
are second order. T, X
axes not shown.

Fig. 31. Partial phase diagram for mixtures
of $DB_{10}ONO_2$+B [65].
The primary N phase
lies ~50K above the
closed reentrant
bubble. The com-
pound B contains a
lateral polar CN
rather than a polar
head group (see
[66]).

VII. CONCLUSION

During the past ten years, a greal deal of satisfying
progress has been made in the discovery and characterization of
polymorphic SmA (and SmC) phases. The essential molecular
architecture is quite well understood empirically, and detailed
x-ray studies have established the long-range structures of most
of the observed phases. However, residual structural questions
still exist for the modulated SmA_{cren} and $Sm\tilde{C}$ phases and
the incommensurate SmA_{ic} phases.

The present focus is on high-resolution investigations of
phase transitions and the evolution of ordering and structure
within a given phase. In parallel with experimental efforts to
characterize the quantitative properties of polymorphic smectics,
there has been an impressive development of a variety of
theoretical models. In many cases, theory and experiment are
already in good agreement, but several challenging problems
remain. The goal is to achieve a universal and microscopic
understanding of SmA phase behavior. This will require (a) the
extension and unification of existing theories, (b) the clear
experimental establishment of several types of universal critical behavior, and (c) the experimental elucidation of systematic trends in non-universal quantities that can be estimated
from microscopic models.

250

ACKNOWLEDGMENTS

Much of the work described here has been the result of pleasant and productive collaborations both formal and informal with many colleagues: P. Barois, A. N. Berker, R. J. Birgeneau, F. Hardouin, P. A. Heiney, G. Heppke, J. O. Indekeu, A. M. Levelut, T. C. Lubensky, P. Pershan, J. Prost, B. R. Ratna, R. Shashidhar, G. Sigaud, Nguyen Huu Tinh. Most of the ac calorimetric results are due to the vigorous efforts of coworkers K Ema, G. Nounesis, and K. J. Stine. This research was supported in part by National Science Foundation Grant DMR 87-02052 and United States-France Cooperative Research Grant INT 85-14185.

REFERENCES

1.   G. Sigaud, F. Hardouin, and M. F. Achard, *Phys. Lett.* **72A**, 24 (1979); G. Sigaud, F. Hardouin, M. F. Achard, and H. Gasparoux, *J. Phys. (Paris)* **40**, C3-356 (1979).

2.   J. Prost, *J. Phys. (Paris)* **40**, 581 (1979).

3.   F. Hardouin, A. M. Levelut, M. F. Achard and G. Sigaud, *J. Chim. Phys.* **80**, 53 (1983) and references therein.

4.   J. Prost and P. Barois, *J. Chim. Phys.* **80**, 65 (1983).

5.   Nguyen Huu Tinh, *J. Chim. Phys.* **80**, 83 (1983).

6.   R. J. Birgeneau, C. W. Garland, G. B. Kasting and B. M. Ocko, *Phys. Rev.* **A24**, 2624 (1981).

7.   R. Shashidhar and B. R. Ratna, *Liq. Cryst.* **5**, 421 (1989).

8.   P. Barois, J. Pommier and J. Prost, in "Solitons in Liquid Crystals", L. Lam and J. Prost, eds., Springer-Verlag, Berlin, chap. 6 (1989).

9.   T. C. Lubensky, *J. Chim. Phys.* **80**, 31 (1983) and references cited therein.

10. J. C. LeGuillon and J. Zinn-Justin, *Phys. Rev.* **B21**, 3976 (1980); *J. Phys. Lett. (Paris)* **46**, L-137 (1985).

11. J. Prost, in "Proceedings of the Conference on Liquid Crystals of One- and Two-Dimensional Order" at Garmisch Partenkirchen, Springer Verlag, Berlin, p. 125 (1980).

12. P. Barois, C. Coulon and J. Prost, *J. Phys. Lett. (Paris)* **42**, L-107 (1981).

13. C. Coulon and J. Prost, *J. Phys. Lett. (Paris)* **42**, L-241 (1981).

14. J. Prost, in "Symmetries and Broken Symmetries in Condensed Matter Physics," N. Boccara, ed., IDSET, Paris, p. 159 (1981).

15. J. Prost, *Adv. Phys.* **33**, 1 (1984).

16. P. Barois and J. Prost, *Ferroelec.* **58**, 193 (1984).

17. P. Barois, J. Prost and T. C. Lubensky, *J. Phys. (Paris)* **46**, 391 (1985).

18. P. Barois, *Phys. Rev.* **A33**, 3632 (1986).

19. J. Prost and J. Toner, *Phys. Rev.* **A36**, 5008 (1987).

20. J. Prost and J. Pommier, private communication.

21. J. Prost, "Exemples de transitions dans les cristaux liquides: la transition nématique-smectique A et les smectiques frustres," to be published.

22. A. N. Berker and J. S. Walker, *Phys. Rev. Lett.* **47**, 1469 (1981).

23. J. O. Indekeu and A. N. Berker, *Phys. Rev.* **A33**, 1158 (1986).

24. J. O. Indekeu and A. N. Berker, *Physica* **A140**, 368 (1986).

25. J. O. Indekeu, A. N. Berker, C. Chiang, and C. W. Garland, *Phys. Rev.* **A35**, 1371 (1987) and references cited therein.

26. A. N. Berker and J. O. Indekeu, in "Incommensurate Crystals, Liquid Crystals and Quasi-Crystals", edited by J. F. Scott and N. A. Clark, Plenum, New York, p. 205 (1987).

27. J. O. Indekeu, *Phys. Rev.* **A37**, 288 (1988).

28. J. O. Indekeu and A. N. Berker, *J. Phys. (Paris)* **49**, 353 (1988).

29. J. F. Marko, J. O. Indekeu, and A. N. Berker, *Phys. Rev.* **A39**, 4201 (1989).

30. V. N. Raja, B. R. Ratna, R. Shashidhar, G. Heppke, Ch. Bahr, J. F. Marko, J. O. Indekeu and A. N. Berker, *Phys. Rev.* **A39**, 4341 (1989).

31. A. N. Berker, K. Hui and J. F. Marko, to be published.

32. J. Wang and T. C. Lubensky, *Phys. Rev.* **A29**, 2210 (1984).

33. J. Wang and T. C. Lubensky, *J. Phys. (Paris)* **45**, 1653 (1984).

34. T. C. Lubensky, S. Ramaswamy and J. Toner, *Phys. Rev.* **A38**, 4284 (1988).

35. Y. Park, T. C. Lubensky, P. Barois and J. Prost, *Phys. Rev.* **A37**, 2197 (1988).

36. Y. Park, T. C. Lubensky and J. Prost, *Liq. Cryst.* **4**, 435 (1989).

37. K. K. Chan, P. Pershan, L. B. Sorensen and F. Hardouin,
    Phys.Rev. Lett. **54**, 1694 (1985); *Phys. Rev.* **A34**, 1420
    (1986).

38. K. Ema, C. W. Garland, G. Sigaud and N. H. Tinh, *Phys. Rev.*
    A39, 1369 (1989).

39. Nguyen Huu Tinh and C. Destrade, *Mol. Cryst. Liq. Cryst.
    Lett.* **92**, 257 (1984).

40. C. W. Garland, M. Meichle, B. M. Ocko, A. R. Kortan, C. R.
    Safinya, J. J. Yu, J. D. Litster and R. J. Birgeneau,
    Phys. Rev. **A27**, 3234 (1983).

41. C. W. Garland, G. Nounesis and K. J. Stine, *Phys. Rev.* **A39**,
    4919 (1989).

42. C. W. Garland, G. Nounesis, K. J. Stine and G. Heppke, *J.
    Phys. (Paris)* **50**, 2291 (1989).

43. P. Das, G. Nounesis, C. W. Garland, G. Sigaud and N. H.
    Tinh, *Liq. Cryst.*, submitted.

44. G. Nounesis and C. W. Garland, unpublished.

45. K. W. Evans-Lutterodt, J. W. Chung, B. M. Ocko, R. J.
    Birgeneau, C. Chiang, C. W. Garland, E. Chin, J. Goodby
    and N. H. Tinh, *Phys. Rev.* **A36**, 1387 (1987).

46. C. Chiang and C. W. Garland, *Mol. Cryst. Liq. Cryst.* **122**,
    25 (1985); C. W. Garland, C. Chiang and F. Hardouin,
    Liq. Cryst. **1**, 81 (1986).

47. D. A. Huse, *Phys. Rev. Lett.* **55**, 2228 (1985); M. A.
    Anisimov, A. V. Voronel and E. E. Gorodetskii, *Sov.
    Phys. JETP* **33**, 605 (1971).

48. A. M. Levelut, *J. Phys. Lett. (Paris)* **45**, L-603 (1984);
    C. R. Safinya, W. A. Varady, L. Y. Chiang and P. Dimon,
    Phys. Rev. Lett. **57**, 432 (1986).

49. G. Sigaud, F. Hardouin, M. F. Achard and A. M. Levelut, *J.
    Phys. (Paris)* **42**, 107 (1981).

50. K. Ema, G. Nounesis, C. W. Garland and R. Shashidhar, *Phys.
    Rev.* **A39**, 2599 (1989).

51. B. R. Ratna, R. Shashidhar and V. N. Raja, *Phys. Rev. Lett.*
    55, 1476 (1985); in "Incommensurate Crystals, Liquid
    Crystals and Quasicrystals", J. F. Scott and N. A.
    Clark, eds., Plenum Press, New York, p. 259 (1987).

52. P. Das, K. Ema, C. W. Garland and R. Shashidhar, *Liq. Cryst.*
    4, 581 (1989).

53. V. N. Raja, Ph.D. thesis, Univ. of Mysore (1988),
    unpublished.

54. E. Fontes, P. A. Heiney, P. Barois and A. M. Levelut, *Phys. Rev. Lett.* **60**, 1138 (1988).

55. E. Fontes, W. K. Lee, P. A. Heiney, G. Nounesis, C. W. Garland, A. Riera and A. B. Smith III, *J. Chem. Phys.*, submitted.

56. F. Hardouin, G. Sigaud, M. F. Achard and H. Gasparoux, *Phys. Lett.* **71A**, 347 (1979); F. Hardouin and A. M. Levelut, *J. Phys. (Paris)* **41**, 41 (1980).

57. R. Shashidhar, B. R. Ratna and S. Krishna Prasad, *Mol. Cryst. Liq. Cryst.* **130**, 179 (1985).

58. R. Shashidhar, B. R. Ratna, V. Surendranath, V. N. Raja, S. Krishna Prasad and C. Nagabhusan, *J. Phys. (Paris)* **46**, L-445 (1985).

59. V. N. Raja, R. Shashidhar, B. R. Ratna, G. Heppke and Ch. Bahr, *Phys. Rev.* **A37**, 303 (1988).

60. J. Prost, private communication.

61. Y. H. Jeong, K. J. Stine, C. W. Garland and N. H. Tinh, *Phys. Rev.* **A37**, 3465 (1988).

62. S. Krishna Prasad, R. Shashidhar, B. R. Ratna, B. K. Sadashiva, G. Heppke and S. Pfeiffer, *Liq. Cryst.* **2**, 111 (1987); C. W. Garland and P. Das, in "Incommensurate Crystals, Liquid Crystals and Quasicrystals", J. F. Scott and N. A. Clark, eds., Plenum Press, New York, p. 297 (1987).

63. R. Shashidhar, B. R. Ratna, S. Krishna Prasad, S. Somasekhara and G. Heppke, *Phys Rev. Lett.* **59**, 1209 (1987).

64. Y. H. Jeong, G. Nounesis, C. W. Garland and R. Shashidhar, *Phys. Rev.* **A40**, 4022 (1989).

65. F. Hardouin, M. F. Achard, Nguyen Huu Tinh and G. Sigaud, *Mol. Cryst. Liq. Cryst. Lett.* **3**, 7 (1986).

66. The structure of compound B is

$$C_8H_{17}O-\underset{NC}{\bigcirc}-COO-\bigcirc-COO-\bigcirc-C_5H_{11}$$

UNIVERSAL BEHAVIOR IN PHOSPHOLIPID MULTIMEMBRANE SYSTEMS

E. B. Sirota, C. R. Safinya, G.S. Smith[*] R. Plano

Corporate Research Science Laboratories
Exxon Research and Engineering Company
Route 22 East, Annandale, NJ 08801, USA

D. Roux

CRPP, CNRS
33405 Talence Cedex, France

N. A. Clark

Dept. of Physics, Univ. of Colorado
Boulder, CO 80309

ABSTRACT

Stacked multimembrane structures of the phospholipid DMPC show two interesting universal behaviors due to geometric concerns. X-ray scattering studies were done to elucidate the structure of the ordered phases where the $L_{\beta'}$ phase was shown to be in fact 3 distinct phases characterized by the direction of chain tilt with respect to the nearest neighbor positions. Two of these phases are lyotropic analogues to the smectic-F and -I phases, while the phase of intermediate symmetry (-L) had not been seen in other systems. Transitions from structures with smectic-F symmetry to those with smectic-I symmetry can be expected in systems where one can reduce the problem to contours on a hexagonal lattice which are effectively repulsive or would like to be straight. In the L_α phase of this system, the layers only dilute with water to ~ 25Å. By adding cosurfactant (pentanol) to the system, the thickness of the bilayer is reduced, thus reducing the bending modulus k_c from $25k_BT$ to ~ k_BT driving the system into a fluctuation-induced universal regime dominated by the repulsive Helfrich interaction which allows the bilayers to dilute to separations >200Å.

I. INTRODUCTION

Dimyristol-phosphotidylcholine (DMPC) is a phospholipid (lecithin) comprised of a hydrophilic zwitterionic head group and two 14-carbon hydrophobic hydrocarbon chains. This amphiphilic molecule, in mixtures with water forms lamellar phases where lipid bilayers are separated by water (Fig. 1). Such systems are of great scientific interest, not only because the lipid bilayer is the basis of all cell membranes [1], but because of their richness in studies of low dimensional physics. [2]

Geometry and Thermodynamics
Edited by J.-C. Tolédano
Plenum Press, New York, 1990

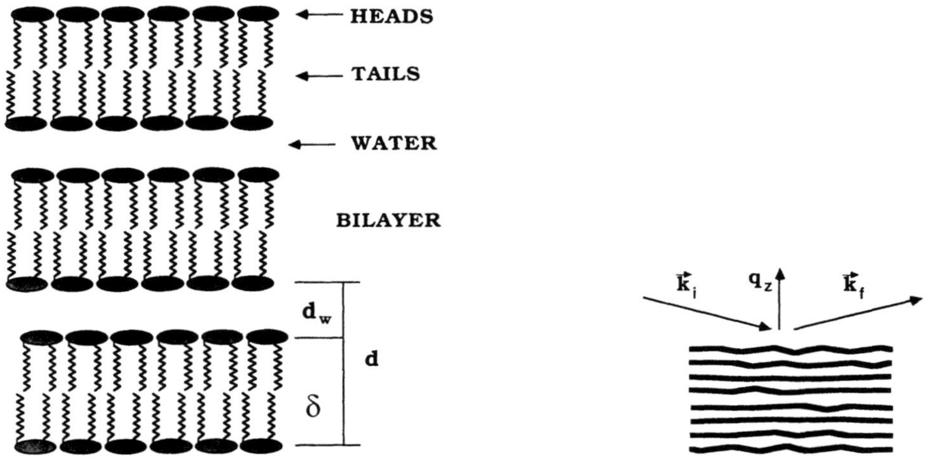

Fig. 1) (left) Schematic of the multilamellar structure formed by amphiphiles. d_w is the thickness of the water layer, δ is the membrane thickness and $d = d_w + \delta$ is the bilayer repeat distance. We define the z direction to be normal to the layers.

Fig. 2) (right) Scattering geometry showing reflection from undulating layers. q_z is the wavevector transfer.

The previously well known lamellar phases in this system are the L_α in which the chains of the molecules are melted and the average chain positions form a 2-dimensional liquid within the bilayer, the $P_{\beta'}$ in which the chains are mostly in the all-trans (frozen) conformation and there is a long wavelength correlated "ripple" of the layers and the $L_{\beta'}$ in which the chains are frozen, tilted with respect to the layers and there is substantial in-plane positional order of the chains.

II. UNDULATIONS IN FLUID MEMBRANES

As in the thermotropic liquid crystalline smectic-A phase, there is only a 1-dimensional density wave in the L_α phase. It was shown by Landau and Peierls[3] that 3-dimensional structures periodic in 1-dimension are at their lower marginal dimensionality and are therefore marginally stable to thermal fluctuations. This means that the r.m.s deviation of the layers from their average position diverges logarithmically with the size of the sample (in a true crystal they would not diverge). Caille [4] showed that the δ-function Bragg peaks arising from scattering from the layers (Fig. 2) are replaced by algebraic singularities with the form:

$$I(q_z) \approx \frac{1}{|q_z - q_m|^{2-\eta_m}} \qquad (1)$$

where $q_m = 2\pi m/d$ (m is the harmonic number) and

$$\eta_m = \frac{m^2 \pi k_B T}{2d^2 \sqrt{BK}} \qquad (2)$$

where B is the bulk modulus form layer compression and K is the bulk modulus for layer curvature. ($K \equiv k_c\, d$). The power las singularity associated with the first harmonic of the smectic peak was first demonstrated in elegant experiments by Als-Nielesen et. al. [5].

In later experiments of lyotropic multilayer membranes, this lineshape and its scaling with harmonic number was demonstrated by Safinya, Roux and collaborators.[6]

For rigid membranes the forces which govern the intermembrane spacing include the hydration force, van der Waals and the electrostatic forces [7] . Helfrich [8] showed that the entropic repulsion of the layers due to undulations can be important when $k_c \sim k_BT$. To understand the origin of this repulsive force in a simple way we first consider an isolated membrane. (We note here that surfactants membranes are usually considered incompressible [9] and therefore surface tension can be neglected) We consider the unperturbed sheet to be in the x,y plane with displacements normal to the layer u(x,y) (Fig. 3) and using equipartition calculate the r.m.s. displacement of the layer using the energy associated with layer curvature.

$$F = \frac{k_c}{2}\left(\frac{\partial^2 u}{\partial x^2} + \frac{\partial^2 u}{\partial y^2}\right)^2 \tag{3}$$

The result is that $<u^2> \sim (k_BT/k_c)A$, or that the displacement diverges strongly with the size of the sheet. If we now confine the layer between 2 flat sheets separated by 2d (Fig. 4) then there would be a collision when $\sqrt{\langle u^2 \rangle} \sim d$. By imposing the constraint that layers cannot cross, the number of states available to the system is reduced. One can see that the area per collision is $(k_BT/k_c)d^2$ which leads to an entropic contribution to the free energy

$$F = \frac{k_c}{2}\left(\frac{\partial^2 u}{\partial x^2} + \frac{\partial^2 u}{\partial y^2}\right)^2 \tag{4}$$

This is a long range repulsive interaction which would overwhelm the exponentially decaying hydration interaction at large distances, and when k_c is small it would overwhelm the attractive van der Waal which has the same d^{-2} dependence.

Helfrich [8] did a complete calculation of the form for this entropic repulsion for the case of finite thickness membranes and determined

$$F_{Helfrich} = 0.23\,\frac{(k_BT)^2}{k_c(d-\delta)^2} \tag{5}$$

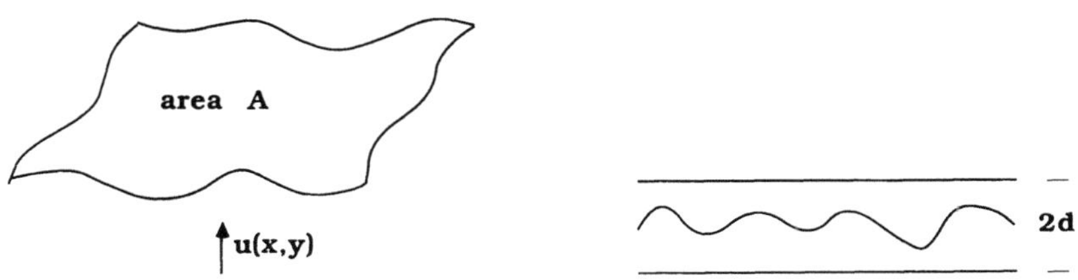

Fig. 3) (left) Tensionless fluid sheet of area A in the x,y plane with with thermally generated curvature waves causing displacements u(x,y).

Fig. 4) (right) Schematic of a fluctuating membrane constrained by 2 adjacent layers.

Returning to the expression for η (eqn. 2) we see that when the Helfrich interaction is dominant, the compressibility is just

$$B_{undulation} = d^2 \frac{\partial^2 \left(F_{Helfrich} / V \right)}{\partial d^2} \tag{6}$$

which results in

$$\eta_1 = 1.33 \left(1 - \delta/d \right)^2 \tag{7}$$

This is a prediction which is independent of the details of the system and only dependent on the membrane thickness and layer spacing. Eqn 7. was experimentally confirmed by Safinya, Roux et al.[6] for a layered microemulsion system of SDS/Pentanol/Water diluted with Dodecane. Roux and Safinya[10] then showed that a water dilution of the same system did not show Helfrich undulations since the long range electrostatic repulsion of the layers dominated and suppressed the undulations, but when the SDS/Pentanol membrane was diluted with brine, the electrostatics were screened and the undulations appeared.

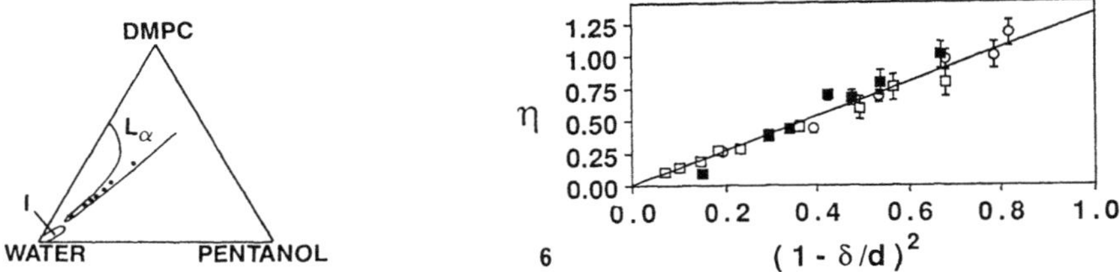

Fig. 5) (left) Schematic phase diagram of the ternary DMPC-pentanol-water system studied along the water dilution line (mixtures studied are indicated as dots).[from ref. 12]

Fig. 6) (right) Power-law exponent h as a function of $(1-\delta/d)^2$ for three dilution systems. The solid squares are for the DMPC-pentanol-water system while the open circles and squares are for SDS-pentanol membranes [6,10] along brine and oil dilution lines respectively. The solid line is the prediction of the Helfrich theory (Eqn 7). [from ref. 12]

For the case of pure DMPC and water it is known that hydration is the dominant force because the membranes are extremely rigid ($k_c \sim 25 k_B T$)[11]. Therefore, in excess water the membranes will only swell apart to 25Å before the van der Waals attraction balances the hydration repulsion. When pentanol is added to the membrane[12] as a cosurfactant (i.e. it also must reside at the interface) it has the effect of thinning the membrane (pentanol has shorter chains than DMPC) which in turn results in a more flexible membrane ($k_c \sim k_B T$). This is the stated condition for Helfrich's undulations to dominate and in fact as can be seen in the ternary phase diagram (Fig. 5) the presence of pentanol allows the membrane to swell with water to a far greater extent (>200Å) than the pure DMPC system. The fact that undulations are causing the swelling is evident from the experimental lineshape and in particular η. Figure 6 shows η_1 plotted versus $(1-\delta/d)^2$ for all 3 systems, and their agreement with Eqn. 7 shows that the undulations are indeed a universal phenomena based only on the geometry of the system.

258

III. IN-PLANE STRUCTURE OF ORDERED MEMBRANES

The in-plane structure of the ordered chains of DMPC in the $L_{\beta'}$ phase was studied using x-ray diffraction [13,14,15]. In 3-dimensional crystals with multiple order and mixed reflections, powder diffraction is usually sufficient to index peaks and determine structure. In more weakly ordered systems such as those studied here, often only 1 or 2 reflections are observable, and these are often broad. It is, therefore, necessary to use some broken symmetry to orient the sample in order to determine the details of the local structure. In the case of layered systems such as smectic liquid crystals the use of freely suspended films [16] has been shown to be very useful in orienting the layer normal to better than .1° and is also a convenient geometry for x-ray scattering studies [17,18] of the in-plane structure.

The water content of the DMPC was varied by directly controlling the chemical potential of water in the system by varying the relative humidity (RH)

$$\Delta\mu = RT\ ln\ (RH) \tag{8}$$

This was done by means of a humidity controller which pumped air at a determined dewpoint through the oven. The oven was also designed to allow scattering in transmission through the film to study the in-plane structure, as well as reflection from the surface, to study the peaks from the layering.

The T-RH phase diagram determined from this study is shown in Fig. 7. What is interesting is that the phase which was previously identified as $L_{\beta'}$ was in fact 3 distinct phases. As we shall see from the scattering patterns, the alignment obtained with the freely suspended films was necessary to distinguish these phases, and therefore they lay undetected until now.

Typical contour plots of the scattered intensity and their corresponding real space structures are shown in Fig. 8 where q_r is in the in-plane direction and q_z is along the layer normal. (While there was good orientation of the layer normal, all our structures were still powders with respect to rotation about the layer normal χ, and structures were interpreted accordingly). The 3 phases are distinguished by the angle made between the direction of chain tilt and the direction between nearest neighbor positions. The low humidity low temperature phase is characterized by chains tilted between nearest neighbors analogous to the thermotropic smectic-F phase ($\phi=0°$), the high humidity high temperature $L_{\beta I}$ phase by tilt towards nearest neighbors ($\phi=30°$), and the intermediate $L_{\beta L}$ phase by tilt at intermediate angles ($0°<\phi<30°$) not previously observed in any liquid crystal system. (We have though denoted this structure smectic-L extending the alphabet in the thermotropic nomenclature).

From the real space representations in Fig. 8, one can see that both the $L_{\beta F}$ and $L_{\beta I}$ have the same reflection symmetry which is absent in the $L_{\beta L}$ phase. Since the $L_{\beta F}$ and $L_{\beta I}$ are of the same symmetry any transition between them would have to be first order as occurs in the smectic-F to smectic-I transition in thermotropic liquid crystals, however the insertion of the $L_{\beta L}$ phase with a lower symmetry allows the occurrence of two 2nd order transitions. A plot of ϕ vs. $\Delta\mu$ (Fig. 9) shows that the transitions are continuous and reversible with no hysterisis.

From the q_r scans through the peaks we determined that the in-plane correlation length was 200Å, and while there was bond orientational order our data was not sufficient to distinguish between a hexatic with finite correlations or a finite size 2D crystal. From q_z scans through the peaks were were able to determine that there was no correlation of the in-plane structure between successive bilayers, however the tails in the two halves of a single bilayer were in registry.

Selinger and Nelson [19] expanded on earlier work by Nelson and Halperin [20] on tilted-hexatics generalizing the mean-field coupling between tilt direction and bond direction to include a term proportional to $\cos(12\phi)$.

$$V(\phi) = -h_6 \cos\left(6\phi\right) - h_{12} \cos\left(12\phi\right) \tag{9}$$

When $h_{12}<0$ the transition is first order and when $h_{12}>0$ the transition is 2nd order.

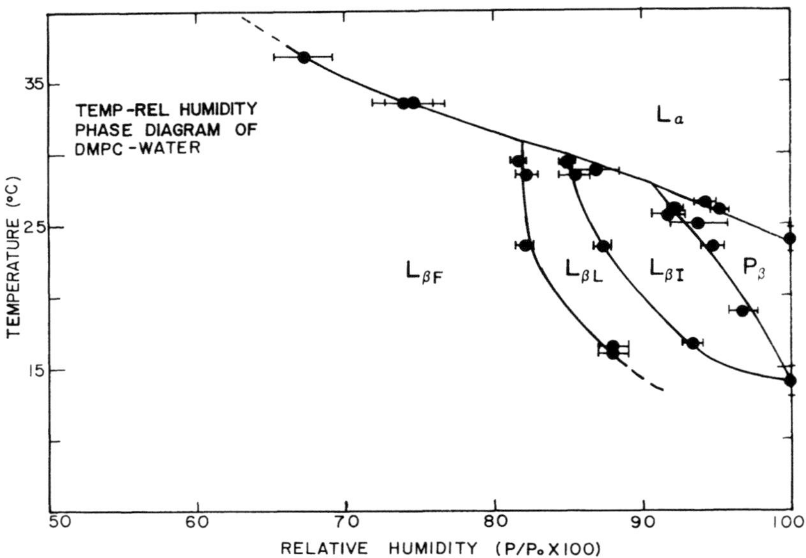

Fig. 7) Phase diagram of DMPC as a function of T and RH including the 3 new $L_{\beta'}$ phases. [from ref 13]

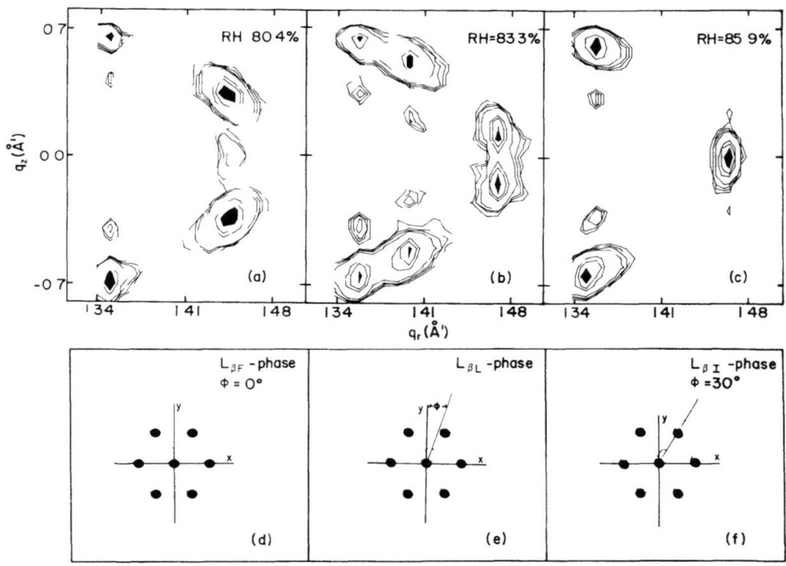

Fig. 8) q_r vs q_z contour plots at T=23.5°C in (a) $L_{\beta F}$, RH=80.4%; (b) $L_{\beta L}$, RH=83.3% and (c) $L_{\beta I}$, RH=85.9%. The contour levels are 100, 113, 125, 150, 300 and 500 counts per 9 seconds. The highest contour levels are filled in to emphasize the fundamental form-factor peaks. Below each is a schematic [(d)-(f)] showing the tilt direction of the molecules relative to the bond direction (defining ϕ). [from ref. 13]

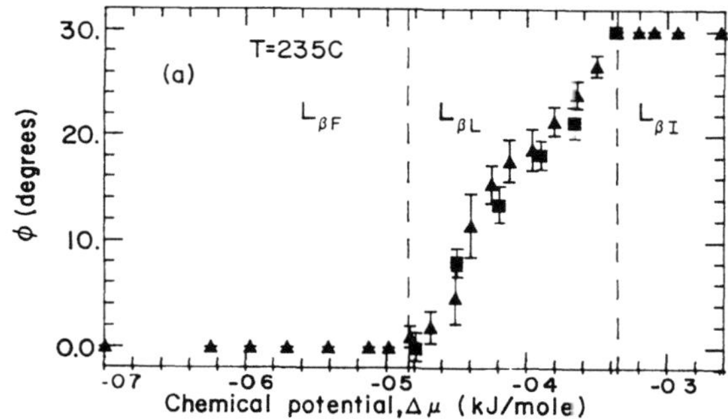

Fig. 9) ϕ vs. $\Delta\mu$ at T=23.5°C. Hydrating data are squares and drying data are triangles showing that hysterisis is negligible.[from ref. 13]

IV. MICROSCOPIC MODEL

One still must ask what microscopics cause the system to prefer $\phi=0$ at low-T and $\phi=30°$ at high-T (or as a function of other potentials). In the case of thermotropics, a model has been developed [21] which can explain this. This model was used to explain the structures and trends of the phase boundaries in the temperature versus thickness phase diagram of freely suspended films of thermotropic liquid crystals.[17] The assumption was that the energy of the microscopic interactions was minimized in the structure of the low temperature phase (smectic-F/crystal-G). Then "turning on" temperature entropy can be allowed to stabilize states with higher energies. (The strength of the model lay in its success in explaining many trends with few assumptions, however only the bulk F to I transition will be described here.)

The assumptions were that adjacent molecules prefer to be displaced normal to their nematic director by a fixed amount (Fig. 10) and that molecules prefer to be positioned such that their center of masses lie in a plane. It is clear that the first assumption cannot be satisfied totally on a hexagonal lattice, leading to frustrated bonds between adjacent molecules which are at the same height. It is convenient to portray the structures projected along the nematic director, and draw contours of the frustrated bonds connecting molecules at the same height. Between contours, an arrow must show which level is higher. From the assumptions of the model, the lowest energy configuration [Fig 11] is where the contours are straight and parallel with arrows pointing in the same direction. The layer normal is then tilted out of the page and the molecular tilt direction is between nearest neighbors ($\phi=0°$). By design, this is the smectic-F structure. If we want the tilt direction to be $0°>\phi\geq30°$ the contours cannot be straight on a hexagonal lattice but must zig and zag. For $\phi=30°$ [Fig 12] there must be an equal number of zigs and zags however they need not be regular. In fact there are of order 2^N choices (where N is the length of a contour). This increase in entropy comes at an energetic cost since the contours are no longer straight and the centers of mass deviate from a single plane. One should note that since neighboring contours will not zig and zag independently, the entropic gain may be much less than one degree of freedom per molecule and the energetic cost must therefore be low in order to stabilize phases with $\phi>0°$. (Experimentally, in thermotropics the smectic-F to -I transition does in fact have a very low latent heat [22]).

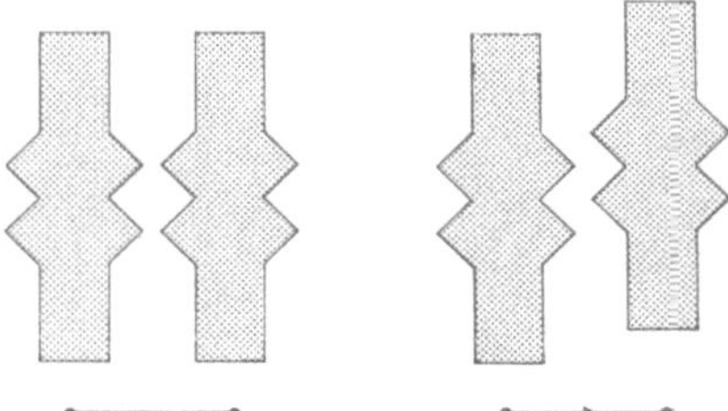

Fig. 10) Schematic showing the assumed prefered local arrangement of nearest neighbors (right), and the unfavorable arrangement (left). The shape portrayed represents the volume swept out by rotation of the molecule about it's long axis. [from ref. 21]

261

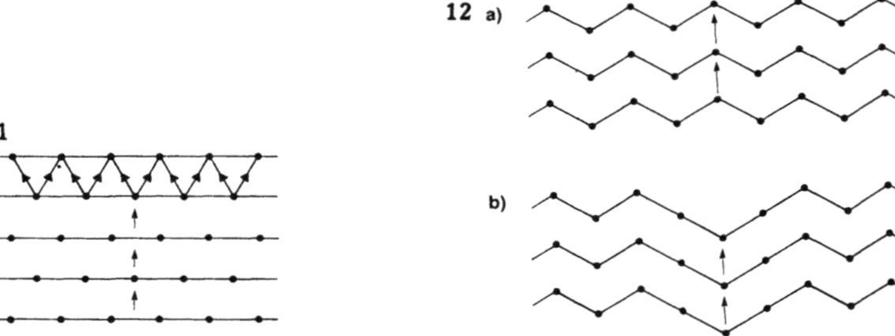

Fig. 11) (left) $\phi=0$ structure. in a projection along the long axis of the molecules. Contours connect molecules at the same height. Arrows show which molecules are higher.[from ref 21]

Fig. 12) (right) $\phi=30$ structure. Average tilt toward a nearest neighbor. (a) and (b) are 2 of a whole class of variants which give this state higher entropy than $\phi=0$.[from ref. 21]

In general then, if a system can be mapped onto a model consisting of contours on a hexagonal lattice, where zig zags cost energy, if the energetic cost is no too high, there should be a transition between a $\phi=0$ state and those with higher ϕ. The presice nature of the transitions being determined by the coefficients of the h_{12} terms when the free energy is computed from a partition function and expanded in terms of harmonics of a coupling potential.

We may then ask how the DMPC system would map onto this model. In this system of double chain molecules, chains were paired in two's on the pseudo-hexagonal lattice and connected by a backbone and dipolar head.. If one expects that there should be some perfered symetry to the pairing, one postulate pairing such as those shown in Fig. 13. As one can see this naturally leads to drawing lines. analogous to the contours in Figs 11 and 12 and the same combinatorics. The energetics would likely be due to the electrostatic forces of the head groups.

V. CONCLUSION

In the L_α phase, through the addition of a cosurfactant, DMPC has been shown to crossover from a hydration dominated regime to one where the interlayer interactions are dominated by the thermally generated Helfrich undulations, entropic in origin and determined only by the geometry of the system. The $L_{\beta'}$ phase, was shown to be 3 distinct phases exhibiting tilt reorientation transitions similar to those in thermotropic systems with different interactions. Such transitions should be expected whenever a system can be mapped onto contours on a hexagonal lattice, independent of the microscopics, and determined only by geometry.

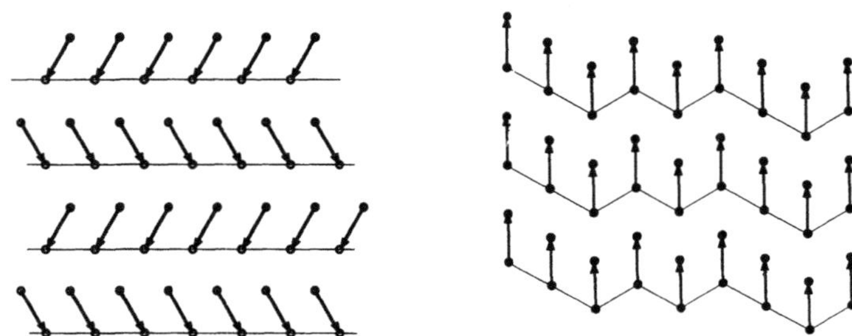

Fig. 13) Speculative pairing of molecules which would lead to a mapping of the double chain lyotropic model onto the contour model.

We acknowledge helpful conversations with S.K. Sinha. Part of this work was done at the Stanford Synchrotron Radiation Laboratory and at the National Synchrotron Light Source (Brookhaven), which are funded by the U.S. Department of Energy. A part of this work was supported by NSF Grant No. DMR-8307157.

REFERENCES

* current address:: LANSC, Los Alamos Natl. Lab., Los Alamos, NM 87545, U.S.A.

1. S.J. Singer and G.L. Nicolson, Science **175**, 720 (1972).

2. C.R. Safinya, in Phase Transitions in Sort Condensed Matter, T. Riste Ed. (Plenum Publishing Corporation, New York, 1989), p249-270; D.R. Nelson, "Statistical Mechanics of Membranes and Surfaces", proceedings of the Jerusalem Winter School, edited by Nelson, Piran and Weinberg (1987).

3. L.D. Landau, in Collected Papers of L.D. Landau, D. Ter Haar Ed. (Gordon and Breach, New York, 1965), p. 209; R.E Peierls, Helv. Phys. Acta. Suppl. **7**, 81 (1934).

4. A. Caille, C.R. Acad. Sci. Ser. B **274**, 891 (1972).

5. C. R. Safinya, D. Roux, G. S. Smith, S. K. Sinha, P. Dimon, N. A. Clark, and A. M. Bellocq, Phys. Rev. Lett. **57**, 2718 (1986).

6. J. Als-Nielsen, J. D. Litster, R. J. Birgeneau, M.Kaplan, C. R. Safinya, A. Lindegaard-Andersen, and S. Mathiesen, Phys. Rev. B **22**, 312 (1980).

7. A. Parsegian, N. Fuller and R. P. Rand, Proc. Natl. Acad. Sci. **76**, 2750 (1979).

8. W. Helfrich, Z. Naturforsch. **33a**, 305 (1978).

9. J.H. Schulman and J.B. Montagne, Ann. N.Y. Acad. Sci., **92**, 366 (1961).

10. D. Roux and C. R. Safinya, J. Phys. (Paris) **49**, 307 (1988).

11. G. S. Smith, C. R. Safinya, D. Roux, N. A. Clark, Mol. Cryst. Liq. Cryst. **144**, 235 (1987).

12. C.R. Safinya, E.B. Sirota, D. Roux and G.S. Smith, Phys. Rev. Lett., **62**, 1134 (1989).

13. G. S. Smith, E. B. Sirota, C. R. Safinya, and N. A. Clark, Phys. Rev. Lett. **60**, 813 (1988).

14. E.B. Sirota, G.S. Smith, C.R. Safinya, R.J. Plano and N.A. Clark, Science **242**, 1406 (1988).

15. G.S. Smith, E.B. Sirota, C.R. Safinya and N.A. Clark, (to be published in J. Chem. Phys.).

16. C.Y. Young, R. Pindak, N.A. Clark, R.B. Meyer, Phys. Rev. Lett. **40**, 773 (1978).

17. E.B. Sirota, P. S. Pershan, L.B. Sorensen and J. Collett, Phys. Rev. Lett., **55**, 2039 (1985); E.B. Sirota, P.S. Pershan, L.B. Sorensen and J. Collett, Phys. Rev. A, **36**, 2890 (1987); E.B. Sirota, P.S. Pershan and M. Deutsch, Phys. Rev. A, **36**, 2902 (1987).

18. D.E. Moncton and R. Pindak, Phys. Rev. Lett. **43**, 701 (1979).

19. J.V. Selinger and D.R. Nelson, Phys. Rev. Lett., **61**,416 (1988).

20. D.R. Nelson and B.I. Halperin, Phys. Rev. B, **21**, 5312 (1980).

21. E.B. Sirota, J. Phys. (Paris), **49**, 1443 (1988).

22. J. Budai, R. Pindak, S.C. Davey and J.W. Goodby, J. Phys. Lett. (Paris) **45**, 1053 (1984).

PATTERN FORMATION DURING THE ORDERING PROCESSES IN NEMATIC LIQUID
CRYSTALS

Yoshihiro Ishibashi, Tomoyuki Nagaya and Hiroshi Orihara

Synthetic Crystal Research Laboratory
Faculty of Engineering, Nagoya University
Furo-cho, Chikusa-ku, Nagoya 464-01, Japan

Dynamics of interfaces was experimentally studied, using nematic
liquid crystals. It was found that the dynamical scaling law holds well
for the dynamics in the orthogonally twisted nematic cell, while in the
non-orthogonally twisted nematic cells the statistical equivalence could
be found for the patterns formed in the cells with different twist
angles. The evolution of patterns observed in the non-orthogonally
twisted nematic cells is analyzed with the use of the theory which takes
into account the effect of the initial distribution of interfaces.

INTRODUCTION

In recent years much progress has been made on the problem of time
evolution of a system quenched from a disordered phase into an ordered
phase.[1,2] In such quenched systems interfaces separating two regions
appear, and move to establish the thermal equilibrium of the whole
system. The motions of interfaces in the non-conserved system, studied
theoretically by many workers, can be most conveniently investigated in
real-time and in real-space by quench experiments of liquid crystals
placed in the orthogonally or the non-orthogonally twisted nematic
cells.[3-5]
The velocity of the curved interface, v, is governed by two
factors; the curvature of an interface and the difference of potentials
in two regions separated by the interfaces. Namely

$$v = \Gamma \kappa \pm v_0 \tag{1}$$

where κ is the curvature, Γ the kinetic coefficient and v_0 is the
velocity caused by the latter factor above mentioned.
An example where v is determined only by v_0 is the wall motion in
ferroelectric domains. In ferroelectrics usually domain walls are flat
planes, where $\kappa = 0$, because the anisotropy in interactions is very
strong. The wall motions take place in such a way that the region made
more stable, for example, by an applied field expands at the cost of the
less stable region.
An opposite case where only the curvature of interfaces plays a
decisive role is seen in the twisted nematic cells and will be described
in §2, and a case where both the curvature of interfaces and the
unbalance of the potential may jointly govern the local velocity of the
interfaces will be discussed in §3.

We prepared a twisted nematic (TN) cell consisting of MBBA placed between two glass plates rubbed in two mutually perpendicular directions. In such a cell, upon quenching from $43°$ C to $35°$ C across the transition temperature there appear two types of state, i.e., left- and right-hand twisted states, being separated by a disclination, a one-dimensional interface in the two dimensional system (Fig. 1). These two regions are distinguishable under a set of crossed polarizers.

The temporal behaviors of disclinations and two states observed under a polarizing microscope are shown in Fig. 2, where the origin of time (0 see) was taken at the time when disclinations became visible. The areas of the right and left-hand twisted regions are almost the same all through the temporal evolution. This is a manifestation of that these two states are thermodynamically equivalent.

To see the dynamical characteristics of the system, we first measured a total length of disclinations per unit area. It has turned out that L(t) decreases with time and can be expressed by a power law as

$$L(t) = L_0 (t - t_0)^{-\mu} \qquad (2)$$

with L_0 = 35 (mm), t_0 = -0.06 (sec) and μ = 0.44.

Next we obtained the spatial correlation function C(x, t) at time t;

$$C(x, t) = \langle S(x, t) \; S(0, t) \rangle \qquad (3)$$

where S(x, t) = -1 in the "dark" region and S(x, t) = 1 in the "bright" region. From the C(x, t) (see Fig. 3), the correlation length at time t, $\ell(t)$, is defined as x at which C(x, t)/C(x, 0) becomes a half. It is seen that $\ell(t)$ increases with time and is well expressed by a power law as

$$\ell(t) = \ell_0(t - t_0)^{\nu} \qquad (4)$$

with ℓ_0 = 17 mm, t_0 = - 0.12 sec and ν = 0.44 (Fig. 4). Note that the exponents μ and ν are exactly the same, though their numerical value itself differ from 0.5 expected by the Allen-Cahn theory.[1]

Let us scale the length with the use of $\ell(t)$, and redraw the relationship between C(x, t)/C(x, 0) and x/$\ell(t)$. It is seen that all data fall on just one universal curve(Fig. 5), implying that the dynamical scaling law holds in the evolution of patterns shown in Fig.2.

INTERFACES IN NON-ORTHOGONALLY TWISTED NEMATIC CELLS [4,5]

Next we describe the observation performed with the use of the non-orthogonally twisted nematic cells. Figure 6 shows the θ-twisted state and the ($\pi - \theta$)-twisted state, where $\theta \neq \pi/2$. It is seen that two states are not energetically equivalent, and the velocity of an interface is written as

$$\frac{dR}{dt} = - \frac{\Gamma}{R} \pm D\varepsilon , \qquad \varepsilon = \frac{\pi}{2} - \theta \qquad (5)$$

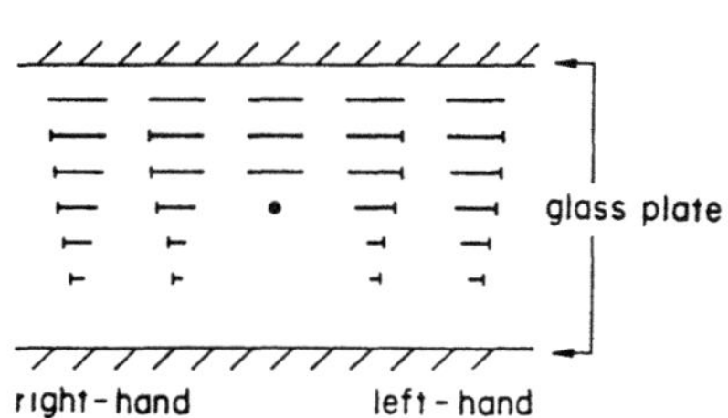

Fig. 1. An orthogonally twisted nematic cell. The molecules are represented by the nails whose points are turned towards readers. The disclination separates the left- and right-hand twisted regions.

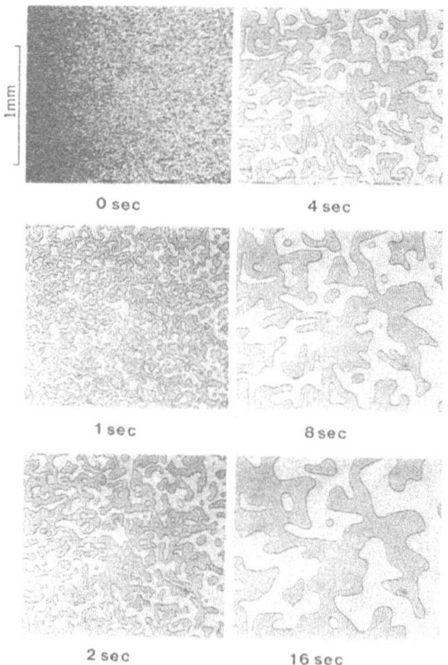

Fig. 2. Temporal evolution of the left- and the right-twisted regions. The boundaries are disclinations.

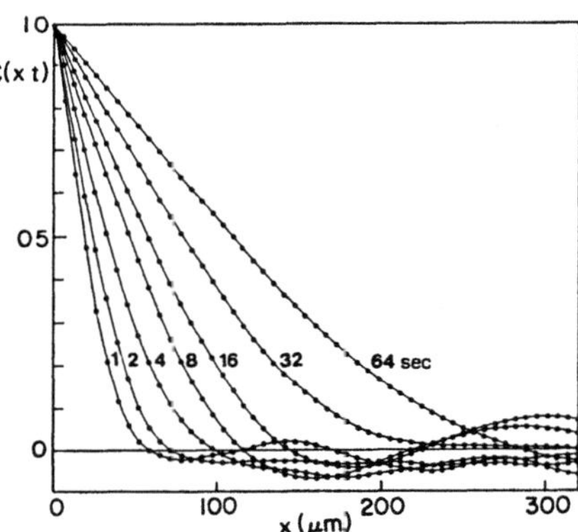

Fig. 3. Dependence of the spatial correlation function on time.

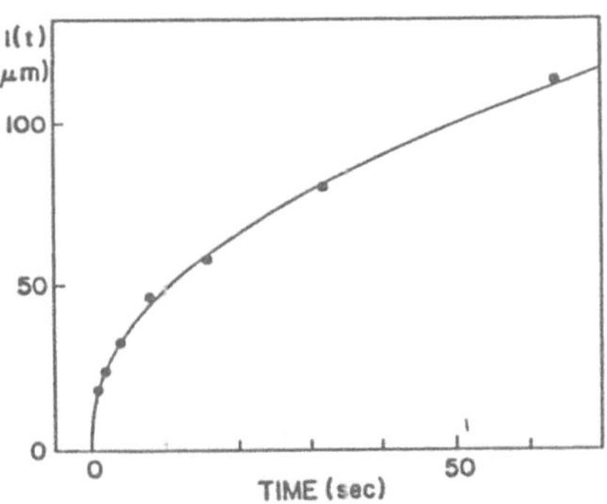

Fig. 4. The time dependence of ℓ(t).

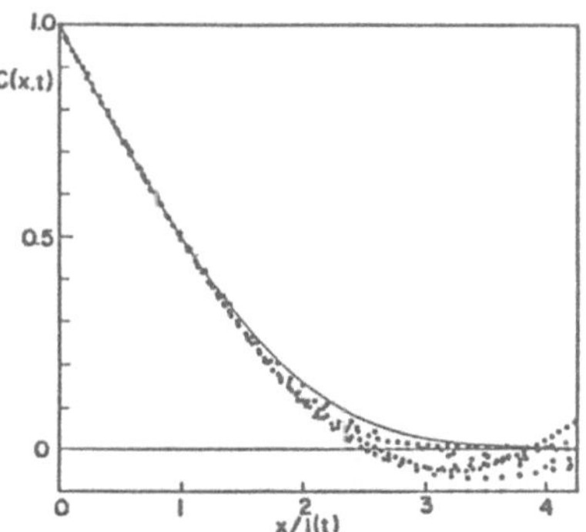

Fig. 5. C(x, t)/C(x, 0) vs. x/ℓ(t).

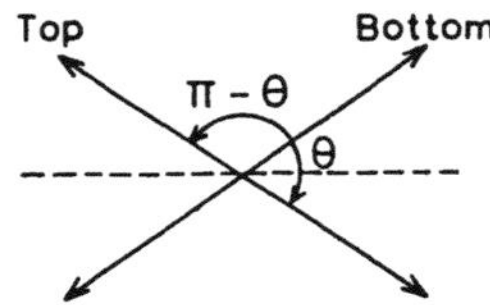

Fig. 6. The non-orthogonally twisted nematic cell. The orientations of molucules at the top and the bottom are shown by the double-ended arrows.

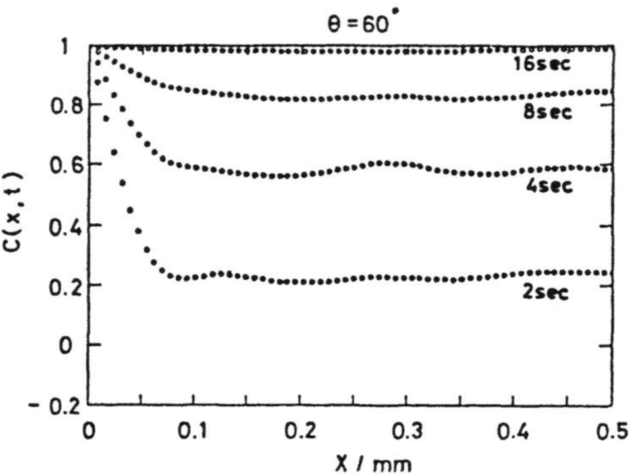

Fig. 8. Dependence of the spatial correlation functions on time for the cell with $\theta = 60°$.

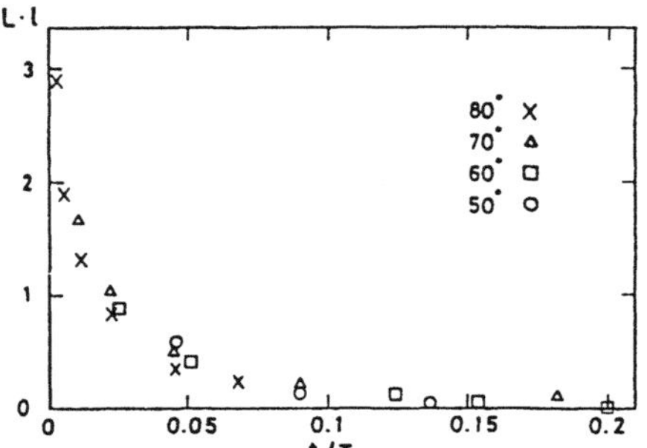

Fig. 9. The scaled total length disclinations vs. the scaled time. All data seem to fall on a universal curve.

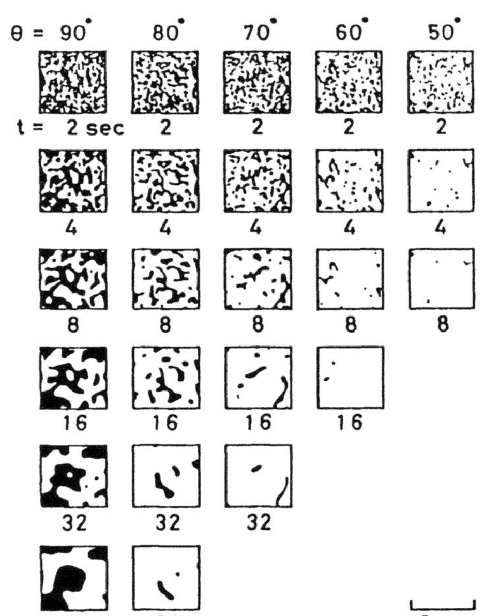

Fig. 7. Temporal evolution of the θ (bright) and the $\pi - \theta$ (dark) twisted regions.

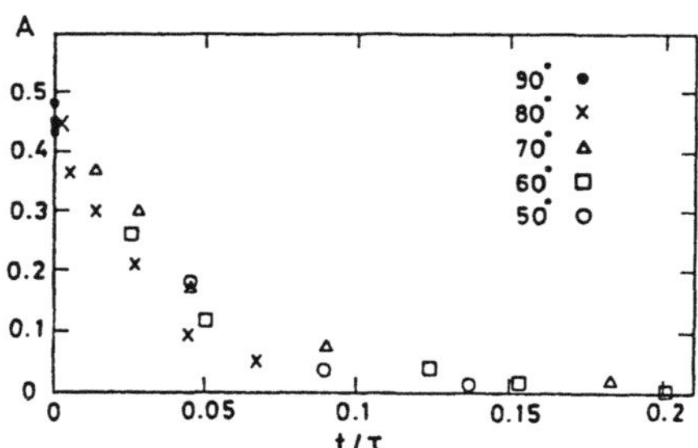

Fig. 10. The area fraction vs. the scaled time. All data seem to fall on a universal curve.

where the $D\varepsilon$ -term represents the velocity due to the difference of the potential energies of the two states. Figure 7 shows the evolution of patterns in various cells with different θ -values. It is seen that the ratio of the θ -twisted regions and the $(\pi - \theta)$ -twisted regions is not unity, and the larger the ε-value, the sooner the unstable regions disappear.

The spatial correlation functions for the cell with $\theta = 60^\circ$ are shown in Fig. 8. It is easily understood that the dynamical scaling law does not hold in this case, because these curves can not be made coincident no matter how the length (abscissa) is scaled.

If time and space are scaled as follows, however, one can see that there is a similarity between patterns in non-orthogonally twisted cells with various θ -values. Let us introduce the scaled quantities;

$$\tilde{R} = R / \ell , \quad \tilde{t} = t / \tau , \quad \tilde{x} = x / \ell , \tag{6}$$

where

$$\ell = \Gamma / D\varepsilon , \quad \tau = \Gamma / (D\varepsilon)^2 . \tag{7}$$

Then, eq. (5) is reduced to

$$\frac{d\tilde{R}}{d\tilde{t}} = - \frac{1}{\tilde{R}} + 1. \tag{8}$$

In this way all scaled quantities can be reduced to be in the form independent of Γ and $D\varepsilon$. Furthermore, let us define the scaled quantities;

$$\tilde{L}(\tilde{t}) = L(t)\ell , \quad \tilde{C}(\tilde{x}, \tilde{t}) = C(x, t),$$

$$\tilde{A}(\tilde{t}) = A(t), \tag{10}$$

where $A(t)$ is the area fraction of the less stable region, and plot $\tilde{L}(\tilde{t})$ and $\tilde{A}(\tilde{t})$ as the functions of $\tilde{x}$ (Figs. 9 and 10). It is seen that these quantities lie on a universal curve, respectively. Furthermore, we plot $\tilde{C}(\tilde{x}, \tilde{t})$ in Fig. 11 and then find that patterns in non-orthogonally twisted nematic cells with different θ -values are of the same structure statistically. For example, the pattern seen in the 60° cell at $t = 2.0$ sec is statistically the same as the one in the 80° cell at $t = 17.6$ sec.

Next, we outline a method of further analyses of patterns in the non-orthogonally twisted nematic cells. Since it is known that for $\varepsilon = 0$ the relation that

$$L(t) \propto t^{-1/2} \tag{11}$$

holds, the scaled function will be written as

$$L(t) = ct^{-1/2} f_L(t), \tag{12}$$

where c is a constant independent of Γ and $D\varepsilon$, and

$$f_L(0) = 1. \tag{13}$$

From this, it is predicted that

$$L(t) = c\Gamma^{-1/2}t^{-1/2} f_L(t/\tau) \qquad (14)$$

holds. At this point, the concrete form of the function $f_L(t/\ )$ is not yet known.

This problem has been approached from another viewpoint by Toyoki and Honda.[6] In an attempt to extend the OJK-theory,[7] they studied the effect of the initial domain distribution in the Ising system on the following temporal evolution. They actually considered the case where a magnetic field is applied only for a short time during quench to set the initial distribution of domains, but the field is removed when observations of domain patterns are started. According to them, the evolution of the area fraction and the length of the interfaces can be expressed as

$$A(t) = \frac{1}{2} - \frac{1}{2} \operatorname{sgn}(\varepsilon) \operatorname{erf}\left[\left(\frac{2\varepsilon^2 t}{B}\right)^{1/2}\right],$$
$$L(t) = \frac{1}{\sqrt{2}}(8\Gamma t)^{-1/2} \exp\left(-\frac{2\varepsilon^2 t}{B}\right). \qquad (15)$$

Our data shown in Figs. 9 and 10 agree well with the formulas (15).

It should be noted, however, that our experimental conditions differ from the ones assumed in the theory. In our experiment the   is not vanishing all through the experiment, corresponding to that the magnetic field is kept applied (but not removed as above mentioned). It is interesting that the agreement is, nevertheless, reasonable. This may be due to the situation that the $D\varepsilon$ - term in (5) plays only a minor role. The agreement seems to suggest, rather, that the initial distribution of domains is decisively important. As is seen from Fig. 12, where domains are schematically depicted as "islands" in "sea", the connectivity of interfaces differs in average in the cases of $A(0) = 1/2$ and $A(0) \neq 1/2$ (The $A(0) \neq 1/2$ case is regarded as the result given rise to by a rise of the sea level). Then, the mean squared radius of curvature is smaller in the $A(0) = 1/2$ case, making the first term in (5) more influential. This is the explanation why our data can be well fitted by the Toyoki-Honda formulas,[6] in spite that the experimental conditions differ from what assumed in the theory.

CONCLUSIONS

In the present study it has been found that the dynamic scaling law holds in the orthogonally twisted nematic cell, and in the non-orthogonally twisted cells, the statistical equivalence could be found for the pattern evolution in the cells with different twist angles. The Toyoki-Honda theory is found applicable to pattern evolutions in the non-orthogonally twisted nematic cells in spite that the situation assumed in the theory is somewhat different from what actually takes place in the present experiments.

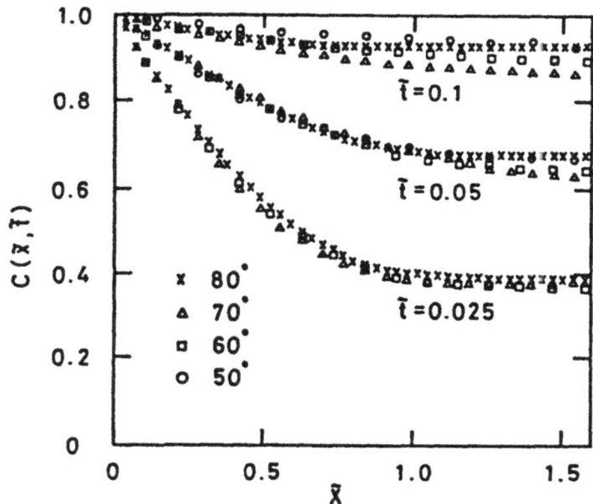

Fig. 11. Dependence of the correlation
functions on the scaled length
$\tilde{x}$ at various scaled times $\tilde{t}$.

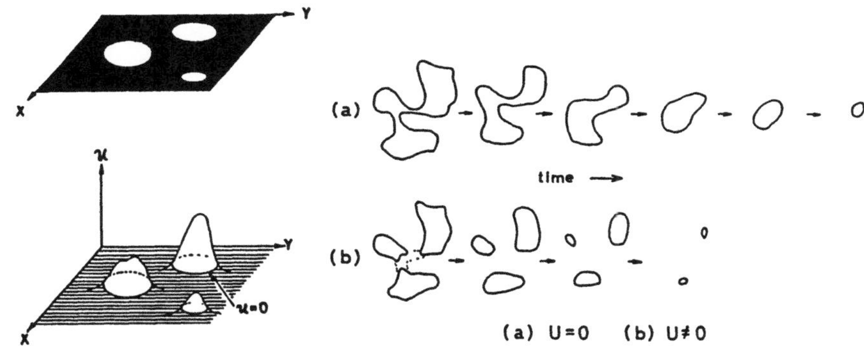

Fig. 12. Evolutions of the interfaces presented by the u-field theory.

References

1. S. M. Allen and J. W. Cahn, Acta. Metall. 27, 1085 (1979).
2. K. Kawasaki and T. Ohta, Prog. Theor. Phys. 67, 147 (1982).
3. H. Orihara and Y. Ishibashi, J. Phys. Soc, Jpn. 55, 2151 (1986).
4. T. Nagaya, H. Orihara and Y. Ishibashi, J. Phys. Soc. Jpn. 56, 1989
    (1987).
5. T. Nagaya, H. Orihara and Y. Ishibashi, J. Phys. Soc. Jpn. 56, 3086
    (1987).
6. H. Toyoki and K. Honda, Phys. Rev. B33, 385 (1988).
7. T. Ohta, D. Jasnow and K. Kawasaki, Phys. Rev. Lett. 49, 1223 (1982).

SPATIALLY MODULATED STRUCTURES IN MODELS
WITH COMPETING INTERACTIONS – SOME NEW RESULTS

Walter Selke

IFF der KFA Jülich
D–5170 Jülich, Federal Republic of Germany

ABSTRACT

Models with discrete symmetry and competing interactions are considered. Among the interesting properties of these models are sequences of distinct, commensurate phases, structure combination branching processes and incommensurate structures. New results include lock–in transitions of the phase factor in the mean–field theory of the ANNNI model and analyses of models for alloys and microemulsions.

INTRODUCTION

Various microscopic models have been proposed and analysed to study spatially modulated structures which have been observed in many materials, such as magnets, ferroelectrics, alloys, polytypes, and adsorbates. One may classify the models according to the range of interactions and the symmetry of the local variables which form the modulated structures. Usually, rather complex patterns can occur already at zero temperature in models where the local variable is continuous and/or in models with long-range competing interactions. Examples are the model of Frenkel and Kontorowa where the local variable is the position of an atom in a harmonic chain subject to an external periodic potential, the chiral XY model with a classical spin vector, and the Ising model with long–range antiferromagnetic interactions.

However, for models with discrete symmetry and short–range competing interactions, entropy plays an essential role in stabilizing complex phases, while the ground states are rather trivial. A prototypical example is the axial next–nearest

Geometry and Thermodynamics
Edited by J.-C. Tolédano
Plenum Press, New York, 1990

neighbour Ising (or ANNNI) model: competing interactions couple near–by Ising spins, $s_\alpha = \pm1$, which are situated in the same, adjacent or next–nearest layers of a crystal. The crucial properties of this and related models have been reviewed recently (1). Among the most prominent features of the models are sequences of distinct, commensurate phases, structure combination branching processes which may accumulate at devil's top steps, incommensurate structures and Lifshitz points. For an introduction and details the interested (non–expert) reader is asked to look at the review (1).

The aim of this contribution is to draw attention to some new advances in the studies of models with discrete symmetry and short–range competing interactions, which have been published or achieved after completion of that review.

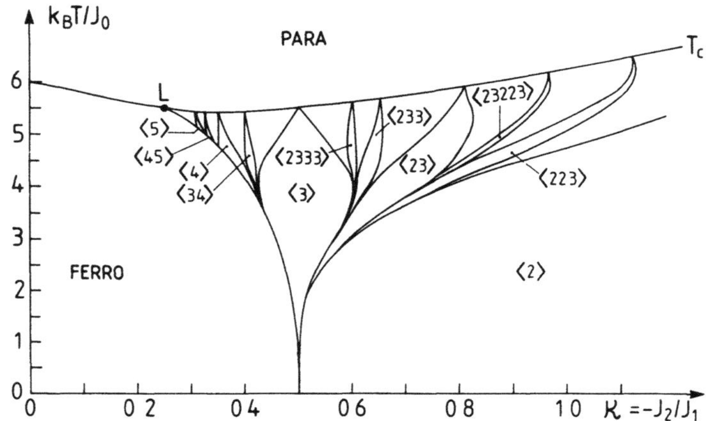

Fig. 1. Mean–field phase diagram of the ANNNI model, $J_0 = J_1$. From (3).

SOME NEW RESULTS

Mean–field theory of the ANNNI model

The mean–field equations of the ANNNI model are

$$M_i = \tanh(\beta H_i) \qquad i = -\infty, \ldots, 0, \ldots \infty$$

M_i is the magnetization of the i–th layer, β is the inverse temperature, and H_i is the effective field

$$H_i = 4J_0 M_i + J_1(M_{i+1} + M_{i-1}) + J_2(M_{i+2} + M_{i-2})$$

assuming the coordination number in a layer to be 4.

The set of coupled equations may be interpreted as a four–dimensional non–linear mapping of the vector $(M_{i-2}, M_{i-1}, M_i, M_{i+1})$. Expanding about the fixed points of that mapping, many properties can be calculated with high numerical precision, as has been done by Siems and Tentrup (2). In particular, they have determined the locations of accumulation points (devil's top steps) of structure combination branching processes (1). For example, the sequence of phases of type $< 23^n >$, $n = 2, 3, 4, \ldots$, which split from the $< 3 >$ phase at successively higher temperatures, accumulates ($n \to \infty$) at $k_B T/J_0 = 2.856331\ldots$ for $J_0 = J_1$, in very good agreement with our original estimate based on an extrapolation (3). For orientation, some of the main commensurate phases are shown in the mean–field phase diagram of the ANNNI model with $J_0 = J_1$, figure 1. — Moreover, additional accumulation points of the phases $< 2^k 3(2^{k-1}3)^n >$, k fixed, $n \to \infty$, have been located. The points move towards lower temperatures with increasing value of k. In the limit $k \to \infty$, the multiphase point at $-J_2/J_1 = \frac{1}{2}$ and $T = 0$ is approached (2).

By weakening the in–layer coupling, J_0, a new type of transition has been observed in mean–field theory. Close to the paramagnetic phase boundary, T_c, the magnetization per layer, M_i, has a sinusoidal form

$$M_i = A\cos(qi + \emptyset)$$

where q is the wavenumber and $\emptyset$ is the phase factor. As had been noted already some years ago by Yokoi, Coutinho-Filho and Salinas (4), the phase factor changes at $-J_2/J_1 = \frac{1}{2}$, where the $< 3 >$ phase with $q = \frac{2\pi}{6}$ is stable, with

$$\emptyset = \begin{cases} 0 & \text{for } J_0 > \frac{1}{3}J_1 \\ \frac{\pi}{6} & \text{for } J_0 < \frac{1}{3}J_1 \end{cases}$$

In the latter case, the magnetization patter is of the form

$$\ldots + +0 - -0 + +0 \ldots$$

0 denotes a disordered layer, while the (non–vanishing) magnetization has the same absolute value in the other layers. More generally, patterns with disordered layers may occur close to T_c at sufficiently small ratios J_0/J_1 for those structures whose wavenumbers $q/2\pi$ have even denominators (5). This interesting behaviour has been compared to experimental results on the magnet $CeSb$ (5).

At low temperatures, patterns with disordered layers are, of course, unstable (in mean–field theory, the ordering temperature of a layer is zJ_0, where z is the coordination number in a layer). Concretely, by solving the mean–field equations on finite lattices numerically, Yokoi has found for the $< 3 >$ phase that the phase factor $\emptyset$ first stays at its critical value, $\frac{\pi}{6}$, as the temperature is lowered, then undergoes a rapid change to 0 (implying a vanishing of the disordered layers), at which value it locks–in at lower temperatures (6). This numerical result is corroborated by exact calculations for an Ising model with competing interactions between spins on nearest and next–nearest levels of a Caylee tree augmented by ferromagnetic interactions on the same level of the tree (7).

The lock–in transition of the phase factor is reflected by the topology of the $< 3 >$phase, as sketched in figure 2.

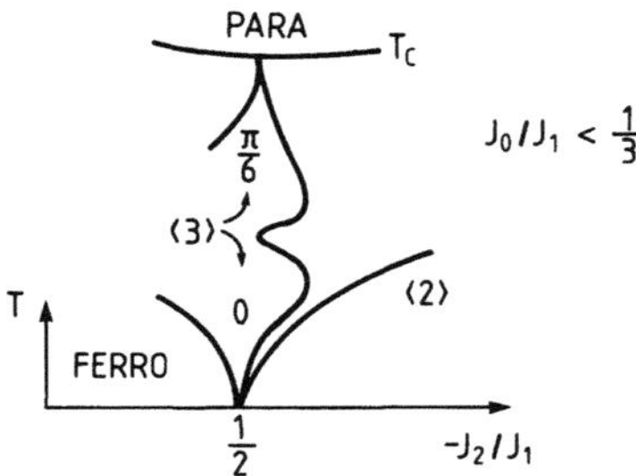

Fig. 2. Topology of the $< 3 >$phase in mean–field approximation for weak in–layer coupling, J_0. After (6).

Of course, it needs to be clarified whether such interesting transitions are artefacts of mean–field theory which becomes less reliable the weaker the in–layer coupling, J_0, becomes.

<u>Ferroelectrics, alloys and microemulsions</u>

Recent experimental findings on the ferroelectrics $BCCD$ show striking similarities to basic features of the phase diagram of the ANNNI model, including sequences of commensurate phase, branching processes, incommensurate structures and (the possible existence of) a Lifshitz point (8,9). Indeed, the experimental phase diagram has been discussed semiquantitatively in analogy to the one of the ANNNI model (2).

Alloys are an active field for experimental studies of long–period spatially modulated superstructures. For example, high–resolution electron microscopy has revealed branched commensurate structures in Cu_3Al (10). Following suggestions by de Fontaine and Kulik (11) these and similar experimental findings (1) have been discussed in the framework of Ising models with short–range competing interactions. Very recently, Upton and Yeomans studied an isotropic Ising model on a cubic lattice with coupling constants (12)

$$J(r) = \begin{cases} J & \text{between nearest–neighbour sites} \\ J' & \text{between next–nearest–neighbour sites along cubic axes} \\ J'' & \text{between next–nearest–neighbour sites across face diagonals} \end{cases}$$

The model may account not only for one–dimensionally modulated superstructures, but also for phases modulated in more than one lattice direction. It would certainly be interesting to see which of the various complex features of the model will persist by going beyond mean–field theory.

A closely related Ising model with the same range of interactions has been proposed by Widom to the study of microemulsions (13). For special choices of the coupling parameters its mean–field theory is identical to the one of the ANNNI model. Low temperature expansions supplied with conjectures have been carried out (14), but the arguments have yet to be made rigorously. — It may be worthwhile to investigate the effect of dilution in that model, mimicing microemulsions in porous media. A first modest attempt towards that aim has been undertaken by doing Monte Carlo simulations on the two–dimensional version of the model (15). As expected for Ising models with competing interactions, dilution may lead to glassy–like structures.

<u>Theoretical results on other models</u>

The study of models with discrete symmetry and short–range competing interactions remains a dynamic field. New models are introduced and phase diagrams are refined using a variety of methods.
Among the senior models the three–state chiral clock model has been analysed furthermore. The model reduces for vanishing chirality parameter to the well-known three–state Potts model, i.e. the local variable can take on three values, $S_\alpha = 0, 1, 2$, with a preference for identical values at neighbour sites. Turning on the chirality parameter, a chiral sequence $\ldots 012 \ldots$ along one axis of the system is favoured. By fixing the boundary conditions against this chiral ordering, wetting phenomena may occur (16). Sequences of layering transitions have been established for the three-dimensional model by applying low–temperature expansions as well as mean–field theory (17,18). — Using Monte Carlo techniques, the transition from the spatially modulated phase to the paramagnetic phase has been demonstrated to be in the universality class of the xy–model for the three–dimensional case (19), in accordance with analoguous results for the two–dimensional model (1). — An exactly soluble variant of the 2d chiral clock model has been proposed and analyzed in detail recently, see, e.g. (20).

Another interesting three–state model is the Blume–Capel model with competing interactions between spins in adjacent and next–nearest layers. Its Hamiltonian may be written in the form (21)

$$\mathcal{H} = \mathcal{H}_{\mathrm{ANNNI}}(S_\alpha) + D \sum_\alpha S_\alpha^2 \qquad S_\alpha = 0, \pm 1$$

The model is expected to display for a specific choice of the coupling constants a tricritical Lifshitz point, as has been confirmed by applying mean–field theory. Due to the competing interactions, sequences of commensurate phases occur. The character of the associated (devil's) staircases, looking at the wavenumber vs. a characteristic parameter of the model, may be investigated by calculating their Hausdorff dimension. Such calculations have been performed for the model on a Caylee tree (21) and also far the straight ANNNI model on that lattice structure (22). In both cases, evidence for complete devil's staircases at low temperatures has been provided.

Models with competing multispin interactions have attracted some interest in the last time (23,24). However, it looks like more detailed analyses are needed to demonstrate clearly the existence of incommensurate or long–period commensurate structures in such models.

In conclusion, this brief and presumably rather incomplete status report on recent activities shows that models with discrete symmetry and short–range competing interactions are still thriving.

<u>Acknowledgement</u>

I thank my colleagues for keeping me informed on their work. In particular, I enjoyed discussions with Silvio Salinas and his group at the University of Sao Paulo this summer which have had a useful impact on this contribution.

REFERENCES

1. W. Selke, Phys. Reports 170: 213 (1988).

2. R. Siems and T. Tentrup, Ferroelectrics (1989), in print.

3. W. Selke and P.M. Duxbury, Z. Physik B57: 49 (1984).

4. C.S.O. Yokoi, M.D. Coutinho-Filho and S.R. Salinas, Phys.Rev.B 24: 4047 (1981).

5. K. Nakanishi, J.Phys.Soc. Japan 58: 1296 (1989).

6. C.S.O. Yokoi, private communication, and 17th International Conference on Thermodynamics and Statistical Mechanics, Rio de Janeiro, August 1989

7. J.G. Moreira, Ph.D. thesis, Universidade Federal de Minas Gerais, December 1987.

8. R.G. Ao, G. Schaack, M. Schmitt and M. Zöller, Phys.Rev.Lett. 62: 183 (1989).

9. H.G. Unruh, F. Hero and V. Dvorak, Solid State Commun. 79: 403 (1989).

10. D. Broddin, G. Van Tendeloo, J. Van Landuyt, S. Amelinckx and M. De Graef, Phil.Mag.A 59: 979 (1989).

11. D. de Fontaine and J. Kulik, Acta Metall. 33: 145 (1985).

12. P. Upton and J. Yeomans, Phys.Rev.B 40: 479 (1989).

13. B. Widom, J.Chem.Phys. 84: 6943 (1986).

14. A. Berera and K.A. Dawson, Preprint (1989).

15. M.J. Velgakis, Preprint, submitted to Physica A (1989).

16. W. Selke, in: Lecture Notes in Physics, Vol. 206, Springer, Berlin (1984), p. 191; S. Dietrich, in: Phase Transitions and Critical Phenomena, Vol. 12, eds. C. Domb and J.L. Lebowitz, Academic Press, New York (1988) p. 1.

17. K. Armitstead and J.M. Yeomans, J.Phys.A 21: 173 (1988).

18. M. Siegert and H.U. Everts, J.Phys.A 22: 117 (1989).

19. M. Siegert and H.U. Everts, J.Phys.A 22: L783 (1989).

20. R.J. Baxter, J.Stat.Phys. 52: 639 (1988) and references therein.

21. T. Tomé and S.R. Salinas, Phys.Rev.A 39: 2206 (1989).

22. C.P.C. Prado and N. Fiedler–Ferrari, Phys.Lett.A 135: 175 (1989).

23. K.A. Penson, Phys.Rev.B 29: 2404 (1984).

24. M.D. Grynberg and H. Ceva, Phys.Rev.B 40 (1989), in print.

WEAKLY PERIODIC STRUCTURES WITH
A SINGULAR CONTINUOUS SPECTRUM

Serge AUBRY

Laboratoire Léon Brillouin (CEA-CNRS)
CEN Saclay F-91191-Gif-sur-Yvette Cedex (France)

Abstract

Any <u>translationaly invariant</u> hamiltonian has always classical ground-states with a minimum kind of order which we called "weak periodicity" or equivalently with "local order at all scales". This property does not imply necessarily periodicity (crystals) or quasiperiodicity (incommensurate structures and quasicrystals). We describe examples of structures which are ground-state of translationally invariant hamiltonians and which are neither periodic nor quasi-periodic but are only weakly periodic. Their Fourier spectrum S(q) (observable spectrum by X-ray, electrons or neutrons scattering) contains neither any Dirac peaks except at q=0 nor any smooth continuous component and is a "singular continuous measure".

INTRODUCTION

During the last decades, a large number of structures in solid condensed matter were observed to be non periodic. However, in most cases, it was possible to describe these structures as standard periodic crystals with superimposed incommensurate periodic modulations in space. These modulations could concern for example the site occupation by different kinds of atoms (antiphases in alloys), the orientation or the average length of spins (Helimagnets and Spin Density Waves), the electronic density (Charge Density Waves) or (and) atomic displacements (Displacive Incommensurate Structures) etc... More recently, the discovery of quasicrystals has shown for the first time, that certain quasi-periodic structures cannot be described as a standard periodic crystal with modulations but could be described directly by quasiperiodic lattices (such as Penrose lattices).

The microscopic interpretation of an observed structure requires a sufficient knowledge of its energy as a function of the microscopic parameters involved in the structure such as the atomic positions, the local magnetization etc.... In addition, it is clear that the energy of such models has to be invariant by any translation (and rotation) in space. We will not discuss here the properties of the ground-states of models with a quenched disorder (e.g.amorphous, spin glasses) which physically occurs when the kinetic of the system becomes partially frozen and thus prevents it from a complete thermal equilibrium at low temperature.

The purpose of this talk is to discuss from a general and theoretical point of view, some aspects of the following symmetry group problem: how the ground-state of an Hamiltonian which is invariant by a translation group T, could break this initial symmetry. If instead of the translation group, this Hamiltonian is invariant by a compact group G (G can be for example a point symmetry group for a crystal Hamiltonian or the local rotation group of spins for a Heisenberg model), it is well-known that the ground-state has to be invariant by a subgroup of G. The theory of irreducible representations of G yields the classification of all possible symmetries for the ground-state.

When the Hamiltonian is invariant by the action of a non compact group such has the translation group, the answer to the problem is by far not so clear. The simplest situation occurs when the ground-state is invariant by a subgroup of the translation group. This well-known situation yields a crystal for the ground-state. But when the ground-state is an incommensurate structure, the ground-state has apparently no <u>translation symmetry</u>. However, we are tempted to say that it has more translation symmetry than a purely random structure! In fact, this intuition is true because an incommensurate structure has to remain invariant by transformations which are <u>uniform</u> limit of translations with diverging lengths while a random structure is not.

The fact that the "ideal" structure of a classical model is the absolute minima of a functional energy invariant by a translation group, implies that the groud-state has a minimal property called "weak periodicity". Let us consider for example, a 3-dimensional system of atoms of different species A,B,C... with given concentrations which interact through a potential energy $\Phi(\{u_i^A\},\{u_j^B\},...)$. $\{u_i^A\}$ is the coordinate vector which determines the position of the atoms i of the specy A, $\{u_j^B\}$ the coordinate vector of the atom j of the specy B, etc... In order to fix the ideas, this potential Φ can be the sum of pair potentials depending on the relative distance of all possible atomic pairs i,j.

$$(1) \qquad \Phi(\{u_i^A\},\{u_j^B\},...). = \sum_{i_1,i_2} V_{AA}(u_{i_1}^A - u_{i_2}^A)$$

$$+ \sum_{j_1,j_2} V_{BB}(u_{j_1}^B - u_{j_2}^B) + \sum_{i_1,j_2} V_{AB}(u_{i_1}^A - u_{j_2}^B) + etc...$$

but we can also assume the existence of 3-body interactions and more. A simple problem which belongs to this class is the close packing of spheres A,B,C with different diameters. One has just to consider hard core potentials in (1) and chemical potentials for the atoms which favors the most compact filling. In most cases, finding the ground-state of model (1) for given pair potentials is an unsolved problem (even for the apparently simpler close packing problem). It depends obviously on the respective concentration of the atoms A, B,C ... but possible phase separation between phases (which are ground-states respectively) with different compositions in A,B,C.... may occur.

In general, the energy functional $\Phi(\{u_i^A\},\{u_j^B\},...)$ has infinitely many local minima (metastable states) which are chaotic but with more energy than the ground-state. In some one dimensional models with short range interactions, the ground-states and the metastable states can be mapped onto a subset of trajectories of a hamiltonian system (for example the Frenkel Kontorowa model maps onto the standard map[1]). In that cases, the energy functional of the structural model has just to be interpreted as the action of the associated dynamical system. Then, the 1d space variable is mapped onto the time in the dynamical system. For most (non integrable) dynamical systems, the existence of chaotic trajectories is now well admitted on the basis of numerical simulations. In addition, this assumption can be rigorously proven in some cases as in the FRENKEL KONTOROWA (FK) model. On this basis of these remarks and also by analogy with problems of turbulence, D. RUELLE has conjectured that the ground-state of some models might be chaotic[2].

In fact, we pointed some years ago[1] that the problem of finding a ground-state is formally different of the problem of finding the behavior of a dynamical system and thus cannot be confused. For Hamiltonian dynamical systems, all the extrema of the action which correspond to trajectories, have to be considered. By contrast, only in the absolute minima of the energy functional corresponds to the ground-state. Indeed, we have shown in ref. 1, that the ground-states of a large class of models invariant under the action of a translation group necessarily have a property which we called "weak periodicity" which prevent these ground-states to be random although the associated dynamical sysyem have random trajectories.

GROUND-STATES AND WEAK PERIODICITY

The systems which we study are imbedded in a d dimensional space (physically d=3). For proving that the ground state is weakly periodic, we need to assume:

<u>Hypothesis 1:</u> The energy functional $\Phi(\{u_i^A\},\{u_i^B\}...)$ is formally invariant by translation, that is for any translation $\mathbf{R}$:

$$(2) \qquad \Phi(\{u_i^A\},\{u_j^B\}...) = \Phi(\{u_i^A + \mathbf{R}\},\{u_j^B + \mathbf{R}\}...)$$

This condition does not require a pair potentials between the atoms.

<u>Hypothesis 2</u> The energy variation of the free energy with respect to the displacement δ of any atom is finite and continuous versus δ.

283

This condition implies not only the continuity of the potentials of the model with respect to the atomic positions, but also the absence of long range interactions between the atoms. If the pair potential interaction as a function the particle distance r, decay faster than $\frac{1}{r^{d+\alpha}}$ with $\alpha > 0$, hypothesis 2 will be fulfilled providing that the space of configurations be restricted to a subspace determined by some DELAUNEY conditions in order that the forces between the atoms are always bounded. Let us describe these conditions.

By definition, a configuration is said to fulfill a DELAUNEY conditions, if there exists two lengths l and L, such that

1) any ball in $\mathbb{R}^d$ with radius l contains at most one atom

2) any ball with radius L contains at least one atom (dense matter)

It is clear that physically, the Delauney conditions are always fulfilled for condensed matter, because

1) the local density of matter in a solid cannot become infinite

2) the global density is defined and cannot be zero.

Let us define now what is weak periodicity:

<u>Definition:</u> A configuration $C = (\{u_i^A\}, \{u_j^B\}, \{u_k^C\}..)...$ is said to be weakly periodic when for any local atomic configuration C_D (that is the piece of the configuration C which lie inside some finite convex domain D) and any given accuracy ε, there exists a length $L(C_D,\varepsilon)$ such that within the accuracy ε, the same local configuration is found translated in any ball with radius $L(C_D,\varepsilon)$.

The physical meaning of this property means that any finite block which already exists in the structure, repeats itself at bounded distance and everywhere. This property can be also called "local order at all scale". It has similitudes with the Harald BOHR definition of "almost periodic functions" except that the uniform topology of functions is here replaced by the "weak topology" in the configuration space (see refs. 1 and 6). Fig.1 exhibits a scheme suggesting intuitively weak periodicity. There is no rules which relates the maximum distance $L(C_D,\varepsilon)$ to the size of the block C_D. Although it has to be finite, this distance might be very large.

Periodic and quasiperiodic structures are necessarily weakly periodic structures, the reverse is not true: weak periodicity does not imply neither periodicity nor quasiperiodicity. By contrast, a purely random configurations cannot be weakly periodic. Indeed any finite block of the structure repeats itself within the accuracy ε with some probability. Thus in a finite ball with radius L, a given block within the accuracy ε, has a non zero probability to be found again and a non zero probability for not being found. Thus, whatever L is large, there will exist balls with radius L which do not contain this given block. Therefore, the condition for weak periodicity cannot be fulfilled.

In 1983, we proved the following theorem [1] which should give a special physical interest for the concept of weakly periodic structure:

<u>Theorem :</u> If for an energy functional which fulfills the above 2 hypothesis, there exists a minimum energy configuration which fulfills a Delauney condition for some l and L, then this model has a "weakly periodic" ground-state.

References 1 and 6 described the proof of this theorem which appears to be an extension of parts of our theories on the FK model. The non-trivial distinction between a "minimum energy configuration" and a ground-state is explained. The conditions for applying this theorem appears to be rather universal, because in physical models, the Delauney conditions must be always fulfilled. In the opposite cases, infinite fluctuations of the density of matter could occur which would simply mean that the considered model is not suitable for describing solid condensed matter. However, we have not determined precisely the mathematical conditions which are necessary, in order that these Delauney conditions be fulfilled with certainty.

Consequently, the ideal ground state of a physical classical model for solid condensed matter, is always weakly periodic.

<u>Figure 1</u> *Scheme showing an example for the generation of a piece of "weakly periodic" structure with black dots and open dots. The structure is not random because any finite block which can be determined in the structure is found again at a finite distance. The same property holds for blocks of blocks and so on... (This construction has been done using a 2d version of the Thue-Morse sequence in 1 dimensions)*

WEAK PERIODICITY AND MINIMAL ORBIT

A more abstract definition for a weakly periodic structure (see refs. 1 and 6) can be given as a property of its orbit $\hat{C}$ By definition, the orbit $\hat{C}$ of the configuration C under the action of the translation group $\mathbb{R}^d$ is the set of configurations $T_{\mathbf{R}}(C)$ obtained by translation of C for all vectors $\mathbf{R}$ completed by all the possible limit configurations (these ones are "accumulation configurations"). $\hat{C}$ is closed for the weak topology. An orbit $\hat{C}$ is minimal when the orbit $\hat{C}'$ of any configuration C' in $\hat{C}$ is identical to $\hat{C}$ The translation group applied to the restricted configuration space $\hat{C}$ does not let invariant any subset of $\hat{C}$ which is not either the empty space or the full space $\hat{C}$. This concept of minimal orbit for a non compact group extends the concept of irreducible representation for a compact group. Unfortunately, it is not possible at the present time to give a full classification of the minimal orbits under the action of the translation group. We have the rigorous result:

A configuration C is weakly periodic if and only if its orbit $\hat{C}$ (under the action of the translation group) is minimal.

When the configuration C is periodic (crystal) in d independant directions of the d dimensional space , the orbit $\hat{C}$ is homeomorphic to a d dimensional torus. When the configuration C is quasi-periodic with n incommensurate modulations with zero gap phason modes, the orbit $\hat{C}$ is homeomorphic to d+n torus.

This theorem on weak periodicity of the ground-state also applies to the FK model. In that model, the energy functional is invariant under the action of a discrete translation group generated by the period of the underlying periodic potential. It has been shown that this invariance is one of the two basic properties which are necessary for developing the exact theory for the ground-state presented in ref.3. In that model, the orbit set $\hat{C}$ of the ground-state are represented as the closure of a trajectory in a dynamical system (the standard map). This set $\hat{C}$ might be a periodic cycle, but also it can be a KOMOGOROV, ARNOL'D MOSER (KAM) torus or non trivially a Cantorus.

By analogy with the FK model, we can call $\hat{C}$ the hull set of the configuration C. It can be a different of a torus which proves that ground-states can be more complex than quasi-periodic structures.

In spite of the possible absence of translation invariance, this hull set $\hat{C}$ allows one to define the point symmetry group G of configuration C. A configuration C is said to be globally invariant by a symmetry g when the hull set $g(\hat{C})$ of $g(C)$ is identical to $\hat{C}$. In other words, there exists a sequence of translations $\mathbf{R}_n$ such that

$$(3) \qquad \lim_{n \to \infty} T_{\mathbf{R}_n}(g(C)) = C$$

Of course, this point symmetry group is not necessarily a crystal point symmetry group and might be for example the icosaedral group.

WEAK PERIODICITY IN PSEUDOSPIN MODELS (LATTICE GAS)

The above theory which has been developed in a general context is applicable to pseudospin models on a periodic d-dimensional lattice Z^d. A pseudospin σ_i with two possible values, is associated with each site $i \in Z^d$ of the lattice (for convenience these values are 0 and 1). The Hamiltonian of this model which is a function of the pseudospin configuration $\{\sigma_i\}$, is supposed to be translationally invariant (i.e the energy functional of

286

configuration $\{\sigma_i\}$ is the same as those of configuration $\{\sigma_{i+n}\}$ for an arbitrary lattice vector $n \in Z^d$. This Hamiltonian can be expanded as a sum

$$(4\text{-}a) \qquad H(\{\sigma_i\}) = \sum_i h\,\sigma_i + \sum_{i,j} J(j\text{-}i)\,\sigma_i\,\sigma_j + \sum_{i,j,k} K(k\text{-}i,j\text{-}i)\,\sigma_i\,\sigma_j\,\sigma_k + \,....$$

of terms involving 1-pseudospin, 2 pseudospins, 3 pseudospins, ... n pseudospins etc... We assume that the sum

$$(4\text{-}b) \qquad B = \sum_j |J(j\text{-}i)| + \sum_{j,k} |K(k\text{-}i,j\text{-}i)| + \,.... \quad < +\infty$$

which does not depend on i, is absolutely convergent. Then, the energy variation of a configuration obtained by switching any single pseudospin is always finite (and its modulus is bounded by B). The energy variation remains finite when switching a finite number of pseudospins.

It is often possible to modelize more complex models by such pseudospin models. For example, if one study the ground-state structure of a simple compound (all the atoms are identical), we can introduce a periodic lattice with lattice spacing small compared to the atomic size. Since one atom at most which can occupy a given site i of this lattice, we set $\sigma_i = 1$ when this site is occupied by an atom and $\sigma_i = 0$ when this site is empty. When there is n kinds of atoms A,B,C..., the structure can be described by an array of pseudospins σ_i with n+1 possible values per site, each of these values corresponding either to an empty site or to a site occupied by one of the n kinds of atoms. (These model are called lattice gas models). A simplified version of the above theorem can be proven for these models. This theorem does not need any Delauney conditions or similar ones but requires:

1) that the Hamiltonian form be translationaly invariant as explicited by the form (4-a)

2) that the energy variation of the system be always finite when a finite number of pseudospins is changed as explicited by condition (4-b).

Then, the ground-states $\{\sigma_i\}$ of these models are weakly periodic. More precisely, for a pseudospin model, for any block of pseudospins $\{\sigma_i\}_D$ defined by a finite domain D (i.e. a finite subset) of the periodic lattice Z^d and $i \in D$, there exists a length $L(\{\sigma_i\}_D)$ which depends on this selected block such that in any d-cube with size L, the same block $\{\sigma_i\}_D$ is found again.

In other words, there exists a set T of lattice vectors n in Z^d

1) which is relatively dense with respect to the parameter $L(\{\sigma_i\}_D)$ (For any cube C of Z^d with size $L(\{\sigma_i\}_D)$, $T \cap C \neq \varnothing$ (the empty set)

2) such that for $n \in T$, we have $\sigma_i = \sigma_{i+n}$ for $i \in D$

In the case of pseudospin models, we can prove a theorem reciprocal of the theorem on the weak periodicity of ground-states. Instead, of finding the ground-state of a translationally invariant Hamiltonian, we reverse the problem. For a given weakly periodic pseudo-spin configuration, we search a translationally invariant Hamiltonian which has this configuration as ground state. More precisely this theorem is

Reciprocal Theorem

Let us consider a weakly periodic array of pseudospins $C = \{\sigma_i\}$ (with $\sigma_i = 0$ or 1) where i belong to $\mathbb{Z}^d$ (d=1,2,3...):

Then, there exists a well defined pseudospin Hamiltonian $H(\{s_i\})$ (in general it involves p-pseudospins interactions with p varying from 1 to ∞, but it can be chosen such that these pseudospin interactions decay for example exponentially at infinity) which is translationally invariant ($H(\{s_i\}) = H(\{s_{i+n}\})$ for any n)

a) such that configuration $\{\sigma_i\}$ be a ground-state

b) all the translated configurations $\{\sigma_{i+n}\}$ and their limits are equivalent ground-states of $H(\{\sigma_i\})$ ("phase degeneracy")

c) $H(\{s_i\})$ has <u>no other</u> ground-states.

Let us note that this Hamiltonian is not unique. The proof of this theorem was given in ref. 6. Unfortunately, although the existence of the Hamiltonian is proven, the method used for constructing it, yields a complex and rather artificial Hamiltonian, which is often unusable in practice.

COMMENTS ABOUT A THEOREM BY MIEKISZ and RADIN[9]

For lattice gas models which have a restricted form (4-a) with only pair interactions ($H = \sum_{i,j} V(i\text{-}j)\, \sigma_i \sigma_j$), it has been recently proven by these authors that for "most" interactions $V(n)$ (i.e. generically), no groundstate configuration is periodic unless it is the vacuum.

This theorem appears to be quite surprising for a physicist because it would suggest that cristalline materials would only exist in nature with probability zero! Concerning the incommensurate structures, the experimental studies use to show the existence of lockings when the wave vector of the modulation becomes commensurate. Theses observations are explained within the standard phenomenological theories by the energy gain of the incommensurate structures when reaching highest symmetries (for example "more" translational symmetry for the commensurate structures or more point symmetry group for the quasicrystals with icosaedral symmetry). The commensurate periodic phases which have extrasymmetry generally extend over finite domains in the parameters space and thus get a finite probability to be observed in contradiction with the above mentionned theorem.

The incommensurate phases with no extrasymmetry which have a given incommensurability ratio are only observable on a zero measure manifold in the parameter space but they can be physically detected because they have globally a finite measure. This situation is confirmed by rigorous studies on the Frenkel-Kontorowa model. The Devil's

Staircase transformation[1] which can be complete or incomplete, is an example which confirm this general feature.

In order to understand the failure in the application of the above mentionned theorem to the physical world, we have to analyse the meaning of the word "generic" and to warn the reader about an easy physical misinterpretations of this word. According to the mathematical definition, in order to say that a given property for a Hamiltonian H is "generic", one must consider that this Hamiltonian belongs to a certain space $\mathbf{H}$, with a certain topology. This property is generic when it is true for H belonging to the intersection of countable set of dense open sets in $\mathbf{H}$. Clearly, the property of genericity crucially depends on the choice of the space $\mathbf{H}$ and on the topology in this space of Hamiltonian.

MIEKISZ and RADIN considered the space $\mathbf{H}$ of pseudospin Hamiltonians with pair interactions with the form $H = \sum_{i,j} V(i\text{-}j)\, \sigma_i \sigma_j$ which have a well defined norm $\| H \| = \sum_n |V(n)|$. This norm defines the topology of this space. As a result, it is straightforward to prove that the Hamiltonians of this space has the <u>generic</u> properties

A) that there exists no constant K and $\alpha > 0$ such that $|V(n)| < \dfrac{K}{|n|^{d+\alpha}}$ (i.e. it is exceptional to find V(n) such that there exist K and α with this property).

B) that the set $\{\text{sign}(V(n)\}$ is random.

Therefore, the question becomes : Does the "physical models" have generically this property? In physical models, the pair potential interactions V(n) decay often exponentially with the distance $|n|$ and thus does not fulfills property A. When these interactions are due to charges, dipoles, quadrupoles etc..., they have to decay <u>smoothly</u> as power law. They might also exist oscillatory long range interactions with a power law decay when there is Fermi surface effects (e.g. RKKY interactions for spins). In physical systems, the long range interactions have a well-defined origin related to first principles and cannot be arbitrarily chosen so that the set of $\{\text{sign}(V(n)\}$ cannot be random. Consequently, property B is never fulfilled. Thus, no physical Hamiltonian has both the properties A and B which proves that the "most" (i.e. generically) Hamiltonians of the space of $\mathbf{H}$ are unphysical.

This remark makes that the theorem of MIEKISZ and RADIN looses most its physical interest. Otherwise, it seems impossible to recover a similar theorem as a generic property in the smaller space of "physical" Hamiltonians (which remains to be properly defined). The proof used for the initial theorem, turns out to be based on the property that one can make small perturbations of the Hamiltonian which can makes negative the energies of certain defects of the ground-state. To do that perturbation, one must introduce perturbations on the long range interactions which are appropriately chosen but are generally "unphysical"!. Thus, unlike the conclusion of this theorem, it is much more reasonable to believe that nature greatly favors crystal structures. In some compounds, the existence of phase diagrams with infinitely many periodic phases and Devil's staircase transformations as well as incommensurate (and perhaps more complicated structures), are suggested by the observations. Although not exceptional, these situations do not occur with probability one.

SINGULAR CONTINUOUS SPECTRUM FOR WEAKLY PERIODIC STRUCTURES

Till recent works[4,5], it was not clear wether non periodic and non quasi-periodic weakly periodic ground-states could exist or not (at least on some theoretical models). What could be the properties of the Fourier spectrum $S(q)$ dq of the density-density correlation function of such structures? This Fourier spectrum which is a positive measure, determines the intensity in the reciprocal space which is observable by neutron, X-ray, electron etc.. scattering on the considered structure. For crystals and standard incommensurate structures which are periodic and quasiperiodic structures respectively, this Fourier spectrum, is the sum of Dirac peaks distributed on a countable set of wave vectors q_n with intensity I_n,

$$S(q) = \sum_n .I_n \, \delta(q-q_n).$$ The same property hold for Penrose lattices and their extensions.

The property of weak periodicity seems to be too "loose" for finding general conclusions on the Fourier spectrum. Particularly, the existence of an absolutely continuous component cannot be excluded.

We are going to show one dimensional examples of weakly periodic structures which have a "singular continuous spectrum" with scaling properties (which thus are neither periodic and nor quasi-periodic). These measures are intermediate between discrete measures which are sum of Dirac peaks and absolutely continuous measures where $S(q)$ is a "good" measurable function. Instead of being concentrated on a countable set of points (discrete measure) or diluted rather uniformly in the reciprocal space (absolutely continuous measures), a singular continuous spectrum $S(q)$ dq is concentrated on a zero measure but uncountable set of points. They are neither concentrated enough to form Dirac peaks nor are "diluted" enough to be described by a smooth density function. Let us describe the simplest example

1-THE THUE-MORSE CHAIN

Let us consider the sequence of pseudospins $\sigma_i = \pm 1$ generated by iteration of the substitution rules:

(5-a)	+1	$\rightarrow$	(+1,-1)
(5-b)	-1	$\rightarrow$	(-1,+1)

Starting from the pseudospin +1, one obtain for example after 6 iterations , the following sequence : + - - + - + + - - + + - + - - + - + + - + - - + + - - + - + + - - + + - + - - + + - - + + - + - - + (where the symbol 1 has been dropped)

After an infinite number of iterations, we obtain a sequence which is completed for negative or zero i with the rule $\sigma_i = \sigma_{-i+1}$. It can easily proven that there is never more than two consecutive pseudospins with the same sign.

If one consider an arbitrary block of p consecutive pseudospins σ_i, σ_{i+1}, σ_{i+2},.... σ_{i+p-1}, it necessarily belongs to some block $B_n = \{\sigma_{-2n+1}, \sigma_{-2n+2}, \sigma_{-2n+3},.... \sigma_{2n}\}$ with $2n+1$ pseudospins . As a consequence of the substitution rule defining the chain and of the remark on the sign alternance given above, this block has to be identical to the blocks $\{\sigma_{-2n+1+p.2n}, \sigma_{-2n+2+p.2n}, \sigma_{-2n+3+p.2n}, \cdots \sigma_{2n+p.2n}\}$ for a sequence of values p_j of p which must fulfill $|p_j - p_{j-1}| \leq 4$. Consequently, the arbitrary initial block is repeated at distance shorter than $L(B) = 4 . 2^n$. The pseudospin chain thus fulfills the condition for weak periodicity.

According to the above theorem, there exists a translationally invariant Hamiltonian which has this pseudospin sequence $\{\sigma_i\}$ as a ground-state (and all its translated configurations $\{\sigma_{i+n}\}$ with their accumulation points). As we mentionned, the Hamiltonian which is constructed is rather complex and unusable. In this special case, we can find much a simpler pseudospin Hamiltonian (see the note ref [16]) which is

$$(6\text{-}a) \qquad H = \sum_{i,n>0} J(n)\ S_i\ S_{i+2n}$$

where the coefficients have to fulfill the condition

$$(6\text{-}b) \qquad J(n+1) > \frac{J(n)}{2} > .... > \frac{J(1)}{2^n} > 0$$

This model which is rather artificial, is an extension of the well studied ANNNI model where the interactions between the pseudospins only exist intermitently for the distances p which are power of 2. In addition, the upper envelop of $J(n)$ has to decay slower than the inverse distance 2^{-n}.

The Fourier spectrum of this model defined as

$$(7\text{-}a) \qquad S(q) = \lim_{N \to \infty} \frac{1}{N} |\sum_{n=1}^{N} \exp(i\,q\,n)\,\sigma_n|^2$$

has been studied in ref.7. It has been shown that $S(q)$ is a singular continuous measure. Let us shortly explain why

We define

$$(8\text{-}a) \qquad A_n(q) = |\sum_{m=1}^{2^n} \exp(i\,q\,n)\,\sigma_n|^2$$

for the pseudospin block with size $N=2^n$ obtained after n substitutions (5) on the initial elementary block +1. Then, a recursion formula can be obtained

$$(8\text{-}b) \qquad A_{n+1}(q) = 4 \sin^2(2^{n-1}\,q) . A_n(q)$$

For most values of q (with probability 1), the set of numbers 2^{n-1} q modulo 2π is uniformly distributed on the interval $[0,2\pi[$. Calculating the average $\ln(4\sin^2 x)$ which is found zero, it comes out

$$(9\text{-}a) \qquad \lim_{n\to\infty} \frac{1}{n} \ln A_n(q) = 0$$

However, there exists many values for q which do not yields uniform distributions for the numbers 2^{n-1} q modulo 2π. Particularly, for the values of q which have the property that their binary expansion

$$(9\text{-}a) \qquad \frac{q}{2\pi} = \sum_{n=1}^{\infty} a_n \frac{1}{2^n} \quad \text{with} \quad a_n = 0 \text{ or } 1$$

generates a sequence $\{a_n\}$ periodic for large enough n, the set of numbers 2^{n-1} q modulo 2π converges to a limit cycle (and thus is not uniformly distributed). Then, it is easy to prove that the limit

$$(9\text{-}b) \qquad \lim_{n\to\infty} \frac{1}{n} \ln A_n(q) = \kappa(q)$$

is defined and generally not zero. Since $\ln(4\sin^2 x) < \ln 4$, we have the strict bound

$$(9\text{-}c) \qquad \kappa(q) < 2\ln 2$$

The intensity of the Fourier spectrum $S_{2^n}(q) = \frac{1}{2^n} A_n(q)$ behaves as n goes to infinity as a power of the system size N^{γ} with $N = 2^n$ and $\gamma(q) = \frac{\kappa(q)}{\ln 2} - 1$. Consequently, when $\gamma(q) > 0$, the limit intensity $S(q) = \lim_{N\to\infty} S_N(q)$ diverges and yields a singularity. However, this singularity cannot correspond to a Dirac peak, because the intensity should diverge proportionally to the system size $N = 2^n$ and which would imply $\gamma(q) = 1$. This situation cannot happens because of the strict bound $\kappa(q) < 2\ln 2$. The set of q values for which the distributions of the numbers 2^{n-1} q modulo 2π is not uniform contains many other numbers (for which $\kappa(q)$ may not be zero) and is uncountable but they have globally a zero Lebesgue measure.

It is easy to show S(q) cannot have any Dirac peaks for any other value of q because the limit product $\prod_{n=1}^{N} \sin^2(2^{n-1} q)$ for N going to infinity, is necessarily zero. Otherwise, for most values of q, $\gamma(q) = -1 < 0$. When $\gamma(q) < 0$, the limit intensity S(q) becomes zero. Consequently the non zero positive measure, S(q) dq cannot have any Dirac peaks and is zero for most q. However, it has divergent singularities at infinitely many q values. The

Lebesgue measure of the support of this measure is zero. These conditions on S(q) dq are those for being a singular continuous spectrum. (see ref.7 for more details)

2-GENERALIZED FIBONACCI TILINGS IN ONE DIMENSION
KESTEN CONDITION

Earlier studies on a different model with a singular continuous spectrum were done in ref.4 and 5. This model is not constructed as an array of pseudospins on a periodic lattice. but to construct it, we generate first a quasi-periodic sequence of pseudospins

$$(10\text{-}a) \qquad \sigma_i = \chi_\Delta(i\, \zeta + \alpha) = 0 \text{ or } 1$$

where ζ is an irrational number, α is an arbitrary phase and $\chi_\Delta(x)$ is a periodic function with period 1 defined in the first unit period by

$$(10\text{-}b) \qquad \chi(x) = 1 \quad \text{for} \quad 0 \leq x < \Delta$$

and

$$(10\text{-}c) \qquad \chi(x) = 0 \quad \text{for} \quad \Delta \leq x < 1$$

Δ is a given parameter. Then, we define a one dimensional chain of identical atoms where the bound lengths between consecutive atoms can take only two possible given values $l_1 \neq l_2$. The position of the atoms i are recursively determined by

$$(11) \qquad u_{i+1} - u_i = l_2 + (l_1 - l_2)\, \sigma_i$$

In other words, this is a one dimensional tiling with two different tiles.

According to the general theorem on weakly periodic structures of pseudospins, there exists a translationally pseudospin Hamiltonian which have configurations (10-a) with arbitrary phases α as ground states. Then, it is easy to construct also a translationally invariant Hamiltonian for the atomic configurations, which has the tilings defined by (11) with arbitrary phase α and arbitrary u_0 as ground states.

Although, the pseudospin sequence σ_i is clearly quasiperiodic, the atomic distribution is generally not quasiperiodic unless Δ and ζ fulfills the KESTEN condition

$$(12) \qquad \Delta = p\, \zeta \text{ modulo 1 where p is some integer}$$

In a closely related model, we have shown in ref. 11 that the Kesten Condition is a necessary and sufficient condition for having an "average lattice". The same arguments applies here.

Let us define the concept of average lattice. The average distance l between consecutive atoms is easily found to be:

(13-a) $\qquad < u_{i+1} - u_i >_i = \Delta \, l_1 + (1 - \Delta) \, l_2 = l$

Then, the fluctuation v_i of the atom i around its average position $i \, l + u_0$ is defined by

(13-b) $\qquad u_i = i \, l + u_0 + v_i$

Using the model definition (10), this fluctuation is given by

(14-a) $\qquad v_n = (l_2 - l_1) \left(\sum_{i=1}^{n} \chi(n \, \zeta + \alpha) - n \, \Delta \right)$

The Kesten theorem[13] asserts that v_n is bounded if and only if (iff) Δ is an integer multiple modulo 1 of ζ, i.e there exists an integer p (positive or negative) such that

(14-b) $\qquad \Delta = p \, \zeta \quad \text{modulo } 1$

Morover, we can prove :

<u>Theorem:</u> When the KESTEN condition (14-b) is fulfilled, the sequence $\{u_i\}$ is quasi-periodic and can be written as

(15) $\qquad u_i = u_0 + [\, f(i \, l + \alpha') - f(\alpha') \,]$

where $\qquad \alpha' = \dfrac{\alpha \, l}{\zeta} \qquad$ redefines a new phase and function $f(x)$ can be written as

$f(x) = x + g(x)$ where $g(x)$ is well defined and periodic with period $\dfrac{1}{\zeta}$.

The Fourier spectrum of this chain defined as

(16-a) $\qquad S(q) = \lim_{N \to \infty} \dfrac{1}{N} \, | \sum_{n=1}^{N} \exp(i \, q \, u_n) |^2$

is a discrete sum of Dirac peaks at the wave vectors $q = Q_{m,n} = m \, Q_1 + n \, Q_2$ generated by the two basis vectors

(16-b) $\qquad Q_1 = \dfrac{2\pi}{l} \qquad\qquad$ and $\qquad\qquad$ (16-c) $\qquad Q_2 = \dfrac{2\pi \, \zeta}{l}$

Conditions generalizing the Kesten condition, for d-dimensional tilings obtained by the cut and projection method, have been recently obtained by DUNEAU and OGUEY[12].

When the Kesten condition is not fulfilled, this hull function g(x) does not exist as a bounded variation function. Then, the sequence u_n is generally not quasi-periodic although it is weakly periodic. (The condition for weak periodicity is easily checked). We found on the basis of a combined analytical and numerical analysis that

When $\dfrac{l_1}{l_2}$ is irrational, S(q) has no Dirac peaks (but for q=0).

When $\dfrac{l_1}{l_2} = \dfrac{r}{s}$ is rational (r and s are irreducible integers), S(q) is the sum of DIRAC

peaks at the integer multiples of $\dfrac{2\pi}{l_1}$ and of a singular continous measure.

CONSTRUCTION BY INFLATION RULES OF GENERALIZED FIBONACCI LATTICES

For finding these results, we used the fact that when the Kesten condition is not fulfilled the sequence $\{u_i\}$ can be constructed recursively by inflation rules using <u>3 tiles</u> . At the initial step, two of the three tiles are equal so that there is only two tiles with length l_1 and l_2 . (This construction involves only two tiles when the Kesten condition is fulfilled). At the step n of the inflation procedure, there is three possible inflation rules J_n which are I_0, I_1 or I_2. The sequence J_n is unambiguously determined according to the knowledge of two sequences of integers $\{a_n\}$ and $\{p_n\}$ (n≥0) which depends on ζ and Δ.

1- The first sequence $\{a_n\}$ is defined by the standard continued fraction expansion of the irrational number ζ:

$$(17\text{-a}) \qquad \zeta = a_0 + \cfrac{1}{a_1 + \cfrac{1}{a_2 + \cfrac{1}{a_3 + \cdots \cfrac{1}{a_i + \cdots}}}}$$

The truncations at order i of this expansion yield the sequence of "best approximations to zero" of ζ

$$(17\text{-b}) \qquad \frac{r_i}{s_i} = a_0 + \cfrac{1}{a_1 + \cfrac{1}{a_2 + \cfrac{1}{a_3 + \cdots \cfrac{1}{a_i}}}}$$

which is used for defining the second sequence $\{p_n\}$.

2. This sequence $\{p_n\}$ is defined by a new kind of expansion which extends the well known concept of binary expansion (basis $\delta_n = 2^{-n}$). Using the base of numbers $\delta_n = s_n \zeta - r_n$ (with alternate signs), Δ has a unique expansion as:

$$(18\text{-}a) \qquad \Delta = \sum_{n=0}^{\infty} p_n \, \delta_n$$

with the conditions

$$(18\text{-}b) \qquad 0 \leq p_n \leq a_{n+1}$$

and

$$(18\text{-}c) \qquad p_{n-1} = a_n \quad \text{if} \quad p_n = 0$$

For arbitrary ζ and Δ, the sequence of inflation rules J_n is generally not periodic. However, it is especially convenient to choose ζ and Δ in order that it be periodic with period r $(J_n = J_{n+r})$ at least for n larger than some n_0. This situation is obtained if and only if both sequences $\{a_n\}$ and $\{p_n\}$ $(n \geq 0)$ become periodic for n larger than n_0 $(a_{n+r} = a_n, p_{n+r} = p_n)$. If periods r=1 and r=2 are chosen, we obtain situations for ζ and Δ where the Kesten condition are necessarily fulfilled.

To fix the ideas, let us consider one of the simplest non-trival case with period 3

$$(19\text{-}a) \qquad \zeta = \tau^{-1} = \frac{\sqrt{5}-1}{2} \quad \text{and}$$

$$(19\text{-}b) \qquad \Delta = \frac{1}{2}$$

Then the continued fraction expansion of ζ has period 1 : $a_0 = 0$ and $a_n = 1$ for $n \geq 1$ which yields $\delta_n = \tau^{-1} \, (-\tau)^{-n}$. Then, the expansion of Δ is :

$$(20) \qquad \Delta = \frac{1}{2} = \tau^{-1} + [\, -\tau^{-2} + \tau^{-3} + \tau^{-5} - \tau^{-6} - \tau^{-8} + \tau^{-9} + \tau^{-11} - \ldots \,]$$

$$= \delta_0 + \delta_1 + \delta_2 + \delta_4 + \delta_5 + \delta_7 + \delta_8 + \delta_{10} + \delta_{11} + \ldots$$

which has period 3 for $n \geq 1$. It is clear that in that case the Kesten condition is not fulfilled. Since, the sequence of inflation rules has period 3, it is more convenient to consider a unique inflation rule which is the product $J = J_3 J_2 J_1$ of three consecutive inflation rules. Then, one obtains for the considered example

$$(21\text{-}a) \qquad A \rightarrow J(A) = C \, B \, A \, C \, A$$
$$(21\text{-}b) \qquad B \rightarrow J(B) = C \, A \, A \, C \, A$$
$$(21\text{-}c) \qquad C \rightarrow J(C) = A \, C \, A$$

With the initial tiles $A = l_1$, $B = l_2$, $C = l_2$, the blocks $A_n = J^n(A)$, $B_n = J^n(B)$ or $C_n = J^n(C)$ converges for n infinite to the tiling defined by (11) with ζ and Δ given by (19) where the phase α and the position of the initial atom u_0 have some value.

A theorem of BOMBIERI and TAYLOR[14] asserts the existence of a discrete component in the Fourier spectrum of a one-dimensional tiling determined by a given inflation rule when some conditions are fulfilled. These conditions concern the characteristic matrix $\overline{\overline{M}}$ of this inflation rule, the columns of which count the number of A,B, C in J(A), J(B), J(C) respectively:

$$(22) \qquad \overline{\overline{M}} = \begin{pmatrix} 2 & 3 & 2 \\ 1 & 0 & 0 \\ 2 & 2 & 1 \end{pmatrix}$$

Their theorem holds when the characteristic matrix has only one eigen value with a modulus larger than 1. In the present example, the eigenvalues of $\overline{\overline{M}}$ are τ^3, -1, τ^{-3}. Thus, we are in a marginal case where the modulus of the second eigenvalue is just equal to 1. (Note that the absence of average lattice noted above requires that the second eigenvalue of the characteristic matrix, has a modulus larger or equal to one.)

The method for studying the Fourier spectrum of this tiling is similar as for the Thue-Morse sequence. The inflation rules (21) can be used for calculating recursively the Fourier transforms of the blocks A_n, B_n, C_n. Instead of a single block recursion relation, a matrix $\overline{\overline{P}}_n(q)$ relates the Fourier transforms of the three blocks of tiles at order n+1 as linear combinations of the Fourier transforms of the blocks of tiles at order n. The results do not depends qualitatively on the locations of the atoms in the elementary tiles. For most q, this matrix product has an average behavior which yields $S(q) = 0$. For special values of q:

$$(23\text{-}a) \qquad q = \frac{2\pi}{1} \, \frac{j + k\,\tau}{4}$$

with

$$(23\text{-}b) \qquad 1 = \frac{l_1 + l_2}{2}$$

$\overline{\overline{P}}_n(q)$ converges to a periodic cycle. Then, the modulus of the Fourier transform modulus of the blocks A_n, B_n and C_n diverge as a power N^α of the block size N. Singularities of $S(q)$ are found ("Quasipeaks") when the exponent α is larger than 1. Unless $\frac{l_1}{l_2}$ is rational, this exponent is strictly smaller than 2 which implies the absence of Dirac peak. Other "Quasi-peaks" may be found for $q = \frac{2\pi}{p\,1} \, (r + s\,\tau)$ where p, r, s are arbitrary integers. Thus, the measure $S(q)\,dq$ has to be singular continuous.

S(q) dq can be approximated by a sequence of measures $S_n(q)$ dq which are obtained by choosing for Δ the truncations of the expansion (20) of $\Delta = \frac{1}{2}$ (which yields the sequence Δ_n of best approximations to Δ by the integer multiples of ζ modulo 1). Since $S_n(q)$ dq is exactly calculable as a series of Dirac peaks, one can check scaling properties in the limit of infinite n. For a given quasipeak at q_∞, it is found that S(q) is invariant by applying a scaling factor $(-\tau)^{-3}$ to $q - q_\infty$ and λ for the peak intensity. Each peak has similar scaling properties but $\lambda(q)$ depends on the peak. This exponent can be calculated explicitely for many values of q by using the inflation rules. When the numerical resolution of the observation is increased by a factor τ^3, one can see that each quasipeak splits into several quasipeaks and so on so that there is no Dirac peak in the limit of an infinite resolution[4]. The multifractal analysis of S(q) dq has been recently performed[15]. (see refs. for details)

ON THE POSSIBILITY OF HAVING FRACTAL GROUND-STATES (NOT WEAKLY PERIODIC)

There exists models where the condition for having weakly periodic ground-states are not fulfilled and which might be of physical interest. For example, the existence of (non summable) long range interactions between the "elements" of the model may be physically essential. We have in mind the long range interactions between domains in "Martensitic Transformations" generated by elastic deformations. These domains correspond to a phase A which grows inside a phase B which has a different lattice parameter. There exists a competition between two forces: a short range interaction between domains which tends to merge these domains together and thus to grow the domains A and a long range interaction which is repulsive. In some temperature range, the domains A cannot grows beyond a finite size. We expect that the situation may be much more complex. Elucidating these situation could perhaps help for understanding the strange structural properties of martensitic transformations.

A first approach consist in modelizing these problems by pseudospin models. The unit cells i of the lattice which are in phase A have the pseudospin σ_i , those which are in phase B have pseudospin 0. Typically, the Hamiltonian $H\{\sigma_i\}$) should be the sum of two types of terms

1- short range interaction terms, for example:

$$-h \sum_i \sigma_i - \sum_{<i,j>} J \sigma_i \sigma_j$$ which are attractive and favors a single domain A with $\sigma_i = 1$

for all i , ($<i,j>$ denotes nearest neighbor sites) and

2- long range interaction terms of elastic origin

$$\sum_{i,j} J_{i-j} \sigma_i \sigma_j$$

were each coefficient J_{i-j} is positive and small compared to h and J and goes to zero as a function of the distance $|i-j|$ as $\frac{1}{|i-j|^3}$.

In the above model, we define the groundstate $\{\sigma_i\}$ as a configuration such that by flipping an arbitrary finite number of pseudospins 1-the energy variation is defined 2- this energy variation is positive. Since the sum $\sum_j J_{i-j}$, diverges, it is clear that the solution $\sigma_i = 1$ for all i cannot provide the ground-state of this model and that we must have the pseudospin average $<\sigma_i> = 0$. A single finite cluster D where $\sigma_i = 1$ for i in D, and $\sigma_i = 0$ for i not in D cannot be the ground-state because far enough from this cluster, the long range repulsive force becomes neglegible, so that a new domain of phase A can grow.

The ground-state configuration must fulfills conditions which seems at first view in contradiction

1- The pseudospin average is zero

2- There is an infinite number of pseudospins which are not zero.

The solution to this problem could be a ground-state with a "fractal structure". The application of this possible result to the physical systems has to be careful because the structure due to long range elastic interactions are obviously very sensitive to the elastic defects such that dislocations and grain boundary. Nevertheless, it is intersting to remark that structures with a fractal character, were already observed in a martensitic microstructure[8]

CONCLUSION

For understanding general aspects of structures in solid condensed matter, we have developped an approach based 1- on energetic considerations 2- on the invariance of the energy under the action of the translation group and proven that ground-states are at least "weakly periodic".

Unlike the predictions of MIEKISZ and RADIN, many ground-states in physical systems are periodic (what means the concept of genericity for physical systems?). As a function of the physical parameters, there may also exist infinitely many phases with "Devil'staircase" transformations. There also exist phases where the standard translational invariance is broken (incommensurate structures, quasicrystals). We know now on some theoretical models that this translational invariance can be broken in a more subtle way. Indeed, we have proven that there exists translationnally invariant Hamiltonians with a ground-state

a) which breaks the translation symmetry

b) which is neither periodic nor quasi-periodic

c) which is not random and has zero entropy

d) which has a Fourier spectrum $S(q)$ which is a purely a singular continuous measure and quasipeaks with scaling properties

In this studies, the use of inflation rules has been very helpful. It can be a simple way for generating weakly periodic structures (However, inflations rules do not always generate structures which are weakly periodic although they are often weakly periodic. The required conditions on the inflations rules for having this property are not yet elucidated .)

There is several questions concerning the physical interpretation of these results.

The first question is: Could such kind of structure exist in nature? In fact, there is many simple physical models in 1 2 or 3 dimensions for which the ground state structures are unknown. The generalized Fibonacci tilings presented here, may have applications for quasiperiodic antiphases in alloys or for quasi-crystals. Thus, the existence of such structures in nature, does not look unreasonable. We conjecture that singular continuous spectrum may be found again with very simple physical Hamiltonians.

The second question is: how a weakly periodic structure with singular continuous spectrum could be observed? In the examples which we described here, the Fourier spectrum of the structure has quasipeaks which may look experimentally as the Dirac peaks of an incommensurate structure. Only a careful analysis with high resolution, may be able to detect that these peaks have anomalous shape and width on good quality samples.

REFERENCES

[1] S. AUBRY "Devil'staircase and order without periodicity in classical condensed matter" J.Physique **44** 147-162 (1983)

[2] D. RUELLE "Five Turbulent Problems" Physica **7D** 40-44 (1983)

[3] S.AUBRY and P.Y. LE DAERON Physica **8D** 381-422 (1983)

[4] S.AUBRY, C. GODRECHE and J.M LUCK: " An Structure Intermediate between Quasiperiodic and Random" Europhysics Letters **4** , pp. 639-643 (1987)

[5] S.AUBRY, C. GODRECHE and J.M. LUCK " Scaling Properties of a Structure Intermediate between Quasiperiodic and Random" J.Stat.Phys. **51** 1033-1075 (1988)

[6] S.AUBRY "Weakly Periodic Structures and Example" in "Nonlinear Coherent Structures in Physics,Mechanics and Biological Systems" ed. J.POUGET Journal de Physique série Colloques

[7] Z.CHENG, R. SAVIT, R. MERLIN Phys.Rev **B37** 4375 (1988)

[8] E. HORNBOGEN, Z. Metallkde. **78** 352-354 (1987)

[9] J.MIEKISZ and C. RADIN, "Are Solid Really Crystalline?" preprint (1988)

[10] J.MIEKISZ Nuclear Physics B(Proc.Suppl.) **5A** 122-124 (1988)

[11] S. AUBRY, C. GODRECHE and F. VALLET J.Physique (Paris) **48** 327 (1987)

[12] M. DUNEAU and C. OGUEY preprint (1989)

[13] H. KESTEN, Acta Arith. **12** 193 (1966/67)

[14] E. BOMBIERI and J.E. TAYLOR J.Physique(Paris) Coll. **C3** (1986) Contemporary Mathematics **64** 241 (1987)

[15] C. GODRECHE and J.M. LUCK in preparation

[16] <u>Note added.</u> After the writing of this paper, C. RADIN pointed to me that the Hamiltonian (6) for the Morse sequence suffers a severe flaw because the set of its ground-states is highly degenerate and contains many other configurations which are periodic and thus are not Morse sequences (private communication). Otherwise, he pointed to me that he can write a Hamiltonian involving only four body interactions which has the Morse sequence as unique ground-state (see C.Gardner,J. Miekisz, C.Radin and A. van Enter to appear in J.Phys.A)

THE ANDERSON METAL - INSULATOR TRANSITION:

INCOMMENSURATE VERSUS DISORDERED SYSTEMS

Spiros N. Evangelou

Physics Department
Division of Theoretical Physics
University of Ioannina
Ioannina 451 10 GREECE

INTRODUCTION

Much work has been devoted to the study of quantum transport or wave propagation in disordered systems in connection with the phenomenon of Anderson localisation[1] and the assosiated metal - insulator transition[2]. The electron properties and lattice dynamics of modulated structures and quasicrystals have also been considered[3]. In this context Aubry and Andre[4] studied the discrete Harper's equation used to describe the quantum theory of an electron confined in a plane with a periodic potential in the plane and a uniform perpendicular magnetic field. These systems are intermediate between periodic and random and they often exhibit an Anderson localization transition with a rich complex scaling behavior[5,6]. From a duality property it was shown[4] that the Anderson transition occurs by breaking of analyticity at a critical value of the incommensurate potential strength. The critical properties of incommensurate systems are mostly understood[4]. The Anderson transition in a random potential is more difficult. Here, I shall demonstrate the extra difficulties, emphasize the common characteristics and point out the differences of the two transitions. I shall also present some recent results by using related techniques.

It is well - known[1] that quasi - periodicity or randomness on a perfect crystalline lattice affect the amplitude of the wave function which decays within a localization length ξ. A mobility edge usually appears in the spectrum separating extended states, which retain their spatial amplitude (ξ=infinite), from non-conducting localized states which decay (ξ=finite), usually exponentially. The Anderson transition is also reflected in the nature of the spectrum which changes from absolutely continuous to a dense pure point - like spectrum for localized states[3]. From a one - parameter (for the averaged conductance) scaling theory[2] of the Anderson transition complete localization was predicted, in the absence of spin effects, for $d \leq 2$ with the occurrence of a transition above two dimensions. The field theoretic formulation[6] and numerical simulation from finite - size scaling techniques[7] produced similar results. At this point one may notice an important difference between incommensurate and disordered systems. In disordered systems the metallic state can be realized for systems with intermediate linear system size L, much greater than the mean free path l and much smaller than the localization length ξ. This regime, which

Geometry and Thermodynamics
Edited by J.-C. Tolédano
Plenum Press, New York, 1990

is often called mesoscopic[8], is absent in incommensurate systems where the metallic state is exactly like that of a pure metal. The mesoscopic systems are further characterized by universal properties of the statistics of fluctuations for the physical quantities (e.g. universal conductance fluctuations reproducible for a given small metallic sample).

WAVE - FUNCTION CORRELATIONS AND CRITICAL EXPONENTS AT THE MOBILITY EDGE

Electronic wave functions in incommensurate or disordered quantum Hamiltonians above and below the Anderson transition[1] are geometrically simple. The localized wave functions display a single dominant maximum followed by an exponential decay of their amplitude characterized by ξ. The extended states in incommensurate systems are perfect Bloch waves[9]. On the other hand, although the amplitude of the extended states in a random potential remains constant on average they exhibit many peaks which indicate the presence of mesoscopic fluctuations. Precisely at the mobility edge the localization length and the characteristic length describing the range of fluctuations diverge. The wave function shares the properties of both the extended and localised states. One might expect in this case that the wave functions exhibit self - similar fluctuations on all length scales larger than the lattice spacing, and have a fractal character. For incommensurate systems both the critical spectra and wave functions are multifractals[5,10]. For random systems indications come from several sources for the absence of conventional diffusive forms of eigenfunction correlations at the mobility edge. A property defined auto - correlation function[11] which describes the spatial fluctuations and the two - particle spectral function $S(q,\omega)$[12,13] have revealed novel dependences. These can be characterized by power - law divergencies described by one number, also regarded as a fractal dimension. Therefore, although the nature of extended states is different for the two kinds of systems at the mobility edge we expect to find wave functions that have structure on all lengths and which scale with a continuous set of non -related exponents D_q. We propose that in this respect the critical behavior is common for disordered and incommensurate systems.

In the rest we present results for the wave functions at the Anderson transition based on the multifractal approach[14]. The method is commonly used in the theory of dynamical systems at the onset of chaos. We utilise the fact that one has to deal with scaling of distributions rather than simple averages. The different moments for the distribution of the wave function amplitude are considered and shown to scale in different ways at the modility edge.

Incommensurate systems

We first consider one - dimensional incommensurate and quasicrystalline systems. The simplest model for electronic propagation in an incommensurate or a quasicrystal lattice is the one - dimensional tight-binding difference equation[4]

$$\Psi_{n+1} + \Psi_{n-1} + V_n \Psi_n = E \Psi_n \quad . \tag{1}$$

The potential in the case of the discrete Harper's equation is $V_n = \lambda.\cos(2\pi\sigma.n+\varphi)$ and V_n takes two values arranged in a Fibbonacci sequence in order to describe the quasicrystal. The basic incommensurability parameter σ is an irrational number (i.e. a ratio of two relatively prime integers). The potential strength is given by λ ,φ is a phase factor and when $\lambda=2$ the parameter σ describes the number of flux quanta through a unit cell of a two – dimensional crystal in a magnetic field. But $\lambda=2$ is special for another reason, as when $\lambda<2$, for all states and for almost every σ solutions of (1) are believed to be extended. Similarly, for $\lambda>2$ they are believed to be localised. Precisely at the transition point the wave function is a curious intermediate between extended and localized states[5]. For small λ, and almost every σ and φ, all states lie in smooth bads of finite measure. As λ approaches 2, the total measure of the bands goes to zero and the spectrum is said to be singular continuous.

At the critical point $\lambda=2$ the nature of the spectrum and the wave functions can be numerically characterized by evaluating the scaling dimensions. In order to understand' the scaling the key point is to consider systematic approximations of the irrational number σ by a series of rational numbers obtained from truncating its continued fraction expansion. The Hausdorff spectral dimension D_O for the spectrum, concerns the support of the set and is $D_O=0.5$ at $\lambda=2$[5,10]. D_O can be also obtained from the scaling law of the sum of bandwidths versus the number of bands for different approximants.

The wave function is a more fundamental measure than the spectrum. The different moments of the wave function also scale in different ways near the mobility edge[5]. Numerical evidence suggests that when $\lambda=2$ the solutions of (1) have a multifractal character[5]. D_q was computed and we obtain $D_0=1$,$D_1=0.7627 \pm 0.0002$, $D_2=0.6123\pm0.0003$, $D_{+\infty}=0.3528\pm0.0001$. D_q may be interpreted as a generalized dimension of the set on which the q'th moment of the wavefunction resides. Related work has also been carried out for the quasi- crystal[15] which corresponds to the case of $\lambda=2$, i.e. is always critical.

<u>Disordered systems</u>

We have recently[16] examined wave functions in the presence of a three-dimensional random potential. The site energies ε_i are independent random variables chosen from a flat probability distribution on the interval $(-W/2, W/2)$. The parameter W describes the strength disorder and the metal-insulator transition is believed to occur at $W_c=16.5$[7]. For $W>W_c$ all states are localized and the conductivity is zero, while for $W<W_c$ mobility edges appear in the band separating localised states near the edges from extended states near the band center. At the critical region $(W=W_c)$ the mobility edges lie at $E_c=0$. Numerical computations in finite samples and a finite–size scaling analysis can be performed. In [16] the critical wave function amplitudes $|\psi_n(r)|$ are evaluated for disordered cubic lattices. Subsequently the critical indices D_q were computed. The procedure is very accurate for small and positive values of q but it becomes not so accurate for larger q and even worse for negative q. In the latter case a larger statistical ensemble is required. The results D_q are compared[16] with the expressions given by Wegner[6]. They were derived from non –linear σ models in $d = 2 + \varepsilon$ dimensions by studying the averaged moments of the localized wave function component $\psi_n(r)$ at site r

$$P_q(E) = \langle \sum_n |\psi_n(r)|^{2q} \; \delta(E-E_n) \rangle \, / \langle \rho(E) \rangle, \qquad (2)$$

where $\langle \varrho(E) \rangle$ is the averaged density of states. The anomalous scaling of P_q was related to the multifractal structure of the wave function[17]. Near the mobility edge E_c, $P_q(E)$ vanishes as $(E_c-E)^{\pi_q}$. One can easily indentify, from π_q, the exponents D_q in powers of ε [17]

$$
\begin{aligned}
D_q &= \nu^{-1}.\ \pi_q\ /(q-1) \\
&= d-\ \nu^{-1}.q\ +\ \nu^{-1}.\ q\ (q^2-q\ +1)\ \zeta(3)\ \varepsilon^3/4\ +\ 0(\varepsilon^4),
\end{aligned} \tag{3}
$$

with $\zeta(3) \approx 1.202$. The results for the fractal dimension is $D_0=d=3$, for the information dimension $D_1=D_I=d-\varepsilon\ +\zeta(3)\varepsilon^4/4\ +0(\varepsilon^5)$ and the correlation dimension $D_2=d-2\varepsilon+6.\zeta(3)\varepsilon^4/4\ +0(\varepsilon^5)$. At $\varepsilon=1$ the ε – expansion up to fourth order can be trusted only for small q since for $q \gtrsim 1.25$ it gives increasing D_q. This result was improved in Ref.17 by means of a Borel resummation of the series. The results of ref.16 are shown in fig.1 with the Borel resummed series.

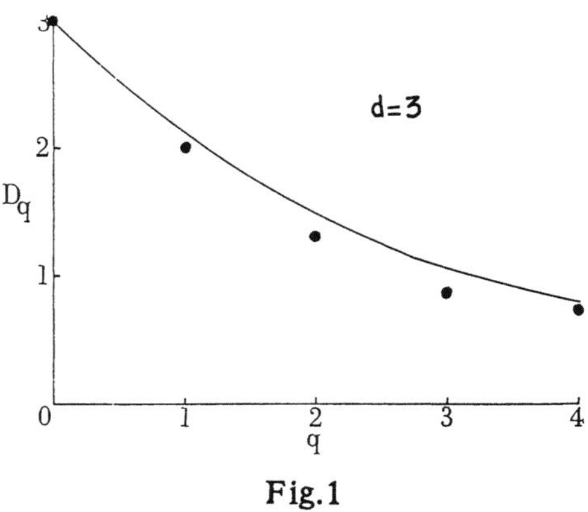

Fig.1

DISCUSSION ON THE VALIDITY OF THE ONE-PARAMETER SCALING THEORY

Localised wave – functions are self – similar only in a trivial may. In the inculating regime the distributions of many local quantities (wave function amplitude, current, etc) become asymptotically log – normal[18]. Consequently two- parameter scaling holds[17] for the first two cumulants which describe the distribution of $\log|\psi_n(r)|$ at site r. This naturally leads again to a multifractal scaling described by a set of related in this case generalized Lyapunov exponents $L(q)=\gamma.q+(\mu/2).q^2$, $q\in[-\infty,+\infty]$[17]. The mean inverse localization length $\gamma=\xi^{-1}$ is defined from the averaged logarithmic response of the wave function and μ is related to its statistical variance. For weak disorder, even in one dimension, deviations from the log – normal law[17] imply a set of unrelated $L(q)$. This can be seen as a precursor of the full critical fluctuation effects which are absent in one – dimension. It was further argued[18] that the distributions in the metallic regime have log – normal tails instead. Then the variance of the conductance appears as a universal number. The role of these tails becomes more pronounced as we enter the metallic regine (e.g. for $l \ll L \ll \xi$. However, we have shown that the description for the critical distribution of the wave function amplitudes at the mobility edge requires an infinite set of un – related exponents D_q. This means that in contrast to the usual critical phenomena the q+1-th moment is described by a scaling exponent which cannot be related to that of the moment q by a simple gap exponent. The one – parameter scaling idea, in this sence, fails as it does in other examples such as aggregate structures and strange objects.

In summary, we have given evidence that critical wave functions both for incommensurate and disordered systems are multifractals. We have numerically demonstrated that the critical wave fuctions in three dimensions cover a space of lower dimensionality (e.g. bi-dimensional in the information sence since $D_1 \approx 2$). Moreover we   show that the broad log-normal distributions valid in the insulating regime cross over to distributions with non – universal tails at the mobility edge. The existence of these tails is responsible for the rich scaling structure. This effect becomes even more pronounced in the region of mesoscopic fluctuations. In the case of incommensurate systems the situation is similar at the transition but simpler in general. The differences lie in the fact that in disordered systems apart from the critical point interesting structure also occurs in the disordered metallic state (mesoscopic) which is not present in incommensurate systems.

REFERENCES

1.   P.W. Anderson, Phys. Rev. 109, 1492 (1958). See also the reviews in : "Chance and Matter" , Les Houches XLVI, 1986,Eds   Souletie J. , Vannimenous J., and Stora R. (North-Holland 1987)
2.   Abrahams E., Aderson P.W., Licciardelo D.C. and Ramakrishnan T.V. Phys. Rev. Lett. 42 673 (1979).
3.   See the reviews by Souillard B. in   ref.1 p.305 and references there in.
4.   Aubry S. and Andre G.,Ann. Israel Phys. Soc. 3, 133(1980)
5.   Evangelou S.N. in "Disordered Systems and New Materials",and references threin. Evangelou S.N., J.Phys. C20, L295 (1987)
6.   Wegner F.J., Z.Phys. B 25 327(1976), B 35 207 (1979) Physics Reports 67, 15 (1980);   ibid Nucl. Phys. B270, 1 (1980),
7.   Mac Kinnon A. and Kramer B. Z.Phys. B 53 1 (1983), Pichard J–L. and Sarma G. Z.Phys. C 14 L 127, L617 (1981)
8.   Atshuler B.L., Proc. 18th Int. Conf. on Low Temperature Phys. Kyoto (1987) Jap. Journal of Appl.Phys. Supp. 26,1938 (1987). Lee P.A. and Stone A.D., Phys. Rev. Lett. 55 1622 (1985)
9.   Aubry S. and Quemerais P. in :"Low Dimensional Electronic Properties of Molybdaenium Bronzes and Oxides" Eds. C.Schlenker Riedel Publ. Co (1988) .
10. Tang C. and Kohmoto M., Phys. Rev. B 34 , 2041 (1986), see also Siebesma A.R. and Pietronero L., Europhys. Lett. 4, 597 (1987)
11. Soucoulis C. and Economou E.N., Phys. Rev. Lett 52, 565 (1984)
12. Chalker J.T. and Daniell G.J., Phys. Rev. Lett 61, 593 (1988),
13. Chalker J.T. Phys. C20, L493 (1987)
14. T.C. Halsey, M.H. Jensen, L.P. Kadanoff, I. Procaccia and B. Shairman, Phys. Rev. A33, 1141 (1986)
15. Evangelou S.N., J.Phys. C20, L295, (1987)
16. Evangelou S.N. (submitted)
17. G.Palabin G. and Vulpiani A., Phys. Rev. B35, 2015 (1987); ibid Physics Reports 156, 147 (1989)
18. Altshuler B.L., Kravtsov V.E. and Lerner I.V.,Phys. Lett A134, 488 (1989)

THEORY OF PHASE TRANSITION BETWEEN TWO INCOMMENSURATE PHASES IN $NbTe_4$

M.B. Walker and Rose Morelli

Department of Physics
University of Toronto
Toronto, Ontario
Canada M5S 1A7

This talk reviews briefly some theoretical predictions[1-3] concerning the phase transition between two incommensurate phases which occurs at approximately 150K in $NbTe_4$. The nature of the high-temperature incommensurate phase, which exists at room temperature, is now well understood both from the theoretical[4] and experimental[5-6] points of view. Only relatively recently, however, in the experimental work of Mahy et.al.,[7] has it become clear that a low-temperature incommensurate phase (which is distinct from the high-temperature incommensurate phase) exists in the temperature range between 50K and 150K. The theory reviewed here yields a precise structural model (characterized by its superspace group symmetry)for the low-temperature incommensurate phase, as well as an understanding of the physical mechanism driving the transition.

The basic tetragonal $NbTe_4$ structure can be viewed as being made up of columns parallel to the c-axis (see Fig. 1). A charge-density wave is formed on each column, and it is the relative phasing of the charge-density waves on the different columns which determines the overall three-dimensional structure. The interaction between nearest-neighbor columns can be shown to be stronger than the interaction between second-neighbor columns, and it is this interaction which stabilizes the structure of the high-temperature phase. On the other hand, it is the second-neighbor interaction which favors the production of the low-temperature incommensurate phase. The key question which must be answered is, ''If the second-neighbor interaction is assumed much weaker than the first-neighbor interaction, how is it that this second-neighbor interaction can ever come into play and cause a phase transition to a new structure?''

Geometry and Thermodynamics
Edited by J.-C. Tolédano
Plenum Press, New York, 1990

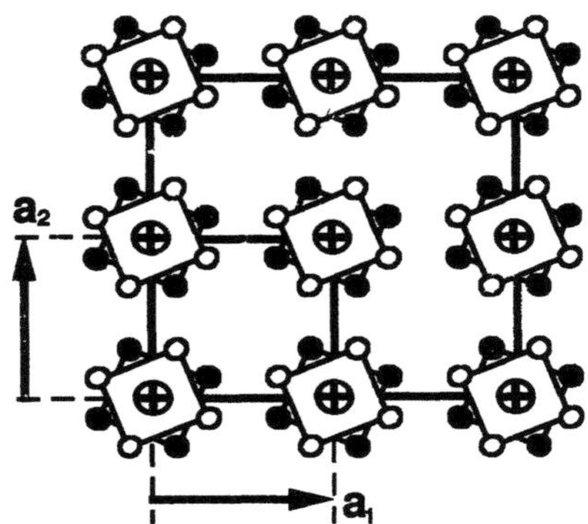

Fig. 1. The presumed structure of the normal phase of $NbTe_4$ projected onto the basal plane. The structure is made up of columns parallel to the c-axis with Nb ions, indicated by open circles containing crosses, at positions $z = \pm c/4$ along the axis of each column. Squares of Te ions at z=0 and z=c/2 are indicated by open circles and closed circles, respectively. The vectors a_1 and a_2 both have length a. The question sometimes raised in the literature of whether or not the structure has a centre of inversion symmetry can be shown not to play a role in the development of the theory.

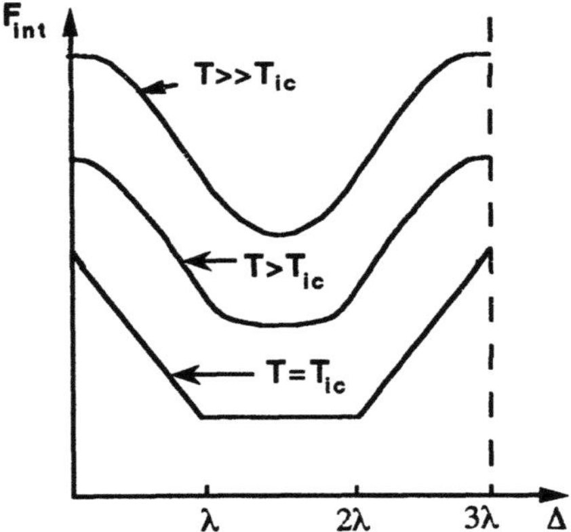

Fig. 2. The interaction energy of the incommensurate charge-density waves on two neighboring columns as a function of the relative c-axis displacement Δ of these columns is shown for three different temperatures. The quantity λ is the domain length and T_{ic} is the incommensurate to commensurate transition temperature.

Details of the theoretical study of the transition between the two
incommensurate phases have been published in Refs. 1-3, and a detailed
review[8] of the properties of the $NbTe_4$-$TaTe_4$ series of compounds is expected to be published shortly. The predicted superspace group symmetry
of the low-temperature incommensurate phase, and hence the necessary
information for the formation of a detailed structural model, is given
in Ref. 9. Thus, although some detail was given in the oral presentation at the Conference, it is perhaps appropriate here to mention only
the main point, and to refer those who wish further details to the references listed above.

Consider two charge-density waves on neighboring columns, and suppose that initially both charge-density waves have the same phase. Now
shift one of the charge-density waves in the direction of the column
axis by an amount Δ relative to the other. The interaction energy as a
function of Δ is shown, for three different temperatures, in Fig. 2.
Note that this interaction between nearest-neighbor columns has its
minimum when Δ has the value $3\lambda/2$. The full three-dimensional structure for the high-temperature incommensurate phase is therefore such
that the charge-density wave on each column is shifted by an amount $3\lambda/2$
in the c-direction relative to all its nearest neighbors.

Now note from Fig. 2 that, as the temperature is lowered, the nearest-
neighbor interaction energy as a function of Δ becomes very flat in the
neighborhood of the minimum. Thus, departures of Δ from its equilibrium value up to a certain point cost very little first-neighbor energy.
In a certain sense, the effective first-neighbor interaction strength
is proportional to the curvature of the curves shown in Fig. 2 at their
minima, and this curvature becomes very small as the temperature is lowered. This phenomenon is caused by the sharpening up of the incommensurate
phase domain walls which occurs as the temperature is lowered. When the
effective first-neighbor interaction becomes sufficiently weak, the
second-neighbor interaction will begin to have non-trivial consequences
and will cause a transition to a new incommensurate phase.

REFERENCES

1. M.B. Walker and Rose Morelli, Phys. Rev. B38, 4836 (1988).
2. Rose Morelli and M.B. Walker, Phys. Rev. Lett. 62, 1520 (1989).
3. Rose Morelli and M.B. walker, Phys. Rev. B (in press).
4. M.B. Walker, Can. J. Phys. 63, 46 (1985).
5. H. Bohm and H.G. von Schnering, Z. Kristallogr. 171, 41 (1985).
6. S. Van Smaalen, K.D. Bronsema and J. Mahy, Acta Cryst. B42, 43 (1986).
7. J. Mahy, J. Van Landuyt, S. Amelinckx, K.D. Bronsema, and S. Van
      Smaalen, J. Phys. C19, 5049 (1986).
8. M.B. Walker to appear in ''Nuclear Spectroscopy on Charge Density
      Wave Systems", Eds., T. Butz and A. Lerf, Kluwer Academic Press.
9. Z.Y. Chen and M.B. Walker, Phys. Rev. B (in press).

THE ORIGIN OF POLYTYPES IN SiC AND ZnS

Volker Heine and C. Cheng*

Cavendish Laboratory, Madingley Rd, Cambridge CB3 OHE, ENGLAND

ABSTRACT

The origin of polytypes is considered, with particular reference to detailed calculations on SiC and ZnS. Why do they occur in these materials ? Are they equilibrium phases, frozen in from some higher temperature, or are they metastable artifacts of the growth mechanism ? The main emphasis will be on the nature of the rather long ranged interaction needed to stabilise polytypes as equilibrium phases.

1. INTRODUCTION

Polytypes can arise in layered materials when the atomic layers can be stacked on top of one another in different ways to form a multiplicity of structures. In order to gain insight into this phenomenon, a lengthy computational and theoretical study of SiC and ZnS has been undertaken with several co-workers [1-13, particularly 4-6 and 12-13]. Of the order of 100 polytypes have been reported for these materials [14-15]. We shall concentrate on SiC, returning to ZnS at the end. There are many layered materials which in principle could support stacking variants but which do not produce regular polytypes, i.e. a superlattice with a regular repetition of some stacking pattern. Hence three basic questions arise. (A) Why do these particular materials form polytypes ? (B) Are the polytypes genuine thermodynamically stable equilibrium phases at some high temperature, from which they are frozen in : or on the contrary are they metastable artefacts of some growth process ? (C) If they are stable phases at some temperature, it requires some fairly long ranged interaction, with range at least equal to the repeat distance, to stabilise a particular polytype compared with others : what is the nature of this long ranged interaction ? In this connection I remember a maxim of a friend of mine, an experimentalist : "It is important to keep theoreticians honest !" The test of any proposed mechanism is whether it can explain which particular polytypes are observed, and which not.

Both SiC and ZnS can, and often do, crystallise in the cubic diamond structure, or rather the GaAs structure with alternate atoms being Si and C (or Zn and S). A (1, -1, 0) section is shown in Figure 1(a). It consists of hexagonal SiC atomic double layers with one of the tetrahedral bonds in the layer lying diagonally in the phane of the paper as shown. Such an atomic double layer is our basic unit, to be referred to as a 'layer'. The stacking direction is (1,1,1). Any layer can be rotated by 180° about the stacking direction and be placed on the layer 'below' (i.e. layer

---

* Now at Chemistry Dept, University of Pennsylvania, PA 19104 - USA

Geometry and Thermodynamics
Edited by J.-C. Tolédano
Plenum Press, New York, 1990

to the left in Figure 1) in either orientation, as shown in Figure 1(b). We shall denote the two orientations as seen in the two directions of the diagonal bonds as + and -. The cubic structure consists of all + or all - stackings. The tetrahedral coordination is retained independent of the stacking, as is evident from Figure 1(b).

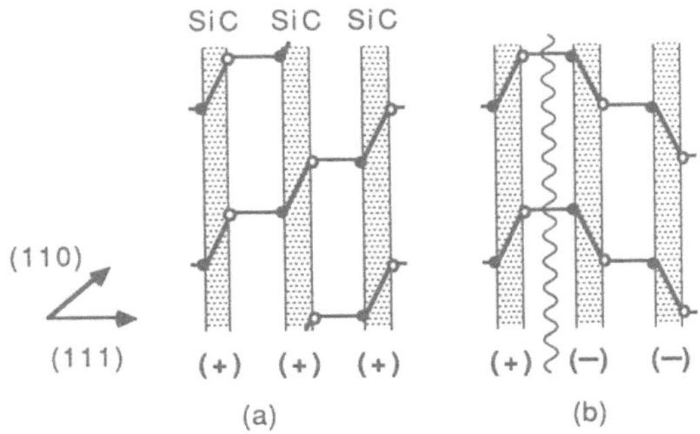

Fig.1 . The stacking of (+) layers and (-) layers in SiC. (a) The cubic structure, with (111) being the stacking direction and (110) the direction of the lines of bonds shown. (b) The configuration +-- with antiphase boundary shown by a wiggly line.

A typical polytype of SiC has the stacking sequence [++---] repeated. In the Zhadanov notation [15 and references there] it is denoted by <23>, and a set of two or three successive layers of the same sign is referred to as a 2-band or 3-band. Table 1 lists some typical polytypes.

Table 1. Some observed polytypes of SiC.

<u>Commonest polytypes</u>

< ∞ >	= cubic = ...++++... = ...----...
<3>	= [+++---] repeated
<2>	= [++--] repeated
<23>	= [++---] = [--+++] repeated

<u>Other typical polytypes</u>

<2333>	<2223>
$<23(33)_n>$	$<(22)_n23>$
$<(23)_n33>$	$<22(23)_n>$

<u>Also under special conditions</u>

wurtzite structure = <1> = ...+-+-+-...

312

The connection with incommensurate modulated structures is now clear from Table 1. A polytype can be seen as a + and - modulation with a half-wavelength equal to some value between 2 and 3 layer spacings. For a regular polytype it has to be a rational fraction between 2 and 3, so that the analogy is with modulated structures where the wavelength has locked on to a rational fraction of the lattice spacing. The wurtzite structure <1> does not conform to this pattern and only occurs under special growth conditions. As we shall see, the cubic phase < ∞ > is also only metastable.

2. THE ZERO ORDER MODEL

The extreme crystal structures are the cubic < ∞ > found in the group IV semiconductors, and the wurtzite <1> found in some of the strongly ionic II-VI and I-VII semiconductor compounds. The SiC is semi-ionic [8] because the carbon is much more electron-attractive than Si. Consideration of electronegativities shows that it and ZnS lie on the boundary between the two structure types. It is therefore not surprising that more complex intermediate stacking structures are found. The results of total energy calculations for simple polytypes are shown in Figure 2 [4,16].

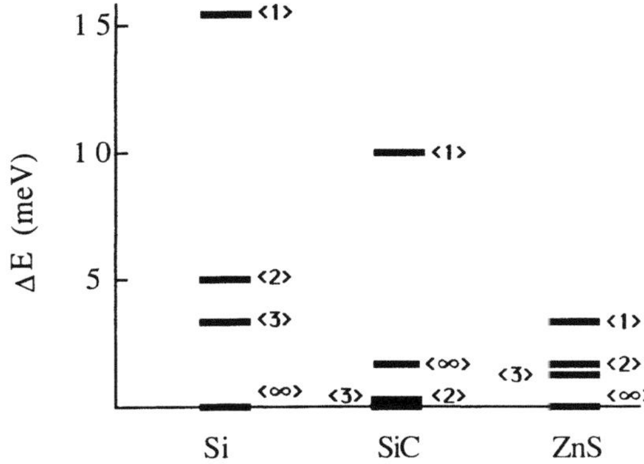

Fig.2 . Relative energies of different structures as given by quantum mechanical total energy calculations, for Si, SiC and ZnS. (Data from Refs 16, 4 and 13 respectively.).

These energies are the results of very careful quantum mechanical calculations with density functional theory, non-local norm-conserving pseudopotentials to represent the atomic cores, a basis of about 3000 plane waves for each electron wave function, and sampling of the wave functions at many points in the Brillouin zone. The relative accuracy is between 1 and 0.1 meV out of 260 eV, i.e. about 1 part in 10^6. All energies are per SiC pair of atoms.

Figure 2 shows that the situation for pure Si is simple : the stable structure is the cubic one < ∞ >, with the wurtzite stacking <1> having the highest energy, and other structures lying in between, all as one could expect. The situation in SiC has begun to turn, with <1> stil having the highest energy but < ∞ > no longer the lowest. Within the accuracy of the calculations, <2> and <3> have the same energy, the lowest energy found. From a more detailed analysis of 5 calculated structures, no other structure with a lower energy is to he expected, to an accuracy of less than 1 meV.

In an approximate sense, we can reasonably infer from Figure 2 that since <2> and <3> have very nearly the same lowest energy, then so will any of the polytypes in Table 1 consisting only of 2-bands and 3-bands. In broadest outline that accounts for the polytypes observed for SiC.

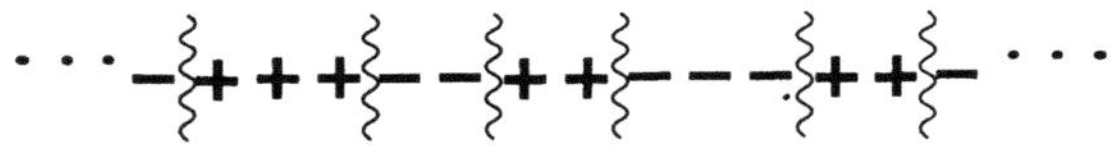

Fig.3 . A polytype configuration of (+) and (-) layers, considered as an array of antiphase boundaries shown as wiggly lines.

We can express our result quantitatively. Figure 3 shows how we can consider a polytype as an array of antiphase boundaries ("boundaries" for short) where the stacking changes form + or - or vice versa. We can write the energy of the polytype (per SiC pair of atoms) as

$$E = E_{ob} + \mu_b x + \Sigma c_r I_r \tag{2.1}$$

where E_{ob} is the bare energy of a layer, μ_b a bare chemical potential or energy for introducing one boundary into the system, x is the concentration of boundaries per layer, and I_r is the boundary-boundary interaction (BBI) r layers apart with appropriate coefficient c_r depending on polytype. For example for polytype <2> we would have

$$E<2> = E_{ob} + \tfrac{1}{2}\mu_b + \tfrac{1}{2}I_2 + \tfrac{1}{2}I_4 + \tfrac{1}{2}I_6 + ... \tag{2.2}$$

For our purposes it is convenient to transform (2.1) slightly because we want to confine ourselves to polytypes consisting of 2-bands and 3-bands only. Let us write

$$E(<2>) = E_o + \tfrac{1}{2}\mu \tag{2.3a}$$

$$E(<3>) = E_o + \tfrac{1}{3}\mu \tag{2.3b}$$

where E_o and μ can be shosen without loss of generality to give E(<2>) and E(<3>) exactly by (2.3) : they include terms such as the interactions in (2.2) and in that sense are "renormalised" quantities compared with the "bare" ones in (2.1), (2.2). Our zero order model is to write for any polytype

$$E = E_o + \mu_x \qquad \left(\tfrac{1}{3} \leq x \leq \tfrac{1}{2}\right) \tag{2.4}$$

which interpolates linearly between the two end points (2.3) and treats the polytype as just a mixture of phases <2> and <3>. That of course is not exact, and a proper transformation of (2.1) is to write

$$E = E_o + \mu_x + \Delta_{BBI} \qquad \left(\tfrac{1}{3} \leq x \leq \tfrac{1}{2}\right) \tag{2.5}$$

314

where Δ_{BBI} is the remainder of the sum of the I_r in (2.1) which is not included in (2.4). For the simplest higher order polytype <23> we have

$$\Delta_{BBI}\langle 23 \rangle = (1/5)(2I_5 - I_4 - I_6) + \ldots \qquad (2.6)$$

which now includes only interactions with second neighbour boundaries and beyond. What (2.6) says is that in <23> the second neighbour boundaries are 5 layers apart instead of 4 or 6 as in <2> and <3> respectively.

To recapitulate, there is a well known transition from the cubic diamond structure $< \infty >$ in the semiconductors to the wurtzite structure <1> with increasing ionicity, where these two structures represent opposite extremes from a stacking point of view. That SiC and ZnS lie close to the borderline is an accident, but one should expect more complex stackings there with bands of width 2 and 3 being the next obvious development from bands of width 1 in wurtzite. It is an accident that the signs and magnitudes of the interlayer interactions work out to give <2> and <3> practically the same energy in SiC, and hence the polytypes of Table 1. Regarding ZnS, see Section 7.

That deals with question (A) of Section 1. As regards questin (B), our view in (2.1) and (2.5) is that we hypothesise that polytypes are equilibrium phases and try to explain them that way, returning to growth mechanisms in Section 5.

3. BOUNDARY-BOUNDARY INTERACTIONS AND THE SIMPLEST POLYTYPES

We turn now to question (C) of Section 1 and the nature of the BBI introduced in (2.1) and (2.5). We can list five sources of interaction that have been discussed.

(i)	Coulomb effects
(ii)	Electronic structure interactions
(iii)	Kinks in the boundaries
(iv)	Atomic relaxations around boundaries
(v)	Phonon free energy

The Coulomb interactions decay exponentially very rapidly with distance and have been shown to the negligible for SiC [3,4] and I think the same conclusion would apply to other materials. By item (ii) I mean interlayer interactions inherent in the electronic structure of an ideal polytype with perfect tetrahedral bond angles and with bond lengths equal to that of the cubic structure. These are the effects included in the energies of Figure 2 and exclude the additional effects of items (iii) to (v). There are indeed interlayer interactions extending to many layers. They are the remnants in the semiconductor of the Friedel oscillations in a metal, and are expected to decay exponentially with distance in a semiconductor [11]. Their magnitude is too small beyond a spacing of 3 layers to be detectable by the total energy calculations of Section 2, but Cheng et al [6] have argued from extrapolation that the interactions are less than those of item (v) in SiC. They dominate in metals where there is a nesting Fermi surface. We mention item (iii) because thermally excited kinks on the boundaries generate the BBI which gives the richness of the phase diagram in the well known Axial Next Nearest Neighbour Ising (ANNNI) model [17,18]. I think this is the relevant mechanism in some alloys but not in SiC where it would mean breaking several bonds : the energy for that is far too high, and it would also cause topological problems due to the details of the crystal structure [2].

Item (iv) deserves a longer discussion than space allows here. At a boundary the cubic symmetry is broken so that there are relaxations of the bond angles and bond lengths from the perfect tetrahedral bonding. The resulting longitudinal relaxations of the interatomic spacings constitute a local strain field, and the overlap between the strain fields of two boundaries leads to a BBI between them. These have been considered for SiC [6] and found to be smaller than the phonon effects, but they might be the dominant mechanism is some metallic alloys [19]. The local strain field arises from the Friedel oscillations in the electron density around the boundary which would be much stronger in a metal than in SiC.

We turn finally to the phonon free energy, item (v). We set up a 10 parameter shell model for the phonons in cubic SiC [9]. The meagre experimental data had to be supplemented by total energy calculation for several phonon frequencies and eigenvectors to fit the model to. Besides Coulomb interactions, the model only contains explicitly couplings to cores and shells on nearest neighbour atoms, so that it can be applied equally well to any polytype. This, incidentally, does not mean one is approximating interatomic force constants to nearest neighbours. The latter are defined by the force experienced by a second atom j when a given atom i is moved, all other atoms being kept fixed. However the shells are allowed to move (polarise) and propagate a force well beyond nearest neighbours [5].

From this model the phonon spectra and phonon free energies were calculated. Let us concentrate specifically on the polytype <23> as the simplest intermediate phase between <2> and <3>. The results are presented as a phase diagram in Figure 4 : at low temperature T the

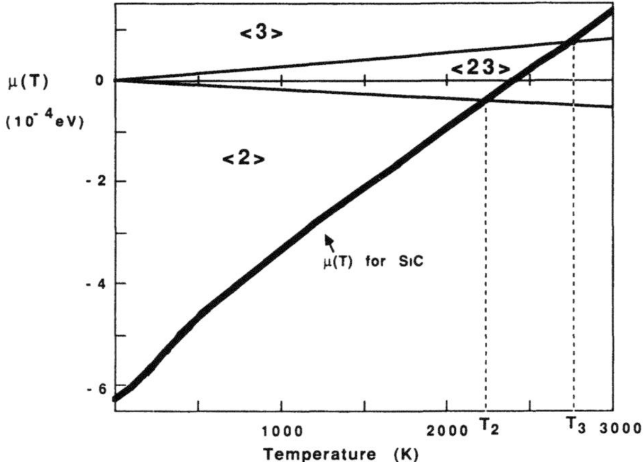

Fig.4 . The variation of chemical potential μ with temperature (heavy line) and consequent regions of stable phases, i.e. phase <2> below T_2, phase <3> above T_3 and <23> in between.

phase <2> is the ground state, whereas at high temperature the phonon free energy stabilises the phase <3>, as seems to be indicated by annealing experiments [15]. In between, the phase <23> has a lower free energy than either, and constitutes an intermediate phase ower a range calculated to be 500 K ! In constructing this diagram there has been one arbitrarily adjusted constant, namely the precise energy difference of 0.1 meV between <2> and <3> at zero temperature to obtain the cross-over at $T_o \approx 2400K$ (Figure 4) at about the observed temperature.

Let us explain in a bit more quantitative detail. The argument of (2.1), (2.5) can be applied just as well to the phonon free energy (suffix ph) as to the electronic structure energy (suffix e). Thus for the total free energy we have

$$F(T) = [E_{oe} + F_{oph}(T)] + [\mu_e + \mu_{ph}(T)]x + \Delta_{BBl}(T). \tag{3.1}$$

The quantities F_{oph} and μ_{ph} could be obtained as functions of T by fitting (2.3) to the phonon free energies calculated for phases <2> and <3>, and then Δ_{BBl} for <23> from the phonon free energy of <23> using (2.5). The μ_e is zero within the accuracy of the total energy calculations of Section 2 and was given a very small negative value as already mentioned, while E_{oe} is an irrelevant additive constant for present purposes. Thus the phase diagram of Figure 4 could be calculated.

Note the simple interpretation of (3.1). The first two terms represent the zero order model of mixing phases <2> and <3>. The important point is that when the total μ is zero

$$\mu(T) \equiv \mu_e + \mu_{ph}(T) = 0, \tag{3.2}$$

then the second term of (3.1) becomes independent of x, i.e. the zero order energy does not depend on the polytype. That is the multiphase degeneracy at temperature $T_o \approx 2400K$ in Figure 4 where all regular and irregular stacking would have equal free energy (in the zero order model). The polytype <23> has negative $\Delta_{BBI}(T_o)$ and hence exists as a stable intermediate phase around T_o. The stability range of temperature is so wide because the $\mu_{p\pi}(T)$ is so small, i.e. the phonon free energies of <2> and <3> differ so little [5]. In principle the phonon free energies of other higher order polytypes can be calculated, and hence their Δ_{BBI} and whether they are stable phases on the boundary between <23> and <2> or <3>. In practice the differences in free energies becomes too small for practical computation.

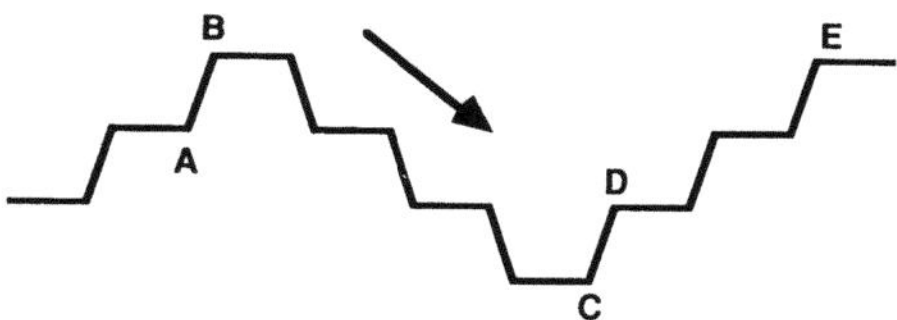

Fig.5 . Longitudinal acoustic phonon travelling down a chain of bonds BC.

Figure 5 illustrates why the BBI due to phonon free energy is moderately long ranged. A longitudinal acoustic phonon travelling down the row of atoms BC will certainly be affected by the boundary at C, being partially reflected and partially transmitted as a transverse wave along CDE. The reflected wave will similarly "notice" the boundary at B and mediate a BBI in the free energy.

In conclusion, all the sources (i) to (v) of possible long ranged BBI have been considered for SiC [5,6,11,12]. Not everything can be calculated as reliably as one would like, but it seems that the phonon free energy mechanism dominates in SiC. It gives <2> as the ground state at low temperature and <3> stable at high temperature, with <23> a stable intermediate phase over a range of the order of 500 K. Higher order polytypes will be discussed in the next two sections by more general arguments. Incidentally the rate of transformation from one structure to another in SiC is so slow that it has been impossible to establish a phase diagram, only to obtain some broad ideas [15].

4. THE DEVIL'S STAIRCASE AND HIGHER ORDER POLYTYPES

In this section we will consider higher order polytypes with repeat lengths of from 5 to, say, 30 layers, particularly those which belong to systematic families as in Table 1. We will outline the general principles which can lead in principle to an infinite sequence of stable intermediate phases (existing over increasingly narrow temperature intervals !). The theory relates to the theory of incommensurate structures. Modified to apply to SiC, it can perhaps explain the observation of some polytypes and not others.

We start with the ideas of Bak and Bruinsma [20], hereafter referred to as BB (not to be confused with the BBI of Sections 2 and 3). They considered some entities spaced along a line, in our case the boundaries as in Figure 3, constrained to occupy lattice sites. These entities interact with some potential I_r when r lattice spacings apart where I_r has the form shown in Figure 6, i.e. it is alsways repulsive, extends to infinity, and is everywhere convex (curved upwards).

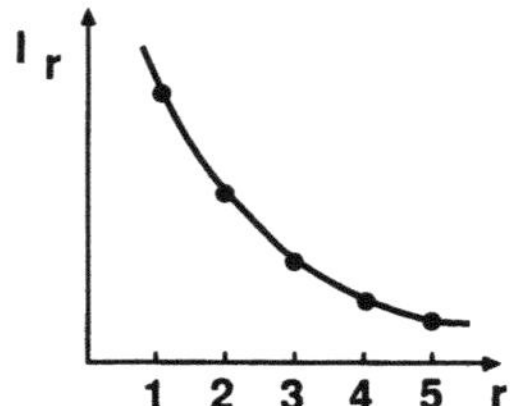

Fig.6 . Interactions I_r which are positive, convex and of infinite range.

BB had a chemical potential μ_b for introducing new boundaries so that their model is identical to (2.1). Under these conditions BB showed that one gets a phase diagram shown in Figure 7 with an infinite number of phases called a devil's staircase. The μ varies with T as in Figure 4 and (3.1) so that a horizontal line in Figure 7 is the temperature range of stability of a phase. The simple structures have broad stability ranges, with higher and higher polytypes having narrower and narrower ones. These phases include the families like those listed in Table 1. They are constructed as follows. Suppose the entities were not confined to lattice sites but interacted with a repulsive potential I(r) given by the curve in Figure 6 where r is now a continuous variable. What would the entities do ? They would space themselves out equally ! This defines a set of equally spaced points z_n. We now more each of these to its nearest lattice site. This construction is like fitting a regular modulation to the lattice. The infinity of phases results when the mean spacing $1/x$, controlled by μ, is varied.

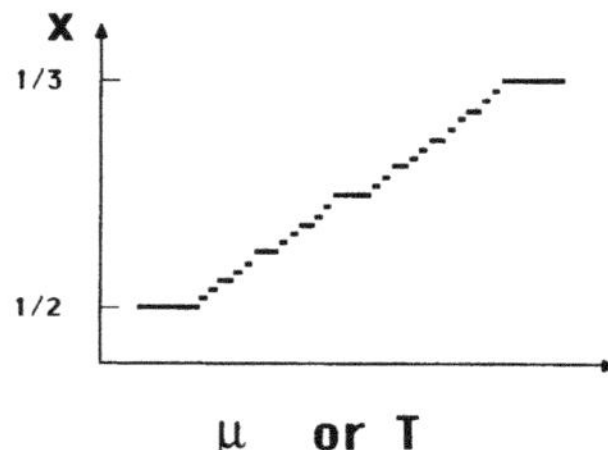

Fig.7 . A devil's staircase of an infinite number of stable phases, characterised by their mean concentration x of boundaries per unit length, occuring successively as μ is varied which in turn depends on T.

There is one problem about applying this model to SiC : we have two types of boundary, those in going from + to - and those from - to +. It is not obvious that is significant but look at Figure 8 and consider the interaction of boundary (1) with boundary (2) of opposite type and boundary (3) of the same types as (1). In one case the crucial interaction is between layers of opposite sign, in the other case of like sign. As a result, we have to distinguish between $\overset{even}{I_r}$ and $\overset{odd}{I_r}$, i.e. between boundaries with an even or odd number of other boundaries in between. In fact we have [5]

$$\overset{even}{I_r} \approx -\overset{odd}{I_r} . \tag{4.1}$$

One can try to rescue the BB argument by noting that the two types of boundary have to alternate and so just treating one of them. They will always have an odd number of boundaries between them, and the $\overset{odd}{I_r}$ was found to be positive from the calculations for <23> in Section 3 and for a few other cases [5]. So far so good. This would explain why the number of bands in the repeating pattern is always even (aside from <2> and <3> themselves). In particular it

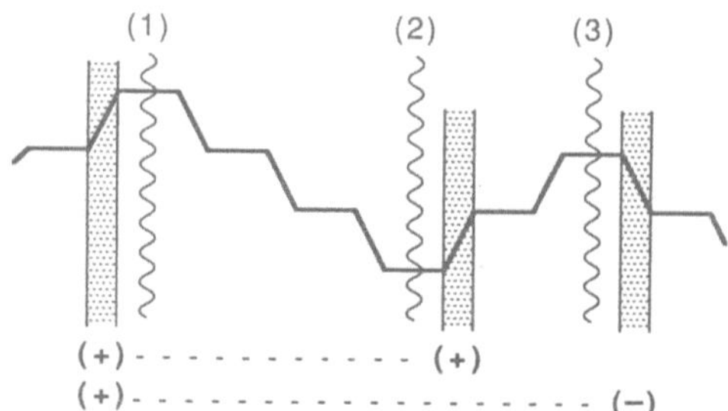

Fig.8 . The leading term in the BBI comes from the layers shown adjacent to the boundaries. Note the +- configuration of the relevant layers for neartest neighbour boundaries (1) and (2), and the opposite ++ configuration for next nearest neighbour boundaries (1) and (3).

explains why the phases <223> and <233> which occur in the BB theory do not seem to be observed polytypes, and why the phases <222323> and <232333> which are not allowed by BB theory are in fact observed polytypes. Of course <223> really has <223223> as the complete repeating unit taking into account + and - layers but would simply be <223> in the BB model : and similarly for <233>. These details reinforce the phonon free energy model because the sign difference (4.1) would not be expected to apply to the longitudinal relaxation mechanism also mentioned in Section 3. However difficulties arise when one looks at adapting the detailed algebra of BB to our case. One needs extra conditions on how slowly the I_r^{odd} fall off with distance r, and it is not clear whether these extra conditions are met [5].

In conclusion, one can make plausible arguments that the phonon free energy BBI can explain the occurence of whole families of polytypes as equilibrium phases and why they have an even number of bands in the repeat unit. However, even if these polytypes are proved in the future to be always metastable at all temperatures, they must be very nearly stable, and factors discussed in this section must be operative to make them so. See also the comment at the end of the next section.

5. GROWTH MECHANISMS

There can be no doubt that the polytypes with very long repeat distances must be growth structures, say repeat distances of 30 to several hundred layers. On any plausible mechanism the free energy difference would have become infinitesimally small so that any model of equilibrium phases does not make sense. Nor do they always form systematic families in the sense of Section 4. In some cases the repeat distance results from a screw dislocation with a giant Burgers vector, which can be observed particularly in whiskers [15]. There may be other plausible mechanisms, more applicable to bulk material [21,22]. In other cases the boundaries are not very regularly spaced, as shown by X-ray streaks rather than spots.

The problem about the dislocation mechanism has long been to see how such a giant Burgers vector can arise. Frank [23] has recently provided an explanation with some experimental corroboration. The whisker starts to growth with some random normal screw dislocations of positive and negative sign. Some will diffuse to the surface of the whisker under the force from its elastic image potential. When a net preponderance occurs of screw dislocations with one sign, a new mechanism operates. The whisker has a net twist which provides an extra force to expel dislocations of opposite (minority) sign and to attract those of majority sign to the centre where they coalesce to form the giant Burgers vector. The core of such a dislocation is hollow.

Although we are dealing with metastable growth structures, the factors discussed in previous sections are still operating and exerting an overall control. The particular stacking sequence of a polytype arises on the screw dislocation mechanism from a spontaneous stacking fault in, say, the structure <3> which is the stable form at high temperature, and this fault is then propagated periodically by winding around the screw. Suppose the stacking fault is a 2-band so that the polytype is $<3^n2>$. Why is it a 2-band and not a 4-band ? I can see no *geometrical* reason favouring one rather than the other, and a few polytypes with a 4-band have been observed [14]. However the considerations of the previous sections would give a 2-band fault a lower (free) energy than a 4-band fault.

In conclusion, the energetics considered previously in explaining polytypes as equilibrium phases, are still relevant in determining which particular stacking faults and metastable growth structures occur.

6. AN ANOMALY : THE CUBIC PHASE OF SiC

The philosopher Popper pointed out that one can never prove a hypothesis is valid under all circumstances, but one serious counter example can destroy it. We saw in Section 2 that the cubic form $<\infty>$ of SiC has a significantly higher energy than the phases <2> and <3> and polytypes derived from them. So why does the stuff grow most easily in the cubic form ? In one sense there is no problem : crystal growth is fast compared with annealing times so that metastable structures can easily arise. But it turns out that we can obtain a more detailed insight. Let us confine ourselves to the energy differences at T = 0 because the phonon free energy effects are smaller in comparison.

Table 2. The energy $\Delta U(+), \Delta U(-)$ for adding one + or - layer to a crystal for different configurations near the surface (in meV per SiC pair of atoms). The surface is on the right, the bulk to the left.

Configuration	$\Delta U(+) - E_{ob}$	$\Delta U(-) - E_{ob}$
.....+++	- 1.58	+ 1.58
.....+-+	- 7.12	+ 7.12
.....-++	- 3.00	+ 3.00
.....--+	- 7.70	+ 7.70

It is first necessary to analyse the results of Section 2 in more detail. For a polytype of N layers we can write the total energy as

$$U = NE_{ob} - \sum_i [J_1 S_i S_{i+1} + J_2 S_i S_{i+2} + (J_3 + K S_{i+1} S_{i+2}) S_i S_{i+3}] \tag{6.2}$$

Here $S_i = \pm 1$ for + and - layers respectively. This form is very general including all interactions betweem layers up to third neighbours [11]. The values fitted to the calculated polytype energies are [4], in meV per SiC pair of atoms,

$$J_1 = 4.85, \qquad J_2 = -2.56, \qquad J_3 = -0.50, \qquad K = -0.21. \tag{6.2}$$

We can now use the form (6.1) for a semi-infinite crystal and calculate the additional energy $\Delta U(+)$ and $\Delta U(-)$ for adding one more layer to the surface of type + or -. Since (6.1) contains only interactions up to third neighbours, there are only four cases, namely when the crystal ends with layers +++, -++, --+ or +-+ (and their opposites). The results for ΔU are given in Table 2. We see that in *all* cases $\Delta U(+)$ is less than $\Delta U(-)$, i.e. it is energetically favourable to add one more layer of the same type as the last one ! Obviously the cubic form $<\infty>$ results, whatever one starts with. This result comes about because we are adding strictly one layer at a time, without allowing lower layers to relax to a achieve a global minimum energy.

320

Of course one can criticise this calculation. Firstly there should be an extra correction energy for the top layer to take into account its contact with the SiC liquid : but that is a constant term which marches upwards as each layer is added on top. Secondly we assume that the topmost layer is not distorted very much so that its interaction with the layers below is given by the same J_n, K as for bulk. This is a more serious assumption, but perhaps the liquid is not so very different from the bulk. Thirdly the expression (6.1) is only valid for bulk material. For a finite or semi-infinite system there are extra terms [11].

In any case the point is that this simple argument, with all its approximations, gives the right answer and removes any mystery about the common growth of the cubic structure.

7. POLYTYPES IN ZINC SULPHIDE

We come finally to ZnS where the story starts the same way as before, namely the material is near the borderline between the cubic $< \infty >$ and wurtzite $<1>$ structures of Table 1. In fact the cubic form is the stable phase at low temperature, and the wurtzite structure becomes stable above a temperature $T_o = 1020°C$, with the material transforming fairly readily between.

The results of total energy calculations are included in Figure 2, with the $< \infty >$ structure lying lowest as expected. The difference from SiC lies in the J_1 in the sense of Section 6. We have for ZnS $J_2 = -0.08$ meV and J_3 and K less than 0.01 meV, only J_1 being significant. The phonon free energy of different types has not been studied but it must be such that the total J_1 is driven negative above T_o, which produces the phase change. Let us assume the phonon effect on J_2, J_3, K is small. Then around T_o we would have all the J_i very small, with no tendency to stabilise any particular polytype stacking, unlike SiC. The observed polytypes of ZnS are also quite different from those in SiC. There are no apparent patterns, or families of polytypes. Quite wide bands occur with 10 or more layers of the same sign i.e. in the cubic arrangement [24].

The ZnS polytypes are in fact metastable structures resulting from incomplete transformation of the material from $<1>$ to $< \infty >$ as the temperature falls. The lattice repetition of the stacking sequence arises from their growth around giant screw dislocations. The experimental indication of this comes from observing different polytypes such as $<11,7,2,2>$, $<9,5,5,3>$ and $<13,9>$ along different parts of a single whisker [25]. Note that they all have the same repeat distance of 22 layers corresponding to the Burgers vector of the screw dislocation. What has happened is that transformation from $<1>$ to $< \infty >$ has been initiated at random sites along the whisker and has propagated around the screw for some distance [24] producing locally a particular polytype.

Engel [13] has accounted for the observed distribution of polytypes in whiskers as follows. We picture the conversion of structure $<1>$ to structure $< \infty >$ as T falls just below T_o. The atomic process is a (+) layer turning locally into a (-) layer or vice versa, the change then propagating throughout that layer and hence around the screw, driven by the small free energy lowering. This process is supposed nucleated randomly at different sites along the whisker. The key point is what happens when two such propagating changes meet : they may reinforce each other or may block one another. Suppose each of them is the conversion of a (-) layer to a (+) layer, i.e. ...+-+... to ...+++... If they meet from propagating around the screw on next nearest neighbour layers their combined effect will be to turn ...+-+... into ...+++++... and we have 5 layers of the $< \infty >$ structure formed. However if one change is (-) to (+) and the other is (+) to (-), then when they meet we obtain configurations like ...++--... which would be energetically a bit more favourable than the original $<1>$ structure but less than the $< \infty >$ pattern. The energy difference depends on the (small) $J_1(T)$ and J_2, the other J_3, K etc still being assumed negligible. Engel [13] has modelled the process, assuming that a particular change propagates in a layer with a speed proportional to the lowering of the free energy obtained. By choosing $J_1/J_2 = -8$ at the effective temperature of polytype formation he found that the polytypes generated in the model resembled remarkably those observed in whiskers [24,25].

To summarise, in ZnS the J_2 in (6.1) is very small and J_3, K effectively zero. The J_1 is temperature-dependent, passing through zero around $T_o = 1020°C$, such that the structure $<1>$

is stable at T above T_o and $< \infty >$ at T below T_o. Thus around T_o all polytypes with any distribution of bands have very nearly equal free energy, corresponding to the great variety of observed polytypes with a wide distribution of band widths. The polytypes are formed as incomplete metastable conversions from $<1>$ to $< \infty >$ as T falls below T_o. Experimental evidence for this from whiskers is coroborated by Engel's theoretical study which accounts well for the observed distribution of polytype structures.

ACKNOWLEDGEMENTS

We wish to acknowledge explicitly the contribution of the other co-workers on this long project, namely R.J. Needs, K. Kunc, I.L. Jones, N. Churcher, J. Smith, J. Yeomans, J.A.A. Shaw and G. Engel. The work has been funded principally by the Science and Engineering Research Council of the U.K.

REFERENCES

[1] J.D.C. Mc Connell and V. Heine, Europhys. Conf. Abstracts A6, 172 (1982)

[2] J. Smith, J. Yeomans and V. Heine in Modulated Structure Materials, T. Tsakalakos, ed. Nijhoff, Dordrecht, (1984) p. 95

[3] C. Cheng, R.J. Needs, V. Heine and N. Churcher, Europhys. Lett. 3, 475 (1987)

[4] C. Cheng, R.J. Needs and V. Heine, J. Phys. C : Solid State Phys. 21, 1049 (1988)

[5] C. Cheng, V. Heine and I.L. Jones, J. Phys. Condens. Matter (submitted)

[6] C. Cheng, V. Heine and R.J. Needs, J. Phys. Condens. Matter (submitted)

[7] C. Cheng, R.J. Needs, V. Heine and I.L. Jones, Phase Transitions 16/17, 263 (1989)

[8] N. Churcher, K. Kunc and V. Heine, Solid State Commun. 56, 177 (1985); J. Phys. C : Solid State Phys. 19, 4413 (1986)

[9] N. Churcher, K. Kunc and V. Heine, in Phonon 1985 Proceedings of the Second International Conference on Phonon Physics Budapest August 1985, J. Kollar, N. Kro eds., World Scientific Publ., Singapore (1985), p. 956

[10] C. Cheng, K. Kunc and V. Heine, Phys. Rev. B 39, 5892 (1989)

[11] J.J.A. Shaw and V. Heine, J. Phys. Condens. Matter (submitted)

[12] C. Cheng, V. Heine and R.J. Needs, to be submitted to Europhys. Lett.

[13] G. Engel and R.J. Needs, to be published.

[14] D. Pandey and P. Krishna, J. Cryst. Growth 31, 66 (1975) ; Current Topics in Materials Science 9, 415 (1982) ; J. Cryst. Growth Charact. 7, 213 (1983)

[15] N.W. Jepps and T.F. Page, J. Cryst. Growth Charact. 7, 259 (1983)

[16] M.Y. Chou, M.L. Cohen and S.G. Louie, Phys. Rev. B 32, 7979 (1985)

[17] J. Yeomans, Solid State Phys. 41, 151 (1988)

[18] W. Selke, Physics Reports 170, 213 (1988)

[19] R. Bruinsma, Phase Transitions 16/17, 275 (1989)

[20] P. Bak and R. Bruinsma, Phys. Rev. Lett. 49, 249 (1982)

[21] D. Pandey and A.P. Sutton (to be published)

[22] P. Pirouz in Proc. Sixth Intern. Symp. on "Structure and Properties of Dislocations in Semiconductors" S.G. Roberts and P.R. Wilshaw eds., Inst. of Phys. Public, Bristol (1989)

[23] F.C. Frank, Phil. Mag. A 56, 263 (1987)

[24] S. Mardix, Phys. Rev. B 33, 8677 (1986)

[25] S. Mardix, A.R. Lang, G. Kowalski and A.P.W. Makepeace, Phil. Mag. A 56, 251 (1987)

STRUCTURAL MODULATIONS IN THE HIGH-TEMPERATURE SUPERCONDUCTOR $YBa_2Cu_3O_{7-\delta}$ AND SEMI CONDUCTING WO_{3-x} ASPECTS OF NON-EQUILIBRIUM BEHAVIOUR

Ekhard K.H. Salje

IRC Superconductivity and Department of Earth
Sciences, Downing Street, Cambridge CB2 3EQ, England

1. INTRODUCTION

A thermodynamic description of non-harmonic structural modulations may be attempted in terms of the energies required to form walls (alias kinks, boundaries, areas of large structural gradients) and the inter-wall interactions. The latter are also called 'boundary-boundary interactions' (BBI), and they are normally repulsive in character. The wall energies can be negative, as in equilibrium incommensurate phases, or positive if the structural state is unstable or metastable. An example for narrow walls in a matrix of a perovskite-type structure are the so-called CS planes in WO_3 (Fig.1). Different wall configurations have been described in great detail for individual chemical phases (review by Sundberg, 1981a, and Sundberg, 1981b) and it was found that these walls occur in equilibrium phases with a high degree of regularity. Here we are concerned only with walls along one of the crystallographic $\{10n\}, n>2$ directions. A unifying, although more simplistic view, of the various structural and chemical features is based on the observation that the essential structural building blocks of the walls are blocks of four edge-sharing octahedra which are aligned within the walls as shown in Fig.2. These structural units represent stable configurations which are accounted for within the framework of Landau-Ginzburg theory by negative Ginzburg energies:

$$G_{Ginzburg} = 1/2 \ g \ (\nabla Q)^2 + 1/4 \ g' \ (\nabla Q)^4 + ...$$

Fig. 1 CS planes of edge-sharing WO_6 octahedra in a matrix of WO_3-type structure.

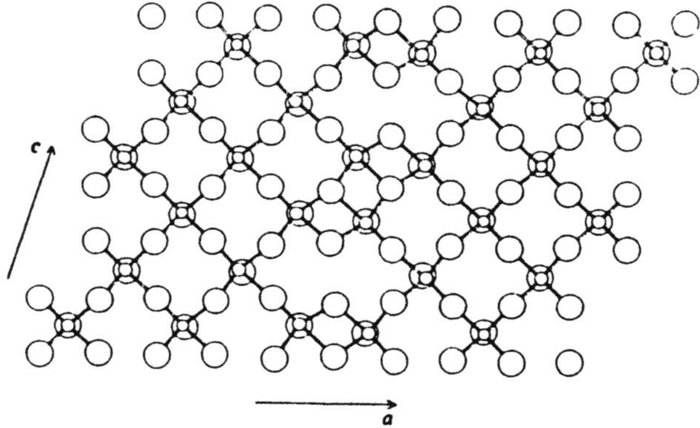

Fig. 2 The structural blocks of edge-sharing octahedra which mainly determine the wall energy. These walls are embedded within a matrix of corner-sharing octahedra.

where Q is the order parameter and g and g' are coefficients. The structural modulation is thermodynamically stable if the total negative wall energy is larger than the total energy related to the BBI. We may now ask what is the physical nature of the BBI which leads to the periodic spacings of the walls. Iguchi, Salje and Tilley (1981) have argued that the electronic energy is a relevant part of the BBI and it was shown by Salje and Güttler (1984) that the semiconductor to metal transition in these materials is related to the breakdown of the structural modulations in the {102} and {103} directions. A second possible contribution is the phonon energy which seems to dominate the BBI in SiC (Heine, this volume). The third, and possibly dominating source of inter-wall repulsion in WO_{3-x} is related to the structural strain within the perovskite-type layer between the walls. This idea has already been advocated by Iguchi and Tilley (1978) and the structural investigations by Viswanathan and Salje (1981) showed that these strain fields do, indeed, exist.

2. STRAIN RELATED BOUNDARY-BOUNDARY INTERACTIONS IN WO_{3-x}

The crystal structure of WO_{3-x} shows that the inter-boundary spacings are topologically closely related to the equilibrium structure of WO_{3-x}. The details of the structural behaviour are, however, rather different in three points:

a) The WO_6 -octahedra are tilted against each other in the fully oxidized WO_6 structure (Diehl et al., 1978). This tilt can be overcome by a structural phase transition with a transition temperature of 740K (Salje, 1977). The tilts do not occur in the modulated structure because they are structurally not compatible with the boundary conditions of two-dimensional walls on either side of the slabs of WO_6 -like material. The total energy of the BBI contains thus a contribution related to the suppression of the octahedra tilts.

b) The tungsten atoms in the walls are shifted along the c-axis. The orientation of the shift is antiparallel for the two nearest W-positions. The next W-atoms in the inter-wall region follow this pattern and transmit it to all adjacent octahedra which leads to a puckering of all W-atoms. The phase of this up-down shift of the W-positions accomodates only even numbers of WO_6 octahedra between two walls. In the case of an odd number, an additional phase shift has to occur which requires an additional energy contribution to the BBI. One finds empirically that crystal structures with odd numbers of WO_6 octahedra between the walls are, indeed, thermodynamically less stable than those with an even number of octahedra.

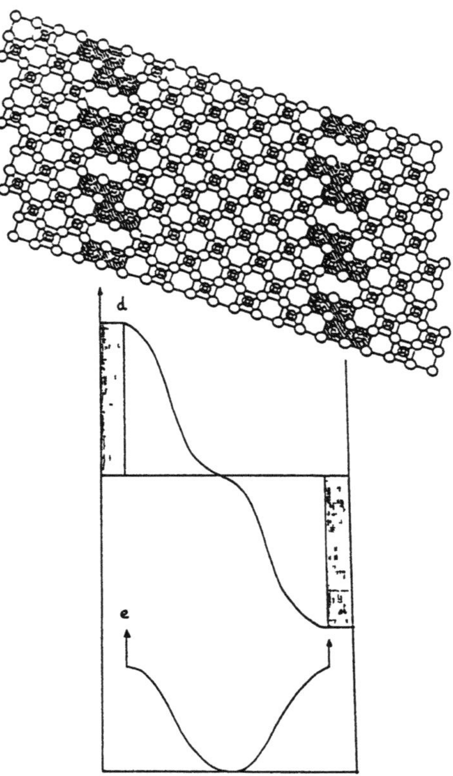

Fig. 3 Crystal structure of $W_{14}O_{41}$ and graph indicating the
decay of the lattice distortion (d) with increasing
distance from the wall; e indicates the elastic energy
related to the BBI.

c) The interatomic distances between tungsten and oxygen deviate
significantly from those of the fully oxidized material. These deviations
are largest for octahedra linked directly to the wall and decrease
systematically for octahedra further away from the walls. The most
regular octahedra are found in the middle between two walls (Fig.3.).
The pattern of these deformations is similar to that which occurs during
the structural phase transition between tetragonal and orthorhombic
WO_3 near 1013K although   similar deformations still exist in the
tetragonal structure and they would only disappear in a hypothetical
cubic structure which was not observed in experiments extending to
temperatures above 1500K. We thus conclude that the energy
contribution from the bond-distance variations exceeds those of the
octahedra tilts.

The variation of bond-distances averaged over each layer of
octahedra is sketched in Fig.3. A quantitative estimate for the related
strain energies is in good agreement with an exponential interaction
energy:

$$E_{BBI} \propto \exp(-r/r_0)$$

where r is the distance from the wall and r_0 is the characteristic length of the interaction (here we measure along the conventional monoclinic axes). In the case of WO_{3-x} we find r_0 to be of the order of 40Å.

Two extreme cases can now be discussed. Firstly, the modulated structures become thermodynamically unstable if the total BBI energy (a positive energy) is larger than the absolute value of the wall energy (a negative energy). This point is reached in WO_{3-x} for spacings of about seven octahedra between the walls, i.e., $r < 0.7\, r_0$. The second case concerns the limit of large inter-wall distances. If r becomes much larger than r_0, the BBI has decayed exponentially and irregular spacings between the walls will occur. This situation has, indeed, been observed in WO_{3-x} for distances larger than 18 octahedra between the walls ($r > 1.7 r_0$). The average inter-wall distance (or the mean value of the wavevector of the modulation) varies, in this case continuously, leading to incommensurations. The total phase diagram (Berglund and Sahle, 1981) can then be described as follows using the oxygen fugacity as the external thermodynamic parameter: at high oxygen fugacities we find the paraphase of fully oxidized WO_3. Reducing the oxygen fugacity, incommensurations occur with small wavevectors k. The first lock-in phase becomes stable at k=1/18 followed by 1/16 etc. The smallest wavevector of a lock-in phase with a large stability range is 1/8, smaller stability ranges might exist for odd values of 1/k. The phase diagram of WO_{3-x} can thus be conveniently described as an incomplete "devils staircase" where the external thermodynamic variable is the oxygen fugacity rather than temperature.

3. STRUCTURAL MODULATIONS IN $YBa_2Cu_3O_{7-\delta}$: Co

The structural modulations in Co doped $YBa_2Cu_3O_{7-\delta}$ are now analysed within the framework developed so far. The structural features of the High Temperature Superconductor are characterized by structurally 'soft' perovskite-like layers of Cu(I) and oxygen and structurally 'hard' Cu(II)-O layers (Schmahl et al.,1989). Oxygen deficiency leads to defect structures which can be mapped on to the same geometrical grid as in WO_{3-x}. The major structural difference between the two structures arises from the two-dimensional character of the perovskite-type layer in the superconductor whereas the same structure has a three-dimensional character in the tungsten oxides. This reduction in dimension is, however, partly compensated for by the fact that the Cu(II)-O layers can structurally relax depending on the structural distortions in the Cu(I)-O layers so that, in fact, the defect structures can be transmitted through the crystal not only within the soft layers but also between them, i.e. along the crystallographic c axis. A clear indication for this behaviour is that the domain structure as observed along the c axis appears equivalent to volume twinning (Fig.4, after Schmahl et al.1989). These domain boundaries (or walls) have thus a three-dimensional character and can be treated in the same way as in WO_{3-x}. The BBI leads again to the formation of a periodic pattern for inter-wall distances of less than ca. 100Å (Fig.4b), domain patterns with larger distances show an irregular distribution of walls (Fig.4a). As the crystal structure transforms from an orthorhombic phase at high oxygen fugacities (or low temperatures, or low Co content) to a tetragonal phase at low oxygen fugacities (or high temperatures, or high Co content) there are two crystallographically equivalent directions along which the walls can develop (Salje et al.1985). These two directions are orthogonal with strongly repulsive BBI between walls of different orientation. Intersections of walls are thus energetically disfavoured and patches of orthogonal domain strucures occur if the structural deviation from tetragonality is small (Fig 4.c).

This interpenetration avoidance no longer applies in the tetragonal phase and the point of the structural phase transition is characterized by the onset of tweed structures replacing the striped orthorhombic structure (Fig. 4d-h). There is, however, no abrupt change of the inter-wall distances or the wavevector of the modulation during the structural phase transition, as already shown by Schmahl et al.1989. It is also important to note that the distribution of modulation vectors in the tetragonal phase does not change as a function of the Co-content even for values equivalent to twice the critical value at which the phase transition occurs. This is in contrast to the assumption that these modulations are directly related to fluctuations in the Ginzburg interval of this structural phase transition for two reasons. Firstly, the extent of the Ginzburg interval (equivalent to a temperature interval between Tc and 2Tc) exceeds the classical estimates. Secondly, the fluctuation-dissipation theorem leads to a divergency of the long wavelength fluctuations at the transition point, in contrast to the observations. We prefer, thus, to describe the modulations as the result of BBIs between walls which are themselves independent of the structural phase transition and may assume an exponential form as described before. The characteristic length, r_o, seems indeed, to be independent of the Co content (and may, presumably, also be independent of the equilibrium temperature).

4. STRUCTURAL MODULATIONS AND NON-EQUILIBRIUM BEHAVIOUR

Tweed patterns also occur amongst the most common microstructures in kinetic experiments (e.g. Salje,1988, Salje and Wruck,1988). Within the context of our present line of arguments, these modulations are the consequence of noise induced fluctuations which create kinks of the order parameters with weak BBIs leading to a wide

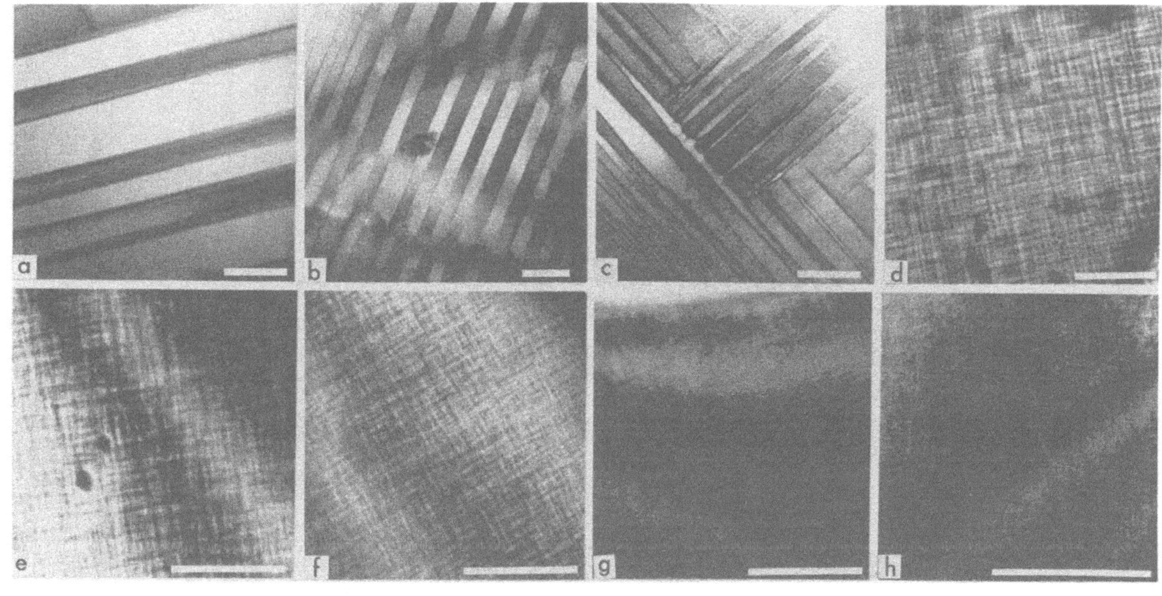

Fig. 4 Microstructures in Co-doped $YBa_2Cu_3O_{7-\delta}$ ($\delta \cong 0$). Striped microstructures occur in orthorhombic crystals (a,b,c). Tweed patterns characterise the tetragonal material (d-h). The scale bar is 0.1μm, the Co-content is 0%(a), 1%(b), 2%(c), 2.5%(d), 2.8%(e), 3%(f), 5%(g), 7%(h) with respect to Cu.

distribution of wavevectors. The lifetime of these modulations can be exceedingly long and tweed structures appear as the stationary solution of the rate equation (Carpenter and Salje, 1989). These fluctuations occur even if the initial and final equilibrium states are uniform and no critical fluctuations occur at the point of the phase transition such as in ferroelastic crystals (Folk et al. 1979). We may thus assume that the actual wall energy is positive and that the formation of structural modulations is related to the kinetics of the time evolution of the order parameter. In the limit of non-conserved order parameters we find that the kinetic equation is:

$$\frac{dQ}{dt} = \frac{1}{\tau} e^{\Delta G_a/kT} \frac{\partial G}{\partial Q}$$

where Q is again the order parameter, ΔG_a is the Gibbs energy of activation and G is the thermodynamic potential which can often be approximated by the Landau potential and the Ginzburg energy. A typical sequence of Gibbs energies as determined experimentally (Salje, 1988) is shown in Fig. 5. Let us consider the course of a kinetic experiment in which a crystal which is in thermodynamic equilibrium at room temperature (position I) is shock-heated to 1300K (position II). If the relaxation time of the order parameter is much longer than the time needed for the heating process, we may assume that the immediate effect of temperature is to excite phonons and relax the lattice with the frozen-in order parameter of the starting material. If the crystal is now annealed, we find that nucleation of the final state can only occur if the relevant interfacial energy is sufficiently low, which is not the case in most ferroelastic materials. The isothermal relaxation of the order parameter is thus determined by the continuous change of Q according to the given rate law. All structural states encountered during the course of the kinetic experiment are then closely related to equilibrium states which would occur at different temperatures. In a simple picture, we find that the space-averaged order parameter can be represented by the 'ball sliding down the hill' along the dotted line in Fig. 5. Stoichiastic fields may then lead to fluctuations of the order parameter on a mesoscopic scale which lower the energy if the structural state of the fluctuation advances those of the stationary solution. It is this additional stabilization of the fluctuations which is the main reason for the formation of kinks. As there are always two orthogonal, elastically soft directions in ferroelastic crystals, kinks may develop perpendicular to these directions and they are then also aligned perpendicular to each other. There is no macroscopic lattice distortion breaking the symmetry of the starting material and the intersection between kinks along different directions are not necessarily points of high excess energies. The most common microstructure under these circumstances is the tweed pattern, where the fine details of the domain structures are determined by the nature of the walls and the BBIs. Fig. 6 shows such long-lived, kinetically induced domains in a Na-feldspar (Harris et al.1989).

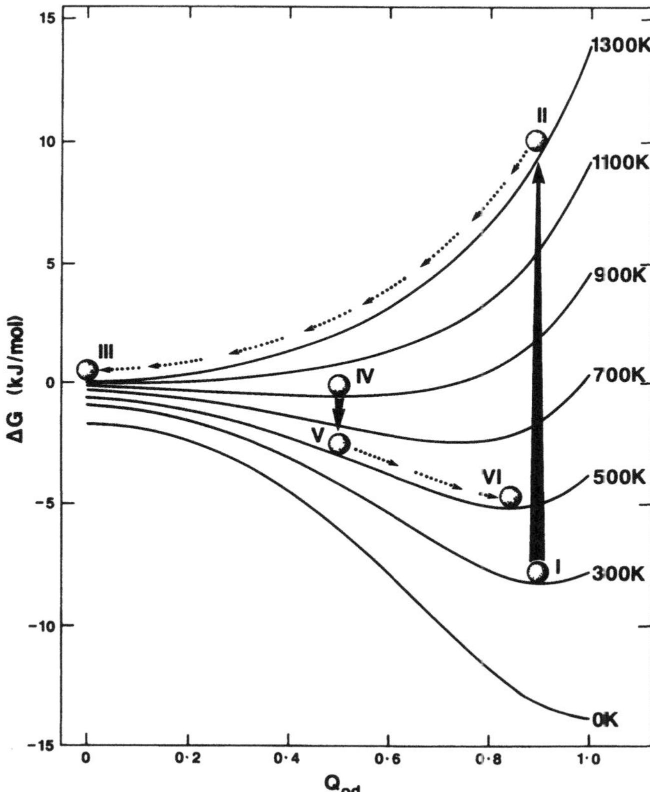

Fig. 5 Experimentally determined excess Gibbs energies of the C2/m-C1 phase transition in Na-feldspar. Typical sequences of kinetic experiments are I → II → III or IV → V → VI.

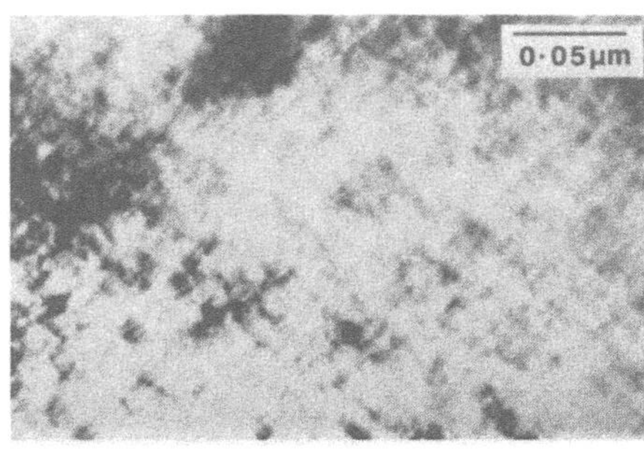

Fig. 6 Tweed microstructure of a Na-feldspar developed during the kinetic path II → III in Fig. 5 (courtesy of A. Putnis, Cambridge)

REFERENCES

Berglund, S., and Sahle, W., 1981, Accommodation of oxygen loss in WO_3 equilibrated with $CO + CO_2$ buffers, J. Sol. State Chem., 36:66.

Carpenter, M.A., and Salje, E., 1989, Time-dependent Landau theory for order/disorder processes in minerals, Min. Mag., 53:483.

Diehl, R., Brandt, G., and Salje, E., 1978, Crystal Structure of Triclinic WO_3, Acta Cryst., B34:1105.

Folk, R., Iro, H., and Schwabl, I., 1979, Critical dynamics of elastic phase transitions, Phys. Rev. B20,:1229-1244.

Harris, M.J., Salje, E., Guttler, B., and Carpenter, M.A., 1989, Structural states of natural potassium feldspar: an infrared spectroscopic study, Phys. Chem. Min. (in print).

Iguchi, E., and Tilley, R.J.D., 1978, The elastic strain energy of crystallographic shear planes in ReO_3-related oxides. 1. The formation energy of isolated CS planes, J. Sol. State Chem. 24: 121.

Iguchi, E., Salje, E., and Tilley, R.J.D., 1981, Polaron Interaction Energies in Reduced Tungsten Trioxide, J. Solid. State Chem., 38:342.

Salje, E., 1977, The structure of the orthorhombic phase of WO_3, Acta Cryst., B33:574.

Salje, E., and Guttler, B., 1984, Anderson transition and intermediate polaron formation in WO_{3-x}: Transport properties and optical absorption, Phil. Mag., B50:607.

Salje, E., 1988, Kinetic rate laws as derived from order parameter theory I: Theoretical concepts, Phys. Chem. Min., 15:336.

Salje, E., Kuscholke, B., and Wruck, B., 1985, Domain wall formation in minerals: I Theory of twin boundary shapes in Na-feldspar, <u>Phys. Chem. Min.</u>, 12: 132.

Salje, E., and Wruck, B., 1988, Kinetic rate laws as derived from order parameter theory II: Interpretation of experimental data by Laplace-transformation, the relaxation spectrum, and kinetic gradient coupling between two order parameters, <u>Phys. Chem. Min.</u>, 16: 140.

Schmahl, W. , Putnis, A., Salje, E., Freeman, P., Graeme-Barber, A., Jones, R., Singh, K.K., Blunt, J., Edwards, P.P., Loram, J., and Mirza, K., 1989, Twin formation and structural modulations in orthorhombic and tetragonal $YBa_2(Cu_{1-x}Co_x)_3O_{7-\delta}$, <u>Phil. Mag.</u> (in press).

Sundberg, M., 1981a, Crystallographic shear in reduced tungsten trioxide, <u>Chem. Commun. 5</u>, University of Stockholm.

Sundberg, M., 1981b, Structure and oxidation behaviour of W24O70, a new member of the {103} CS series of tungsten oxides, <u>J. Solid State Chem.</u>, 35:120.

Viswanathan, K., and Salje, E., 1981, Crystal structure and charge carrier behaviour of $(W_{12.64}Mo_{1.36})O_{41}$ and its significance to other related compounds, <u>Acta Cryst.</u>, A37:449.

INCOMMENSURATE MODULATIONS IN BISMUTH-BASED

HIGH-Tc SUPERCONDUCTORS

J.C. Tolédano, J. Schneck, L. Pierre

C.N.E.T., 196 avenue Henri Ravera, 92220 Bagneux - FRANCE

ABSTRACT

We describe the incommensurate modulations which are observed in the various superconducting phases of the bismuth-based high Tc materials, both for the nominal compositions and for the substituted ones. The origin of these modulations is discussed with reference to standard physical models available for 2D-incommensurate systems. The dependence of the modulation wavevector observed for various types of substitutions (Pb for Bi or Y for Ca) agrees with the values deduced from the mismatch between the natural and constrained Bi-O distances.

INTRODUCTION

Satellite diffraction peaks revealing the occurence of incommensurately modulated structures have been observed in several families of materials displaying high-Tc superconductivity. Up to now the best characterized situation, in this respect, is that of the bismuth-based superconductors, having the generic formula $Bi_2Sr_2Ca_{n-1}Cu_nO_{2n+4}$ in which the modulation gives rise to sharp and intense satellites. In the other families of superconductors, the satellite reflections are often weak and diffuse, and they mostly appear in compounds shifted in composition with respect to the standard superconducting material (e.g. in oxygen defficient material).

In order to analyze the main features of the incommensurate modulation of the bismuth based superconducting phases a convenient schematization of the crystal structures of these phases consists in regarding the structures as stacking of slabs[1] whose thickness along c depends on the superconducting phase considered, e.g. ~ 9 Angström for n=1 in the above formula, and ~ 15.5 Angström for n=3 (Fig.1).

Each slab consists in a pair of BiO planes enclosing an atomic configuration resembling the perovskite structure, and involving the remaining atoms (Sr, Ca, Cu, O).

Though no complete structural determination of the modulation is yet available, the broad lines of its characteristics are known from X-ray investigations for the n=1[2,3], and n=2[4,5] phases, and further information has been obtained, for n=1, 2, 3, 4, from observations in high resolution electron microscopy[6]. These studies show that the modulation consists in a complex set of atomic displacements concerning all the constituants of the structure (no direct data exists for the oxygen atoms but these are most likely to be affected as well). These displacements are unusually large for modulated structures (up to 0.5 Angström), and they are perpendicular to the a-axis of the structure. Their wavevector is parallel to the b-axis and the period is approximately equal to 5b unit cells of the basic unmodulated structure. Several experiments have shown that the modulation wavevector is temperature independent both below[7] and above[8] room temperature, up to 600°C.

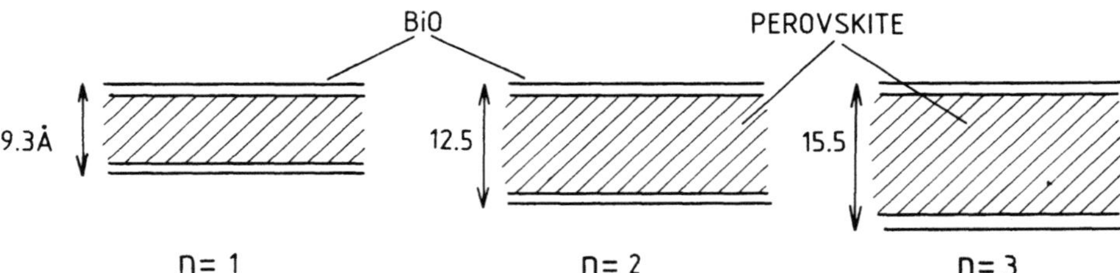

Fig.1 . Schematic representation of the basic building blocks of the series of bismuth-based superconductors.

Considering the complexity of the pattern of atomic displacements and the lack of temperature dependence of the wavevector, the modulation of bismuth-based superconductors appears, at first sight, difficult to relate to one of the standard models of incommensurate systems. However, the key to the interesting aspects of this modulation and to the sorting out of its simple features resides, not in the effect of temperature, but in that of compositional changes.

In this paper, we consider briefly four aspects of the above modulation. First, we develop simple energetic arguments supporting the idea that the modulation is triggered by an incommensurate compression wave existing in the sole Bi-O planes. In the second place, we discuss the characteristics and origin of the various stacking modes observed for the slabs. Thirdly, we consider the possible origin of the 2D-modulation in the Bi-O planes by reference either to an electronic model, or to a structural one. In the latter case, we compare the predictions of the model, to the experimental data obtained for various types of chemical substitutions. Finally, we discuss the recent data relative to the coexistence of two incommensurate modulations in lead-substituted compounds, for a certain range of lead content.

TWO-DIMENSIONAL NATURE OF THE "PRIMARY" MODULATION

Several observations suggest that the existence of the modulation and its wavelength are not significantly influenced by the interaction between consecutive slabs. Thus, the wavelength λ has been noted to possess the same value in the bulk of a sample and on a surface cleaved between the two Bi-O planes of consecutive slabs[9]. Similarly, this wavelength has also been found to keep its standard value in a single slab obtained isolated by an ionic-thinning procedure[10]. It is therefore licit to restrict the first step of an analysis of the modulation to the situation in a single slab.

For all values of n, the modulation can be schematically described as the superimposition of 3 types of collective atomic displacements[11] (Fig.2b).

i) an in-plane longitudinal compression-wave in the Bi-O planes. This wave forms, in each plane, bismuth-concentrated regions and bismuth-depleted ones. Note that, in the two Bi-O planes limiting a slab, the compression-waves are shifted by half a wavelength along the modulation direction $\vec{b}$ (i.e. the concentrated region of one plane faces the depleted region of the other).

336

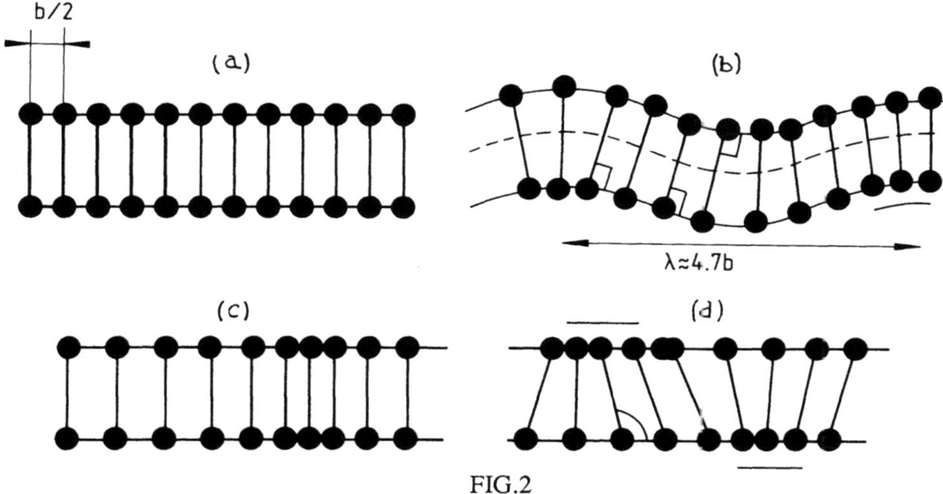

FIG.2

ii) A buckling of the Bi-O planes associated with modulated c-displacements of the atoms. In contrast with the situation in i), the buckling waves of the two Bi-O planes of a slab are in-phase with each other.

iii) Within the slab, the perovskite columns parallel to $\vec{c}$ experience quasi-rigid modulated rotations.

It has previously been noted[1] that the latter rotations within the perovskite layers, were likely to be mechanical consequences of the waves affecting the Bi-O planes. Assuming that a slab is a continuous elastic medium, let us establish this inference and extend it by showing that the entire set of features i) - iii) can be considered as simple consequences of the occurence of a compression wave in each Bi-O plane.

One is naturally led to this idea by the observation that the modulation wavelength has the same value $\lambda \approx 4.7$ b in slabs of different thicknesses[6], i.e. for different values of n. Moreover, the longitudinal Bi-displacements defining the amplitude of the compression wave are, at least, twice as large as any other atomic displacement within the slab[4].

This assigment of the triggering component of the modulation to the Bi-O compression wave differs from certain previous assignments[1] which rather considered the buckling wave as the primary effect.

The temperature independence of the modulation, and its large amplitude, entitle us to assume that we have to deal with a "zero temperature" physical problems for which energetic considerations are sufficient.

As starting point, we assume that each Bi-O plane experiences the "spontaneous" onset (the mechanism of which will be discussed below) of a compression-wave, while the perovskite layers are "passive" with respect to this onset. As attested by the occurence of short bonds between a Bi-atom and the neigbouring oxygen atom of the perovskite layer[2], the constituants of the Bi-O plane are strongly bonded to this layer. As a consequence, the Bi-O compression-wave will impose a deformation to the perovskite layer.

If the compression-waves in the pair of Bi-O planes limiting a slab were not shifted with respect to eachother (fig.2c), the deformation of the perovskite layer would also consist in a compression-wave. As the layer is assumed to be passive in the formation of the modulation, such a deformation corresponds, in a continuous approximation, to an increase of the elastic energy of the slab. Per unit volume, this increase is of the order of $\sim (C\epsilon^2/2)$, where C is a longitudinal elastic constant and ϵ is the relative contraction or expansion of the lattice along $\vec{b}$. We have $\epsilon \sim (u/a_0)$, with u the amplitude of the compression-wave and a_0 the distance between Bi-atoms in the undistorted lattice.

For a given amplitude u of the compression-wave in the Bi-O planes, a shift by half a wavelength of one of the two limiting Bi-O planes has the effect of decreasing the elastic energy of the slab by replacing the compression of the perovskite-layer by a shear of this layer (fig.2d). Per unit volume, the elastic energy of the slab is then $\sim (C'\varepsilon'^2/2)$ where C' is the shear elastic constant and $\varepsilon' \sim (u/d)$, with d the thickness of the slab. For the n=2 phase, $a_o = 2.7$ Å and d = 12.5 Å. Since C' is usually significantly smaller C, the elastic energy of the slab will be two orders of magnitude smaller in the case of a shear, for given amplitude u of the triggering Bi-O compression-wave. In the actual situation, it is not the amplitude u which is likely to be imposed by the onset of a modulation, but rather, the amount of elastic energy of the perovskite layer which can be compensated by the triggering displacements in the Bi-O planes. In this case, we deduce from the above considerations, that the resulting amplitude u' of the Bi-O wave corresponding to a shear of the slab will be an order of magnitude larger than u.

In either of its preceding formulations, this result provides an explanation for the observed π-phase-shift of the compression waves in the two Bi-O planes limiting a slab. It explains, as well, the rotations of the perovskite columns which are directly induced by the π phase-shift.

As a final step of the argumentation let us show that the periodic buckling of the slab corresponds, either to a further decrease of its elastic energy (for given u-displacement) or to a further increase of the amplitude of the Bi-O compression wave (for given elastic energy). Indeed, we can see (fig.2b), that the buckling has the effect of locally cancelling the shear of the slab by restoring the 90°-angles between the perovskite columns and the Bi-O planes. In agreement with the experimental observations, the buckling of the two Bi-O layers limiting a slab have to occur in-phase in order to preserve the length of the perovskite columns.

In the framework of the preceding description, the amplitude B of the buckling is determined by the value of the shear angle cancelled by its onset. Assuming, for instance, a sinusoidal buckling-wave, we have $(2\pi B/\sigma) \sim (u'/d)$. For the n=2 phase the available data yield $B \sim 0.25$ Å in good agreement with the experimental value $B \sim 0.2$ Å [4]. Furthermore, the preceding expression relating B to u' leads to the expectation that the ratio (B/u') of the buckling-component, and of the compression-component of the Bi-modulation should decrease when the thickness of the slab increases, i.e. for increasing values of the n-index. Consistently with this expectation, the experimental data[2-4] indicate that the ratio (B/u') decrases by a factor $\sim (0.75)$ between the phases n=1 and n=2, while the reduction factor deduced from the respective thicknesses of the slab for n=1 and n=2 is $\sim (0.81)$. Furthermore, high-resolution electron microscopy data[6] indicate that the buckling amplitudes relative to phases with different n-values can be directly compared on the images, in samples where phases having different slab thicknesses are intergrown and can be observed simultaneously. The trend of decrease with increasing n is qualitatively found to be valid.

It is worth pointing out that one also expects a decrease of the longitudinal amplitude u of the Bi-O compression wave, with increasing n, since for given value of u, the elastic energy of a slab increases with its thickness. Experimental confirmations of this feature exist[3]. A quantitative estimation of the expected decrease is difficult because it relies on the calculation, for each n-value, of the residual value of the elastic energy which is left after the shearing and buckling of the slab have taken place.

STACKING MODES OF THE SLABS

Experimental data indicate that there is a phase relationship between the modulations of consecutive slabs. This relationship depends of the n-index and also of the type of chemical substitution which is performed in the compounds.

In the standard (unsubstituted) compounds, the stacking of the slabs is such that there is a modulation of the distance separating neighbour Bi-O planes of consecutive slabs. A simple situation prevails for $n \geq 2$ in which the buckling waves of consecutive slabs are π-out of phase (fig.3a). This phase-shift can be directly observed on images of high-resolution electron microscopy[6,13]. It can also be deduced from measurements in reciprocal space.

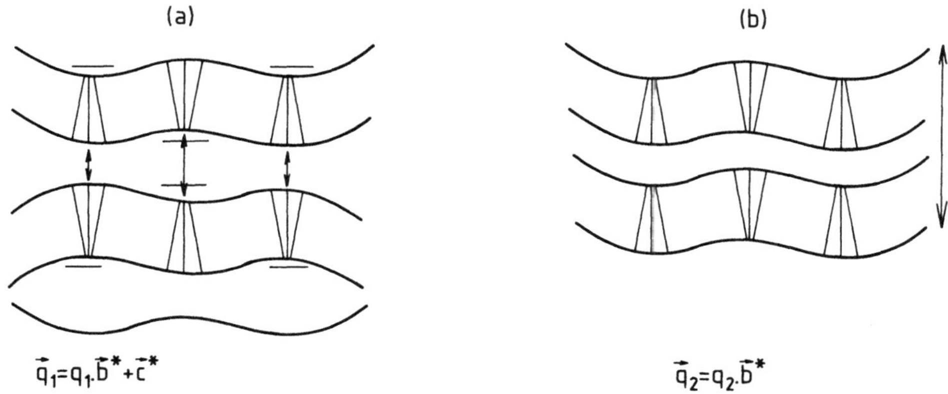

FIG.3. Stacking mode in standard and Pb-substituted compounds.

These measurements reveal[7,14] that the wavevector $\vec{q}_1$ of the modulation has a commensurate component along $\vec{c}^*$, corresponding to a twofold superstructure in the slab-ordering along $\vec{c}^*$. This wavevector is $\vec{q}_1 = \left(q_1.\vec{b}^* + \vec{c}^*\right)$, where $q_1 \approx (2\pi/\lambda_1) \approx 0.21$ is the wavenumber of the modulation within a slab. For n = 1 the situation is more complex, the wavevector of the modulation being $\vec{q}_1 = q_1\vec{b}^* + \alpha\vec{c}^*$, with α varying from sample to sample in the range 0.33 - 0.62[2,3]. In this case, the phase-shift of the buckling waves in consecutive slabs are out of phase by angles varying between $(\pi/3)$ and $(2\pi/3)$,, the most frequent value being $(\pi/2)$.

Chemical substitutions which concern essentially constituants of the perovskite layers, i.e. $Ca \rightarrow Y$ or $Sr \rightarrow Y$ or $Sr \rightarrow La$ seem to preserve the above type of stacking[15-18].

By contrast, the introduction of lead, which substitutes bismuth in the planes limiting a slab[19], give rise to a different stacking-mode. This is revealed by diffraction measurements[10,20-23] which show a modulation with wavevector $\vec{q}_2 = q_2\vec{b}^*$, deprived from commensurate component along $\vec{c}^*$ for n = 1,2,3,4. The latter feature is consistent with a type of stacking in which the buckling waves of consecutive slabs are in phase with one-another[23] (fig.3b) and in which the distance between slabs is not modulated. Very recently[24] high-resolution imaging in electron microscopy, has confirmed this stacking characteristics for n = 3.

The difference between the two above types of stacking can be correlated to a number of other differences between the lead-free and the lead-substituted compounds. Hence the single crystals of the standard compound can be cleaved easily between the Bi-O planes of consecutive slabs[9]. These crystals display a twinning habit consisting in a relative rotation around c of the two members of a twin, by an angle θ, the value of which seems to vary in an arbitrary way from sample to sample[22] (fig.4a) (θ −twist domains). The compounds also show an extensive intergrowth of distinct phases[6] where phases corresponding to different n-values occupy different slabs stacked upon eachother along c. All these effects are strikingly absent from the lead-substituted compounds. Their cleavage normal to the c-axis is difficult[22]. Their twinning consists exclusively in 90°-twist domains[22] consistent with the pseudo-tetragonal symmetry of the structure (fig.4b). Intergrowth is drastically reduced[20-23] even for a small proportion of lead substitution[20-23].

These features suggest that there is a qualitative difference in the strength of the bonding between consecutive slabs.

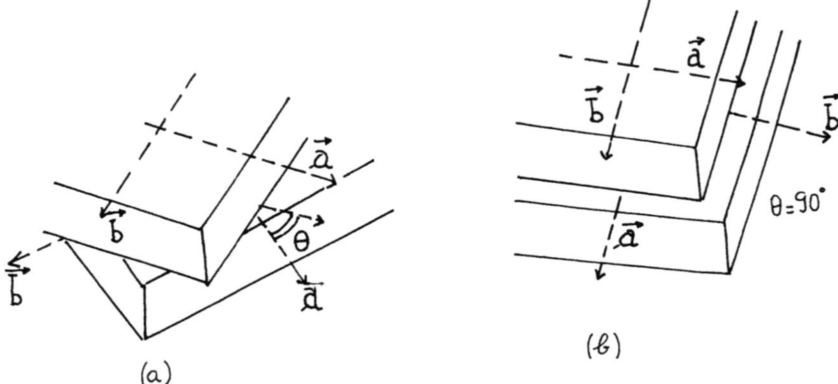

FIG.4

In the lead-free compounds, the experimental data are consistent with a very weak bonding between the slabs. More precisely, there seems to be no covalent-bonding between these structural units, as attested, from band structure calculations, by the absence of electronic density between the slabs[25]. On this basis, we can understand that the interval between consecutive Bi-O planes can be strongly modulated (2.5 - 3.2 Å) without raising significantly the energy of the system. The precise relative shift of consecutive slabs relies on secondary types of interactions, the nature of which is not clear at present.

In the lead-substituted compounds, the anisotropy of the bonding of slabs, disclosed by the twinning habit, suggests that there is a covalent-type of bonding between the BiO-planes. Such a bonding will oppose efficiently variations of the inteval between slabs and explain the observed type of stacking in which the buckling of the slabs are in-phase. Additional structural data and a more detailed-chemical analysis of the lead-substituted compounds are needed in order to elucidate the origin of the new covalent bonds which arise from the introduction of lead.

POSSIBLE MICROSCOPIC MECHANISMS OF THE 2D-MODULATION

Two possible microscopic origins of the modulated displacements in the BiO planes are worth discussing ; an electronic, and a structural one.

In two-dimensional systems, such as the Bi-O planes, a well known mechanism of formation of a modulation is the Peierls mechanim which gives rise to a charge-density-wave and an associated structural distortion. The relevance of such a mechanism to the situation in the bismuth-based superconductors was discarded by several authors on the basis of the observation that the Bi-amplitudes (~ 0.5 Å) are much larger than the usual amplitudes realized in charge density waves systems (~ 0.1 Å). In spite of this difficulty and of other difficulties pointed out hereunder, it is useful to underline that certain important requirements for a Peierls mechanim are realized in the considered compounds.

This mechanism, valid for a metal, involves the opening, at the Fermi level, of a gap in the electronic density of states, and the correlated onset of a modulated structural distortion, the wavevector of which is $2\vec{k}_F$, twice the vector locating the intersection of the considered electronic band with the Fermi level. In a two-dimensional system, the activation of this mechanism requires : i) that the undistorted system be metallic. ii) that parallel sections exist in the Fermi surface, allowing an efficient "nesting".

In the bismuth-based compounds, the undistorted (unmodulated) system of reference has the approximate structure, with tetragonal symmetry, described by Tarascon et al[26]. For this structure, a number of electronic band-structure-calculations have been performed for the n = 2 phase[25]. They all agree on the following conclusions : i) The electronic band structure has pronounced characteristics of a two-dimensional system. ii) Each BiO-plane is associated, in the electron density of states, to a partly filled band, i.e. these planes are metallic. iii) The portion of the Fermi surface corresponding to the BiO-band contains planar and parallel portions allowing the possibility of "nesting".

The necessary conditions of the Peierls mechanism are therefore satisfied. More precisely, by referring to the available results one finds that the nesting conditions defines two possible wavevectors for the modulation at $2k_f \sim 0.2$ and $2k_f \sim 0.4$, the magnitude of the first one being in good agreement with the experimental value (~ 0.21). Besides, recent experimental tunneling-microscope mesurements of cleaved single crystals have shown that in the modulated n=2 system, consistently with the opening of a gap by the Peierls mechanism, the BiO-planes are insulating, the measured gap being ~ 0.3 eV[27].

However, examination of the Fermi-surface of the undistorted structure reveals that the speculative $2\vec{k}_F$ vectors of the modulation are rotated by 45° with respect to the actual modulation wavevector. This qualitative discrepancy may indicate that the Peierl mechanism is irrelevant. It may also be interpretated as an indication that, in agreement with the large amplitude of the modulation, we are dealing with a situation of strong coupling between the lattice distortion and the electronic energy, for which the determination of the modulation wavevector is expected to be less straigthforward.

A number of possible structural mechanisms have been proposed for the onset of the modulation in the bismuth-based superconductors[28,29]. Their chemical and structural consistency has been thoroughly discussed. The most promising one is based on a remark by Zandbergen et al[29] that the Bi-O planes may be structurally frustrated. More precisely, there is a large difference between the average Bi-O distance existing in these planes, and the standard Bi-O distance which is usually found in bismuth oxides. The average Bi-O distances actually realized in the superconductor are believed to be imposed by the perovskite-layer to which the bismuth and oxygen atoms are bonded. In other terms, there is a mismatch between the "natural" Bi-O distance in the considered planes, and the Bi-O distance constrained by the presence of the relatively rigid perovskite layer (fig.5).

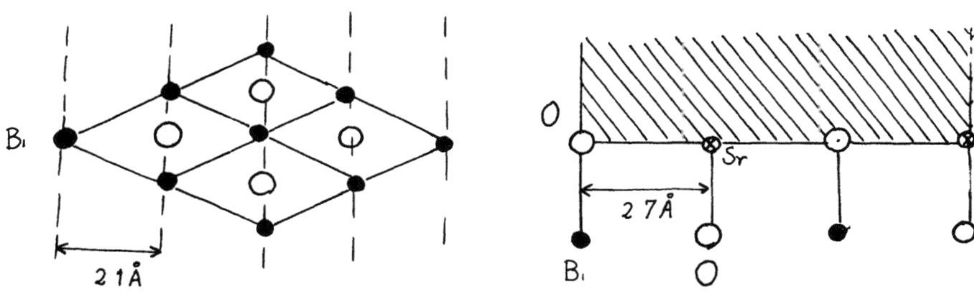

FIG.5

In a similar manner as in the well studied situation[30] of adatoms (the Bi-O planes) physically adsorbed on a substrate (the perovskite layer), the mismatch will give rise to a modulated configuration of atoms in the BiO-plane. In this framework, this modulation would coincide with the Bi-O compression-wave discussed above.

In the framework of the standard adsorption model[30], it is possible to relate the value of the modulation wavelength to the value of the "natural" mismatch between the BiO-planes and the perovskite layers. Note, however, that this model is not fully relevant to the physical situation realized in the superconducting system considered. In the latter system, each bismuth atom is strongly bonded to an oxygen atom in the perovskite layer, and there is a one to one correspondance between the available sites in the SrO plane of the perovskite layer, and the number of bismuth atoms. By contrast, in the adatom model, an incommensurate modulation will only arise when the "coverage" of the lattice is incomplete, i.e. when the number of adatoms differs from the number of lattice sites[31].

In spite of this inadequacy of the adatom model, we can use the relationship established by this model, in the limit of a "floating phase"[30], in order to estimate the approximate value of the modulation wave number in the BiO-planes as a function of the "natural" mismatch. Indeed, this relationship expresses the occurence of a superstructure common to the two lattices. This simple geometrical relationship is likely to be independent of the specific model considered. It yields :

$$q = \frac{b_0 - a_0}{b_0} \tag{1}$$

where a_0 is the "natural" distance between bismuth rows in the BiO-plane, and b_0 is the corresponding distance in the perovskite lattice, equal to half the b-lattice constant of the average structure of the material.

In the standard (unsubstituted) compounds, the lattice constant along the b-axis is ~ 5.39 Å. Hence b_0 ~ 2.695 Å. On the other hand a_0 is the typical bismuth-oxygen distance, i.e. a_0 ~ 2.1 Å [12]. We deduce from (1), q ~ 0.22 in good agreement with the experimental value of the modulation wavenumber q_1 ~ 0.21.

In the case where calcium is substituted by Yttrium, we must consider that a_0 is unchanged since the BiO-planes are unaffected. The available experimental data indicate that the b-lattice constant increases with Y-substitution, and corresponds, for total substitution, to (b_0) ~ 2.74 Å [16]. Equation (1) then determines an expected q ~ 0.235, while the experimental value is q ~ 0.25.

The case of lead-substitution requires more elaborate considerations. Experimental observations have shown that the wavenumber of the modulation decrases when the ratio (Pb/Bi) increases. For the maximum achievable substitution (Pb/Bi ~ 25 %), q is reduced[32] by a factor of 2. As stressed in the preceding paragraph, the lead substitution also induces a change of the stacking mode, associated to a strong chemical bonding between the slabs. In order to understand the correlation between these two effects, we have to refer to the fact that the Pb^{2+} ion, as well as the Bi^{3+} ion possess a chemically inactive pair of electrons[33]. Zandbergen et al[29] have emphasized that this pair points in the direction of the neighbouring slab, and that the large separation between slabs, in the lead-free compounds, (3.2 Å), corresponds to the normal Bi-O distance in the direction of the inactive electron pair.

As explained above, the substitution of Bi^{3+} by Pb^{2+} establishes locally a covalent bond between consecutive slabs, at the lead-site or in its neighborhood. Hence, the local distance between slabs will be the ordinary Bi-O distance ~ 2.1 Å. The occurence of these shorter local distances has two consequences. Firstly, there will be a decrease of the average separation between slabs, leading to a decrease of the c-lattice constant. On the other hand, there will be a reorientation of the inactive electron-pair which does not have sufficient space between the slabs. This pair will point within the Bi-O plane. Accordingly, the average Bi-O distance in this plane will increase, thus reducing the mismatch with the perovskite layer (fig.6).

In order to make the preceding description quantitative, let us assume that each substitution of a Bi^{3+} ion by a Pb^{2+} ion determines one "short" distance (2.1 Å) between slabs, and one "large" Bi-O distance (3.2 Å) within the BiO plane. Conversely, each unsubstituted Bi^{3+} ion remains associated to one "large" distance (3.2 Å) between slabs, and one "short" Bi-O bond (2.1 Å) within the plane. In this framework, we calculate, for (Pb/Bi) = 25 %, an average Bi-O "natural" distance -a_0 ~ 2.32 Å, within the Bi-O planes. Assuming that b_0 is unchanged

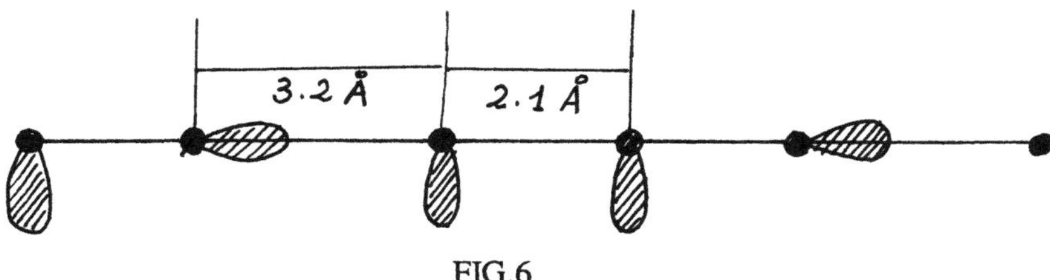

FIG.6

in eq.(1), we deduce, from the new mismatch, the wavenumber q ~ 0.14, the experimental value being q_2 ~ 0.1. On the same basis, we expect a decrease by 1.5 % of the c-lattice constant, in good agreement with the recently measured value[32].

These quantitative evaluations show that it is reasonable to assign the origin of the incommensurate modulation to the lattice mismatch between the Bi-O natural distance and the distance imposed by the perovskite layers.

Note however, that the value of the modulation wavenumber has been related, in the framework of a different description[1], to the number of oxygen atoms, in excess of stoichiometry, which are assumed to be present in the Bi-O planes. Such a relationship has indeed been checked experimentally for the n=2 unsubstituted phase[1]. It is not yet established for the other phases, or in the substituted compounds. Hence, for the Pb-substituted material, this relationship is claimed[34] to determine a disappearance of the modulation ($\lambda = \infty$) for (Pb/Bi) = 25%, while recent experimental data show, instead, that for this level of substitution, sharp modulation satellites with q ~ 0.1 are still observed [32]. We believe, in agreement with Zandbergen[29] that the excess of oxygen in the BiO planes is a secondary effect of the occurence of the modulation.

INTERMODULATION OF TWO STACKING-MODES IN Pb-SUBSTITUTED COMPOUNDS

We have already stressed that Pb-substitution gave rise to an "in-phase" stacking mode of slabs characterized by the modulation wavevector $\vec{q}_2 = q_2 \vec{b}^*$, which differs from the wavevector $\vec{q}_1 = q_1 \vec{b}^* + \vec{c}^*$ of the modulation in the standard compound.

In most diffraction experiments, the $\vec{q}_1$ and $\vec{q}_2$ wavevectors are simultaneously present in the diffraction data, though some observations[23,24] have detected $\vec{q}_2$ alone. The coexistence of $\vec{q}_1$ and $\vec{q}_2$ has first been considered as inessential, and assigned to an inhomogeneous state of the samples, where $\vec{q}_1$ and $\vec{q}_2$ correspond to different areas of a sample, in agreement with observations in direct space[23].

Recent X-ray diffraction data[35], performed on a series of single crystal samples with (Pb/Bi) content varying between 5 % and 25 %, have shown a more complicated situation. For (Pb/Bi) ~ 10 %, they have revealed the occurence of diffraction spots corresponding to the combined harmonics $(\vec{q}_1 + \vec{q}_2)$ and $(\vec{q}_1 - \vec{q}_2)$ of the two types of modulations. This result supports the idea that, for the considered level of substitution, the two types of modulations (differing by their wavelength and by the stacking mode of the slabs) are superimposed in the same microscopic area of the sample, and that they interact. It has also been observed that the intermodulation occurs for a particular ratio of the wavelength of the two types of modulations: $(\lambda_2/\lambda_1) \approx (3/2)$. Hence, the intermodulation (undetected for other values of (λ_2/λ_1) seems to be associated to a mutual commensurability of the two modulations, which both remain incommensurate with respect to the basic crystal lattice.

Such a circumstance can be given a qualitative explanation in the framework of a phenomenological approach. Indeed, the intermodulation requires some coupling between the two modulated displacements, of the form $f(\eta_1,\eta_2) \propto \eta_1^m . \eta_2^p$. The spatial homogeneity of f implies the condition :

$$m\vec{q}_1 + p\vec{q}_2 = 0 \qquad\qquad (2)$$

Considering the fact that $\vec{q}_1$ has a commensurate $\vec{c}*$ component distinct from a reciprocal lattice vector ($2\vec{c}*$ is such a vector) and that $q_1 \approx 0.22$, and $0.1 \leq q_2 \leq 0.2$, we find easily that eq.(2) can only be satisfied for m = 2 and p = 3. We then deduce $q_2 \approx 0.14 - 0.15$, in good agreement with the experimental data[35]. In this case, the simultaneous onset of the two types of modulation can lower the energy of the system by setting a non-zero negative value for the above coupling term[36].

For other values of q_2 (e.g. 0.16 or 0.12), equation (2) cannot be satisfied and the spatial average of the coupling term vanishes. A single modulation $(\vec{q}_2)$ should constitute the stable state of the system.

CONCLUSIONS

We have shown that the apparent complexity of the pattern of modulated displacements was not essential, and that this pattern was triggered by a simpler set of displacements concerning the sole BiO-planes. The mechanism of the 2D-modulation has not been fully clarified. However, we have pointed out that an electronic mechanism could not be discarded since the necessary conditions of a Peierl's instability are realized.

We have also examined, semi-quantitatively, the implications of an assignment to the structural mismatch between the BiO- and perovskite atomic layers. This assignment has still to be confirmed by an accurate quantitative fit to the results of a specific physical model differing from the standard adatom-model. Several models have been recently investigated[37], with features similar to those of the systems considered here. There adequacy remains to be analyzed.

REFERENCES

1. Y. Le Page, W.R. Mc Kinnon, J.M. Tarascon and P. Barboux, Phys. Rev. B40, 6810 (1989)

2. M. Onoda and M. Sato, Solid State Comm. 67, 799 (1988)

3. Y. Gao, P. Lee, J. Ye, P. Bush, V. Petricek, P. Coppens, Physica C160, 431 1989)

4. Y. Gao, P. Lee, P. Coppens, M.A. Subramanian, A.W. Sleight, Science 24, 954 (1988)

5. N. Yamamoto, Y. Hirotsu, Y. Nakamura and S. Nagakura, J.J.A.P. 28 L598 (1989)

6. T.M. Shaw, S.A. Shivashankar, S.J. La Placa, J.J. Cuomo, T.R. Mc Guire, R.A. Roy, K.H Kelleher and D.A. Yee, Phys. Rev. B 37, 9856 (1988). Y. Matsui, H. Maeda, Y. Tanaka, S. Horiuchi, J.J.A.P. 27, L372 (1988). Y. Matsui, S. Takekawa, H. Nozaki, A. Umezono, E. Takayama, S. Horiuchi J.J.A.P. 27 L1241 (1988). S. Ikeda, J. Sato, K. Nakamura, J.J.A.P. 28, L1398 (1989). R. Ramesh, C.J.D. Herington, G. Thomas, S.M. Green, C. Jiang, M.L. Rudee, H.L. Luo, APL 53, 615 (1988)

7. S. Horiuchi, K. Shoda, M. Iwatsuki, Y. Harada, Y. Matsui, J.J.A.P. 28, L386 (1989), P.A. Albouy, R. Moret, M. Potel, P. Gougeon, J. Padiou, J.C. Levet, H. Noel, J. Physique Comm. Brèves Comm. 49, 1987 (1988)

8. J. Schneck (unpublished)

9. Y. Matsui, H. Maeda, Y. Tanaka, S. Horiuchi, J.J.A.P. 28, L946 (1989)

10. E.A. Hewat, J. Microsc. Spectrosc. Electron. 13, 297 (1988)

11. Y. Matsui, S. Horiuchi, J.J.A.P. $\underline{27}$, L2306 (1988)

12. P. Bordet, J.J. Capponi, C. Chaillout, J. Chenavas, A.W. Hewat, E.A. Hewat, J.L. Hodeau, M. Marezio in Studies of high T_c superconductors edit. E.V. Narlikar (NOVA SCIENCE NY 1989)

13. H.W. Zandbergen, P. Groen, G. Van Tendeloo, J. Van Landuyt, S. Amelinckx, Solid State Comm. $\underline{66}$, 397 (1988)

14. K. Kawaguchi, S. Sasaki, H. Mukaida, M. Nakao, J.J.A.P. $\underline{27}$, L1015 (1988)

15. J.M. Tarascon, P. Barboux, G.W. Hull, R. Ramesh, L.H. Greene, M. Giroud, M.S. Hedge, M.R. Mc Kinnon, Phys. Rev. $\underline{B39}$, 4316 (1989)

16. H. Niu, N. Fukuohima, S. Takeno, S. Nakamura, K. Ando, J.J.A.P. $\underline{28}$, L784 (1989)

17. H.W. Zandbergen, W.A. Groen, G. Van Tendeloo, S. Amelinckx, Appl. Phys. $\underline{A48}$, 305 (1989)
S. Olivier, W.A. Groen, C. Van der Beek, H.W. Zandbergen Physica $\underline{C157}$, 531 (1989)

18. J. Darriet, C.J.P. Soethout, B. Chevalier, J. Etourneau Solid State Commun. $\underline{69}$, 1093 (1989)

19. H. Nobumasa, T. Arima, K. Shimizu, Y. Otsuka, Y. Murata, T. Kawai, J.J.A.P. $\underline{28}$, L57 (1989)

20. R. Ramesh, G. Van Tendeloo, G. Thomas, S.M. Green and H.L. Lao, Appl. Phys. Letters $\underline{53}$, 2220 (1988)

21. R. Kawaguchi, S. Sasaki, H. Mukaida and M. Nakao, J.J.A.P. $\underline{27}$, L1015 (1988)

22. J. Schneck, L. Pierre, J.C. Tolédano and C. Daguet, Phys. Rev. $\underline{B39}$, 9624 (1989); M. Clin, K. El Boussiri, J. Repszki, D. Morin, and J. Schneck Journ. Physique to be published.

23. C.H. Chen, D.J. Werder, G.P. Espinosa and A.S. Cooper, Phys. Rev. B$\underline{39}$, 4686 (1989)

24. Y. Hirotsu, O. Tomioka, N. Yamamoto, Y. Nakamura, S. Nagakura, Y. Iwai and M. Takata, J.J.A.P. $\underline{28}$, L1783 (1989)

25. H. Krakauer and W.E. Pickett, Phys. Rev. Lett. $\underline{60}$, 16656 (1988)

26. J.M. Tarascon, Y. Le Page, P. Barboux, B.G. Bagley, L.H. Greene, W.R. Mc Kinnon, G.W. Hull, M. Giroud and D.M. Hwang, Phys. Rev. $\underline{B37}$, 9382 (1988)

27. M. Tanaka, T. Takahashi, H. Katayama-Yoshida, S. Yamazaki, M. Fujinami, Y. Okabe, W. Mizukani, M. Ono, K. Kajimura, Nature $\underline{339}$, 691 (1989)

28. E.A. Hewatt, J.J. Capponi and M. Marezio, Physica $\underline{C157}$, 502 (1989)

29. H.W. Zandbergen, W.A. Groen, F.C. Mijlhoff, G. Van Tendeloo and S. Amelinckx, Physica $\underline{C156}$ 325 (1988)

30. P. Bak, Reports on Progress in Physics $\underline{45}$, 587 (1982)

31. S. Aubry (Private communication)

32. L. Pierre, J. Schneck, D. Morin, J.C. Tolédano, J. Primot, C. Daguet, H. Savary, Ferroelectrics to be published

33. A. F. Wells, Structural Inorganic Chemistry, Clarendon Press (Oxford 1962) p.23

34. N. Fukushima, H. Niu, S. Nakamura, S. Takeno, M. Hayashi, K. Ando, Physica $\underline{C159}$, 777 (1989)

35. L. Pierre, J. Schneck, J.C. Tolédano, C. Daguet, Phys. Rev. B to be published 1st January 1990

36. J.C. Tolédano and P. Tolédano, The Landau Theory of Phase Transitions, World Scientific (Singapore 1987)

37. S. Aubry, K. Fesser, A.R. Bishop, J. Phys. $\underline{A18}$, 3157 (1985)
P.G. de Gennes, J. Physique $\underline{44}$, L-657 (1983)

INCOMMENSURATE - COMMENSURATE PHASE TRANSITION

OF $Cu_{2-x}Te$. ($x < 0.05$)

N. Vouroutzis and C. Manolikas

Department of Physics
Solid State Section
Aristotle University of Thessaloniki
54006 Thessaloniki - Greece

INTRODUCTION

In the temperature range 25°C to 600°C, $Cu_{2-x}Te$ ($x<0.05$) undergoes a sequence of phase transformations schematically shown in Fig. 1.[1] The highest temperature phase (ε-phase) and the one stable at room temperature (α_I-phase) are both

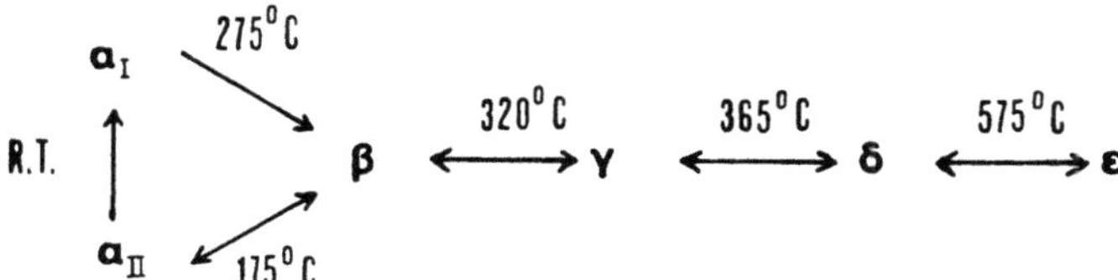

Fig. 1. Phase transition paths of $Cu_{2-x}Te$
($x < 0.05$).

based on a f.c.c. substructure of Te atoms. However between them a number of phases exist (δ, γ, β, α_{II}) all of which exhibit the same hexagonal basic structure and their "average" structure can be modeled after the model proposed by Novothy.[2] In this model Te atoms form trihedral prisms stacked along the c-axis with every second layer of them filled by pairs of Cu atoms located on lines parallel to the c-axis (Fig. 2).

We present here results obtained by means of electron diffraction and microscopy which indicate that the sequence of the "hexagonal" phases presents a normal-incommensurate-commensurate phase transformation.

OBSERVATIONS

Exploration of the reciprocal space by means of electron diffraction, has allowed for the reproduction of the

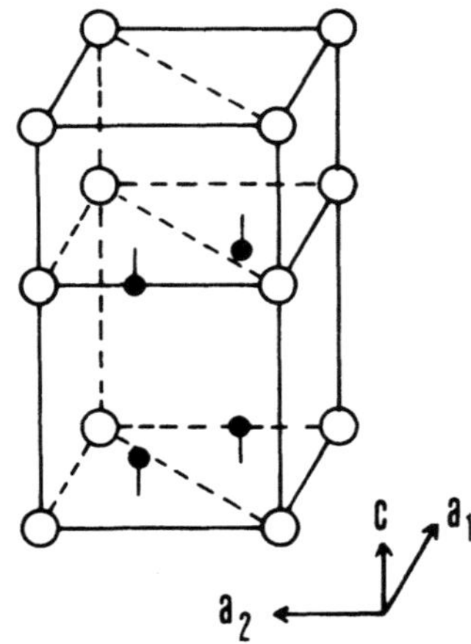

Fig. 2. Novotnys model for the structrure of Cu2Te.

reciprocal lattice of the three phases δ, γ and β shown schematically in Fig. 3. The large circles correspond to reflections of the basic hexagonal structure whereas the small ones correspond to sattelites. The (1010)* and (1230)* sections of these reciprocal lattices can be directly compared with the observed ones depicted in Fig. 4.

It is easily seen that only streaks of diffuse scattering parallel to the c*-axis, exist in δ-phase. Maxima of this diffuse intensity are observed at certain non-commensurate positions. The corresponding incommensurability decreases on approaching from above the transition temperature to the γ-phase.

In γ-phase the streaks of diffuse intensity tranform into rows of sharp irrational sattelite spots parallel to c*-axis. Simultaneously, fragmentation of the crystal into domains takes place, which differ 120° in the orientation of the structure on the basal plane. The latter confirms the orthorombic symmetry of γ-phase.

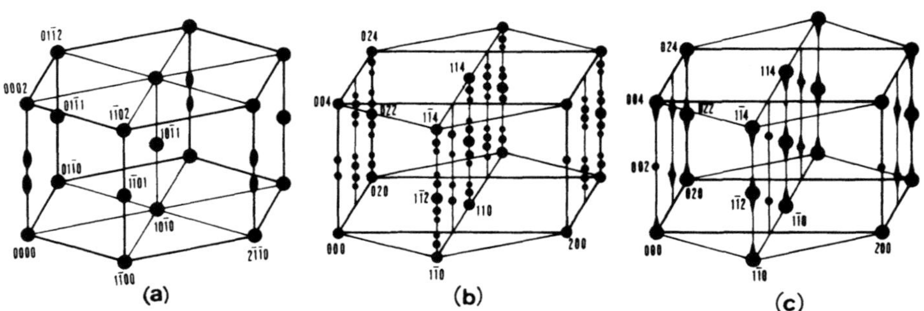

Fig. 3. Reconstruction of reciprocal space as obtained from electron diffraction analysis, a) δ-phase, b) γ- phase, c) β-phase. Large dots correspond to reflections of the basic hexagonal structure whilst small ones correspond to satellites.

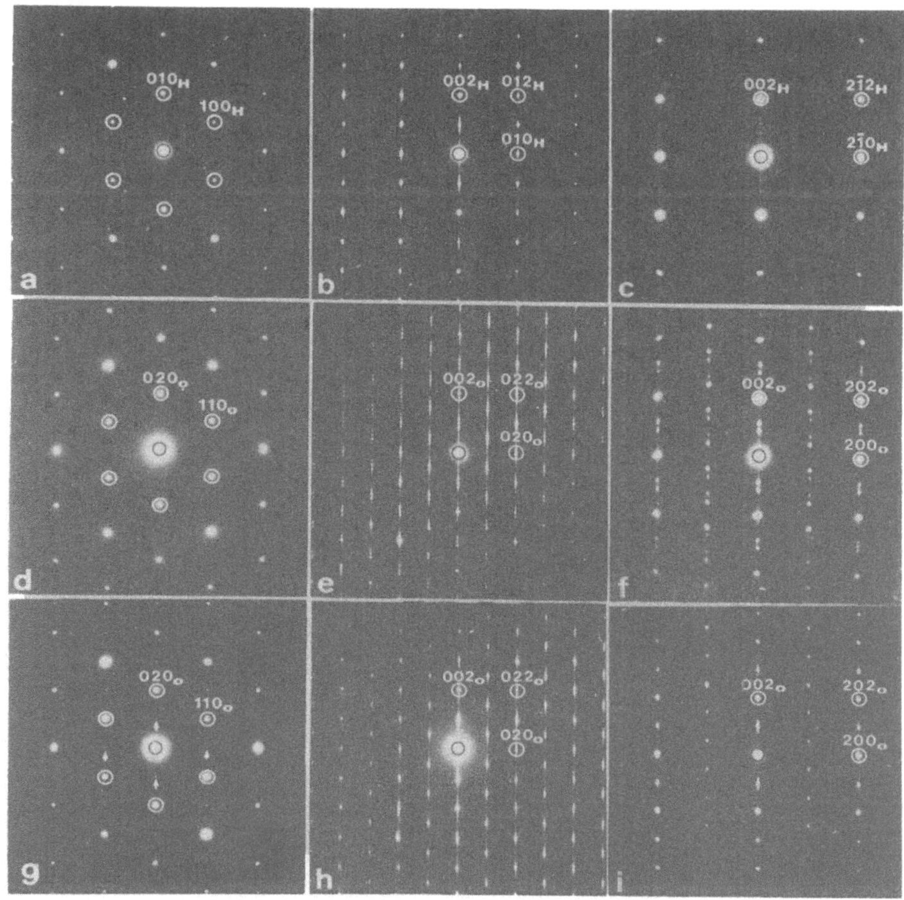

Fig. 4. Electron diffraction patterns of the three high
temperature "hexagonal" phases corresponding to
the [0001]-zone (first column), [1010]-zone
(second column) and [1230]-zone (third column).
(a,b,c) δ-phase, (d,e,f) γ-phase, (g,h,i)
β-phase.

On the other hand dark-field images made by use of the
irrational sattelite reflections, reveal that the modulation
along the c-axis is due to regularly spaced interfaces,
perpendicular to the c-axis (Fig. 5a). Occasionally adjacent
pairs of these interfaces end inside the crystal suggesting
that they are translation interfaces with a displacement
vector equal to 1/2 [111]$_o$. Similar configurations of
interfaces were observed in numerous of other materials
(TaSe$_2$, NbTe$_4$ etc)[3,4,5,6] and described as discommensuration
(DC) arrays.

Upon cooling into the β-phase very few irregular,
immobile DCs remain (Fig. 5). Most probably they are pinned
by point defects in the crystal.[6] The irregularity in the DCs
array has as a result the streaking of the spots along the
c*-axis (Fig. 4). Also the appearence of residual
incomensurability of the order of 0.01 to 0.05 in the β-phase

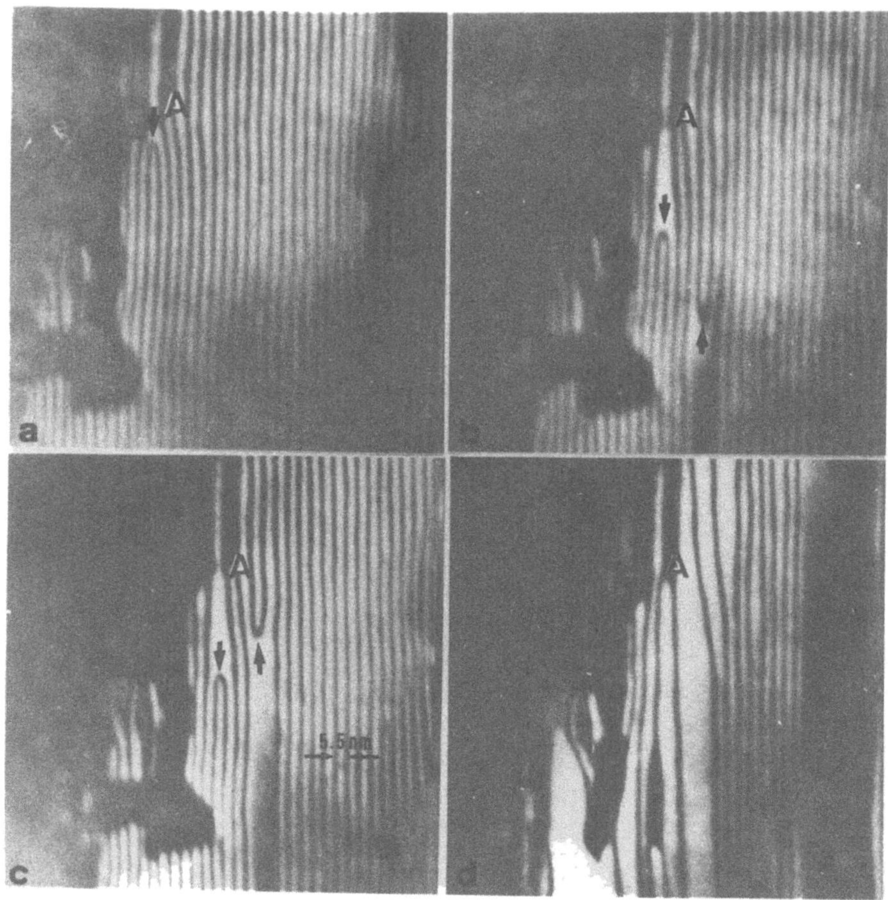

Fig. 5. Four successive patters of DCs array upon cooling
in the vicinity of γ -> β phase transition.

must be attributed to them, in the sense that an existing
commensurate structure of β-phase with c = 2c₀ becomes
slightly incommensurate by the presence of DCs. Similar is
the case of Ba₂NaNb₅O₁₅ which has a quasi-commensurate
structure at room temperature.[8]

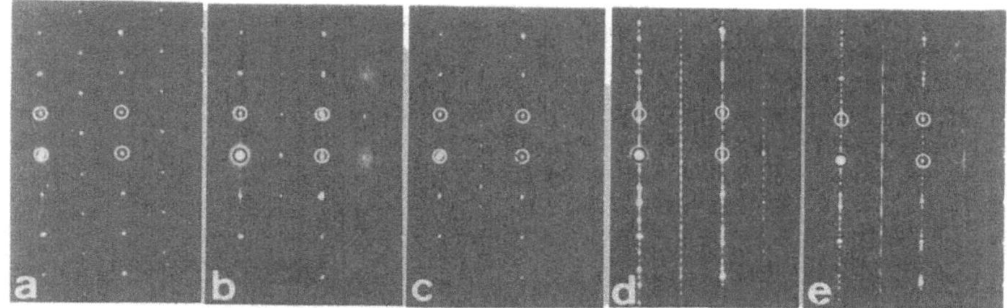

Fig. 6. Diffraction patterns of various room temperature
phases: a) commensurate β-phase, b) 2c₀ phase, c)
4c₀-phase, d) 5c₀-phase, c) 7c₀-phase.

By cooling the β-phase below 175°C, numerous of commensurate modulated along the c-axis phases are formed. As shown in Fig. 6 superstructures with the c-axis equal to $2c_o$, $4c_o$, $5c_o$ and $7c_o$ exist at room temperature. In addition, a superstructure corresponding to the commensurate β-phase is occasionally observed.

DISCUSSION AND CONCLUSIONS

The electron diffraction and microscopy observations reveal that $Cu_{2-x}Te$ is a typical one-dimensional modulated system. Most probably the modulation is due to clustering of the copper atoms along the c-axis and formation of charge density waves. However since neither exact structural data nor the temperature dependence of physical properties are known, at present one can only speculate on this point.

The sequence of phase transformation can be included in the following general schema[7]:

Commensurate	T_L	Incommensurate	T_I	Normal
phase	<-----	phase	<----→	phase
(a_{II}-phase)	170°C	(β- and γ-phases)	365°C	(δ-phase)

The normal phase corresponds to the high temperature δ-phase, which upon cooling below T_I transforms into the incommensurate phase. The peculiarity of the latter phase is that it consists of two phases with a transition between them at 320°C. One of them (γ-phase) is characterized by regularly spaced DCs, whereas the other one (β-phase) contains a number of irregular DCs which are pinned rigidly to sites of the substrate structure. The posibility of such splitting of incommensurate phases has been prooved out theoretically by Pokrovsky.[8]

The presence of numerous of commensurate a_{II}-phases is questionable at moment. More observations particularly by high resolution electron microscopy are needed in order to clarify this point.

REFERENCES

1. N. Vouroutzis and C. Manolikas. "Phase transformations in cuprous telluride" Phys. St. Sol. (α) 111, 491 (1989).
2. H. Novotny, "Die krystallstructur von Cu_2Te" Z. Metallkunde, 37, 40 (1946).
3. C.H. Chen, J.M. Gobson and R.M. Fleming. "Microstructure in the incommensurate and the commensurate charge-density-wave states of 2H-TaSe2: A direct observation by electron microscopy". Phys. Rev. B26, 184 (1982).
4. H. Bestgen, "Direct observation of discommensurations in the incommensurate superlattice of ferroelectric Rb2ZnCl4 by transmission electron microscopy", Sol. Stat. Comm., 58, 197 (1986).
5. J. Mahy, J. Van Landuyt, S. Amelinckx, Y. Uchida, K.D. Bronsema and S. van Smaalen, "Direct observation of discommensuration arrays in NbTe4 by means of low-temperature electron microscopy", Phys. Rev. Letters, 55, 1188 (1985).

6. G. Van Tendeloo, S. Amelinckx, C. Manolikas and Wen Shulin, "The direct observation of "discommensurations", In Barium Sodium Niobate (BSN) and its homologues", Phys. St. Sol. (a) 91, 483 (1986).

7. J. D. Toledano, "Defauts et transitions des phases structurales", Ann. Telecomm. 39, 278 (1984).

8. V.L. Pokrovsky, "Splitting of comensurate-incommensurate phase transition", J. Physique, 42, 761 (1981).

REVIEW OF THE COMMENSURATE AND INCOMMENSURATE TRANSITIONS IN BCCD

M.Renata Chaves and Abílio Almeida

CFUP (INIC) Universidade do Porto
Praça Gomes Teixeira, 4000 Porto Portugal

INTRODUCTION

Several addition compounds of the amino acid betaine and inorganic acids were found in last years to possess structural ferroelastic, ferroelectric or antiferroelectric phase transitions. These materials are betaine phosphate (antiferroelectric), betaine arsenate (ferroelastic and ferroelectric), betaine borate (ferroelastic) and betaine phosphite (ferroelectric).[1]

Betaine calcium chloride dihydrate $(CH_3)_3NCH_2 COO.CaCl_2.2H_2O$ (BCCD) contains no inorganic acid but a salt. BCCD is orthorhombic at room temperature in the centrosymmetric space group Pnma (a=10.97; b=10.15; c=10.82 Å), with four molecules per unit cell.[2] It was reported that dielectric measurements on this material along the **b** axis revealed anomalies at eight different temperatures $T_1=T_i=164K$, $T_2=127K$, $T_3=125K$, $T_4=116K$, $T_5=75K$, $T_6=51K$, $T_7=47K$ and $T_8=43K$. Along the **a** axis one pronounced peak occurs at $T_4=116K$.[3] These anomalies were associated with different phases and a hysteresis loops study showed that some of them are polar along the **b** axis while other ones are polar along the **a** axis; BCCD also exhibits non-polar phases.[3,4]

These phases were investigated by conventional X-ray and the satellite reflections show modulated structures in **c** direction.[5,6] The strongest superstructures reflections occur in the planes (C k1) and (1 k1).

Between 164K and 127K the modulation wave number $\delta(T)$ $(\mathbf{k}=\delta(T)\mathbf{c}^*)$ changes from 0.32 to 0.285 (INC1). Between 127K and 125K there is a commensurate region $\delta(T)=2/7$. Between 125K and 116K there is a further continuous change of the modulation wave number, $0.285 > \delta(T) > 0.25$ (INC2). Below 116K phases with $\delta(T)=1/4$ (116K > T > 73K), $\delta(T)=1/5$ (73K > T > 47K) and $\delta(T)=1/6$ (60K>T > 47K and T < 43K) were found.

Between 43K and 47K the intensity of the satellites drops to zero without change of the modulation wave number outside this region. When we go down in the temperature the half width of the satellites increases strongly near the phase transitions and apparently the range of incommensurate phases is expanded by an electric field (E=6kV/cm).[6]

Geometry and Thermodynamics
Edited by J.-C. Tolédano
Plenum Press, New York, 1990

The present work is a survey of recent experiments which clearly evidence an especially rich variety of different commensurate phases (C) and incommensurate phases (INC) in BCCD; this material is at present the best illustration of an incomplete devil's staircase behaviour.

Some theoretical concepts and ideas relevant to the discussion of the experimental studies will be briefly reviewed in the next paragraph. The last sections deal with some aspects of the theoretical models used to describe the observed phase transition sequence in BCCD.

SYMMETRY ANALYSIS OF THE PHASE TRANSITION SEQUENCE IN BCCD

By analogy with other compounds with INC phases, Perez-Mato[7] assumed that the commensurate modulated phases in BCCD correspond to the eventual lock-in of the modulation wave number into simple rational values while the symmetry of the structural distortion is kept essentially unchanged except for its wavelength.

In the language of Landau theory an order parameter could then be defined for each distorted phase, its symmetry corresponding to the same irreducible representation of the point group C_{2v}.

The point group of k is C_{2v} and the star of the vector k contains only the two non-equivalent vectors k and $-k$. In consequence the order parameter must be transformed according to one of the four possible physically irreducible representations of the group Pnma generated by each of the four irreducible representations of the point group of the vector k.

As BCCD reflections $(0,k,1,m)$ satisfy the condition $k + 1 =$ even while in the plane $(h,0,1,m)$ no satellites are observed, Perez-Mato concluded, using superspace group theory and by taking into account the selection rules of the satellites that the active irreducible representation is Λ_3 ($\chi(E)=1$, $\chi(C_{2z})=-1$, $\chi(\sigma_y)=-1$, $\chi(\sigma_x)=1$).[7]

If it is assumed that the symmetry of the two component order parameter $(Q= r\, e^{i\phi}, Q^*= r\, e^{-i\phi})$ is kept constant as the temperature is decreased, the possible space groups of these commensurate phases can be deduced from the condition:

$$[\{R\,|\,t\}]\begin{bmatrix} Q \\ Q^* \end{bmatrix} = \begin{bmatrix} Q \\ Q^* \end{bmatrix}$$

where $\left[\{R\,|\,t\}\right]$ is an element of the respective space group. The calculus of these space groups are rather trivial and will depend on the phase ϕ .[7,8] By writing $\delta(T)$ as the ratio of two prime integers, the three sets of the possible space groups for the corresponding commensurate phases are indicated in Table 1 (t,m integers).

Table 1.   Space groups for the commensurate phases in BCCD

$\delta(T)$	$\phi = 0$	$\phi = \pi/2$	$\phi =$ arbitrary	$\phi=(2t+1)\pi/4m$
odd/odd	P 11 2_1/n	P $2_1 2_1 2_1$	P 112$_1$	——
even/odd	P 2_1/n 11	P n 2_1 a	P n11	——
odd/even	P 1 2_1/c 1	——	P 1 c 1	P 2_1 c a
(2t+1)/2m				

By assuming that the space groups with ϕ arbitrary are non relevant[8], we should point out that the polar phases with δ(T)=odd/even have the electric polarization (**P**) parallel to the **a** axis while the polar C phases with δ(T)=even/odd have **P** parallel to the **b** axis; the δ(T)=odd/odd C phases are necessarily non polar. An electric field along the **b** axis should increase the range of stability of the phases polar in **b** direction. Similar effects are expected in **a** direction.[7]

EXPERIMENTAL RESULTS

<u>General remarks</u>

In the following we present a review of some experimental results concerning pyroelectric effect, dielectric constant, measurements of thermal expansion and elastic properties, electron paramagnetic resonance and Raman scattering.

As we have seen previously, the commensurate phases have a definite polar character in BCCD and so the pyroelectric and dielectric measurements are particularly adequate to identify the high order commensurate phases in this system. The existence of high order C phases in a narrow temperature range of stability have been predicted by microscopic models developed for incommensurate systems.[9,10,11]

<u>Macroscopic technics; influence of external parameters</u>

<u>Pyroelectric effect.</u> Pyroelectric current was measured with a short-circuited technic using an electrometer with a resolution of about 10^{-14} A.[12]

Figure 1 shows the pyroelectric coefficient (p_b) along the **b** axis as a function of temperature for zero d.c. field (E).[13,14] Anomalies of p_b(T) occur at T_2=127K, T_3=125K, T_4=116K, T_4'=115K, T_4''=114.7K, T_5=77K, T_5'=75.5K, T_6'=56k, T_6=55K, T_7=48K and T_8=45K. Figure 2 (a,b,c) depicts these anomalies in a enlarged scale.[14] The limits of the commensurate phase δ(T)=2/7 are marked by small pulses of opposite sign of the pyroelectric current at T_2=127K and T_3=125K, revealing the polar nature of that phase. Below this phase the integration over temperature of the pyroelectric coefficient shows that INC2, δ(T)=1/4, δ(T)=1/5 and δ(T)=1/6 are nonpolar along **b**. At T_7=48K, p_b(T) shows an intense peak reaching values of the order 0.3 $\mu C/(cm^2 K)$ followed by a secondary peak at T_8=45k. The anomaly at T_7 marks the onset of a ferroelectric phase (P ~ 1 $\mu C/cm^2$), also observed by hysteresis loop measurements.[3]

In spite of the fact that no spontaneous polarization along **b** is ascribed to the three commensurate phases occuring between T_4 and T_7, the pyroelectric coefficient exhibits clear anomalies in the vicinity of the transition points. On this basis and relying on the correspondence between polarization direction and commensurate wave number stressed above, it has been conjectured that three narrow commensurate phases could exist between the phases INC2 and δ(T)=1/4, δ(T)=1/4 and δ(T)=1/5 and δ(T)=1/5 and δ(T)=1/6 with δ(T)=m/n (m even; n odd). The values δ(T)=4/15, δ(T)=2/9, and δ(T)=2/11 were tentatively assigned to these phases. The anomaly observed at T_4''=114.7K can be associated with the phase δ(T)=6/23 sandwiched between the phases δ(T)=4/15 and δ(T)=1/4.[15]

In what concerns the effect of a d.c. bias field along **b** the results obtained are also in agreement with the predictions of the symmetry analysis[8] reported by Perez Mato.[7] The temperature range of stability of the polar phases along **b** increases as can be seen in figure 2(d,e,f).

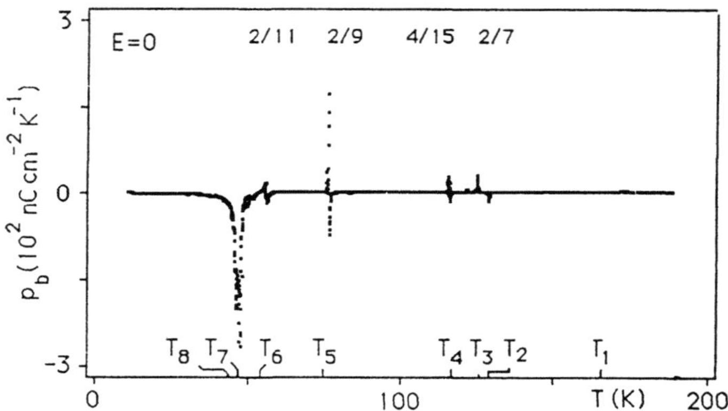

Figure 1. Pyroelectric coefficient along the **b** axis versus the temperature.

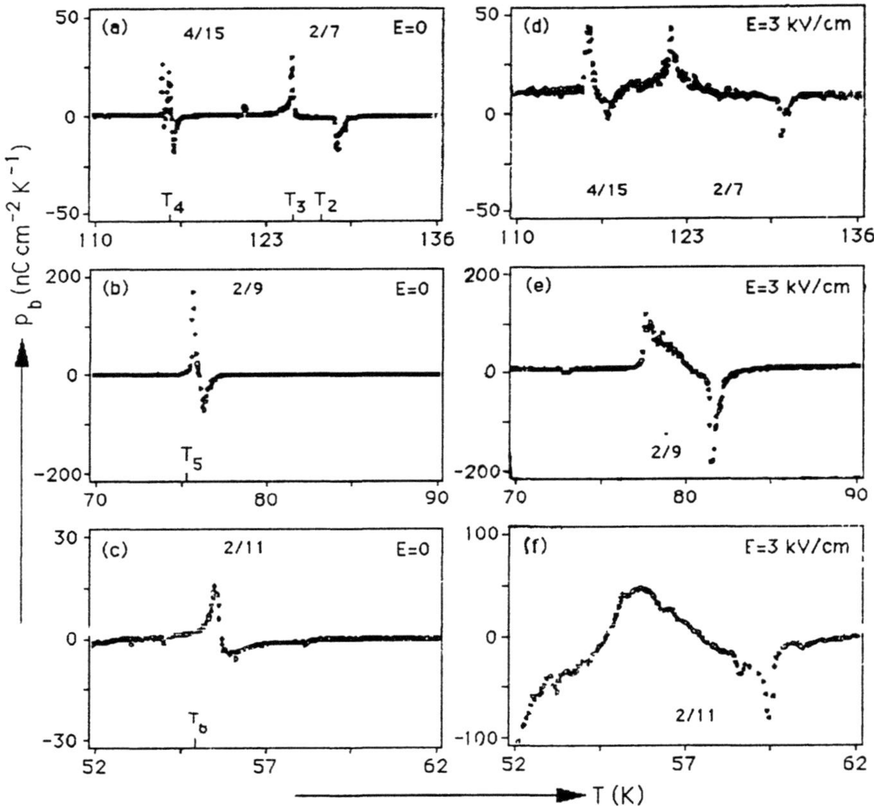

Figure 2. Temperature dependence of pyrolectric coefficient: (a), (b) and
(c) show in detail results depicted in figure 1; (d),(e) and (f)
show the effect of a d.c. electric field on the high-order
commensurate phases.

The measurement of the pyroelectric coefficient for zero d.c. field versus temperature along the **a** axis revealed clearly the existence of four anomalies which limit the onset and anihilation of the $\delta(T)=1/4$ and $\delta(T)=1/6$ C phases (see figure 3).[13,16] A very small anomaly is seen below the range of stability of the $\delta(T)=1/6$ phase. Under an applied d.c. electric field in the **a** direction, the temperature ranges of stability of the polar phases along this direction ($\delta(T)=1/4$, $\delta(T)=1/6$) seem to be slightly increased by the bias field. The small anomaly referred above is more explicit under a bias electric field as can be seen in figure 4; relying on the correspondence between polarization direction and wavevector parity it may be associated with the $\delta(T)=1/8$ C phase.[16]

<u>Dielectric constant</u>. As we shall report in the following a dielectric constant study confirms the main conclusions drawn by the analysis of the experiments concerning the pyroelectric effect.[17]

Dielectric constants along **a** (ε_a) and **b** (ε_b) axes is measured at a constant frequency of 10 kHz using an a.c. voltage of 1.5 volts. The temperature dependence of ln ε_b is shown in figure 5. As can be seen ε_b shows anomalies at $T_1=164K$, $T_2=126K$, $T_3=124K$, $T_4=116K$, $T_4'=115K$, $T_4''=114.8K$, $T_5=77K$, $T_6=56K$, $T_7=51K$ and $T_8=47K$. $\varepsilon_a(T)$ shows a single clear anomaly at $T_4'''=114K$. The higher temperature anomalies ($T>114K$) shown in more detail in figures 6 and 7 reflect the sequence of phase transitions from the reference phase down to the commensurate $\delta(T)=1/4$ phase.

The general form of the Landau free energy density for a given modulated phase in BCCD must verify the same symmetry constrains as of Rb_2ZnCl_4 family[18,19] and so it is expected that the anomalies near the lock-in transitions may be described by the same equations:[17]

$$\varepsilon = \bar{\chi}_0 + \frac{\bar{\chi}_0^2 \xi^2}{\bar{\chi}_0 \xi^2 - \gamma}\left[\frac{E(m)}{(1-m)^2 K(m)} - 1\right],$$

$$\frac{m^{1/2}}{E(m)} = \frac{4 r_o^{n-1}\sqrt{\chi(\bar{\chi}_0 \xi^2 - \gamma)}}{\pi \delta};$$

(eqns.1)

we shall represent

$$A = \frac{\bar{\chi}_0^2 \xi^2}{\bar{\chi}_0 \xi^2 - \gamma}.$$

Just above the lock-in transitions, the distance between solitons becomes large and its repulsive interaction weakens. Hence, in this region, an electric field can change easily the relative size of positive and negative domains, wich explains the large susceptibilities observed (T_2, T_3, T_4).

The anomalies in ε_b at $T_2=126K$ and $T_3=124K$ (see figure 6) reflect the lock-in of the modulation wave number at the value $\delta(T)=2/7$. For this commensurate phase the coupling between the primary order parameter and P_y is allowed by symmetry and therefore ε_b is expected to be sensitive to the corresponding lock-in transition. The high value of the dielectric constant observed between T_2 and T_3 suggests that the soliton density remains non-zero in the whole of the temperature range of stability of the phase

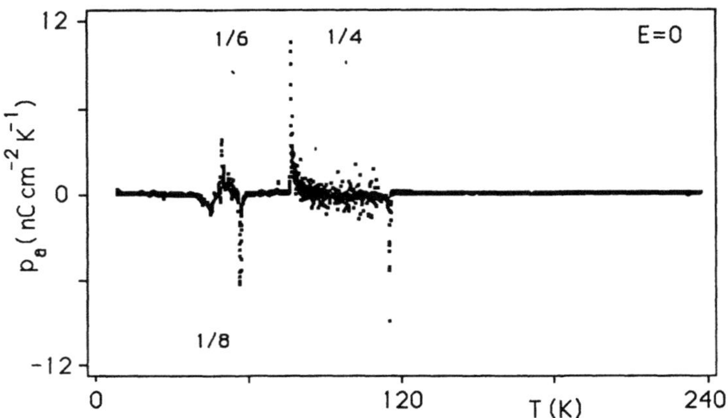

Figure 3. Pyroelectric coefficient along the **a** axis versus temperature.

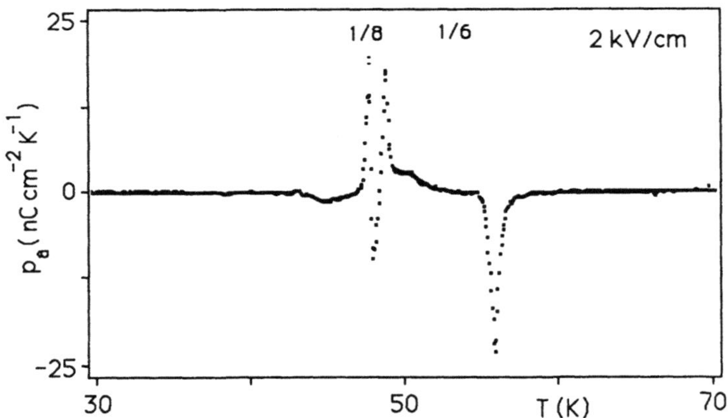

Figure 4. Pyroelectric coefficient along the **a** axis versus temperature under a bias field $E_a = 2$ kVcm^{-1}.

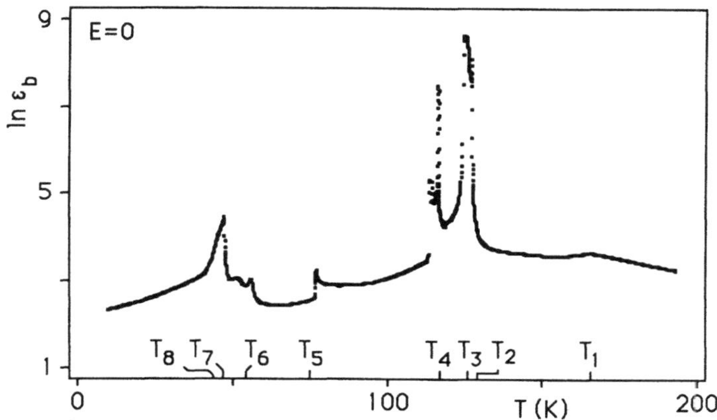

Figure 5. ln ε_b as a function of the temperature.

$\delta(T)=2/7$. Figure 6 depicts experimental data and the fitting with a theoretical curve calculated from eqns.1 with n=7 and A=0.45.

As referred above $\varepsilon_b(T)$ (figure 7) displays three distinct anomalies at T_4=116K, T_4'=115K and T_4''=114.8K. These three anomalies are well resolved and are separated by a sharp decrease of the value of $\varepsilon_b(T)$ at about T=115.6K. In contrast, $\varepsilon_a(T)$ shows a relative maximum at T_4'''=114k and displays only a small shoulder at T_4=116k.[17] This situation clearly suggests a more complex phase sequence than a single transition between the phases INC2 and $\delta(T)=1/4$. Relying on the previous analysis it can be assumed that the anomalies observed in $\varepsilon_b(T)$ at T_4=116K and T_4'=115K mark the onset and disappearence of a commensurate phase of the type $\delta(T)$=even/odd; ($\delta(T)=4/15$). The existence of this lock-in transiticn can be checked by the study of the critical behaviour of dielectric constant by fitting eqns.1 to the experimental data. The figure 7 shows the curve calculated from these equations in the range between 116K and 120K, by assuming a lock-in transition to a phase $\delta(T)=4/15$ and considering A and $\overline{X}_o$ as adjustable parameters. The theoretical curve can describe the experimental data in the region between 116K and 118K, that is, in the vicinity of the lock-in temperature. The anomaly in ε_a at T_4''=114K marks the onset of the commensurate phase $\delta(T)=1/4$ as P_x is a possible secondary order parameter. From the difference in temperature between T_4'' and T_4''' it can be conjectured that an additional phase, may exist sandwiched between the phase $\delta(T)=4/15$ and $\delta(T)=1/4$: $\delta(T)=6/23$.[15,17]

The temperatures T_5 and T_6 have been associated with the transition from $\delta(T)=1/4$ to $\delta(T)=1/5$ and from $\delta(T)=1/5$ to $\delta(T)=1/6$ respectively. For these transitions P_y cannot be coupled to the primary order parameter and therefore no anomalies in $\varepsilon_b(T)$ should be expected at these transition points. Considering the correspondence between the polarization direction and commensurate wave number parity it was conjectured that the anomalies observed at those temperatures could be associated with the existence of the two C phases $\delta(T)=2/9$ and $\delta(T)=2/11$.

Unruh et al.[15] did a very detailed experimental study of the temperature dependence of dielectric constant. Along the **b** axis the results obtained are in close agreement with those referred above, while the dielectric constant measured along the **a** axis depicts a large number of anomalies in this direction.

With an applied electric bias field parallel to the **a** axis no remarkable influence could be detected on the temperature range of stability of the polar phases in **a** direction. The widths of C phases, if spontaneously polarized along **b**, increase at the expense of neighbouring C or INC phases under a bias field along **b**.[16]

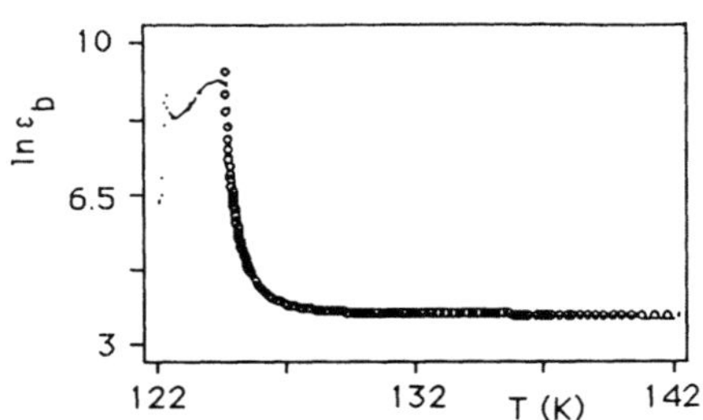

Figure 6. Logarithm of ε_b as a function of the temperature in the vicinity of the transition INCl to the $\delta(T)=2/7$ phase. The experimental data (circles) obtained from eqns.1 with A=0.45 and n=7.

Similarly to other modulated structures an anomalous thermal hysteresis at the lock-in transitions in BCCD is observed.[20] Figure 8(a) show plots of $\ln \varepsilon_b(T)$ near the lock-in transitions. Figure 8(b) clearly shows two anomalies at 126.9K and 124.5K for increasing temperature and at 122.3K and 124.6K for decreasing temperatures. As it has been seen before anomalies define the stability range of temperatures of the phase $\delta(T)=2/7$. The different shapes of the experimental curves suggest that the pinning of the discommensurations may induce different structures on heating and on cooling runs.

At lower temperatures (figure 8c), the plot of $\ln \varepsilon_b$ versus T shows two sharp anomalies at 115.6K and 115.1K and a shoulder at 114.0K on the cooling run while only two anomalies are observed at 114.3K and 115.1K on the heating run. The two sharp anomalies observed on both runs can be associated with the onset of two commensurate phases $\delta(T)=4/15$ and $\delta(T)=6/23$. The small shoulder, observed only for decreasing temperature measurements, seems to mark the stabilization of the phase $\delta(T)=1/4$. This anomaly in $\varepsilon_b(T)$, which is not expected in thermal equilibrium as the coupling between the primary order parameter and the polarization P_b is not allowed by symmetry for the phase $\delta(T)=1/4$, seems to be a direct consequence of non equilibrium configurations: this transition is "labeled" in $\varepsilon_b(T)$ by the metastable persistence of discommensurations[21] below the range of stability of the phases $\delta(T)=4/15$ and $\delta(T)=6/23$.

The existence of memory effects in BCCD can also be put in evidence by creating local non-equilibrium configurations and by observing that the "memory" of these chaotic states can persist in subsequent measurements. Dielectric constant and pyroelectric measurements clearly have shown such effects.

Additional C and INC phases have been observed by measuring the dielectric constant $\varepsilon(T)$ under hydrostatic pressure at 10 kHz.[22] Branching processes were clearly observed near the phase transitions $\delta(T)=1/4$ to $\delta(T)=1/5$, $\delta(T)=1/5$ to $\delta(T)=1/6$ and $\delta(T)=1/6$ to $\delta(T)=1/7$. At increasing

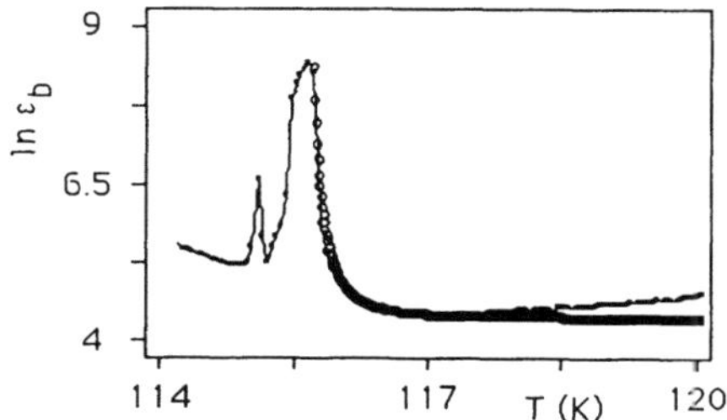

Figure 7. $\ln \varepsilon_b$ versus T near the transition INC2 to the $\delta(T)=4/15$. The experimental data (dots) are compared with a fitted curve (circles) obtained from eqns.1 with A=0.165, $\overline{X}_o$=77 and n=15.

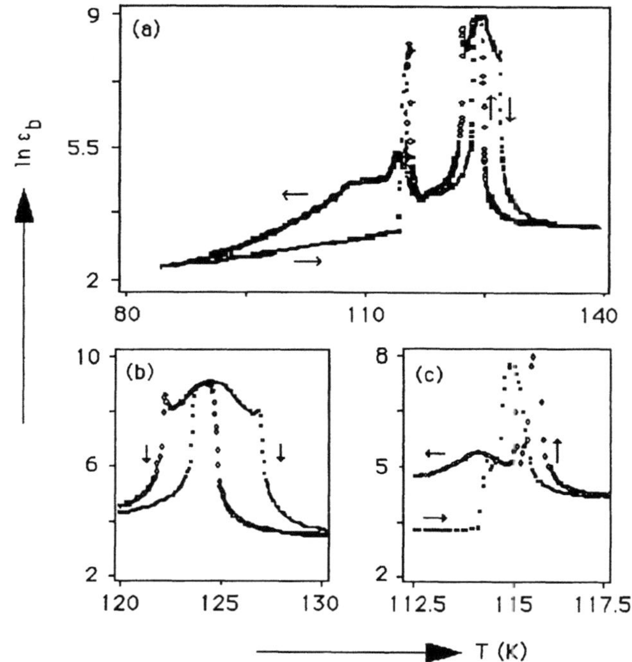

Figure 8. (a): $\ln \varepsilon_b$ versus T for increasing and decreasing temperatures. (b) and (c): details of the anomalies displayed in (a).

pressure, from 0.1 MPa to 489 MPa, the single anomalies observed at these phase transitions increase and broaden, shifting to higher temperatures and separating into several new very sharp anomalies.[23] Finally the anomalies decrease again and some of them disappear completely in the dielectric background.

Thermal expansion and elastic properties. An experimental study of thermal expansion along **c** direction between 100K and room temperature clearly reveals anomalies at 127.7K, 124.4K, 117.4K, 116.2K, 115.8K and 115.4.[15] Within C phases the linear expansion coefficient has a definite smaller value as compared to adjacente INC phases. Measurements between 40K and 80K of thermal expansion confirm C phases to exist with δ (T)=1/4, 2/9, 1/5, 1/6, 1/7 and 0/1. Phases with δ(T)=2/11, 2/13 and 1/8 were not detected in a study of thermal expansion.[24]

The temperature dependence of elastic constants was studied between 100 and 300K and a sharp softening of all ultrasonic waves having longitudinal components of the displacement vector was observed.[25]

Microscopic technics

EPR in Mn^{2+} doped BCCD. These results concern Mn^{2+} (S=5/2, I=5/2) doped BCCD crystals where Mn^{2+} ions replace Ca^{2+} in the crystalline structure. The doped crytals were grown from a solution with a molar ratio on Mn^{2+}/Ca^{2+} of the order 10^{-3}. The measurements were done using an X-band frequency of 9.45 GHz and a magnetic field in the range $0-10^4 G$.[26]

Mn^{2+} replacing Ca^{2+} in the crystal structure is a good EPR probe for the study of the structural phase transitions in BCCD. The room temperature spectra (figure 9A,9B) can be understood by considering the lower order quadrupolar terms for the crystal field and the principal magnetic axes of the Mn^{2+} defects can be reasonably well related with chemical axes in the molecular structure.

By decreasing the temperature the spectra show important changes in correlation with phase transition sequence as can be seen in figure 9C. In the first incommensurate phase the analysis of the hyperfine lines shows that a pure sinusoidal regime is observed over a large temperature range indicating that the pinning of the distortion wave by the Mn impurities is not relevant. So at 158K, each line of the hyperfine sextuplet gives rise to "two edge singularities" as expected. For T < 140K this structure cannot be described assuming a pure sinusoidal modulation; a multisoliton regime is observed. For 119K < T < 121K the distortion wave locks into the commensurate value δ(T)=2/7. The second incommensurate phase appears in the temperature range 112 < T < 119K. A typical line shape corresponding to a pure sinusoidal modulation is never observed clearly and some additional singularities in the spectral density are observed. This can be due to metastable δ (T)=2/7 regions. As the temperature decreases the spectrum changes indicating a multisoliton regime percursor of the commensurate δ (T)=1/4 phase.[26]

The analysis of the hyperfine structure allows also to identify other commensurate phases. When the modulation wave number takes a rational value δ(T)=m/n each non-equivalent center splits in, at most n different centers. We can account for the position of the lines satisfactorily by assuming four magnetically equivalent sets of seven centers for δ(T)=2/7 phase when **H//b**. The δ(T)=1/4 and δ(T)=1/5 phases were identified. At lower temperatures the progressive overlap between the different adjacent hyperfine lines, already observed within the phase δ(T)=1/5, prevents a simple evidence of the transition to the phase δ (T)=1/6.

362

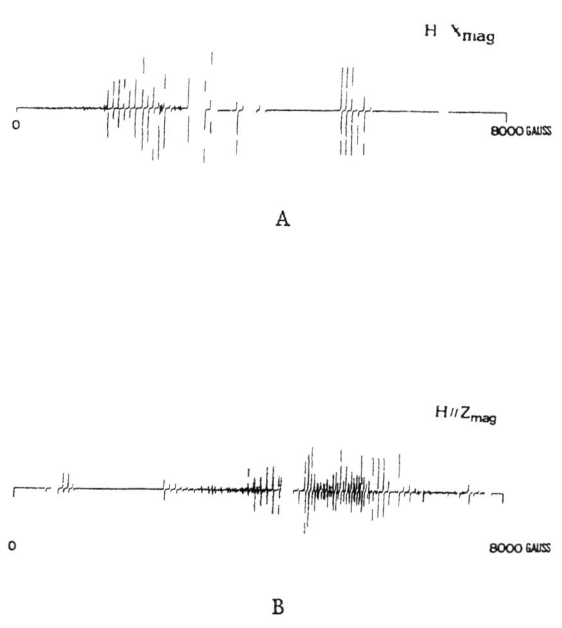

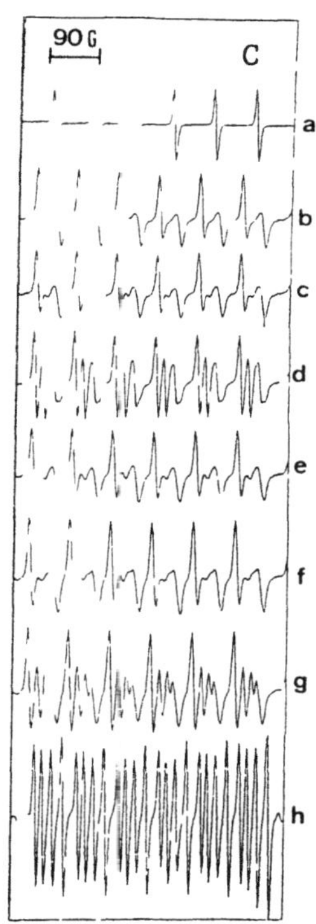

Figure 9. Experimental spectra at room temperature for **H//b** (A) and for **H//Z**$_{mag}$ (B). Sequence of hyperfine structures of the resonance centered at 3 850 G for **H//b**, between 170K and 100K (C). Spectra a) to h) correspond to T=170K (P-phase). T=152K, T=135K (INC1), T=120K (δ(T)=2/7), T=118K, 116K, 113K (INC2) and T=100K δ(T)=1/4), respectively.

An additional transition to a non-modulated phase at low temperatures was clearly observed. This new phase produces a sharp change of the spectrum and each intense hyperfine line reveals an underlying quadruplet. This behaviour shows the non-equivalence of the four sites in the restored unit cell.

Raman Spectroscopic investigation of various commensurate and incommensurate phases. The 333 optical phonon modes of BCCD have to be assigned as $46A_g$, $38B_{1g}$, $46B_{2g}$, $38B_{3g}$, $38A_u$, $45B_{1u}$, $37B_{2u}$, $45B_{3u}$.[27] The point group D_{2h} comprises eight symmetry types, four are Raman-active $A_{1g}(xx,yy,zz)$, $B_{1g}(xy)$, $B_{2g}(xz)$, $B_{3g}(yz)$, three are infrared active $B_{1u}(z)$, $B_{2u}(y)$, $B_{3u}(x)$ and A_u is optically silent.

No complete phonon softening (amplitudon) is found near T_i. It appears that the observed phonon-renormalization are due to coupling effects with another temperature dependent excitation, probably of relaxational type.[27]

The phase diagram which was observed spectroscopically differs from the one determined by X-ray in the low temperature range: it was observed an additional C-phase (45.4K - 46.4K), probably with δ(T)=1/7. No modulation in the ferrolectric phase was found. This result is in contradiction to the results obtained by Volkov et al.[28] in the dielectric measurements of BCCD in the submillimeter wavelength range and with X-ray study.[6] The δ(T)=2/7 phase was not retrieved by spectroscopic techniques.[27]

<u>Concluding Remarks</u>

The results presented in Table 2 summarise the main aspects stated in the last paragraphs in what concerns C phases. Macroscopic technics allowed the identification of high order commensurate phases exhibited by BCCD in narrow temperature ranges of stability while microscopic technics have not detected them up to now. It is an important question to know up to what point the a.c. fields used in the study of dielectric constant and hysteresis loops can induce these phases. Hysteresis loops are obtained with high electric fields whose amplitudes are of the order of magnitude of 1 kV cm^{-1}.[3] The amplitude of a.c. fields in the measurements of ε in the **a** direction is about 20 Vcm^{-1} which is a rather high value.[15]

The results obtained by X-ray study are not in agreement with Raman and EPR measurements, in which concerns the low temperature phase. According to the (p-T) phase diagram a pressure determines new modulated phases in BCCD[22] and so the δ(T)=1/6 phase observed with X-ray below T< 43K, could have been induced, unexpectedly, by some internal stress determined by cooling down the sample.

In adittion we should like to point out that the spatial modulation of BCCD structure possibly consists of a transverse periodic warping of the (010) planes of the betaine zwitterions and the octahedrons. This behaviour is consistent with the occurence of the spontaneous polarization in the **b** direction. The spectroscopic results support a libration soft mode along the **a** axis.[23,29]

Table 2.   Commensurate phases in BCCD.

δ(T)= m/n	Temperature range (K)	**P** along	Experimental study used to identify the C phase*	
2/7	127.8 - 124.5	**b**	(p,HL,ε,α)	(X,EPR,-)
3/11	118.4 - 117.4	0	(-,HL,ε,α)	(-, - ,-)
4/15	116.0 - 115.7	**b**	(p,HL,ε,α)	(-, - ,-)
6/23	115.4	**b**	(p,HL,ε,α)	(-, - ,-)
1/4	115.3 - 75.8	**a**	(p,HL,ε,α)	(X,EPR,R)
2/9	75.8 - 75.2	**b**	(p,HL,ε,α)	(-, - ,-)
1/5	75.2 - 53.3	0	(p,HL,ε,α)	(X,EPR,R)
2/11	53.3 - 53.0	**b**	(p,HL,ε,-)	(-, - ,-)
1/6	53.0 - 47.1	**a**	(p,HL,ε,α)	(X,EPR,R)
2/13	47.1 - 46.9	**b**	(-,HL,-,-)	(-, - ,-)
1/7	46.9 - 46.2	0	(-,HL,ε,α)	(-, - ,?)
1/8	46.2 - 46.0	**a**	(-,HL,ε,-)	(-, -, -)
0/1	46.0 and below	**b**	(p,HL,ε,α)	(?,EPR,R)

(* p: pyroelectric effect; HL: hysteresis loops; ε =dielectric constant; α : thermal expansion; X: X-ray; EPR; R: Raman scattering). The temperatures are those referred by Unruh et al.[15]

As we have seen BCCD exhibits a sequence of modulated phases, as well a reentrant incommensurate phase. In the following we shall see that a phenomelogical model based on a one dimensional order parameter and involving a term of the form $P^2(\partial P/\partial z)^2$ is adequate to account for this complex behaviour.

In the scope of Landau model and by adopting as reference the low temperature non-modulated phase which is ferroelectric with an electric polarization along $\mathbf{b}$ direction, we take the expression below for the free energy density in which the polarization ($\mathbf{P}$) corresponds to the primary order parameter in the phase of reference.[30]

$$f(z) = \tfrac{1}{2}\alpha P^2 + \tfrac{1}{4}\beta P^4 - \tfrac{1}{2}\sigma\left(\frac{\partial P}{\partial z}\right)^2 + \tfrac{1}{4}\gamma\left(\frac{\partial^2 P}{\partial z^2}\right)^2 + \upsilon P^2\left(\frac{\partial P}{\partial z}\right)^2 .$$

The dispersive term $\upsilon P^2(\partial P/\partial z)^2$ is possible by symmetry because it is the product of two trivial invariants (P^2 and $(\partial P/\partial z)^2$). As usual we take $\alpha = \alpha_o (T-T_o)$ and α_o, β as positive constants. A negative $(-\sigma)$ and positive (γ) coefficients are chosen in order to describe the occurence of modulated structures since the terms associated with these coefficients are responsible for a minimum in the dispersion curve $\omega(q)$ at an arbitrary point of the Brillouin zone.

The dispersive term ($\upsilon P^2(\partial P/\partial z)^2$) imposes a temperature dependence of the modulation wavevector. By taking $\upsilon > 0$ we have a term which favours energetically small values of the modulation wavevector and can therefore offset lock-in terms corresponding to larger values of $\delta(T)$. This term is therefore the key to account for the reentrant behaviour.

By using the following set of adimensional variables:

$$t = \frac{\alpha_o \gamma}{8\sigma^2}(T - T_o) ; \quad \mu = \frac{2\sigma\upsilon}{\beta\gamma} ; \quad x = \left(\frac{2\sigma}{\gamma}\right)^{\frac{1}{2}} z ; \quad k = \left(\frac{2\sigma}{\gamma}\right)^{-\frac{1}{2}} q ;$$

$$\xi = \frac{(\beta\gamma)^{\frac{1}{2}}}{2\sigma} P ; \quad g(x) = \frac{\gamma^2\beta}{16\sigma^4} f(z) .$$

we have:

$$g(x) = t\xi^2 + \tfrac{1}{4}\xi^4 - \tfrac{1}{4}\left(\frac{\partial\xi}{\partial x}\right)^2 + \tfrac{1}{4}\left(\frac{\partial^2\xi}{\partial x^2}\right)^2 + \mu\xi^2\left(\frac{\partial\xi}{\partial x}\right)^2 .$$

This latter equation, which depends only on two parameters (t and μ) simulating respectively the temperature and the strength of the special term considered, can be taken as the starting point for the discription of the phase transition sequence in BCCD.

If we assume a pure plane approximation and the existence of an additional umklapp term, we have:

$$U_{umk} = -b\xi^{2p} \approx -b\left(\xi^{2p-4}\right)\xi^4 = -\tfrac{1}{4}\beta_{eff}\xi^4 \quad \text{and} \quad \xi = \xi_0\cos(kx) .$$

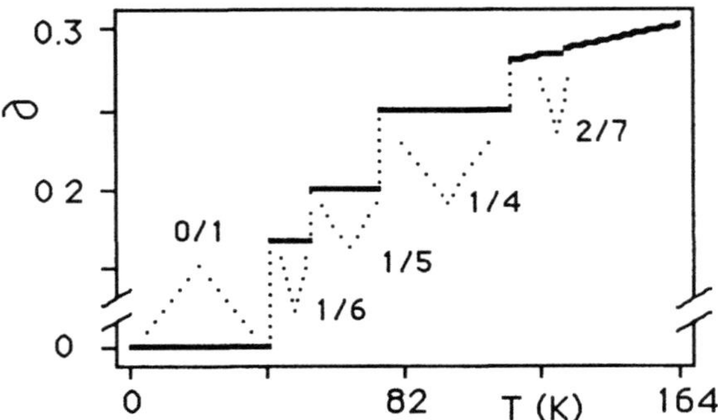

Figure 10. Calculated temperature dependence of the modulation wave number.

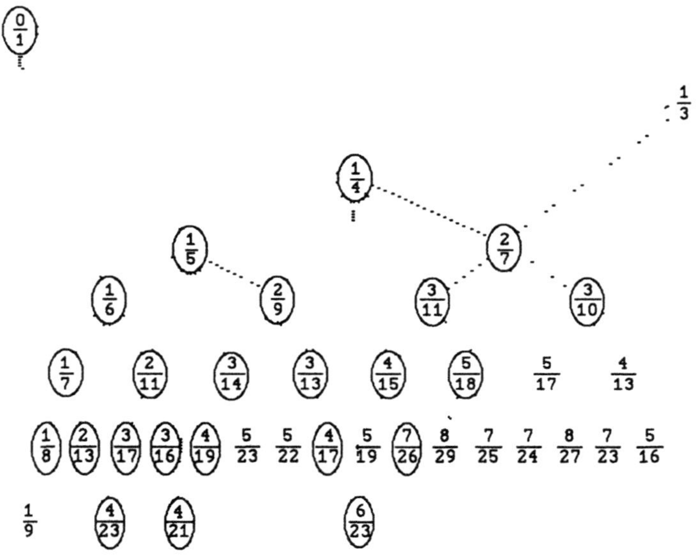

Figure 11. Farey-tree between 0/1 and 1/3. The observed C-phases in BCCD are
indicated.

The average free energy density (F) over a period (x_0) of the modulated distortion is

k=0:
$$F_0 = t\,\xi_0^2 + \frac{1}{4}\,\xi_0^4;$$

k=INC:
$$F_k = \frac{1}{2}\xi_0^2\left(t - \frac{k^2}{4} + \frac{k^4}{4}\right) + \frac{1}{4}\xi_0^4\left(\frac{3}{8} + \mu\,\frac{k^2}{2}\right);$$

k=COM:
$$F_k = \frac{1}{2}\xi_0^2\left(t - \frac{k^2}{4} + \frac{k^4}{4}\right) + \frac{1}{4}\xi_0^4\left(\frac{3}{8}(1 - \beta_{eff}) + \mu\,\frac{k^2}{2}\right).$$

By imposing thermal equilibrium condition $(\partial F/\partial\xi_0 = 0)$ and by adjusting the dispersive coefficient ν, the value of the Curie-Weiss temperature T_0 and the values of the umklapp coeficients β_{eff}, we have the sequence of the phases observed experimentally and depicted in figure 10.[33]

ANNNI-MODEL

Statistical models with Ising-type pseudospins were sucessfully used in the study of some betaine adducts. Tentrup and Siems[31] have shown recently how to adjust the assumptions of the simple ANNNI-model to the BCCD by introducing an adequate dependence of the pseudospin interactions on pressure and temperature. They obtained a phase diagram for BCCD which is in a good agreement with the experimental results, including the ocurrence and sequence of modulated, incommensurate phases and structure branchings.

The splitting of dielectric anomaly with increasing pressure and temperature is caused by the structure branching (bifurcation) processes which have been predicted by Selke and Duxbury[32,33] in their numerical studies of the ANNNI-model. There is a striking similarity between the (p-T) phase diagram of BCCD obtained by Ao et al. and the "devil's flower".[23] At the branching points the boundaries between neighbouring C-phases with modulations $\delta_1 = m/n$ and $\delta_2 = t/p$ become unstable against a higher order commensurate structure with $\delta = (m+t)/(n+p)$. The Farey-tree construction predicts the $\delta(T)$ values of the new C phases. Figure 11 shows the Farey-tree between 0/1 and 1/3 and the observed C-phases in BCCD are indicated.[23]

RECENT EXPERIMENTAL RESULTS; NEW PROSPECTS

The symmetry analysis associated with softening of just one mode explained successfully the experimental results we have reported.[7]

Dvõrák developed a model in which the condensation of a second mode is proposed.[34,35] This idea might be supported by some experimental results reported very recently by Kroupa et al.[36] They observed that the temperature dependence of birefringence displays two anomalies close to each other in the region of T_i and that a small rotation of the order of minutes of the optical indicatrix occurs just below T_i. It seems also that the number of modes observed in the infrared and Raman spectra could not be explained by the freezing of only one soft mode.[35,37,38] Neverthless, as Dvõrák pointed out, there is no clear experimental evidence of the condensation of a second mode.[35]

BCCD is again an open question and supplementary work is needed in order to understand this material which displays an exceptional rich variety of different commensurate and incommensurate phases.

ACKNOWLEDGEMENTS

The authors are deeply indebted to Prof. Dr. H.E. Müser, Dr. J. Albers and Dr. A. Klöpperpieper, University of Saarland, for their cooperation in a project on betaine compounds and for helpful discussions, to Prof. J.C. Tolédano, CNET, for stimulating discussions and his encouragement and to J.L. Ribeiro for help with the preparation and a critical reading of the manuscript.

REFERENCES

1. J. Albers, Ferroelectrics $\underline{78}$, 3 (1988)
2. W. Brill, W. Schildkamp and J. Spilker, Zeitschrift für Kristallographie $\underline{172}$, 281 (1985)
3. H.J. Rother, J. Albers and A. Klöpperpieper, Ferroelectrics $\underline{54}$, 107 (1984); H.J. Rother, J. Albers and A. Klöpperpieper, Z. Kristallogr. $\underline{170}$, 158 (1985)
4. A. Klöpperpieper, H.J. Rother, J. Albers and H.E. Müser, Jpn. J. Appl. Phys. $\underline{24}$, 24-2, 829 (1985)
5. W. Brill and K.H. Ehses and H. Schenk-Strauss, Z. Kristallogr. $\underline{170}$, 24 (1985)
6. W. Brill and K.H. Ehses, Jpn. J. Appl. Phys. $\underline{24}$, 24-2, 826 (1985)
7. J.M. Perez-Mato, Solid State Comm. $\underline{67}$, 1145, (1988)
8. J.C. Tolédano and P. Tolédano, The Landau theory of phase transitions, World scientific Lect. Notes in Phys. Singapore (1988).
9. P. Bak, Rep.Prog.Phys. $\underline{45}$, 587 (1982)
10. P. Bak and J. Boehm, Phys.Rev. B $\underline{21}$, 5297 (1980)
11. F. Axel and S. Aubry, J.Phys.C: Solid State Phys. $\underline{14}$, 5433 (1981)
12. M.R. Chaves, M.H. Amaral and S. Ziolkiewicz, J. Phys. (Paris) $\underline{41}$, 259 (1980)
13. J.L. Ribeiro, M.R. Chaves, A. Almeida, J. Albers, A. Klöpperpieper and H.E. Müser, Ferroelectrics Lett. $\underline{8}$, 135 (1988)
14. J.L. Ribeiro, M.R. Chaves, A. Almeida, J. Albers, A. Klöpperpieper and H.E. Müser, Phys. Rev. B $\underline{39}$, 12320 (1989)
15. H.G. Unruh, F. Hero and V. Dvõrák, Solid State Comm. $\underline{70}$, 403 (1989)
16. J.L. Ribeiro, M.R. Chaves and A. Almeida, J. Albers, A. Klöpperpieper and H.E. Müser, to be published in Ferroelectrics - IMF7 (1989)
17. J.L. Ribeiro, M.R. Chaves, A. Almeida, J. Albers, A. Klöpperpieper, H.E. Müser, J. of Phys. Condensed Matter (1989)
18. P. Prelovšek, J.Phys.C: Solid State Phys. $\underline{16}$, 3257 (1983)
19. P. Prelovšek and R. Blinc, J.Phys.C: Solid State Phys. $\underline{17}$, 1577 (1984)
20. J.L. Ribeiro, M.R. Chaves, A. Almeida, H.E. Müser, J. Albers and A. Klöpperpieper, to be published Ferroelectrics - IMF7 (1989)
21. J.C. Tolédano, Ann. Télécommun. $\underline{39}$, nº7-8 (1984)
22. R. Ao, G. Schaack, M. Schmitt and M. Zöller, Phys.Rev. Letters $\underline{62}$, 183 (1989)
23. G. Schaack, to be published in Ferroelectrics - IMF7 (1989)
24. O. Freitag and H.G. Unruh, to be published in Ferroelectrics - IMF7 (1989)
25. S. Haussühl, J. Liedke, J. Albers, A. Klöpperpieper, Z.Phys. B - Condensed Matter $\underline{70}$, 219 (1988)
26. J.L. Ribeiro. J.C. Fayet, J. Emery, M. Pezeril, J. Albers, A. Klöpperpieper, A. Almeida and M.R. Chaves, J. Phys. $\underline{49}$, 813 (1988)
27. R. Ao, G. Schaack, Indian J. Pure Appl.Phys. $\underline{26}$, 124 (1988)

28. A.A. Volkov, Yu.G. Goncharov, G.V. Kozlov, J. Albers, and J. Petzelt, JETP Lett. <u>44</u>, 603 (1986)
29. M. Fujimoto and Y. Kotake, J.Chem.Phys. <u>90</u>, 532, (1989)
30. J.L. Ribeiro, J.C. Tolédano, M.R. Chaves, A. Almeida, H.E. Müser, J. Albers and A. Klöpperpieper, to be published in Phys.Rev.B
31. T. Tentrup and R. Siems, to be published in Ferroelectrics - IMF7 (1989)
32. W. Selke and P.M. Duxbury, Z. Physik B<u>57</u>, 49 (1984)
33. W. Selke, Phys.Reports <u>170</u>, 213 (1988)
34. V. Dvořák, J. Holakovský and J. Petzelt, Ferroelectrics <u>79</u>, 15 (1988)
35. V. Dvořák, to be published in Ferroelectrics - IMF7 (1989)
36. J.K. Kroupa, J. Albers, N.R. Ivanov, to be published Ferroelectrics - IMF7 (1989)
37. R. Ao and G. Schaack, Ferroelectrics, <u>80</u>, 105 (1988)
38. S. Kamba, J. Petzelt, V. Dvořák, Yu.G. Goncharov, A.A. Volkov, G.V. Kozlov and J. Albers, to be published in Ferroelectrics - IMF7 (1989)

AMORPHOUS AND QUASICRYSTALLINE Al-Cr AND Al-Cr-Si PHASES PRODUCED
BY SOLID STATE DIFFUSION OF ALTERNATING THIN LAYERS

I.Levi and D.Shechtman

Materials Engineering, Technion, Haifa, Israel

ABSTRACT

Amorphous, icosahedral and decagonal phases were produced by
vacuum furnace annealing and by rapid thermal annealing (RTA) of
alternating Al-Cr and Al-Cr-Si thin layers. The microstructure of
the quasicrystalline phases and the periodic phases formed by
these solid-state heat treatments was examined. The optimal
conditions for the formation of the icosahedral phase by solid
state diffusion were found. It was also found that the decagonal
phase is formed in the Al-Cr-Si system and not in the Al-Cr
system, and that the addition of about 5 at% Si to the binary
system stabilizes the icosahedral phase, similar to the Al-Mn
system.

INTRODUCTION

Since the discovery of the icosahedral phase (1) extensive
investigation has been carried out on this phase and other
quasi-periodic phases like the decagonal phase (2). The
quasi-crystalline phases which exhibit long range orientational
order without long range translational order were first produced
by rapid solidification methods, particulary melt-spinning (1).
Another way of obtaining the quasi-periodic phases is by solid
state transformation of thin films. Many methods of producing
quasi-periodic thin films by solid state transformation have been
used until now: ion-beam mixing of alternating thin layers or of
coevaporated alloys (3-8), heat treatment of alternating thin
layers or of coevaporated alloys (8-13), implantation of Mn^+ ions
into Al film at high temprature (150°C) (7,14) and sputtering of
alloys at high temperatures (15). Most of the research was done on
the Al-Mn system (7,8,13-15), but also on other systems like
Al-Mn-Si (6,8,15), Al-Cr (3,4), Al-Ru (7) and Al-Fe (5,13).

In this study we compared the structure of thin films of
alternating Al-Cr and Al-Cr-Si thin layers before and after heat
treatments in a vacuum furnace and by a heat pulse.

Rapid thermal annealing (RTA) by the heat pulse enables to
perform heat treatments at relatively high temperatures for very
short times, and with a very good control of the time and
temperature.

Geometry and Thermodynamics
Edited by J.-C. Tolédano
Plenum Press, New York, 1990

In this study we have found the conditions needed for producing the quasicrystalline phases in the Al-Cr and Al-Cr-Si systems. The phase transformations and microstructures obtained by the various heat treatments are studied in detail.

EXPERIMENTAL

Alternating thin layers of Al and Cr with and without Si were sputtered at room temperature on a cleaved NaCl substrate. Four alloys were examined. Their compositions and thicknesses are described in table 1.

Table 1. Thickness (in nm) and composition of the various alloys

SAMPLE	Al	Cr	Si	TOTAL THICKNESS
T2 - 1	4.0	0.5	——	40
T2 - 3	8.0	1.0	——	40
T2 - 6	6.0	1.0	0.5	40
T2 - 7	4.5	1.0	1.5	40

COMPOSITIONS OF THE VARIOUS LAYERS (at%)

T2-1, T2-3:   Al-15Cr

T2-6:   Al-17Cr-5Si

T2-7:   Al-19Cr-19Si

Heat treatments were performed in a vacuum furnace of 10^{-6} torr in the temperature range of 300°C to 600°C for 1 to 1.25 hr, and by a heat-pulse in the temperature range of 550°C to 700°C for 10 and 60 seconds in a N_2 atmosphere. The microstructure was examined using Jeol 120CX and Jem 2000FX transmission electron microscopes (TEM).

RESULTS

As-Deposited Structures

When the alternating layers are very thin (alloy T2-1), the as-deposited structure is amorphous as can be seen in fig. 1a. Thicker alternating layers (alloys T2-3,6,7) produce a polycrystalline structure of the various elements as shown in figs.1b,c. The dark field image in fig.1d taken from the two strongest rings of fig. 1b shows very small grains (about 10-15 nm) of Al and Cr.

<u>Heat Treatments in a Vacuum Furnace</u>

Amorphous as-deposited structures (sample T2-1) produce the icosahedral phase at relatively high temperatures (440°C), but without any residual Al as can be seen in the diffraction pattern of figs.2a,b. If we compare these results to sample T2-3, which has the same composition but thicker alternating layers, it is clear that the icosahedral phase in this alloy is formed at lower temperature (300°C), but always appears with a considerable amount of α-Al phase as indicated by the diffraction pattern in figs.2c,d.

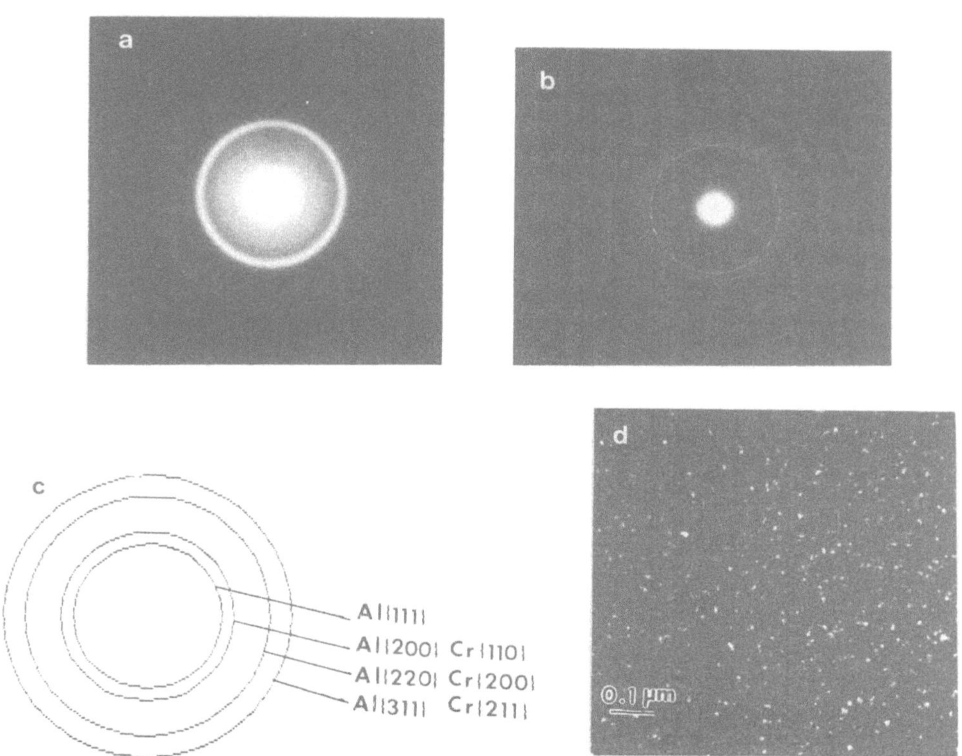

Fig.1. As-deposited structures: a.diffraction pattern from the very thin alternating layers (T2-1);amorphous phase. b.diffraction pattern from thicker alternating layers (T2-3);Al and Cr rings. c. schematic diagram of fig.1b. d.dark field image taken from the two strongest rings of fig.1b;Al and Cr grains.

Heat treatments at higher temperatures produce periodic phases in all the alloys examined. Since the crystal structures of the intermetallic compounds in the Al-Cr system are not known for certain (16-19), we did not analyze the periodic diffraction patterns obtained in this study. Fig.3a is a bright field image of the periodic phase of alloy T2-3, formed following annealing at 500°C for 1 hr. The corresponding periodic diffraction pattern is seen in fig.3b.

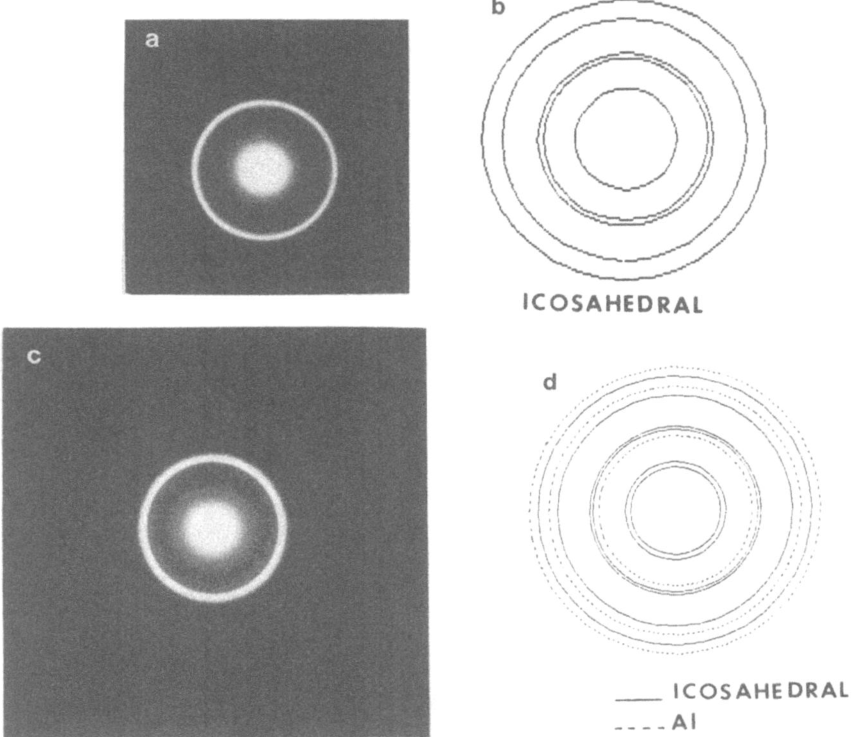

Fig.2. a.diffraction pattern from specimen T2-1 following
annealing at 440°C for 1hr.;icosahedral phase. b.schematic diagram
of fig.2a. c.diffraction pattern from T2-3 after annealing at
300°C for 1 hr;icosahedral and Al phases. d.schematic diagram of
fig.2c.

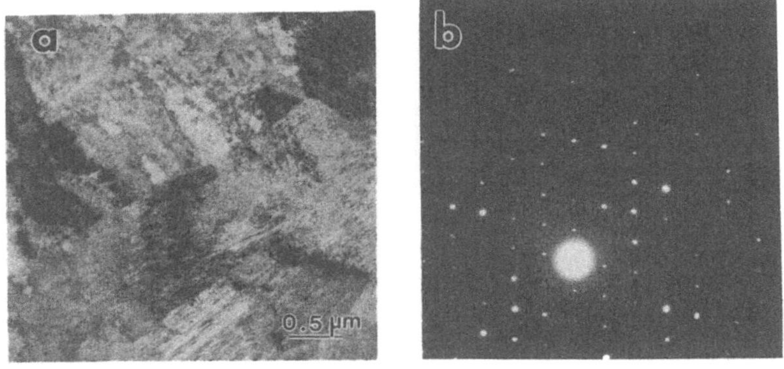

Fig.3. T2-3 after annealing at 500°C for 1 hr. a.bright field
image showing a periodic phase. b.diffraction pattern from the
periodic phase.

An addition of a small amount of Si (about 5 at%) (alloy
T2-6), stabilizes the icosahedral phase. It forms at a lower
temperature and transforms to a periodic phase at a higher
temperature compared to the amorphous as-deposited film (alloy
T2-1). The alloy is almost entirely icosahedral without any
residual Al. Annealing of this film (T2-6) at 400°C for 75 min
produces a polycrystalline icosahedral phase. A typical electron
diffraction pattern of this phase is seen in fig.4a. The
diffraction pattern indicates that there is a small amount of
residual Al in addition to the icosahedral phase. A comparison
between this diffraction pattern and the diffraction pattern in
fig.2b reveals that the ternary alloy contains much less Al than
the binary alloy. A dark field image taken from the bright doublet
of the icosahedral phase is seen in fig.4b. Very small grains of
about 10 nm are observed.

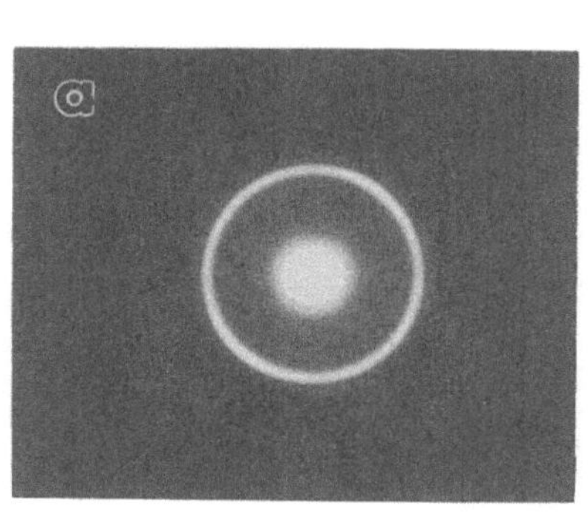
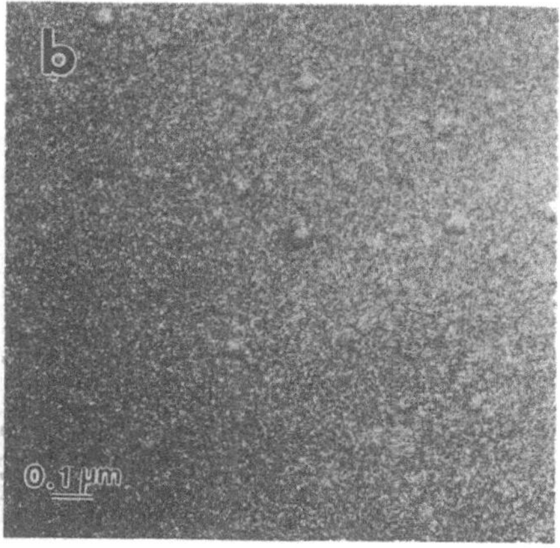

Fig.4. Al-17Cr-5Si alloy (T2-6) after annealing at 400°C for 75
min.;a.diffraction pattern of icosahedral phase. b.dark field
image taken from the bright doublet of fig. 4a;small grains of the
icosahedral phase.

After annealing the Al-17Cr-5Si (sample T2-6)at 440°C for
1 hr periodic intermetallic compounds start to form in the
icosahedral matrix. Two different morphologies of the periodic
phase appear; Fig.5a is a bright field image of a spherolite
particle surrounded by an icosahedral matrix. The corresponding
diffraction pattern can be seen in fig.5b. The other morphology of
platelets is shown in fig.5c. The plates are faulted along their
length as shown by the bright field image and the diffraction
pattern in fig. 5d.

After annealing this alloy at 600°C for 1 hr, all the
icosahedral phase is transformed to the periodic phases and the
microstructure is composed entirely of the periodic spherolites
and plates.

The addition of a larger amount of Si (alloy T2-7,
Al-19Cr-19Si), prevents the formation of the icosahedral phase.
Annealing these polycrystalline alternating layers at 300°C for 1
hr causes an amorphization of the film as can be seen in the
diffraction patterns taken before and following the heat treatment
(figs.6a-c). Annealing this film at a higher temperature (350°C,
1 hr) produces the ternary periodic $Cr_4Si_4Al_{13}$ phase shown in
figs.6d-f.

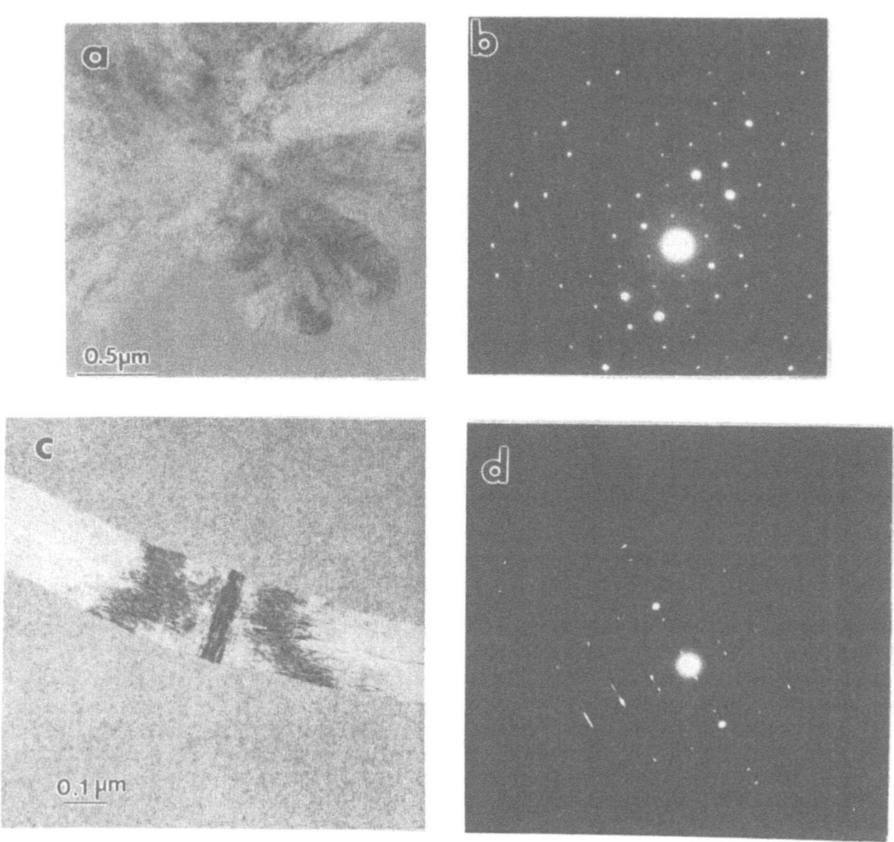

Fig 5 Al-17Cr-5Si alloy after annealing at 440°C for 1 hr, a bright field image showing a spherolite particle in the icosahedral matrix b diffraction pattern of the spherolite particle c plates grow within the icosahedral matrix d diffraction pattern of a plate

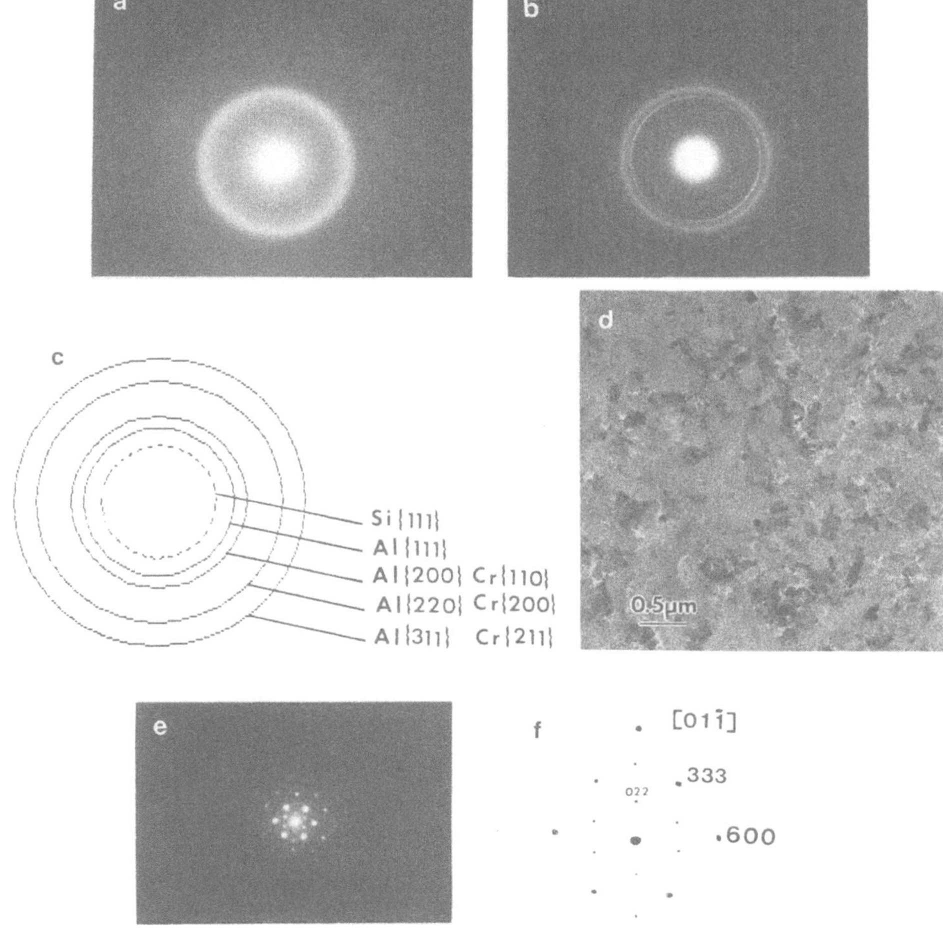

Fig.6. Al-19Cr-19Si alloy. a.annealing at 300°C for 1 hr; diffraction pattern showing an amorphous phase. b.diffraction pattern from the as-deposited layer;Al, Cr, and Si rings. c.schematic diagram of fig.6b. d. bright field image showing $Cr_4Si_4Al_{13}$ phase obtained after annealing at 350°C for 1 hr. e.microdiffraction from $Cr_4Si_4Al_{13}$ particle. f.diffraction pattern analysis of fig.6d.

Heat Treatments by Heat Pulse

Heat treatments by heat pulse were performed at temperatures between 550°C to 700°C for 10 and 60 sec. At these temperatures only alloys T2-3 and T2-6 transform to quasicrystalline phases. The other as-deposited layers (alloys T2-1 and T2-7) produce ring diffraction patterns of some other phase(s), probably periodic.

Al-Cr

Annealing at 550°C for 60 sec produces polycrystalline icosahedral grains. The icosahedral phase appears with residual Al, the same as after annealing in vacuum. After annealing at higher temperatures, periodic phase(s) were formed. Annealing at 700°C for 60 sec produce large grains of periodic phase(s). As a result of this heat treatment (700°C for 60 sec) two different morphologies of periodic phase are formed: In fig.7a grains of about 0.1-0.2 µm are observed. Their periodic microdiffraction is seen in fig.7b. The second morphology is shown in fig.7c and its

diffraction pattern in fig 7d Both the bright field image and the
diffraction pattern show a faulted structure

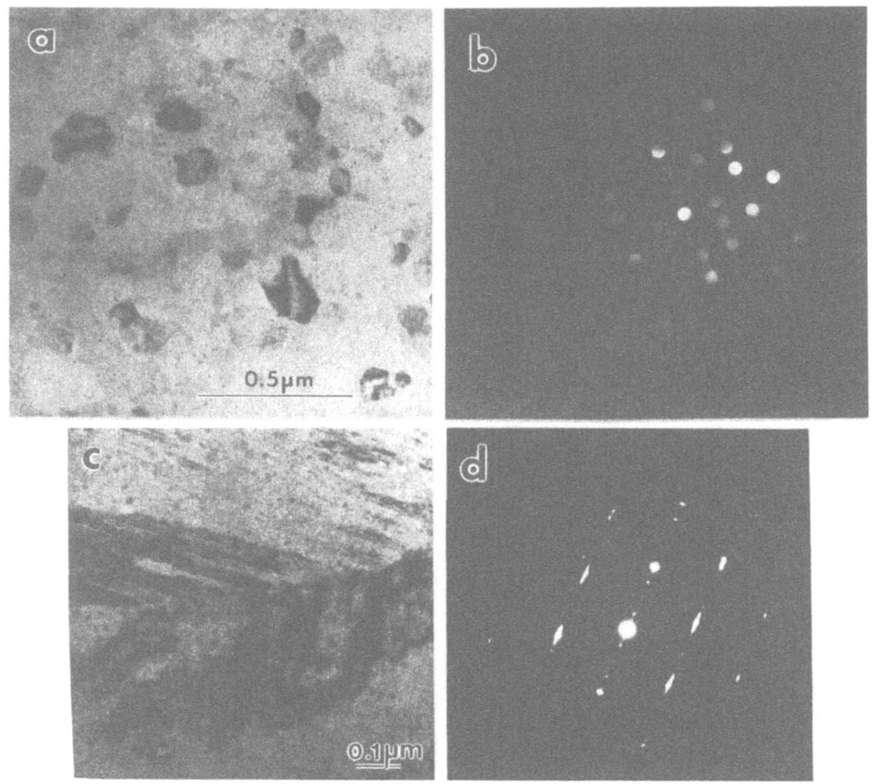

Fig 7 12-3 after RTA at 700°C for 60 sec a bright field
micrograph showing grains of a periodic phase b microdiffraction
from a periodic grain c bright field image of another morphology
of a periodic phase d diffraction pattern from the periodic phase
shown in fig 7c

Al-Cr-Si
     Heat treatments at 550°C to 620°C for 10 sec and 60 sec
produce a polycrystalline icosahedral phase   The resulting
diffraction pattern and morphology are the same as after the
vacuum annealing

     Annealing at 650°C for 60 sec produce the two morphologies of
periodic phase(s) the plates and spherolites which were observed
also after the vacuum annealing, nucleate and grow in the
icosahedral matrix
     At these annealing conditions (650°C for 60 sec) large grains
of the decagonal (T) phase start to nucleate and grow in the
icosahedral matrix Heat treatments at 650°C to 670°C cause
additional growth of the T phase grains Fig 8 shows bright field
images of the decagonal grains in the icosahedral matrix The
growth of the grains with the increasing temperature can be
observed   The electron diffraction patterns of the T phase can be
seen in fig 9 After annealing at 670°C for 60 sec, all the
icosahedral matrix is consumed by the T phase grains, and the
microstructure, shown in fig 10, is composed entirely of the
decagonal grains and the plate-like periodic phase

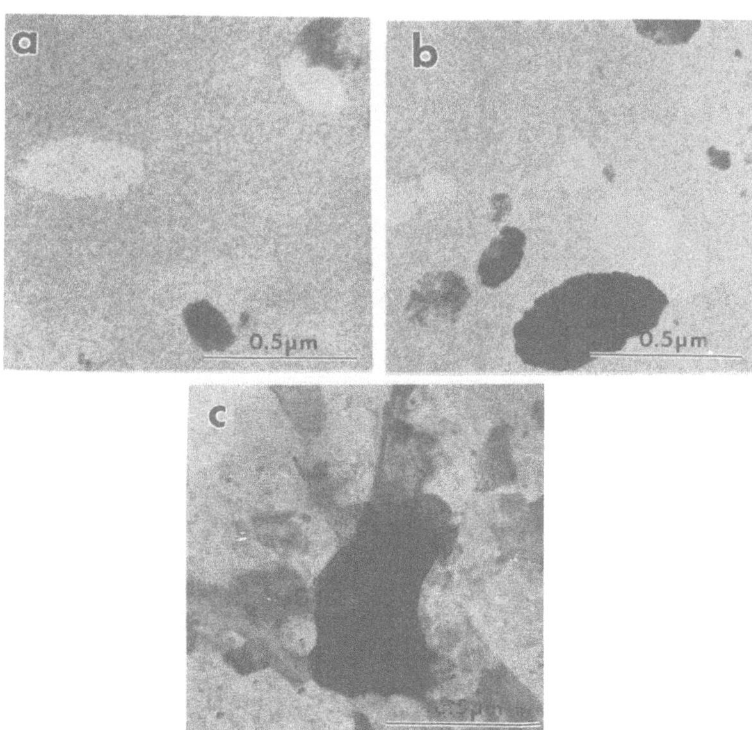

Fig 8   Al-17Cr-5Si alloy (T2-6), bright field   micrographs   showing growth   of the decagonal phase in the icosahedral matrix   a RTA at 650°C for 60 sec   b RTA at 665°C for 60 sec   c RTA at 670°C for 60 sec

Fig.9. Al-17Cr-5Si alloy (T2-6) after RTA at 650°C TO 700°C; diffraction patterns of the decagonal phase.

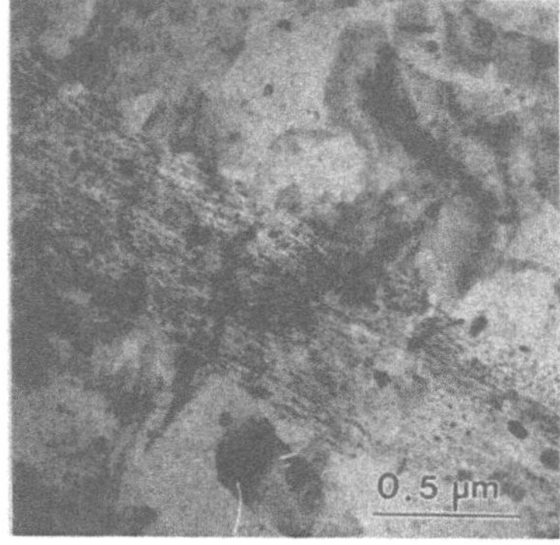

Fig.10. bright field micrograph of T2-6 after RTA at 700°C for 60 sec; T phase grains and periodic plates.

CONCLUSIONS

1. When the Al-Cr icosahedral phase is produced by heat treatments
of an amorphous phase, it appears as a single phase without
residual Al. When it forms from polycrystalline alternating layers
of the same average composition it always appears with Al.

2. The addition of a small amount of Si (about 5 at%) to the Al-Cr
alloy stabilizes the icosahedral phase: It raises the temperature
of the transformation to the periodic phase compared to the binary
alloy, and reduces the amount of residual Al.

3. An addition of about 5 at% Si to the Al-Cr alloy enables the
formation of the decagonal phase. Large grains of the decagonal
phase nucleate and grow in the icosahedral matrix. The T phase
does not form in the binary alloy, as was also found by Kuo (20)
in melt spun ribbons.

4. An addition of about 19 at% Si to the binary alloy results in
the amorphization of the polycrystalline alternating layers after
the heat treatments. No quasicrystalline phases are formed in this
film.

REFERENCES

1. D.Shechtman, I.Blech, D.Gratias, J.W.Cahn. Phys. Rev. Lett.,
    53, (1984), 1951.
2. L.Bendersky. J. Physique Colloq., 47, (1986), C3-457.
3. D.A.Lilienfeld, M.Nastasi, H.H.Johnson, D.G.Ast, J.W.Mayer.
    Phys. Rev. Lett., 55, (1985), 1587.
4. D.A.Lilienfeld, J.W.Mayer, M.Nastasi, E.Rimini, B.M.Ullrich.
    IL Nuovo Cimento, 7D, (1986), 134.
5. D.A.Lilienfeld, Cornell University, Diss. Abstr. Int., 47(2),
    (1986), 143.
6. D.M.Follstaedt, J.A.Knapp. Mat. Res. Soc. Symp. Proc., 74,
    (1987), 333.
7. D.M.Follstaedt, J.A.Knapp. Nucl. Instr. Meth. Phys. Res.,
    B24/25, (1987), 542.
8. D.M.Follstaedt, J.A.Knapp. J. Less. Comm. Met., 140, (1988),
    375.
9. D.M.Follstaedt, J.A.Knapp. Phys. Rev. Lett., 56, (1986),
    1827.
10. D.A.Lilienfeld, M.Nastasi, H.H.Johnson, D.G.Ast, J.W.Mayer. J.
    Mat. Res., 1(2), (1986), 237.
11. W.Malzfeldt, P.M.Horn, D.P.Divincenzo, B.Stephenson,
    R.Gambino, S.Herd. J. De Phys. Colloque, 47, (1986), C3-379.
12. K.Urban, M.Bauer, A.Csanady, J.Mayer. Mat. Sci. Forum, 22-24,
    (1987), 517.
13. A.Csanady, P.B.Barna, G.Radnoczi, K.Urban. Mat. Sci. Forum,
    22-24, (1987), 617.
14. J.D.Budai, M.J.Aziz. Phys. Rev. B., 33, (1986), 2876.
15. K.G.Kreider, F.S.Biancaniello, M.J.Kaufman. Scr. Mett., 21,
    (1987), 657.
16. W.B.Pearson. Handbook of Lattice Spacing and Structure of
    Metals and Alloys, 1967, p.321, Perganon Press.
17. J.N.Pratt, G.V.Raynor. J. Ins. Met. 80, (1951-1952), 449.
18. T.Ohnishi, Y.Nakatani, K.Okabayash. Bull. Univ. Osaka Prefect,
    24, (1975), 183.
19. M.J.Cooper. Acta Crystall., 13, (1960), 257.
20. K.H.Kuo. Mat. Sci. Forum, 22-24, (1987), 131.

PHASONS AND AMPLITUDONS IN MODULATED STRUCTURES

R. CURRAT

Institut Laue-Langevin
BP 156X, 38042 Grenoble, France

ABSTRACT

The normal modes of modulated crystals are discussed using the superspace and Landau approaches. The dynamical aspects which characterize the modulated state, i.e., the phase and amplitude fluctuation spectrum, are emphasized. Available inelastic neutron scattering results on displacively-modulated insulators (ThBr$_4$, biphenyl, K$_2$SeO$_4$) are reviewed.

1. PHONONS IN MODULATED CRYSTALS

Many quasiperiodic systems may be viewed as modulated crystals. This implies the existence of an average (or "basic") reference structure with 3-dimensional periodicity. With respect to this crystalline reference state, the "modulation" can be described as a static distorsion characterized by its own independent periodicity. In what follows we discuss the dynamics of displacively modulated crystals, when the modulated atomic displacements may be regarded as a small perturbation on the basic structure.

In a crystalline solid, normal modes are labelled by a wavevector in the first Brillouin zone and a branch index $\vartheta = 1,2,...3s$. The corresponding eigenfrequencies generate dispersion curves with gaps at the zone-boundaries. An inelastic scattering experiment at fixed momentum transfer $\mathbf{K}$, probes excitations with wavevectors $\mathbf{k}$ defined by $\mathbf{K} + \mathbf{k} = \mathbf{G}$, where $\mathbf{G}$ is a reciprocal lattice vector.

In incommensurately modulated crystals the Brillouin zone has collapsed to zero volume and an infinite number of gaps have been introduced. In the limit of small modulation amplitudes most of the gaps can be ignored and the eigenfrequencies generate pseudo-dispersion curves in the Brillouin zone of the basic structure, with only a few gaps at low-order "zone-boundary" points :

$$k_{G,m} = 1/2 \ (G+mk_i) \qquad m = 0 \pm 1, \pm 2,... \qquad (1.1)$$

where $\mathbf{k_i}$ stands for the modulation wavevector.

The superspace formalism yields general expressions for the fluctuating displacements associated with the normal vibrational modes of the modulated crystal[1]. These are obtained as the restriction to 3D physical space of the normal modes of a higher dimensional periodic structure. When the modulation periodicity is characterized by a single pair of modulation wavevectors $\pm k_i$, the dimensionality of the superspace structure is 4 :

$$(n + r_j + f_j (k_i \cdot n + \tau), \tau) \qquad (j = 1, \ldots, s) \qquad (1.2)$$

and the additional coordinate τ corresponds to the phase of the modulation function f, with respect to the basic lattice $n + r_j$. Because of lattice periodicity, the displacement fields associated witht the normal modes of the higher dimensional structure may be written as Bloch waves :

$$u_{njz}^k = M_j^{-1/2} \exp(-ik \cdot n) U_j^k (\tau + k_i \cdot n) \qquad (1.3)$$

where U_j^k is periodic :

$$U_j^k(\tau) = \sum_m A_{jm}^k \exp(im\tau) \qquad (1.4)$$

Without loss of generality, both vectors u and k can be taken to be parallel to 3D physical space. In particular k is restricted to the first Brillouin zone of the basic lattice.

The atomic displacements of the modulated crystal are obtained by setting $\tau = 0$ in (1.3) :

$$u_{njo}^k = M_j^{-1/2} \sum_m \exp[-i(k+mk_i) \cdot n] A_{jm}^k \qquad (1.5)$$

Expression (1.5) suggests that an excitation of nominal wavevector k can be described as a linear superposition of normal modes of the basic structure, with wavevectors $k, k \pm k_i, k \pm 2k_i, \ldots$ As such, it can participate in inelastic scattering processes at $K = G - k \pm k_i, G - k \pm 2k_i, \ldots$, with scattering strengths proportional to the appropriate mixing coefficients. Physically, the mixing expresses Bragg scattering of the phonons of the basic structure by the periodic potential of the modulation.

The superspace formalism can be used to derive a number of other results (Debye-Waller factors, one-phonon structure factors) connected with the vibrational spectrum of modulated crystals[2].

2.LANDAU THEORY OF THE NORMAL-TO-INCOMMENSURATE TRANSITION

At high temperature, most modulated crystals transform continuously into a "normal" crystalline phase, whose structure is closely related to the basic structure defined in the modulated state[3]. Thus, the correspondence between normal modes of the modulated ($T < T_i$) and unmodulated ($T > T_i$) structures can be further explored, in the framework of the Landau theory of second-order phase transitions[4].

Writing the modulated atomic displacements in terms of a single pair of frozen normal coordinates $<Q(k_i)>=\eta\exp(i\phi)$ and $<Q(-k_i)>=\eta\exp(-i\phi)$, the Landau free-energy for the normal-to-incommensurate phase transition may be derived[5]. It is found to be of the same form as that of the isotropic planar magnet ($n = 2$; $d = 3$), with the amplitude η and phase ϕ of the modulation playing the role of the modulus and orientation angle of the magnetization, respectively. The linearized order-parameter fluctuation spectrum is obtained from the q-dependent static susceptibilities :

$$\chi^{-1}(\mathbf{q}) = \alpha(T-T_i) + \mathbf{q}\overline{\overline{\Lambda}}\mathbf{q} + ... \qquad (T>T_i) \tag{2.1a}$$

$$\left. \begin{aligned} \chi_L^{-1}(\mathbf{q}) &= 2\,\alpha(T_i-T) + \mathbf{q}\overline{\overline{\Lambda}}\mathbf{q} + ... \\[4pt] \chi_T^{-1}(\mathbf{q}) &= \ 0 + \mathbf{q}\overline{\overline{\Lambda}}\mathbf{q} + ... \end{aligned} \right\} \quad (T<T_i) \begin{aligned} \tag{2.1b} \\[6pt] \tag{2.1c} \end{aligned}$$

where the longitudinal and transverse susceptibilities $\chi_L(\mathbf{q})$, $\chi_T(\mathbf{q})$ are associated with fluctuations in the amplitude and phase of the modulation.

The divergence of $\chi_T(\mathbf{q})$ in the limit $q \to 0$ reflects the invariance of the free-energy with respect to an homogeneous change in ϕ. However, as discussed by Zeyher and Finger[6], this does not imply Goldstone behaviour since the anharmonic microscopic Hamiltonian itself is not phase-invariant. As a consequence, *homogeneous* phase fluctuations are expected to be damped through normal anharmonic processes[6,7], due to the *inhomogeneous* (optic-like) nature of the atomic motions involved. Under the assumption of a damped-harmonic-oscillator (DHO) lineshape, with a q-independent damping factor, the quasi-harmonic phase- and amplitude-mode frequencies are obtained as :

$$\omega_\phi^2(\mathbf{q}) = \chi_T^{-1}(\mathbf{q}) \; ; \; \omega_A^2(\mathbf{q}) = \chi_L^{-1}(\mathbf{q}) \tag{2.2}$$

One may readily verify that the atomic motions associated with each type of modes is of the general form (1.5).

There are many inherent limitations to the above description, particularly in low-dimensional systems (non-mean field behaviour) or in systems close to the commensurate limit (pinning by lattice discreteness or lattice defects). In addition, coupling to other degrees of freedom, such as acoustic modes, may lead to deviations from simple DHO-type lineshapes[8,9]. Due to these compounded difficulties, direct experimental evidence for long-wavelength "phasons" is available for a very limited number of modulated systems[2].

3. EXPERIMENTAL RESULTS ON PHASONS AND AMPLITUDONS

The detection of long-wavelength phase- and amplitude-modes via inelastic neutron scattering is discussed and various possible coupling mechanisms are considered : mode mixing, diffraction harmonics and mixed processes. Available experimental results on $ThBr_4$[10], biphenyl[11,12] and K_2SeO_4[13], are reviewed. The discussion follows closely that given in Ref. 2, Sect.VI, to which the reader is referred.

4. CONCLUDING REMARKS

Current developments aim at exploring the dynamics of a broader variety of modulated systems than presently available.

Away from the sinusoidal regime, the excitation spectrum of the modulated crystal becomes more and more pathological when referred to the Brillouin zone of the parent structure and the concept of "pseudo-dispersion curves" becomes progressively irrelevant. In the discommensuration (DC) limit, however, a new picture emerges[14,15], in which phase fluctuations take the form of collective vibrational modes of the DC array. Discrete models predict a phason gap due to pinning of the DC positions by the underlying crystal structure[16]. All these features have been well studied theoretically but await detailed experimental confirmation.

Progress has been made in the field of charge-density-waves, where the specific difficulties associated with the observation of CDW excitations by inelastic neutron scattering are now better appreciated[17,18]. The generalization of the concepts of phase and amplitude fluctuations to more complex modulated structures, with several degrees of freedom, or to systems with purely relaxational dynamics is the object of current investigations[19,20].

So far, direct observation of phasons in modulated crystals has only been achieved by inelastic neutron scattering. Indirect evidences are also available from light and resonance spectroscopy and ultrasonic measurements[21-23]. One challenging task lying ahead is to achieve consistency between the results obtained with these various techniques (see lectures by Copic, Milia and Cummins, this Volume).

REFERENCES

1. T. Janssen, J. Phys. C 12 (1979) 5381
2. R. Currat and T. Janssen, Solid State Phys. 41 (1988) 201
3. V. Heine and E.H. Simmons, Acta. Cryst. A 43 (1987) 289
4. J.C. Toledano and P. Toledano, "The Landau Theory of Phase Transitions : Application to Structural, Incommensurate, Magnetic and Liquid Crystal Systems", World Scientific Lecture Notes in Physics, Vol. 3 (World Scientific Publ. Co., 1987)
5. R.A. Cowley and A.D. Bruce, J. Phys. C11 (1978) 3577
6. R. Zeyher and W. Finger, Phys. Rev. Lett. 49 (1983) 340
7. V.A. Golovko and A.P. Levanyuk, Sov. Phys. JETP 54 (1981) 1217
8. M.B. Walker and R.J. Gooding, Phys. Rev. B 32 (1985) 7412
9. A.P. Mayer and R.A. Cowley, J. Phys. C 19 (1986) 6131
10. R. Currat, L. Bernard and P. Delamoye, in "Incommensurate Phases in Dielectrics", Vol. 2, R. Blinc and A.P. Levanyuk eds. (Elsevier Science, Amsterdam 1986)
11. H. Cailleau, same Vol. as for Ref. 10
12. P. Launois, M.H. Lemée, H. Cailleau, F. Moussa and J. Mons, Ferroelectrics 78 (1988) 137
13. M. Quilichini and R. Currat, Solid State Comm. 48 (1983) 1011
14. B. Sutherland, Phys. Rev. A 8 (1973) 2514
15. A.D. Bruce and R.A. Cowley, J. Phys. C 11 (1978) 3609
16. D.A. Bruce, J. Phys. C 13 (1980) 4615
17, J.P. Pouget, in : Low Dimensional Conductors and Superconductors, D. Jerome and L.G. Caron eds. (Plenum, New York, 1987)
18. B. Hennion, J.P. Pouget, C. Escribe-Filippini and M. Sato, to be published
19. P. Launois, F. Moussa, M.H. Lemée-Cailleau and H. Cailleau, to appear in Phys. Rev. B
20. D. Durand, R. Currat, F. Mezei, L. Bernard and J.M. Kiat, Phys. Rev. B 39 (1989) 2453
21. H. Poulet and R.M. Pick, in "Incommensurate Phases in Dielectrics" (R. Blinc and A.P. Levanyuk eds.), Vol. 1, North-Holland, Amsterdam, 1986
22. R. Blinc, same Vol. as for Ref. 21
23. R. Zeyher, Ferroelectrics 66 (1986) 217

PHONONS IN QUASICRYSTALS AND RELATED STRUCTURES

T. Janssen

Institute for Theoretical Physics, University of Nijmegen
6525 ED Nijmegen, The Netherlands

INTRODUCTION

Quasiperiodic structures are generalisations of periodic structures.[1] Whereas in the latter the support of the Fourier spectrum is a lattice, the positions of the diffraction peaks in a quasiperiodic structure are integral linear combinations of an arbitrary finite number of basis vectors. Therefore, the diffraction vectors are of the form

$$H = \sum_{i=1}^{n} h_i \, \mathbf{a}_i^{*} \qquad ; \; h_i \text{ integers}$$

If n is equal to the dimension of the space considered, the system is periodic. If the number n of rationally independent vectors (the rank) is larger, the system is no longer periodic but still quasiperiodic. In the last decades it has become clear that condensed matter may have perfect long range order, without periodicity but with quasiperiodicity. Examples of such systems are incommensurate modulated phases. The positions of the atoms in such a system are displaced with respect to a system with ordinary space group symmetry. The positions of the atoms in a displacively modulated crystal phase are

$$\mathbf{x}_{nj} = \mathbf{n} + \mathbf{r}_j + \mathbf{f}_j(\mathbf{q} \cdot \mathbf{n}), \qquad \mathbf{f}_j(x) = \mathbf{f}_j(x + 2\pi)$$

where $\mathbf{n}$ is a lattice vector, $\mathbf{r}_j$ the position of the j-th particle in the unit cell of the basic structure and $\mathbf{f}_j$ the periodic displacement. If $\mathbf{q}$ has irrational components with

Geometry and Thermodynamics
Edited by J.-C. Tolédano
Plenum Press, New York, 1990

respect to the reciprocal lattice of the basic structure, the system is no longer periodic, but, since the diffraction spots are at positions which are linear integral combinations of reciprocal lattice basis vectors and the modulation wave vector $\mathbf{q}$, it is still quasiperiodic, the rank n being the dimension of the space plus one. Modulations with more such wave vectors, such that the rank is even higher, occur as well. Other examples are quasicrystals, by which we mean here quasiperiodic structures that deviate only little from a tiling. So for these structures the atoms are in positions

$$ x_{nj\alpha} = \sum_{i=1}^{n} n_i \mathbf{e}_i + \mathbf{r}_{j\alpha} + \sum_{\mathbf{q} \text{ in } M^*} f_{j\alpha}(\mathbf{q}) \exp(i\mathbf{q}.\mathbf{n}) $$

where the $\mathbf{e}_i$ are the basis vectors of the tiling, n_i are n-tuples of integers, $\mathbf{r}_{j\alpha}$ are positions inside a tile (a so called decoration of tile α) and in which the last term describes a displacement from these ideal tiling positions such that its Fourier components fall on the same positions as those of the tiling. One could call such a structure a modulated tiling. Still other examples include helicoidal spin structures and composite (or intergrowth) structures. Because the Fourier spectrum of such a quasiperiodic structure may be seen as the projection of a lattice in an n-dimensional space (n being the rank), the real physical structure in d dimensions is the intersection of a d-dimensional hyper- plane V_E with a periodic structure in n-dimensional space V_S which is the sum of V_E and an additional (n-d)-dimensional internal space V_I. As an example, the incommen- surate modulated structure, mentioned above, is the restriction to the hyperplane $\varphi = 0$ of the (d+1)-dimensional array of lines

$$ (\, \mathbf{n} + \mathbf{r}_j + f_j(\mathbf{q}.\mathbf{n} + \varphi),\, \varphi\,) $$

where the phase φ is a continuous variable. Quasicrystals may be embedded in a similar way, but then the periodic pattern in V_S does not consist of continuous lines but of disjunct line elements, called supermolecules or atomic surfaces. An example is given in Fig. 1 for the Fibonacci chain, a quasiperiodic sequence of long and short intervals on the line, such that the Fourier module (the support of the Fourier transform) has rank two and is generated by two vectors with length ratio $\tau = (\sqrt{5}-1)/2$. The super- molecule in that case is the projection of the 2-dimensional unit cell on V_I.

If one considers (n-d)-dimensional supermolecules parallel to V_I, positioned in the vertices of the n-dimensional lattice (generalisation to more supermolecules per unit cell is then straightforward) the algorithm to construct the vertices of the d-dimensional tiling is as follows.

Take an element $\mathbf{n}$ from $\mathbf{Z}^n$. If $\mathbf{a}_{is}$ is a basis of the n-dimensional lattice, the supermolecule Ω in the vertex

$$ \mathbf{x} = \sum_{i=1}^{n} n_i\, \mathbf{a}_{is} $$

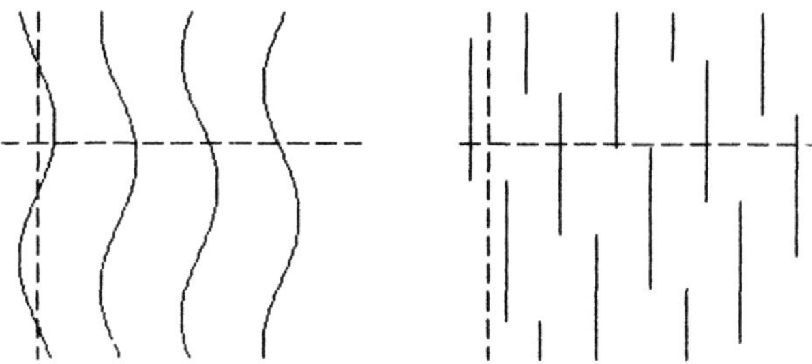

Fig.1 Embedding of a 1-dimensional IC phase (left) and and a 1-dimensional Fibonacci chain (right) into 2-dimensional superspace

gives a vertex of the d-dimensional tiling if it intersects the hyperplane φ = constant. In other words:

$$\mathbf{x}_E \quad \text{is vertex if and only if } \mathbf{x}_I \in \varphi - \Omega$$

Therefore, $\mathbf{x}$ gives a vertex if $\varphi - \mathbf{x}_I$ belongs to the domain Ω in V_I. For example, for the standard octagonal tiling Ω is a 2-dimensional octagon in 4 dimensions, the projection of the 4-dimensional unit cell on V_I.

PHONONS

If one has a quasiperiodic structure as equilibrium configuration of a set of atoms, one may consider the equations of motion if the particles are displaced from these equilibrium points over small distances. In that case one can assume that the forces that drive the particles back to the equilibrium points are harmonic. For simplicity we shall consider systems with only one type of particles. Of course, it is not known whether in such a situation one may find a quasiperiodic ground state, but the formalism is easily extended to the more general case. We shall denote the particle by the vector $\mathbf{n}$, which is a d-dimensional lattice vector of the basic structure in the case of an IC phase[2] and an n-dimensional lattice vector for quasicrystals, and the (n-d)-dimensional vector φ, which belongs to V_I and specifies the position of the physical space which is parallel to V_E. The displacements are

$$\mathbf{u}(\mathbf{n},\varphi)\, e^{i\omega t} + \text{c.c.}$$

where the displacement vector $\mathbf{u}$ is chosen parallel to V_E. Evidently, for quasicrystals one has $\mathbf{u}(\mathbf{n},\varphi) = 0$ if $\varphi - \mathbf{n}_I \notin \Omega$, because then there is no particle at $\mathbf{n}_E$. The equations of motion for the displacements then become

$$m \omega^2 u(n,\varphi) = \sum_{n'} \Phi(n\,n'\,\varphi)\, u(n',\varphi)$$

The dynamical matrix Φ satisfies a number of properties.
- Because the n-dimensional system is periodic, Φ has the symmetry of the n-dimensional lattice:

$$\Phi(n\,n'\,\varphi) = \Phi(n+a\,\,n'+a\,\,\varphi+a_|) = \Phi(n-n',\varphi-n_|)$$

- Because the interaction is only active when both particles n and n' are present, one has

$$\Phi(n\,n'\,\varphi) = 0 \quad \text{if}\quad \varphi - n_| \nsubseteq \Omega \quad \text{or}\quad \varphi - n'_| \nsubseteq \Omega$$

- Because of the n-dimensional lattice periodicity, the solutions of the equations of motion have the Bloch-Floquet property. They are of the form

$$u(n,\varphi) = \exp(i k_s . a_s)\, u(n-a,\varphi-a_|) = \exp(i k_s . n)\, U(\varphi - n_|),$$

where $U(x)$ is a function on the supermolecule Ω in the origin, and because of the lattice translation invariance on each supermolecule. Contrary to the ordinary 3-dimensional case this function is not necessarily smooth. For IC phases the analogue is a periodic function on the internal space. There this function is not smooth either. A smooth function U implies an extended state and it is well known that in quasiperiodic structures localized and critical states can occur as well.

In terms of the function U on Ω the equations of motion become an eigenvalue problem. Substitution of the Bloch Ansatz in the equations gives

$$\omega^2 U(\varphi) = \sum_{n} \Phi(n,\varphi)\, m^{-1} \exp(-i k_s . n)\, U(\varphi + n_|)$$

This is an infinite set of coupled linear equations. For quasicrystals the function Φ has a value different from zero for a small number of vectors n only and for each such vector only in a restricted area of Ω, the region of the supermolecule for which both φ and $\varphi + n_|$ belong to Ω. Because U is a function with the periodicity of the n-dimensional lattice one can decompose it in a Fourier series

$$U(\varphi) = \sum_{K} A(K)\, e^{i K . \varphi} \quad ; \quad \Phi(n,\varphi) = \sum_{K} F(K,n)\, e^{i K . \varphi}$$

Here the sum runs over the vectors that belong to the projection of the n-dimensional reciprocal lattice on V_I. Then one defines the dynamical tensor B as

$$B(K\ K') = \sum F(K' , n) \, m^{-1} \exp(-i\,k_s.n)\,\exp(\,i\,K.n_I)$$

where the sum runs over the lattice vectors n , such that the eigenvalue problem in reciprocal space is just $B\ A = \omega^2\ A$. Therefore the squared eigenfrequencies of the system are the eigenvalues of the matrices B.

SOLUTIONS OF THE EIGENVALUE PROBLEM

The use of the lattice translation invariance does not lead to the same degree of simplification as it does in the 3-dimensional periodic case, where the infinite-dimensional problem is reduced to a problem with dimension which is 3 times the number of particles in the unit cell. To reach a managable problem in the present case one has to make further approximations.

Approximation of the interaction tensor Φ

Consider as an example the case of the Fiboracci chain, a quasiperiodic sequence of intervals with length ratio τ discussed in the first section. Here Ω is a line interval of length $1+\tau$. If one supposes that atoms across a long interval are connected by a harmonic spring constant α_L and those across a short interval by springs with constant α_S the interaction function is given in the following table where a $=$ $(1+\tau)/2$ and $b = (1-\tau)/2$:

n :	0	a_1	$-a_1$	a_2	$-a_2$
$-a < \varphi < -b$:	$-\alpha_L - \alpha_S$	0	α_L	α_S	0
$-b < \varphi < b$:	$-2\,\alpha_L$	α_L	α_L	0	0
$b < \varphi < a$:	$-\alpha_L - \alpha_S$	α_L	0	0	α_S

The simplest approximation is to consider only the lowest Fourier components of the interaction function. If one takes only $F(0 , n)$ into consideration the eigenvalues for the vector $k_s = (k_1, k_2)$ are of the form

$$\omega^2 = 4\ (\ \alpha_L \sin^2(k_1/2) + \alpha_S\,\tau\,\sin^2(k_2/2))/m$$

For long wave lengths this gives a squared frequency which is the average value $(\ \tau\,\alpha_L + (1-\tau)\,\alpha_S\ )/m$ of the force constant times the square of the wave vector k_s. Therefore the acoustic modes are determined by the average spring constant.

Rational approximants

The equations of motion couple an infinite number of points inside the super-molecule Ω . However, when the basis vectors of the n-dimensional lattice are changed in such a way that there are d linearly independent lattice vectors in V_E, the intersection becomes periodic. In nature such commensurate approximants have been observed. By making the unit cell of the periodic structure going to infinity one supposes that the phonons in such a system will tend to those of the quasiperiodic system. Such a procedure has been applied also to incommensurate modulated phases. The commensurate approximation means that the number of points that is coupled by the equations of motion becomes finite. If one considers the Fibonacci chain and approximates τ by 2/3, there

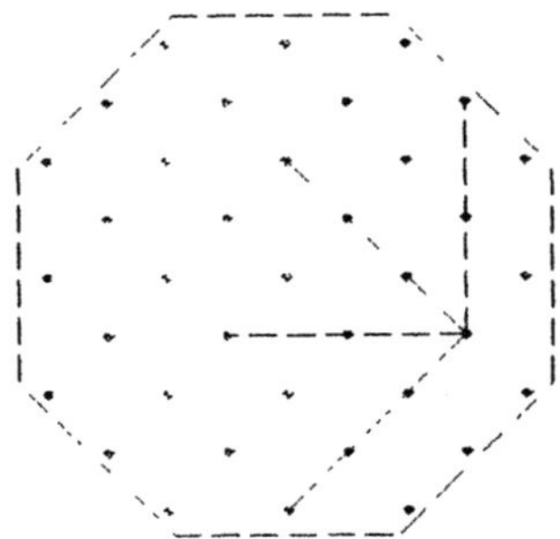

Fig.2 When $\sqrt{2}$ is approximated by 3/2 there are 37 coupled points in Ω. The interactions of one such point with 4 coupled points are indicated by dashed lines.

are 5 coupled points in the interval $[-(1+\tau)/2,(1+\tau)/2]$. Another example is given in Fig. 2, the 2-dimensional octagonal tiling with approximation $\sqrt{2} \approx 3/2$. Then in Ω there are 37 such points. Each of these 37 points is coupled to 4 others.

In the commensurate approximation the phonon problem is reduced to the determination of the eigenvalues of a finite matrix. If the limit of these spectra for the period tending to infinity exists, it is supposed to be the spectrum of the incommensurate structure.

CHARACTERISATION OF THE TYPES OF SPECTRA AND WAVE FUNCTIONS

In ordinary 3-dimensional systems, the spectra of phonons always consist of bands, and the eigenstates are all extended. In quasiperiodic systems there is much more variety. Spectra can be point spectra, or have a point spectrum as component. Spectra may even be singular continuous. Wave functions (or eigenvibrations) can be localized or critical.

An indication of the character of spectra and wave function may be obtained from the analysis of the total spectral measure and the band widths of commensurate approximants, as discussed above. If the irrational number α is approximated by α_p , which is the result of truncation of the continued fraction expansion in the p-th step, this information is obtained for one-dimensional models by plotting the logarithm of the total measure and of band widths as function of p. For many examples one has found a linear dependence, in particular for quasiperiodic systems, like the modulated spring model with incommensurate modulation[3,4], the almost-Mathieu equation and the tight-binding Fibonacci chain[5]. However, other models, like the Thue-Morse[6] and the Cantor tight-binding models show another behaviour, indicating the existence of extended states . A number of examples is given in Fig.3.

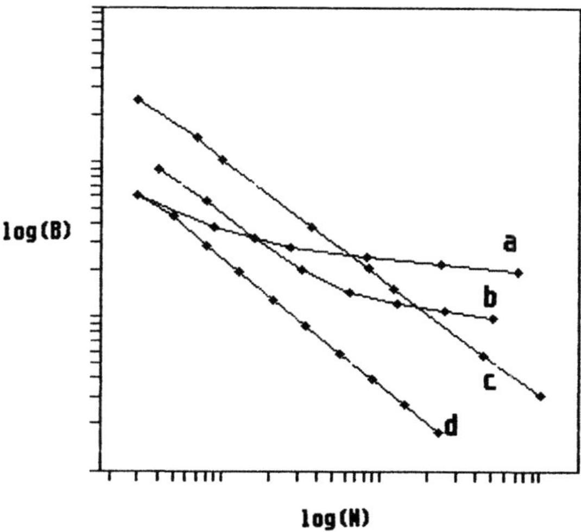

Fig. 3 Logarithm of the total spectral measure against system size :
a) Cantor chain $N_p = 3^p$; b) Thue-Morse chain $N_p = 2^p$; c) $\alpha = (\sqrt{37} - 4)/7$;
d) Fibonacci chain : $\alpha = (\sqrt{5} - 1)/2$.

One way to analyze these scaling properties has been developed by Kohmoto[7]. If the irrationality α is approximated by the rational number $\alpha_p = L_p/N_p$, the size N_p of the system in case of scaling goes as a^p for some number a. If the width of the i-th band scales as $\Delta_j \sim \exp(-p\ \varepsilon_j)$, one may characterize the distribution of scaling exponents by the density function $\Omega(\varepsilon)$, which is suppsoed to scale as $\exp[\ p\ S(\varepsilon)]$. This function $S(\varepsilon)$ can in the p-th approximation easily be calculated as follows. Define for a given approximation the function

$$Z(\beta) = \sum_j \Delta_j^{\beta}$$

Then the function $S(\varepsilon)$ is obtained from the formulas:

$$S(\varepsilon) = F(\beta) + \beta\varepsilon \quad ; \text{ with } F(\beta) = \frac{1}{p} \log Z(\beta) \text{ and } \varepsilon = -\frac{dF}{d\beta}$$

Actually one obtains the function $S_p(\varepsilon)$. If the limit $S(\varepsilon)$ of this sequence of functions exists for p going to infinity the distribution of scaling exponents is known. If $S(\varepsilon)$ is a smooth function this is an indication of a singularly continuous spectrum. It has been shown numerically for simple models that $S(\varepsilon)$ exists. One example is the modulated spring model, a linear chain with harmonic nearest neighbour interaction and with potential energy V expressed in the displacements u_n by

$$V = \sum_n \alpha_n (u_n - u_{n-1})^2$$

where α_n is a quasiperiodic function of n. By way of example one may take the function $\alpha_n = \alpha [1 - \Delta \cos (Q n + \varphi)]$. It has been shown for $\Delta = 1$ that for all irrational values of Q investigated the limit $S(\varepsilon)$ exists. This is even so if one changes the form of the modulation function. Therefore, it seems that phonon spectra in one-dimensional systems may show a behaviour that deviate from those in lattice periodic systems.

If one switches to systems in higher dimensions the analysis becomes more complicated. Preliminary experiments on inelastic neutron scattering from quasi-crystals show that it is difficult to distinguish well determined dispersion branches. Therefore, one should look first to an more easily accessible quantity as the density of states. This quantity, however, is more difficult to analyze in terms of scaling properties. For the moment, the question whether in 3 dimensions there are also sufficiently many localized or critical states to have an influence on the physical properties is unanswered.

REFERENCES

1. T. Janssen and A. Janner, Incommensurability in Crystals, Adv.Phys. 36:519 (1987)
2. R. Currat and T. Janssen, Excitations in Incommensurate Crystal Phases, in: Solid State Physics, Vol.41, H. Ehrenreich and D. Turnbull, eds., Academic Press, San Diego (1988)
3. C. de Lange and T. Janssen, Incommensurability and recursivity: lattice dynamics of modulated crystals , J.Phys.C, 14: 5269 (1981)
4. T. Janssen and M. Kohmoto, Multifractal spectral and wave-function properties of the quasiperiodic modulated spring model, Phys. Rev. B, 38: 5811 (1988)
5. T. Fujiwara, M. Kohmoto and T. Tokihiro, Multifractal wave functions of Fibonacci lattice, Technical Report ISSP , October (1988)
6. F. Axel, J.P. Allouche, M. Kleman, M. Mendes-France and J. Peyriere, Vibrational modes in a one-dimensional quasi-alloy, J. de Physique Colloque , C3-1986:181
7. M. Kohmoto , Entropy function for multifractals, Phys. Rev. A, 37: 1345 (1988)

PHASONS IN FERROELECTRIC LIQUID CRYSTALS

M. Čopič, I. Drevenšek, I. Muševič and R. Blinc

J. Stefan Institute, University of Ljubljana

61000 Ljubljana, Yugoslavia

In ferroelectric liquid crystals the orientational order fluctuations, which are of relaxational type, can be decomposed into a phason and amplitudon mode, and were observed with photon correlation spectroscopy. The phasons show no gap and have parabolic dispersion relation with the minimum at a finite wave vector. In the Sm A phase, both modes become a degenerate soft mode. Its dispersion also has a minimum at a finite wave vector. At wave vectors oblique with respect to the helical axis the phason dispersion relation splits at points $q_z=0$ and $q_z=2q_0$.

1. Introduction

Ferroelectric smectic C^* liquid crystals form a very interesting spatially modulated structure. Like in all smectic systems, the molecules are on average arranged in planes, with the long axis of the molecules tilted with respect to the normal of the smectis planes by an angle θ. Thew tilt direction slowly precesses around the normal and this additional structure is incommensurate with the periodicity of the smectic layers. The helical structure gives rise to strong Bragg reflection of light, which correspond to the satelite reflections in the incommensurate crystals. So, in principle, ferroelectric smectic C^* systems are analogous to solid incommensurate crystals.

There are, however, also important differences. The pitch of the helix, is two to three orders of magnitude larger than the separation of the smectic planes, so that on the scale of the helical pitch the smectic periodicity does not really matter. Also in smectic C^* crystal a lock-in transition does not occur and therefore the analogy is somewhat formal.

The fluctuations of the director field can be decomposed into the fluctuations of the magnitude of the tilt angle, and the small local rotations of the director. The latter can also be described as the fluctuations of the phase of the helical structure and therefore the two modes have been called the amplitudon and the phason mode [1]. Alternatively, they are also called the soft and the Goldstone mode of the transition [2,3]. A homogeneous rotation of the helix around its axis, which can also be represented as a sliding of the helix along its axis, does not perturb the thermal equilibrium. Therefore, in the long wavelenght limit the relaxation time of the overdamped phason mode is expected to be inversely proportional to the square of the wave vector. The amplitudon, representing a local departure from equilibrium, has a finite relaxation time at all wave vectors. These two modes have been observed by light scattering [1]. In the light scattering spectrum, the long wavelenght fluctuations of the helical structure appear in the vicinity of the Bragg scattered spot, corresponding to the periodicity of the helix.

We have observed with light scattering the dispersion relations of the amplitude and phase mode, both at wave vectors along the helical axis and at wavevectors having some perpendicular component. The difference between the

Geometry and Thermodynamics
Edited by J.-C. Tolédano
Plenum Press, New York, 1990

smectic C^* crystal and the solid incommensurate crystal is reflected in the phason dispersion relation. In the solids one always gets a finite gap at the minimum of the phason frequency which is due to pinning on the defects and to lock-in interactions. The smectic C^* phase, however, can on the scale of the pitch of the helix be taken as a continuum, so that the periodicity of the smectic layers and the associated defects are at most very weakly coupled to the phason mode. Therefore, no gap in the dispersion relation is expected.

We have also observed the wave vector dependence of the fluctuations above the transition in the smectic A phase, where the two modes of the Sm C^* phase become a doubly degenerate soft mode.

2. Theory

The collective fluctuations of the orientational order in the smectic C^* phase can be conveniently derived from a Landau thermodynamic potential describing the A to C^* transition. For our purposes it is not necessary to explicitly include the spontaneous electric polarization. Then the smectic C^* phase is described by a two component vector describing the tilt of the molecular director away from the normal to the smectic layers:

$$\vec{\xi}=(\xi_1 , \xi_2)=(n_x n_z, n_y n_z) \tag{1}$$

By writing down the invariants with respect to the D_∞ symmetry of the smectic A phase composed of optically active molecules, one obtains a free energy density [4]

$$g=\frac{1}{2}a\left(\xi_1^2+\xi_2^2\right)+\frac{1}{4}b\left(\xi_1^2+\xi_2^2\right)^2-\Lambda\left(\xi_1\frac{\partial\xi_2}{\partial z}-\xi_2\frac{\partial\xi_1}{\partial z}\right)+$$
$$+\frac{1}{2}K_3\left\{\left(\frac{\partial\xi_1}{\partial z}\right)^2+\left(\frac{\partial\xi_2}{\partial z}\right)^2\right\}+\frac{1}{2}K'\left(\frac{\partial\xi_1}{\partial x}+\frac{\partial\xi_2}{\partial y}\right)^2+\frac{1}{2}K''\left(\frac{\partial\xi_2}{\partial x}-\frac{\partial\xi_1}{\partial y}\right)^2 \quad . \tag{2}$$

Here $a=\alpha(T-T_c)$, b is positive constant, Λ is the Lifshitz constant and K_3 , K', K'' are orientational elastic constants analogous to the twist, splay and bend constants of a nematic liquid crystal.

The condition for minimum of the free energy gives the following expression for the equilibrium value of the order parameter below the transition :

$$\xi_1=\theta_0 cos(q_0 z+\phi_0)$$
$$\xi_2=\theta_0 sin(q_0 z+\phi_0) \tag{3}$$

where $q_0=\Lambda/K_3$ is the inverse pitch, $\theta_0^2=(K_3 q_0^2-a)/b$ the tilt angle and the transition temperature is given by $T_c=T_0+K_3 q_0^2/\alpha$. The constant initial phase ϕ_0 of the helical structure can be set to 0.

To obtain the dynamics of the order parameter fluctuations one also needs an expression for the dissipation function density :

$$\rho=\frac{1}{2}\gamma(\dot{\xi}_1^2+ \dot{\xi}_2^2) \tag{4}$$

where γ is a viscosity coefficient [5]. Before expanding the free energy around equilibrium it is very convenient to transform the free energy expression in the following way. Let us take two new unit vectors

$$\vec{e_u}=\vec{e_x}cos(q_0 z)+\vec{e_y}sin(q_0 z) \quad ,$$

$$\vec{e_v}=-\vec{e_x}sin(q_0 z)+\vec{e_y}cos(q_0 z) \tag{5}$$

where $\vec{e_x}$ and $\vec{e_y}$ are two unit vectors in the directions x and y . Then we can write the order parameter in the form

$$\vec{\xi}=u\vec{e_u}+v\vec{e_v} \quad . \tag{6}$$

Remembering that $q_0 = \Lambda/K_3$ we get the free energy density in terms of u and v

$$g = \tfrac{1}{2}(a - K_3 q_0^2)(u^2 + v^2) + \tfrac{1}{4}b(u^2 + v^2)^2 +$$
$$+ \tfrac{1}{2}K_3\left\{\left(\frac{\partial u}{\partial z}\right)^2 + \left(\frac{\partial v}{\partial z}\right)^2\right\} + \tfrac{1}{2}K_+\left\{\left(\frac{\partial u}{\partial x} + \frac{\partial v}{\partial y}\right)^2 + \left(\frac{\partial v}{\partial x} - \frac{\partial u}{\partial y}\right)^2\right\} + \qquad (7)$$
$$+ \tfrac{1}{2}K_-\left\{\left(\left(\frac{\partial u}{\partial x} + \frac{\partial v}{\partial y}\right)^2 - \left(\frac{\partial v}{\partial x} - \frac{\partial u}{\partial y}\right)^2\right)\cos(2q_0 z) - 2\left(\frac{\partial u}{\partial x} + \frac{\partial v}{\partial y}\right)\left(\frac{\partial v}{\partial x} - \frac{\partial u}{\partial y}\right)\sin(2q_0 z)\right\}$$

where $K_+ = 1/2(K' + K'')$ and $K_- = 1/2(K' - K'')$. The advantage of (7) is that the Lifshitz term has been transformed away. By either taking into account (3) or by minimizing the free energy resulting from (7) we get for the equilibrium values of u and v : $u_0^2 = \theta_0^2$, $v_0 = 0$ below T_c and $u_0 = 0$, $v_0 = 0$ above T_c.

2. a) Fluctuations above T_c

Let us write the order parameter, fluctuating around the equilibrium state, as $u = u_0 + \tilde{u}$ and $v = v_0 + \tilde{v}$. From (7) we can immediately get the equations of motion for the fluctuations. We drop the quartic terms and require that the restoring generalized force acting on the order parameter equals the functional derivative of the free energy. We neglect the inertial terms, so that the restoring force is balanced by the frictional force, obtained from the dissipation function density (4). In terms of u and v we get the equations of motion

$$\gamma\dot{\tilde{u}} = -\frac{\partial g}{\partial \tilde{u}} + \nabla \cdot \frac{\partial g}{\partial \tilde{u}_i}$$
$$\gamma\dot{\tilde{v}} = -\frac{\partial g}{\partial \tilde{v}} + \nabla \cdot \frac{\partial g}{\partial \tilde{v}_i} \qquad (8)$$

where $\tilde{u}_i$ and $\tilde{v}_i$ denote the derivatives with respect to x, y and z . As x and y are equivalent, we may retain only the derivatives with respect to x and z, so that we get from (8) and (7)

$$\gamma\dot{\tilde{u}} = -\alpha(T - T_c)\tilde{u} + K_3\left(\frac{\partial^2 \tilde{u}}{\partial z^2}\right) + K_+\left(\frac{\partial^2 \tilde{u}}{\partial x^2}\right) + K_-\left(\frac{\partial^2 \tilde{u}}{\partial x^2}\cos(2q_0 z) - \frac{\partial^2 \tilde{v}}{\partial x^2}\sin(2q_0 z)\right)$$

$$\gamma\dot{\tilde{v}} = -\alpha(T - T_c)\tilde{v} + K_3\left(\frac{\partial^2 \tilde{v}}{\partial z^2}\right) + K_+\left(\frac{\partial^2 \tilde{v}}{\partial x^2}\right) + K_-\left(-\frac{\partial^2 \tilde{v}}{\partial x^2}\cos(2q_0 z) - \frac{\partial^2 \tilde{u}}{\partial x^2}\sin(2q_0 z)\right)$$

$$(9)$$

The solution of this system is complicated by the spatially dependent K_- terms. These couple fluctuations at wave vector $\vec{q}$ with those at $\vec{q} \pm 2\vec{q_0}$. At small q_x we can treat them as a perturbation. To obtain a nonzero contribution to the relaxation rate, one must go to the second order term. We then get a doubly degenerate soft mode with the inverse relaxation time

$$\frac{1}{\tau} = \frac{\alpha}{\gamma}(T - T_c) + \frac{K_3}{\gamma}q_z^2 + \frac{K_+}{\gamma}q_x^2 + \frac{K_-^2}{\gamma K_3}\frac{q_x^4}{(q_z^2 - q_0^2)} \ . \qquad (10)$$

This approximation breaks down when q_z is close to q_0. The reason is the following. The unperturbed relaxation rate is equal for opposite directions of the wave vector. This degeneracy is harmless except at $q_z = \pm q_0$. These points are coupled by the perturbation term and this causes a splitting in the dispersion relation with a gap equal to $(2K_-/\gamma)q_x^2$.

2. b Fluctuations below T_c

In the smectic C^* phase at $T < T_c$, we have to expand the free energy density around the static equilibrium values $u_0 = \theta_0$ and $v_0 = 0$. Again retaining only terms quadratic in fluctuations and proceeding as before, we get the equations of motion

$$\gamma \dot{\tilde{u}} = -2\alpha(T_c\text{-}T)\tilde{u} + K_3\left(\frac{\partial^2\tilde{u}}{\partial z^2}\right) + K_+\left(\frac{\partial^2\tilde{u}}{\partial x^2}\right) + K_-\left(\frac{\partial^2\tilde{u}}{\partial x^2}cos(2q_0z) - \frac{\partial^2\tilde{v}}{\partial x^2}sin(2q_0z)\right)$$

$$\gamma \dot{\tilde{v}} = K_3\left(\frac{\partial^2\tilde{v}}{\partial z^2}\right) + K_+\left(\frac{\partial^2\tilde{v}}{\partial x^2}\right) + K_-\left(-\frac{\partial^2\tilde{v}}{\partial x^2}cos(2q_0z) - \frac{\partial^2\tilde{u}}{\partial x^2}sin(2q_0z)\right) \qquad . \qquad (11)$$

The terms with K- can again be treated perturbatively and we get two modes, with inverse relaxation times

$$\frac{1}{\tau_a} = \frac{2\alpha}{\gamma}(T_c\text{-}T) + \frac{K_3}{\gamma}q_z^2 + \frac{K_+}{\gamma}q_x^2 + \frac{K_-^2}{2\gamma K_3}\frac{q_x^4}{(q_z^2-q_0^2)} \quad ,$$

$$\frac{1}{\tau_p} = \frac{K_3}{\gamma}q_z^2 + \frac{K_+}{\gamma}q_x^2 + \frac{K_-^2}{2\gamma K_3}\frac{q_x^4}{(q_z^2-q_0^2)} \quad . \qquad (12)$$

The soft mode from the Sm A phase splits into a fluctuation in u, that is in the tilt angle which can be called an amplitude mode, and into a fluctuation in v, which corresponds to the fluctuations of the phase of the helical structure. The relaxation rate of the amplitude mode rapidly increases below the transition temperature, whereas the relaxation rate of the phase mode remains zero at zero wave vector. The perturbation correction is again inappropriate at $q_z=\pm q_0$, where the dispersion relation has a gap equal to $(2K\text{-}/\gamma)q_x^2$.

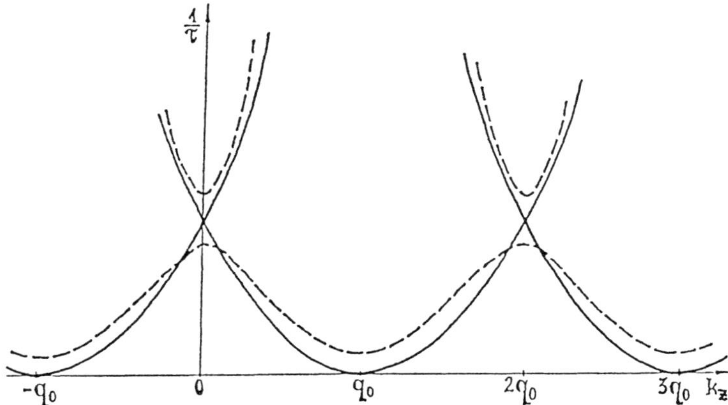

Figure 1. Theoretical phason dispersion relation for scattering wave vectors along the axis of helical modulation (——) and at oblique wave vectors (-----).

In light scattering one observes the fluctuations in the components of the dielectric tensor. These have the same symmetry properties as the products of the components of the molecular director n_in_j. In particular we have

$$\epsilon_{xz}\propto\xi_1 \quad , \quad \epsilon_{yz}\propto\xi_2 \quad . \qquad (13)$$

So we have to transform from the u, v variables back to ξ_1, ξ_2. The main effect of this transformation is that the dispersion relations, given by eq.(10) and (12) are shifted by $\pm q_0$ so that we expect two minima in the inverse relaxation time at the scattering vectors $\pm q_0$, that is in the directions of the Bragg reflections. The form of the dispersion relation for the phase mode in the repeated zone scheme is shown in Fig. 1.

3. Experiment

The experiments were performed on CE8 (S−(+)−[4-(2'-methylbutyl) phenyl 4'-n-octylbiphenyl-4-carboxylate]) which exhibits a smectic C* phase in the temperature interval from 70.3C to 81.0C [6,7,8]. Samples of $75\mu m$ thickness were prepared between clean glass plates. The liquid crystal was aligned with smectic planes perpendicular to the glass plates (bookshelf geometry) by slowly cooling (4K/h) from the isotropic phase to the smectic A phase in a magnetic field of 6.3T. Some samples were also prepared in homeotropic alignment by treating the glass plates with lecithin. In this case the sample thickness was $50\mu m$.

The sample was placed in a cell with a two-stage temperature control, with the temperature stable to a few mK. The cell was filled with glycerine which served both for index matching and for thermal contact. The power of the incident light beam from a He-Ne laser was about 1mW and did not cause any discernible heating of the illuminated sample area. The position of the Bragg diffraction spots gave the value of the helix wave vector q_0. The incident light was polarized as extraordinary wave and scattered light as ordinary wave. In choosing the magnitude and direction of the scattering wave vector one must take into account the relatively large difference between the ordinary and extraordinary index of refraction, so that the Bragg peak appears at an asymmetric position, i.e. the angle of the Bragg scattered light is different from the incident angle. The approximate values for indices of refraction of CE8, obtained from the measurements of conoscopic fringes, are n_o=1.5 and n_e=1.7 .

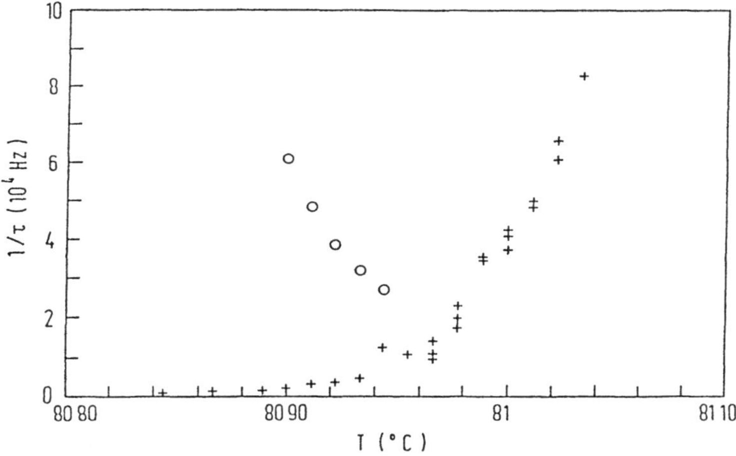

Figure 2. Temperature dependence of the inverse relaxation times at fixed
scattering angle. + - ordinary-extraordinary scattering
o - extraordinary-extraordinary scattering.

4. Results and Discussion

The temperature dependence of the measured inverse relaxation times at fixed scattering angle is shown in Fig.2. Above the transition temperature we get a single inverse relaxation time which decreases towards the transition point T_c . Below T_c the mode splits into two branches, one of which remains slow whereas the relaxation rate of the other increases with decreasing temperature. This increasing, amplitudon, branch was observed in extraordinary-extraordinary (e-e) polarized scattering, in which the wave vector of the observed fluctuations is different from the one in e-o scattering. The e-e scattering geometry was chosen because it has greater scattering cross section for amplitudons. At fixed polarizations the scattering wave vector also changes slightly with changing temperature due to temperature dependence of the indices of refraction. This phenomenon together with the temperature dependence of the helical wave vector q_0 causes a finite, slightly changing value of relaxation rate of the slow phason and prevents direct comparison of the measured data with the expressions (10) and (12). The general features, i.e. the splitting of the degenerate soft mode below the transition into two modes, the slow phason and fast amplitudon, are however evident and in agreement with the results of dielectric measurements in other ferroelectric liquid crystalline materials [9,10,11]

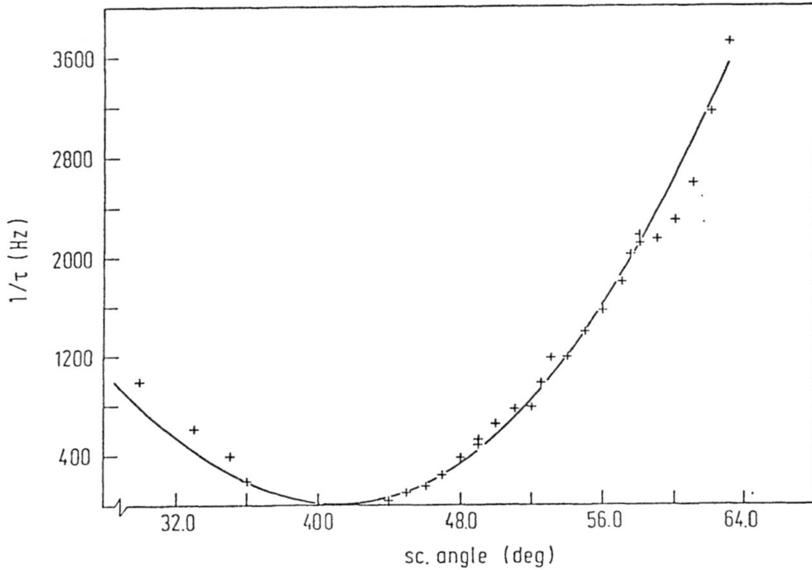

Figure 3. The dependence of the phason inverse relaxation time on q_z, with $q_x=0$. The full line is the least squares fit with the best value of K_3/γ. $T=T_c-0.07K$.

The wave vector dependence of the phason inverse relaxation time below the transition is shown in Figs. 3 and 4. In Fig. 3 the incident and scattering direction were varied in such a way that the scattering vector remained perpendicular to the smectic layers and only changed in magnitude. The resulting wave vector dependence of the relaxation rate is well described by a quadratic function, as expected from Eq. 12b and allows us to determine the ratio of the twist elastic constant to viscosity K_3/γ. In Fig. 4 only the scattering direction was varied. Then the scattering vector in the sample varies not only in magnitude but also in direction. Knowing the indices of refraction and assuming a parabolic dependence of the relaxation rate on the wave vector components, expected by Eq. 12b, we can fit the data and obtain from Fig. 4 both K_3/γ and K_+/γ.

It should be also noted that in both figures the inverse relaxation time goes to an unmeasurable small value at q_0, in agreement with Eq. 12b. In a nonideal sample the fluctuations of the phase of helix could be hindered by defects, leading to a finite relaxation rate even at q_0 [12]. Our results shows that either the defect concentration is very small or that pinning of the helix on defects is weak, in agreement with our expectations.

The dependence of the inverse relaxation time on the scattering direction in the smectic A phase slightly above the transition is shown in Fig. 5. Again, the scattering vector varies both in magnitude and direction in a complicated fashion so that the angular dependence is not simply $sin^2(\vartheta/2)$. Also, the minimum value is no longer zero, in agreement with theoretical expectations (10). A fit to the data gives the values of K_3/γ and K_+/γ. Besides, the position of the minimum in the dispersion curve gives us the value of the helical wave vector above the transition and represents, to our knowledge, the first such measurement.

400

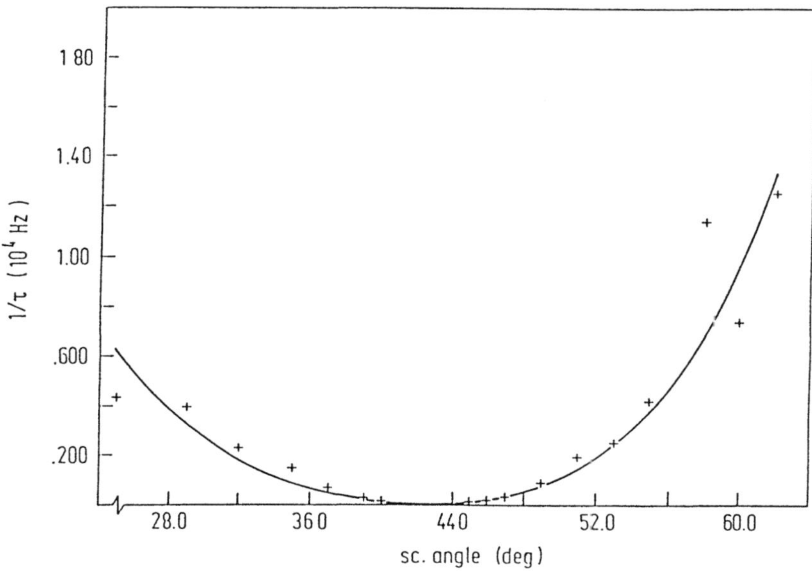

Figure 4. The dependence of the inverse phason relaxation time on the scattering direction. Both q_z and q_x vary. Full line is the least squares fit with the best values of K_3/γ and K_+/γ. $T=T_c-0.72K$.

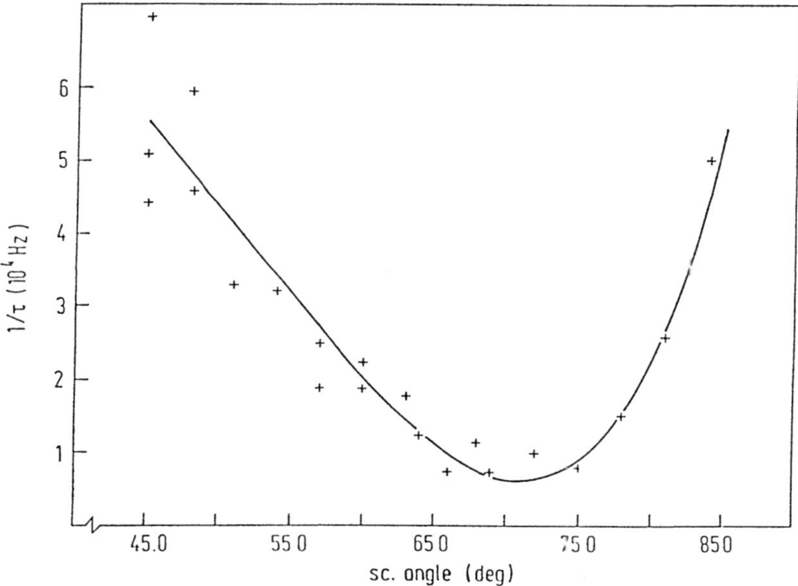

Figure 5. The wave vector dependence of the inverse soft mode relaxation time in the Sm A phase. $T=T_c+0.02K$. The position of the minimum in the dispersion curve gives a value for $q_0=\Lambda/K_3$ even above T_c.

As higher order corrections should be negligible above the transition, the value of q_0 above T_C should be accurately given by the ratio of the Lifshitz parameter Λ and K_3. The observation of this pretransitional influence of the helical structure also clearly shows that q_0 does not vanish at T_C .

The values of the ratio of the orientational elastic constants K_3 and K_+ to viscosity γ at different temperatures below and above the transition are collected in the Table. Within experimental accuracy they do not strongly depend on temperature. The elastic constants K', K'' and viscosity can also be deduced from light scattering studies in a freely suspended film [13,14] . The values obtained in similar materials are comparable to ours.

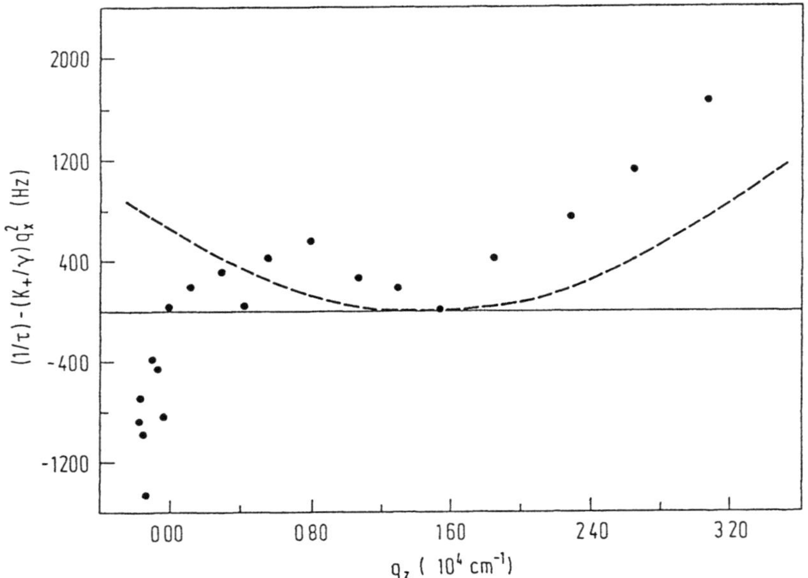

Figure 6.   The difference between the phason relaxation rate and $(K_+/\gamma)q_x^2$ in a mixture of 60 % racemic and 40% right handed material. $T - T_C$=-0.7K.

The experimental data are not accurate enough to allow observation of the perturbative correction of eq. (12) or (10). We believe that the effect of the inhomogeneous terms in the equations of motion can only be observed at q_z=0 or $\pm q_0$, that is at the points where there are gaps in the dispersion relation. This gap is expected to be relatively small, amounting to perhaps one tenth of the relaxation rate at that point. It is experimentally difficult to separate two nearly equal exponentials so that we can unfortunately not measure the gap. We can only observe a marked departure from parabolic dispersion relation close to q_z=0 with $q_x \neq 0$. Fig. 6 shows a set of data, similar to the one in Fig.4 (with the estimated value of $K_+q_x^2$ subtracted), plotted as a function of q_z. These data were taken on a mixed sample of 60% racemic and 40% dextrorotatory material. This mixture was chosen to allow us to measure at both q_z=q_0 and q_z=0 in the same scattering geometry. Around the point q_0 the relation is parabolic, as expected. Close to q_z=0 there is a strong departure from quadratic dependence and the subtracted value even becomes negative. We believe that this is only due to inaccuracy of our value for K_+ which was taken from the Table for a pure material. This value is considerably larger than K_3 so that the data in Fig. 6 can only be taken as a qualitative indication that the dispersion relation is no longer quadratic close to q_z=0. Above the transition temperature, the soft mode dispersion is more difficult to observe and the quality of our data does not allow us to see any departure from the quadratic dependence of the relaxation rate on the wave vector.

402

Table 1   The ratios K_3/γ and K_+/γ in CE8 at different temperatures.

$T-T_c$ [K]	K_+/γ [cm^{2}s^{-1}]	K_3/γ [cm^{2}s^{-1}]
-0.95	-	$2.3\cdot10^{-6}(1\pm0.10)$
-0.72	$3.6\cdot10^{-5}(1\pm0.25)$	$1.6\cdot10^{-6}(1\pm0.40)$
-0.70	$3.3\cdot10^{-5}(1\pm0.05)$	$2.1\cdot10^{-6}(1\pm0.10)$
-0.09	-	$3.1\cdot10^{-6}(1\pm0.09)$
-0.07	-	$2.7\cdot10^{-6}(1\pm0.02)$
-0.05	-	$2.7\cdot10^{-6}(1\pm0.06)$
-0.05	$4.5\cdot10^{-5}(1\pm0.10)$	$6.4\cdot10^{-6}(1\pm0.20)$
$+0.02$	$3.3\cdot10^{-5}(1\pm0.02)$	$3.0\cdot10^{-6}(1\pm0.40)$

References

[1] I.Muševič, R.Blinc, B.Žekš, C. Filipič, M.Čopič, A.Seppen, P.Wyder, A.Levanyuk, *Phys. Rev. Lett.*, 60, 1530 (1988).
[2] R.Blinc, B.Žekš, *Phys. Rev. A*, 18, 7q40 (1978).
[3] T.Carlsson, B.Žekš, C.Filipič, A.Levstik, R.Blinc, *Mol. Cryst. Liq. Cryst.*, 163, 11 (1988).
[4] S.A.Pikin, *Strukturnye prevrasceniya v zidkih kristallah*, Nauka, (1981).
[5] E.M.Lifshitz, L.P.Pitaevskii, *Landau and Lifshitz Course of Theoretical Physics, Vol. 5, Statistical Physics, 3rd Ed.*, Pergamon Press, (1980)
[6] H.R.Brand, P.E.Cladis; *J. Phys. Lett.*, 45, L-217 (1984).
[7] M.F.Bone, D.Coates, A.B.Davey, *Mol. Cryst. Liq. Cryst.*, 102(Letters), 331 (1984).
[8] S.Gierlotka, J.Przedmojski, B.Pura, *Liq. Crystals*, 3, 1535 (1988).
[9] C.Filipič, T.Carlsson, A.Levstik, B.Žekš, R.Blinc, F.Gouda, S.T.Lagerwall, K.Skarp, *Phys. Rev. A,* 38, 5833 (1988).
[10] A.Levstik, T.Carlsson, C. Filipič, I.Levstik, B.Žekš, *Phys. Rev. A,* 35, 3527 (1987).
[11] S.U.Vallerien, F.Kremer, H.Kapitza, R.Zentel, W.Frank, *Phys. Lett. A,* 138, 219 (1989).
[12] P.A.Lee, T.M.Rice, *Phys. Rev. B,* 19, 3970 (1979).
[13] C.Y.Young, R.Pindak, N.A.Clark, R.B.Meyer, *Phys. Rev. Lett.*, 40, 773 (1978).
[14] D.H.VanWinkle, N.A.Clark, *Phys. Rev.A,* 38 , 1573 (1988).

ATOMIC DEBYE-WALLER FACTORS AND PHASONS IN MODULATED INCOMMENSURATE STRUCTURES

J.M. Pérez-Mato and G. Madariaga

Departamento de Física de la Materia Condensada, Facultad de Ciencias
Universidad del Pais Vasco
Apdo 644, Bilbao, Spain

INTRODUCTION

Incommensurately modulated (IC) phases are known to possess low energy vibrational excitations (a zero-gap branch in the ideal case) corresponding to the fluctuations of the phase of the static modulation (see ref. 1 and references therein). This fourth branch of acoustic type is expected to influence drastically the characteristics of the atomic Debye-Waller (DW) factors. To take account of these effects Overhauser proposed a global unconventional thermal factor [2], which was corrected subsequently [3] and used in some experimental studies [4-7]. Other approaches considered as unique peculiarity of the IC structures with respect to thermal factors the possible existence of a modulation of the D-W thermal parameters [8,9]. This last assumption was supported by some derivations in the frame of Landau theory [3] and was rigorously confirmed for a general case in a recent analysis of the problem of dynamics and scattering in an IC structure in the harmonic approximation [10]. The connection of this rigorous approach to the ones which propose special phason thermal factors is unclear.

In the present work, the present theoretical and experimental situation refering this problem is discussed. After reviewing the general expressions for atomic D-W parameters in IC structures, we particularize them for the specific Landau model of phasons and amplitudons in a pure sinusoidal regime. The anomalous contribution of phasons and amplitudons to the atomic thermal tensors is derived. The lack of degeneracy between the amplitudon and phason branches inside the IC phase is shown to cause the build-up of a second harmonic modulation of the thermal parameters. The behaviour of the atomic thermal tensors in a formal soliton regime is also discussed. The basis of a phason thermal factor of the type proposed by Overhauser [2] or its corrected version by Axe [3] are then discussed showing their incompatibility with the results

Geometry and Thermodynamics
Edited by J.-C. Tolédano
Plenum Press, New York, 1990

presented previously. The present experimental situation lacking conclusive results is then briefly described.

GENERAL EXPRESSIONS

We review first some expressions of general validity, refering to the dynamics and the resulting atomic D-W factors of modulated incommensurate structures in the harmonic approximation. Further details can be found in ref. 10.

The equilibrium atomic positions in a modulated incommensurate structure of displacive type are modulated with respect to some reference structure, which has Bravais periodicity, and can be usually identified with a real or imaginary parent structure, in the sense used in the theory of structural phase transitions. In the following, we will refer to this reference as the basic structure. The atoms can then be labelled by a basic lattice vector l and an index μ $(1,...,s)$ inside the basic unit cell. The instantaneous displacement $\delta u(l,\mu)$ of an atom μ belonging to a cell l can then be written in terms of the normal modes λ and their corresponding normal coordinates $C(\lambda)$ as:

$$\delta u(l,\mu) = (Nm_\mu)^{-1/2} \sum_\lambda C(\lambda) \; e(\lambda l l,\mu) \tag{1}$$

where N is the number of cells of the basic lattice included in the crystal, m_μ the mass of atom μ and $e(\lambda l l,\mu)$ is a 3xNxs-dimensional vector describing the structure of the normal mode λ. The normal modes $e(\lambda l l,\mu)$ of the IC structure can be expressed in terms of the normal modes of the basic structure (hereafter refered as basic modes):

$$e(\lambda l l,\mu) = \sum_{kj} a_\lambda(k,j) \; e^{ik \cdot l} \; e^\mu(k,j) \tag{2}$$

where $e^\mu(k,j)$ represents the polarization vector of the basic mode (k,j), j being a branch index. The coefficients $a_\lambda(k,j)$ satisfy [10]

$$\omega_\lambda^2 \, a_\lambda(k,i) = \sum_n \sum_j D^{(n)}{}_{ij}(-k) \; a_\lambda(k-nq,j) \tag{3}$$

where q is the wave vector of the static modulation and the sum extends to all integers n and all branches. The coefficients $D^{(n)}{}_{ij}(k)$ are the coefficients of the Fourier series of a generalized modulated dynamical matrix [10].

The D-W factor of each atom can be shown to satisfy

$$W(l,\mu)(Q)= \sum_{\alpha\beta} Q_\alpha Q_\beta \; B_{\alpha\beta}(l,\mu) \tag{4}$$

with $\alpha,\beta=x,y,z$ and

$$B_{\alpha\beta}(l,\mu) = (1/2) < \delta u_\alpha(l,\mu)(t) \; \delta u_\beta(l,\mu)(t)> \qquad (5)$$

The essential difference with the commensurate case is that here the atomic D-W factors depend on l. In fact, it can be shown that the parameters $B_{\alpha\beta}(l,\mu)$ are in general modulated according to a Fourier series:

$$B_{\alpha\beta}(l,\mu) = \sum_n B^\mu_{\alpha\beta}(n) \; e^{inq\cdot l} \qquad (6)$$

with the nth Fourier amplitude given by [10]

$$B^\mu_{\alpha\beta}(n) = \sum_\lambda (\hbar/4Nm_\mu\omega_\lambda) \; (\; 2 <n_\lambda> +1 \;) \; x$$

$$x \sum_{kjj'} a_\lambda(k,j) \; a_\lambda^*(k\text{-}nq,j') \; e_\alpha^\mu(k,j) \; e_\beta^\mu{}^*(\; k\text{-}nq,j') \qquad (7)$$

where $<n_\lambda>$ and ω_λ denote the Bose thermal occupation factor and the frequency of mode λ, respectively.

LANDAU MODEL

According to the Landau model of the phase transition between the basic and the IC structures [11], the static modulation describing the structural distortion in the IC phase is determined in a first approximation by the structure of the so called "soft-mode" or unstable mode, its normal coordinate being identified with the order parameter. In terms of its polarization vector e_0^μ , the mean atomic positions are then:

$$u_0(l,\mu)= (m_\mu)^{-1/2} \; [\; (1/2) \; Q \; e_0^\mu \; e^{iq\cdot l} \; + c.c. \;] \qquad (8)$$

For further reference we can also write the sinusoidal static modulation in the form:

$$u_{0\alpha}(l,\mu)= \rho \; |\varepsilon_{0\alpha}{}^\mu| \; \cos(q.l +\psi^\mu_{0\alpha} + \Phi) \qquad (9)$$

where $|\varepsilon_{0\alpha}{}^\mu|$ and $\psi^\mu_{0\alpha}$ are the modulus and phase of the α-component of the mass-unweighted polarization vector $(m_\mu)^{-1/2} \; e_0^\mu$ of the soft-mode. ρ and Φ are the amplitude and phase of the modulation order parameter Q. The phase Φ is arbitrary and is taken hereafter as zero ($\Phi=0$).

Following also a Landau approach to describe the system dynamics, the amplitudon and phason modes are related to the basic modes by the relations [12,13]:

$$e_\rho(k\|,\mu) = (1/2)^{1/2} [\, e_0{}^\mu \, e^{ik \cdot l} + e_0{}^{\mu*} \, e^{i(k-2q) \cdot l}] \tag{10}$$

$$e_\Phi(k\|,\mu) = (1/2)^{1/2} [\, e_0{}^\mu \, e^{ik \cdot l} - e_0{}^{\mu*} \, e^{i(k-2q) \cdot l}] \tag{11}$$

where $\mathbf{k}$ is a wave vector close to the static modulation wave vector $\mathbf{q}$. The polarization vector $e_0{}^\mu$ is considered to be the same for the whole set of relevant wave vectors $\mathbf{k}$. This approximation neglecting the variation of the polarization vector along the soft-mode branch must be added when passing from the usual continuous field approach of Landau theory (see for instance refs. 11-13) to the discrete atomic description. It should be stressed that the approximate expressions (10) and (11) are in principle only valid in a strict sinusoidal regime without secondary order parameter distortions [12].

Under these conditions, the contribution of the phason modes to the Fourier amplitudes of the thermal tensor, $B^\mu{}_{\alpha\beta}$, can be readily derived from (7) and (11). The only non-zero terms are :

$$[B^\mu{}_{\alpha\beta}{}^{(0)}]_\Phi = (1/2N) \, \mathrm{Re}[(\mathcal{E}_0{}^\mu)_\alpha (\mathcal{E}_0{}^\mu)_\beta{}^*] \sum_k < C(\Phi,k)C^*(\Phi,k)> \tag{12}$$

$$[B^\mu{}_{\alpha\beta}{}^{(2)}]_\Phi = - (1/4N)(\mathcal{E}_0{}^\mu)_\alpha (\mathcal{E}_0{}^\mu)_\beta \sum_k < C(\Phi,k)C^*(\Phi,k)> \tag{13}$$

where $C(\Phi,k)$ denotes the normal coordinate of the phason mode (Φ,k) (see eq. (11)), and we have used the general property of harmonic normal coordinates:

$$< C(\Phi,k)C^*(\Phi,k) > = (\hbar/2\omega_\Phi(k)) \, (2 <n_\Phi(k)> +1) \tag{14}$$

The contribution of the amplitudon modes (10) have analogous expressions, with the thermal averages substituted by those corresponding to the amplitudon normal coordinates. However, the second order Fourier amplitude $[B^\mu{}_{\alpha\beta}{}^{(2)}]_\rho \ (=[B^\mu{}_{\alpha\beta}{}^{(-2)}]_\rho{}^*)$ have in this case opposite sign than the one in (13).

Considering eqs. (1), (10), (11), the local fluctuations of amplitude and phase of the static modulation (8) can be expressed in terms of the amplitudon and phason normal coordinates:

$$\delta\rho(l) = (2/N)^{1/2} \sum_k C(\rho,k) \, e^{i(k-q).l} \tag{15}$$

$$\rho\,\delta\Phi(l) = - i(2/N)^{1/2} \sum_k C(\Phi,k) \, e^{i(k-q).l} \tag{16}$$

Assuming uncorrelation between the different modes, the thermal averages in eqs. (12) and (13) can then be identified with the local mean fluctuations of the modulation amplitude and phase:

408

$$(2/N) \sum_{k} < C(\rho,k)C^*(\rho,k)> = <\delta\rho^2> \tag{17}$$

$$(2/N) \sum_{k} < C(\Phi,k)C^*(\Phi,k)> = \rho^2<\delta\Phi^2> \tag{18}$$

Despite the modulation of the structure, the mean values $<\delta\rho^2>$ and $\rho^2<\delta\Phi^2>$ do not depend on the position cell l.

Using (17) and (18), the contribution of both phasons and amplitudons to the Fourier amplitudes of the atomic thermal tensor coefficients reduces to

$$B^{\mu}_{\alpha\beta}(0) = (1/4)\, \mathbf{Re}[(\varepsilon_0{}^{\mu})_{\alpha}(\varepsilon_0{}^{\mu})_{\beta}{}^*]\, [<\delta\rho^2> + \rho^2<\delta\Phi^2>] \tag{19}$$

$$B^{\mu}_{\alpha\beta}(2) = (1/8)\,(\varepsilon_0{}^{\mu})_{\alpha}(\varepsilon_0{}^{\mu})_{\beta}\, [<\delta\rho^2> - \rho^2<\delta\Phi^2>] \tag{20}$$

Eqs. (19) and (20) have important practical consequences in X-ray structural analysis of IC phases. According to this model, the phase and amplitude fluctuations give place to a modulation of the atomic thermal parameters with a wave vector $2q$. The Fourier amplitudes for each parameter are determined by the polarization vector $\varepsilon_0{}^{\mu}$ of the static modulation (see eq. (9)), except for a common global factor depending on either the sum or the difference between the mean square fluctuations of both types of fluctuations. In particular, the phase of the second Fourier term $B^{\mu}_{\alpha\beta}(2)$ is given by the sum of the phases $\psi^{\mu}_{o\alpha}+\psi^{\mu}_{o\beta}$ of the static modulation. To include this model of D-W factor modulation in a structural analysis only implies a single additional refinement parameter with respect to the usual X-ray structure determinations, which assume non-modulated atomic thermal tensors. The possible correlation between the phases of the thermal tensor modulation and those of the static atomic modulation was earlier pointed out in ref. 14

As both amplitude and phase modes become degenerate at T_i (the basic-incommensurate transition), the factor $[<\delta\rho^2> - \rho^2<\delta\Phi^2>]$ in eq. (20) becomes zero in this limit (see eqs. (17),(18) and (14)). Accordingly, the only "peculiar" contribution of the phasons is the progressive build-up of a second harmonic modulation of the thermal tensor coefficients with amplitudes starting from zero at T_i and linear with the difference between the phase and amplitude mean square fluctuations. The phase modes also contribute to the large increase of the homogeneous thermal tensors near T_i, which is usual in any second order structural inestability.

SOLITON REGIME

Eqs. (10) and (11) describing the phason and amplitudon modes are only valid if the modulation is purely sinusoidal as given in (9). The presence of secondary order parameter distortions and in general higher harmonics in the modulation would modify the results of the preceding section. The analysis can however be easily generalized to a specific anharmonic modulation corresponding to the so-called soliton regime [11,12], if the description of the IC structure by means of atomic modulation functions (AMF) [9] is introduced.

Neglecting again secondary order parameters but including higher order harmonics of the order parameter, the atomic displacements associated to the primary distortion can be written in general as [11-13] :

$$u_0(l,\mu) = (1/2)Q(l)\,\varepsilon_0{}^\mu\,e^{iq_c.l} + c.c. \tag{21}$$

where only the configuration of the space-dependent order parameter $Q(l)$ is assumed to vary with temperature. In (21), a commensurate value, q_c, (normally corresponding to the expected "lock-in" value) is taken as reference for the modulation wave vector. The order parameter $Q(l)$ (in a continuous approximation, instead of the discrete l-dependence) is the one considered in the usual Landau model for the approach to a lock-in phase through a soliton regime [12].

The AMF along the superspace internal coordinate v, $u^\mu(v)$, are then simply defined by the general rule $u_0(l,\mu) = u^\mu(v = q.l \ (\text{mod. } 2\pi))$, where $\mathbf{q}$ denotes as before the IC modulation wave vector [9]. Following these definitions, the AMF describing the static atomic displacements can be written as

$$u^\mu{}_\alpha(v) = \rho(v)\,|\varepsilon_{0\alpha}{}^\mu|\,\cos(\theta(v)+\psi^\mu{}_{0\alpha}) \tag{22}$$

The v-dependent amplitude $\rho(v)$ and phase $\theta(v)$ in (22) are related to the order parameter configuration $Q(l)$ in (21) by the defining equation: $Q(l)=\rho(v)\exp(i\theta(v))$ with $v=(q-q_c).l$. Expression (21) reduces to the pure sinusoidal case treated in the preceding section for the particular order parameter configuration: ρ=constant and $\theta(v)=v$. The full consistency between eqs. (21) and (22) is guaranteed by the following relations, satisfied by the defined $\rho(v)$ and $\theta(v)$: $\rho(v)=\rho(v+q_c.l)$, $\theta(v+q_c.l) = \theta(v)+q_c.l$. Note that between the phase description along the internal coordinate $\theta(v)$ and the usual continuous approximation $\theta(x)$ for $\theta(l)$, exists a temperature dependent scale transformation, so that the period of $\theta(v)$ is constant and the limit of null soliton density will correspond to zero-width discommensurations in the coordinate v [15]. In the following, we will limit ourselves to the constant amplitude approximation and we assume $\rho(v)=\rho$ to be v-independent.

We consider now the fluctuations of the atomic positions for an arbitrary "cell" v, due to local small fluctuations of the order parameter amplitude $\delta\rho$ and of the modulation phase $\delta\Phi$ (Note that a phason shift should be taken as a global translation of the modulation functions along the internal coordinate v):

$$\delta u^{\mu}{}_{\alpha}(v) = |\epsilon_{o\alpha}{}^{\mu}| \{\delta\rho \cos(\theta(v)+\psi^{\mu}{}_{o\alpha}) - \delta\Phi \, (d\theta(v)/dv) \, \rho \sin(\theta(v)+\psi^{\mu}{}_{o\alpha})\} \tag{23}$$

Assuming that phase and amplitude fluctuations are uncorrelated ($< \delta\rho \, \delta\Phi > = 0$) and that the mean square values $<\delta\rho^2>$ and $<\delta\Phi^2>$ are v-independent (l-independent), it is straightforward to derive for $B^{\mu}{}_{\alpha\beta}(v)$, defined as $(1/2)<\delta u^{\mu}{}_{\alpha}(v)\delta u^{\mu}{}_{\beta}(v)>$, the following expression

$$B^{\mu}{}_{\alpha\beta}(v) = (1/4) \, |\epsilon_{o\alpha}{}^{\mu}| \, |\epsilon_{o\beta}{}^{\mu}| \, \{\cos(\psi^{\mu}{}_{o\alpha} - \psi^{\mu}{}_{o\beta}) \, [< \delta\rho^2> + \rho^2 < \delta\Phi^2> (d\theta(v)/dv)^2] +$$
$$+ \cos(2\theta(v) + \psi^{\mu}{}_{o\alpha} + \psi^{\mu}{}_{o\beta}) \, [< \delta\rho^2> - \rho^2 < \delta\Phi^2> (d\theta(v)/dv)^2] \} \tag{24}$$

Analogously to the preceding section, we obtain a modulation of the thermal tensor coefficients, expressed in this case in terms of a modulation function along the internal coordinate v. The results from section (2) for the pure sinusoidal cases are reproduced just by restricting eq. (24) to the case $\theta(v)=v$, so that $d\theta(v)/dv = 1$.

Expression (24) describes clearly a expected effect in a strong soliton regime where the function $\theta(v)$ adquires a step-like form: the mean atomic displacement resulting from phase fluctuations will become only relevant in the discommensuration regions where $d\theta(v)/dv$ is large (it tends to infinity for zero soliton density), while they are quite negligible in the quasi-commensurate regions where $d\theta(v)/dv \approx 0$. The formal divergency of the phason term at the limit point of null soliton density is irrelevant from a practical point of view because the discommensuration region width (and the number of atoms involved) becomes simultaneously zero. At this point, only the amplitudon modes are relevant, and the modulation of the thermal tensor parameters also becomes step-like, the steps corresponding to the values associated to each of the independent atoms in the commensurate lock-in phase. Therefore, as the soliton regime builds-up, the phason branch is maintained, but its practical influence concentrates steadily in the discommensuration regions of decreasing width.

The influence on the diffraction properties of an expression such as (24) for the thermal tensor can be easily simulated using a general expression for the structure factor as given in ref. 9. However, the resulting effects cannot be reduced to a single simple rule and are rather subtle and/or weak.

A more realistic model would require to take into account secondary distortions and to go beyond the constant amplitude approximation. In principle, this is rather straightforward and requires a direct and simple generalization of the arguments above. However, considering the smallness of the effects and the experimental sensitivity, the resulting expressions for the thermal tensor parameters become too complex to be useful in an experimental situation.

In constrast with the preceding analysis, early in 1971 Overhauser argued that the thermal factor resulting from phason excitations essentially differs from the usual D-W factor [2]. Namely, phase fluctuations were supposed to give place in the diffraction pattern to a systematic decrease of satellite intensities by a factor of the form $\exp(-(1/2)n^2<\delta\Phi^2>)$, where n is the order of the satellite. The exponent of this factor lacks the familiar quadratic dependence on the modulus of the diffraction vector of a normal D-W factor. In principle, this peculiarity should be easily detected experimentally.

Although the argument leading to this factor was originally developed for a monoatomic crystal in a pure sinusoidal regime, it can be easily extended for an IC poliatomic structure with an arbitrary anharmonic modulation by means of the following simple reasoning. The static structure factor of an arbitrary IC structure only suffers a phase shift if the modulation is globally translated along the internal coordinate (modulation phase shift), such that for a modulation shift of $\delta\Phi$, the structure factor of the transformed structure, $F'(H)$, is related with the original one in the form [16]

$$F'(H) = F_0(H) \exp(-in\delta\Phi) \tag{25}$$

where n is the order of the satellite. From this expression it is then straightforward, assuming small local fluctuations of the modulation phase, to derive as effective value of the structure factor:

$$<F(H)> \approx F_0(H) (1 - (1/2) n^2 < \delta\Phi^2>) \approx F_0(H) \exp(-(1/2)n^2<\delta\Phi^2>) \tag{26}$$

where $F_0(H)$ refers to the static configuration. For satellite diffraction vectors, $F_0(H)$ is expected to be proportional to ρ^n. As pointed out by Axe [3], the phase fluctuations in principle also introduce a correction of the mean amplitude of the modulation with the approximate form $\rho \exp(-(1/2) <\delta\Phi^2>)$. Therefore if the experimental structure factor refers to the mean structure, expression (26) should be modified to

$$<F(H)> \approx F_m(H) \exp(-(1/2)n(n-1) <\delta\Phi^2>) \tag{27}$$

Despite some previous attempts [3], such thermal factors as the one in (26) or its corrected version (27) are practically impossible to reconcile in general with the more rigorous predictions of previous sections. Note for instance that according to (19) and (24), phase fluctuations, apart from giving place to a modulation of the atomic thermal tensors, also contribute to a "normal" homogeneous D-W factor, common for all atoms which are lattice equivalent in the basic structure. In contrast, according to (26) or (27), the atomic displacement fluctuations resulting from phasons must not be included in the normal atomic D-W factors, since their effect is completely described by the special factor. On the other hand, according to (19) and (20), any special phason effect (apart form large normal atomic D-W factors) disappears at T_i, while in the Overhauser picture, these effects should be especially enhanced since $<\delta\Phi^2>$ is expected to diverge at T_i. That the Overhauser factor should not be valid close to T_i is not strange since it includes the assumption of small phase fluctuations. However, also at the other limit of the soliton regime, its predictions are in contradiction with the discussion in section 4. There, we showed that phason fluctuations should be relevant only for atoms on the discommensurations, while the argument leading to the Overhauser factor is, as shown above, independent of the form of the modulation function.

The problem of the validity of the Overhauser factor can be clarified if we compare the argument above, leading to it, with the analogous one we can develop for the "fluctuations of the origin" in a commensurate crystal structure. The origin in three-dimensional space of a crystalline structure is arbitrary in the same sense that the origin of the modulation phase is arbitrary in an IC structure. When translating the space origin a vector $\mathbf{u}$, the static structure factor transforms in the form $F(\mathbf{H})\exp(i\mathbf{H}.\mathbf{u})$. Following the same reasoning as above, we then obtain $<F(\mathbf{H})> = F(\mathbf{H})\exp(-(1/2)<(\mathbf{H}.\mathbf{u})^2>)$.

Although this simple argument leads to a qualitative correct $\mathbf{H}$-dependence of atomic thermal factors, it is obvious that we cannot take the result too literally. A correct model requires to include thermal factors for each particular atom, and as a consequence, the global temperature dependence of each structure factor can differ considerably from the one given by the above global factor. Therefore, the argument is formally not rigorous in both cases.

EXPERIMENTAL RESULTS AND CONCLUSIONS

The two incompatible models of a global phason thermal factor given in (26) and (27) have been claimed to have been observed by means of the analysis of the temperature dependence of the intensity of a small set of satellite diffraction satellites [4-7]. However, these studies, following the original work of Overhauser [2], assume a proportionality of $<\delta\Phi^2>$ with temperature, while, according to (18) and (14), this behaviour can only be associated (neglecting the strong temperature dependence of the phason branch frequencies) with $\rho^2<\delta\Phi^2>$, where ρ is in principle temperature dependent. In addition, the temperature variation of the static structural modulation and its influence on the satellite intensities is ignored in refs. 4-6. Also, in ref, 7 a thermal factor of the type (27) is assumed for the whole range of the IC phase even close to T_i, where this type of factors should be clearly not valid.

On the other hand, attempts to introduce either global phason thermal factors or a modulation of the atomic thermal parameters in the full structure refinement of IC structures have been unconclusive or unsuccessful. In fact, almost all IC structure determinations published up to now have been successfully performed assuming "normal" non-modulated atomic D-W factors. This indicates that in general any peculiar phason effect, in particular the expected thermal tensor modulation discussed in the previous sections 2-4, are rather weak.

Nevertheless, we have recently detected the presence of a second order modulation of atomic thermal tensors in an X-ray determination of the modulated structure of thiourea[17]. Although the measurements were done in a commensurate phase with a nominal commensurate wave vector, by continuity and given the high order of the commensurate wave vector value (1/9) we do not expect strong differences with the neighbouring IC configuration. The structure of the determined modulation is not fully consistent with the model of section 3. Its amplitude is however very small, typically one to two orders of magnitude smaller than the non-modulated thermal parameters. More precise and careful measurements are required to arrive to clear conclusions about this point.

Summarizing, we can conclude that the contribution of phasons to the atomic thermal tensors can be divided in two parts: one strong with no modulation, analogous to the anomalous contribution of any zero-gap branch in a commensurate structure, and a weaker one giving place to a modulation of the thermal parameters, whose amplitude depends on the relative weight of amplitude and phase fluctuations.

Acknowledgements: This work has been supported by the DGICYT (Spain), Project nº PB87-0744.

REFERENCES

1. R. Currat and T. Janssen, Solid State Physics, Vol. 41, ed. H. Ehrenreich, D. Turnbull (Academic Press, New York) (1987).

2. A.W. Overhauser, Observability of Charge-Density Waves by Neutron Diffraction, <u>Phys. Rev. B</u> 3, 3173 (1971).

3. J.D. Axe, Debye-Waller factors for incommensurate structures, <u>Phys. Rev. B</u> 21, 4181 (1980).

4. L.D. Chapman and R. Colella, Experimental evidence from X-Ray Diffraction for Phase Excitations in solids, <u>Phys. Rev. Lett.</u> 52, 652 (1984).

5. L.D. Chapman and R. Colella, X-ray scattering studies of charge-density waves in tantalum disulfide, <u>Phys. Rev. B</u> 32, 2233 (1985).

6. R. Colella, Thermal attenuation of satellite reflections in modulated structures, <u>Phys. Rev B</u> 39, 1501 (1989).

7. K.H. Ehses, Phase Fluctuations in the IC-Phase of Rb_2ZnCl_4, <u>Jap. J. Appl. Phys.</u> 24, 793 (1985).

8.   A. Yamamoto, Structure Factor of Modulated Crystal Structures, <u>Acta Crystallogr. Sect. A</u> **38**, 87 (1982).

9.   J.M. Perez-Mato, G. Madariaga, F.J. Zuñiga, and A. Garcia Arribas, On the Structure and Symmetry of Incommensurate Phases. A Practical Formulation, <u>Acta Crystallogr. Sect. A</u> **43**, 216 (1987).

10. A.Garcia, J.M. Pérez-Mato and G. Madariaga, The Dynamics of incommensurate structures and inelastic Neutron scattering, <u>Phys. Rev. B</u> 39, 2476 (1989).

11. R. Blinc and A.P. Levanyuk (Eds.), "Incommensurate Phases in Solids.Fundamentals", North-Holland 1986.

12. A.D. Bruce and R.A. Cowley, The theory of structurally incommensurate systems: III. The fluctuation spectrum of incommensurate phases, <u>J. Phys. C</u> **11**, 3609 (1978).

13. V. Dvorak, in "Modern Trends in the theory of Condensed Matter", ed. A.Pekalsky and J. Przystawa (Berlin-Springer), 447 (1980).13.

14. W.A.Paciorek, D.Kucharczyk, Structure Factor Calculations in Refinement of a Modulated Crystal Structure, <u>Acta Cryst. A</u> 41,462 (1985).

15. J.M. Perez-Mato, R. Walisch, and J. Petersson, Static quadrupolar perturbed NMR in structurally incommensurate systems I. Electric-field-gradient tensor, <u>Phys. Rev. B</u> **35**, 6529 (1987).

16. J.M. Pérez-Mato, G. Madariaga, F.J. Zúñiga, Determination of incommensurate structures: choice of atomic parameters, <u>Phase Transitions</u> (1989),to be published.

17. G. Madariaga, F.J. Zúñiga, W.A. Paciorek, J.M. Pérez-Mato, J.M. Ezpeleta, I.Etxebarria, A New X-Ray Determination of the Modulated Structure of Thiourea, <u>Ferroelectrics</u> (1989),to be published.

THE LANDAU FREE ENERGY AND ACOUSTIC ANOMALIES IN

INCOMMENSURATE INSULATORS

H.Z. Cummins, Gen Li and N. Tao
Department of Physics, City College of the City University
of New York
New York, NY 10031

R.M. Pick, C. Dreyfus and M. Hebbache
Département de Recherches Physiques, Universite P & M Curie
4 Place Jussieu, 75230 Paris Cedex 05, France

Incommensurate phases of insulating crystals have been studied extensively during the last fifteen years, with some 66 different materials already identified.[1] The phenomenological Landau free energy has played a central role in both the general theoretical understanding of these phases and in the analysis of experimental data.

The Landau free energy can be viewed as a "joining" approach, providing a connection between microscopic theories (such as the theory of $NaNO_2$ discussed in this workshop by Karl Michel) which can predict numerical values for the free energy coefficients, and experiments whose analysis also provide numerical values for these coefficients. Unfortunately, theoretical studies often ignore the issue of numerical values, while experimental investigations usually treat the coefficients as fitting parameters, adjust them to provide an optimum fit to a particular set of experimental data, and ignore the question of agreement with the results of other experiments.

In 1984, Sannikov and Golovko presented an analysis of K_2SeO_4 in which the results of a number of different experiments were analyzed self-consistently to extract values for the free energy coefficients of this material.[2] In the spirit of their analysis, we have undertaken a new study of the elastic constants of K_2SeO_4 as a test of the Landau approach. The study, which is part of an international research collaboration supported by NATO Scientific Affairs Division, will be described in detail in a forthcoming publication. Here we present a preliminary description of the results in which the free energy, with coefficients determined from the existing experimental literature, is analyzed to predict the temperature-dependent complex elastic constants which are then compared with our experimental data.

K_2SeO_4 is an orthorhombic pseudohexagonal crystal with space group D_{2h}^{16} = Pnam from the hexagonal-orthorhombic transition at 745K to the incommensurate transition at T_I = 129.5K. Between T_I and the lock-in transition at T_L = 93K it is structurally modulated with the modulation

Geometry and Thermodynamics
Edited by J.-C. Tolédano
Plenum Press, New York, 1990

wavevector $q_0 = (1-\delta)a^*/3$; the misfit parameter δ decreases continuously from ~ 0.07 at T_I to ~ 0.02 at T_L where it drops discontinuously to zero. Below T_L, K_2SeO_4 is orthorhombic with space group C_{2v}^9 = Pna2$_1$ and is an improper ferroelectric with spontaneous polarization P_z. A crucial neutron scattering study by Iizumi et al[3] established that the intermediate phase is incommensurate and showed that there is a soft optic mode with wavevector q_0 on a Σ_2 branch whose frequency approaches zero as $T \rightarrow T_I^+$, showing that K_2SeO_4 is a soft-mode driven displacive transition material.

THE LANDAU FREE ENERGY

The Landau free energy functional $f(x)$ for K_2SeO_4 can be expressed as a sum of three parts:

$$f(x) = f_Q(x) + f_\epsilon(x) + f_P(x) \tag{1}$$

where $f_Q(x)$ includes terms only in the order parameter Q which is the complex position-dependent amplitude of the Σ_2 mode at the commensurate wavevector $q_c=a^*/3$, $f_\epsilon(x)$ includes terms in the strains ϵ_i (i=1-6) both alone and in combination with Q, and $f_P(x)$ includes terms in the polarization P_3 both alone and in combination with Q.

$$f_Q(x) = \frac{1}{2}\alpha QQ^* + \frac{1}{4}\beta'(QQ^*)^2 + \frac{1}{6}\gamma_1'(QQ^*)^3 - i\frac{\sigma}{2}(Q\frac{dQ^*}{dx}-Q^*\frac{dQ}{dx}) + \frac{\kappa}{2}\frac{dQ^*}{dx}\frac{dQ}{dx}$$
$$+ \frac{1}{2}\gamma'(Q^6+Q^{*6}) \tag{2}$$

$$f_P(x) = \frac{1}{2\chi_0}P^2 + \frac{\eta}{2}P^2QQ^* + i\xi P(Q^3-Q^{*3}) - PE \tag{3}$$

$$f_\epsilon(x) = \frac{1}{2}\sum_{ij=1}^{6} C_{ij}\epsilon_i\epsilon_j - \sum_{j=1}^{3} h_j\epsilon_j QQ^* + \sum_{i,j=1}^{3} g_{ij}\epsilon_i\epsilon_k QQ^* + \sum_{i=4}^{6} g_{ij}\epsilon_i^2 QQ^*$$
$$+ \frac{1}{2}a_5\epsilon_5(Q^3+Q^{*3}) - \sum_{i-1}^{6} \epsilon_i\sigma_i \tag{4}$$

To simplify $f(x)$, we first make the usual transformation to polar coordinates $Q=\rho e^{i\phi(x)}$, and invoke the continuum constant-amplitude approximation $d\rho/dx=0$. Since we will not consider the polarization P we minimize the average free energy density $F = (1/L)\int_0^L f(x)dx$ with respect to P to find its equilibrium value, and eliminate it from $F(x)$ assuming that the electric field E and the stresses σ_i are zero.

Minimization of the resulting expression with respect to ϕ shows that near T_I, where ρ is small, $d\phi/dx$ is a constant independent of x which we take as $q-q_c$. Dropping the terms in ρ^6 which are small near T_I and all strains except ϵ_1, ϵ_2, ϵ_3 then gives an approximate form for $f(x)$:

418

$$f_1(x) = \frac{1}{2}[\alpha - 2\sigma(q - q_c) + \kappa(q - q_c)^2]\rho^2 + \frac{1}{4}\beta\rho^4 + \frac{1}{2}C_0\epsilon^2 - \sum_{i=1}^{3} h_i \epsilon_i \rho^2$$

$$+ \sum_{i=1}^{3} g_i \epsilon_i^2 \rho^2 \tag{5}$$

This is just the incommensurate plane wave limit; ρ is the amplitude of a Σ_2 mode at the (still unspecified) wavevector q, and $f_1(x)$ of eq. (6) is thus the free energy density of a single mode. Minimization of $f_1(x)$ with respect to q gives

$$q = q_0 = q_c + \frac{\sigma}{\kappa}$$

so that with $q=q_0$,

$$f_1(x) = \frac{1}{2}(\alpha - \frac{\sigma^2}{\kappa})\rho^2 + \frac{1}{4}\beta\rho^4 + \frac{1}{2}C_0\epsilon^2 - \sum_{i=1}^{3} h_i \epsilon_i \rho^2 + \sum_{i=1}^{3} g_i \epsilon_i^2 \rho^2 \tag{6}$$

Note that the Lifshitz invariant term $-i\frac{\sigma}{2}(Q\frac{dQ^*}{dx} - Q^*\frac{dQ}{dx})$ in eq. (2) shifts the minimum in the part of $f_1(x)$ quadratic in ρ from $q_c = a^*/3$ to q_0, and also increases the transition temperature from T_0 where $\alpha = \alpha_0(T-T_0) \rightarrow 0$ (the virtual para-commensurate transition) to $T_I = T_0 + \frac{\sigma^2}{\alpha_0\kappa}$ where $(\alpha - \frac{\sigma^2}{\kappa}) \rightarrow 0$.

Finally, replacing $(\alpha - \frac{\sigma^2}{\kappa})$ by A and β by B and noting that $f_1(x)$ is no longer dependent on x and is therefore identical to the average free energy density F, we have (approximately)

$$F = \frac{1}{2}A\rho^2 + \frac{1}{4}B\rho^4 + \frac{1}{2}\sum_{ij}C_{ij}\epsilon_i\epsilon_j - \sum_{i=1}^{3}h_i\epsilon_i\rho^2 + \sum_{i=1}^{3}g_i\epsilon_i^2\rho^2 \tag{7}$$

This is the form we will use to predict the elastic constants near T_I.

EVALUATION OF K_2SeO_4 FREE ENERGY COEFFICIENTS

Many of the coefficients appearing in the approximate free energy density expression of Eq. (7) and the full free energy functional of Eqs. (1)-(4) can be evaluated from the existing experimental literature on K_2SeO_4.

The coefficient $A=A_0(T-T_I)$ can be determined from the soft mode frequency Ω_0 determined in the neutral scattering experiment of Iizumi et al[3], since in the Landau approximation $A = 2\Omega_0$. The constants B and h_i can be found from the specific heat $C_P = -T(\frac{\partial^2 F}{\partial T^2})$ and thermal expansion coefficients $\alpha_j = (\frac{\partial \epsilon_j}{\partial T})_{\sigma=0}$. From the free energy expression of Eq. (7), the specific heat is

$$C_P(T>T_I) = C_P^0.$$

$$C_P(T<T_I) = C_P^0 + A_0^2\ T/2[B-2h_i(C^{-1})_{ij}h_j] \tag{8}$$

where C_P^0 is the background specific heat due to other degrees of freedom. Eq. (8) predicts that with decreasing temperature the specific heat should exhibit a positive jump ΔC_P at T_I of

$$\Delta C_P = A_0^2\ T_I/2[B-2h_i(C^{-1})_{ij}h_j] \tag{9}$$

The thermal expansion coefficients $\alpha_j = (\frac{\partial \epsilon_j}{\partial T})_{\sigma=0}$, with the equilibrium strains determined by $\epsilon_j^0 = (C^{-1})_{ji}h_i\rho^2$, are given approximately by

$$\alpha_j(T>T_I) = \alpha_j^0 \tag{10}$$

$$\alpha_i(T<T_I) = \frac{\alpha_j^0 - (C^{-1})_{ji}h_iA_0}{[B-2h_i(C^{-1})_{ij}h_j]}$$

which also predicts a jump, $\Delta\alpha_j$, at T_I. Combining these results for C_P and α_j, we find

$$h_i = -A_0T_IC_{ij}\ \frac{\Delta\alpha_j}{2\Delta C_P}$$

$$B = \frac{1}{2}\left[\frac{T_IA_0^2}{\Delta C_P} + 4h_kC_{ki}^{-1}h_i\right] \tag{11}$$

We have evaluated B, h_1, h_2 and h_3 from the specific heat data of Atake et al[4] and the thermal expansion data of Flerov et al.[5] These results, and the elastic constants determined in ultrasonic experiments, determine all the important coefficients related to C_{33} in Eq. (7) except for g_3 which we have estimated from the spontaneous strain data of Kudo and Ikeda.[6]

DERIVATION OF THE ELASTIC ANOMALY

The elastic constants of K_2SeO_4 exhibit anomalies in the vicinity of the phase transitions due to coupling of strains to the order parameter. The three longitudinal strains ϵ_1 , ϵ_2 , and ϵ_3 all couple to QQ^* through the terms $h_j\epsilon_jQQ^*$ in Eq. (4) although the only major anomaly near T_I (approximately 25%) occurs in the C_{33} mode.[7] Such anomalies are often found in incommensurate and other structural phase transitions.

The static value of the linear elastic constant C_{33} above and just below T_I, found from the reduced free energy density of Eq. (7), is

$$C(\omega=0) = \delta\sigma/\delta\epsilon = (C_0+2g\rho_0^2) - 4(h-2g\epsilon_0)^2/2B \quad (T<T_I) \tag{12}$$
$$C(\omega=0)=C_0 \qquad\qquad\qquad\qquad\qquad (T>T_I)$$

The temperature-dependent value of the first part of Eq. (12), $C_0+2g\rho_0^2$, is shown in Fig. 1a. The second part of Eq. (12), which arises below T_I from the stress produced by $\delta\rho$, depends on the acoustic frequency ω and can also be analyzed dynamically through a coupled mode analysis. The dynamical contribution to (the complex) $C(\omega)$ is

$$-\left[\frac{4(h-2g\epsilon_0)^2}{2B}\right]\frac{\Omega_0^2}{\Omega_0^2-\omega^2-i\omega\Gamma_0} \tag{13}$$

where Ω_0 is the amplitudon frequency.

In the limit $\Omega_0 \gg \omega$, Eq.(32a) reduces to

$$- [4(h-2g\epsilon_0)^2/2B] \;/\; [1-i\omega(\Gamma_0/\Omega_0^2)]$$

Eq. (13) reduces to the second term in Eq. (12) in the static limit $\omega \to 0$. This coupled mode contribution to C is shown in Fig. 1b. Note that this bilinear coupled mode contribution to $C(\omega)$ in the limit $\Omega_0 \gg \omega$ is indistinguishable from coupling to a simple Debye relaxational mode with relaxation time $\tau = \Gamma_0/\Omega_0^2$, a case first discussed in 1954 by Landau and Khalatnikov.[8] (A similar bilinear coupling to the phason is allowed for C_{11}, but not for C_{22} or C_{33}.)[9,10]

The linearization procedure leading to equation (12) eliminates the effect of fluctuations. The contribution of the leading (cubic) coupling term $-h\epsilon\rho^2$ to the stress is $-2h\rho_0\delta\rho - h(\delta\rho)^2$. The first term, $-2h\rho_0\delta\rho$, is the induced bilinear coupling discussed above, while the second term, $-h(\delta\rho)^2$, couples strain to fluctuations in the order parameter and is important near T_I where these fluctuations become large.

This fluctuation term, which has been discussed by several authors[11-15], can be viewed (to leading order) as a renormalization of the acoustic mode frequency due to one acoustic phonon coupling to a pair of optic phonons on the soft Σ_2 branch at $\pm$ q where q can lie anywhere in the Brillouin zone.

The cubic and quartic coupling terms together produce a fluctuation-induced stress term $\delta\sigma = -(h-2g\epsilon_0)(\delta\rho)^2$ which produces downward curvature of the elastic constants near T_I on both sides of the transition. As noted by Yao et al, the fluctuation contribution above T_I can be computed from the soft-mode dispersion curve determined by inelastic neutron scattering.[16] Below T_I, the fluctuation effect has two components: one due to pairs of amplitudons which should look essentially like the effect above T_I (except for a scale change in the temperature dependence), and a second component due to coupling to pairs of phasons which should be essentially temperature independent.

Following essentially the same analysis as Yao et al,[16] the contribution of the fluctuation term to the complex elastic constant is found to be

$$-\frac{4(h-2g\epsilon_0)^2k_BT}{(2\pi)^3m^*}\int_q d^3q\{\Omega^2(q)[4\Omega^2(Q)+2i\omega\Gamma(q)]\}^{-1} \tag{14}$$

where $\Omega(q)$ and $\Gamma(q)$ represent the frequency and damping of soft modes in the normal phase, or of amplitudons and phasons in the incommensurate phase, of wavevector q.

Finally, we can collect all contributions to the complex elastic constant $C(\omega) = C'(\omega)-iC''(\omega)$, and relate the imaginary part $C''(\omega)$ to the damping constant or Brillouin linewidth (half width at half maximum) γ by

$$\gamma = -(q^2/d)C''/\omega \tag{15}$$

so that

$$C'(\omega) = [C_0+2g\rho_0^2] - [4(h-2g\epsilon_0)^2/2B]\frac{\Omega_0^2(\Omega_0^2-\omega^2)}{(\Omega_0^4+\omega^2\Gamma^2)}$$

$$-\frac{4[h-2g\epsilon_0]^2k_BT}{(2\pi)^3m^*}\int_q\frac{d^3q}{4\Omega(q)^2+\omega^2\Gamma(q)^2} \tag{16}$$

and

$$\gamma(\omega) = \gamma_0 + \frac{q^2}{d}\frac{4(h-2g\epsilon_0)^2}{2B}\frac{\Omega_0^2\Gamma_0}{(\Omega_0^4+\omega^2\Gamma_0^2)}$$

$$+\frac{4[h-2g\epsilon_0]^2k_BT}{2(2\pi)^3m^*}\int_q\frac{\Gamma(q)d^3q}{\Omega_q^2[4\Omega(q)^4+\omega^2\Gamma(q)^2]} \tag{17}$$

Evaluation of $C'(\omega)$ and $\gamma(\omega)$ in Eqs. (16) and (17) requires the parameters of h, g, B, and C_0 (as discussed above) as well as Ω_0 and Γ_0 of the amplitudon for the bilinear coupling term, and $\Omega(q)$ and $\Gamma(q)$ of the soft branch above T_I, and of both the amplitudon and phason branches below T_I, for the fluctuation integral.

The temperature-dependent $\Omega(q)$ on the soft Σ_2 branch for $T > T_I$ can be taken from the inelastic neutron scattering data of Iizumi et al[3] from which

$$[\hbar\Omega(q)]^2 = [\hbar\Omega_0]^2 + \beta_1'(q_x-q_0)^2 + \beta_2'q_y^2 + \beta_3'q_z^2 \tag{18}$$

where $[\hbar\Omega_0]^2 \cong 0.071(T-T_I)[(meV)^2]$ and $q_0 = 0.31\ a^*$.

The amplitudon and phason parameters needed to evaluate the fluctuation integrals below T_I were taken from Unruh et al[17] and Quilichini and Currat.[18]

The contribution of the fluctuation integrals to C' and γ are shown in Fig. 1(c).

In Fig. 1(b) we show both the static limit $\omega=q=0$ (solid line) and the predicted $90°$ scattering result ($\omega = 2\pi \times 0.54$ cm^{-1}) (broken line). Finally, Fig. 1(d) shows the total theoretical anomaly in C'_{33} and γ_{33} which are the sum of (a), (b) and (c).

BRILLOUIN SCATTERING EXPERIMENTS

Previous studies of the elastic anomalies in K_2SeO_4 have been undertaken by several authors with ultrasonic[7,19-21], acoustic resonance[22] and Brillouin scattering[7,23-28] methods. The K_2SeO_4 crystals used in our

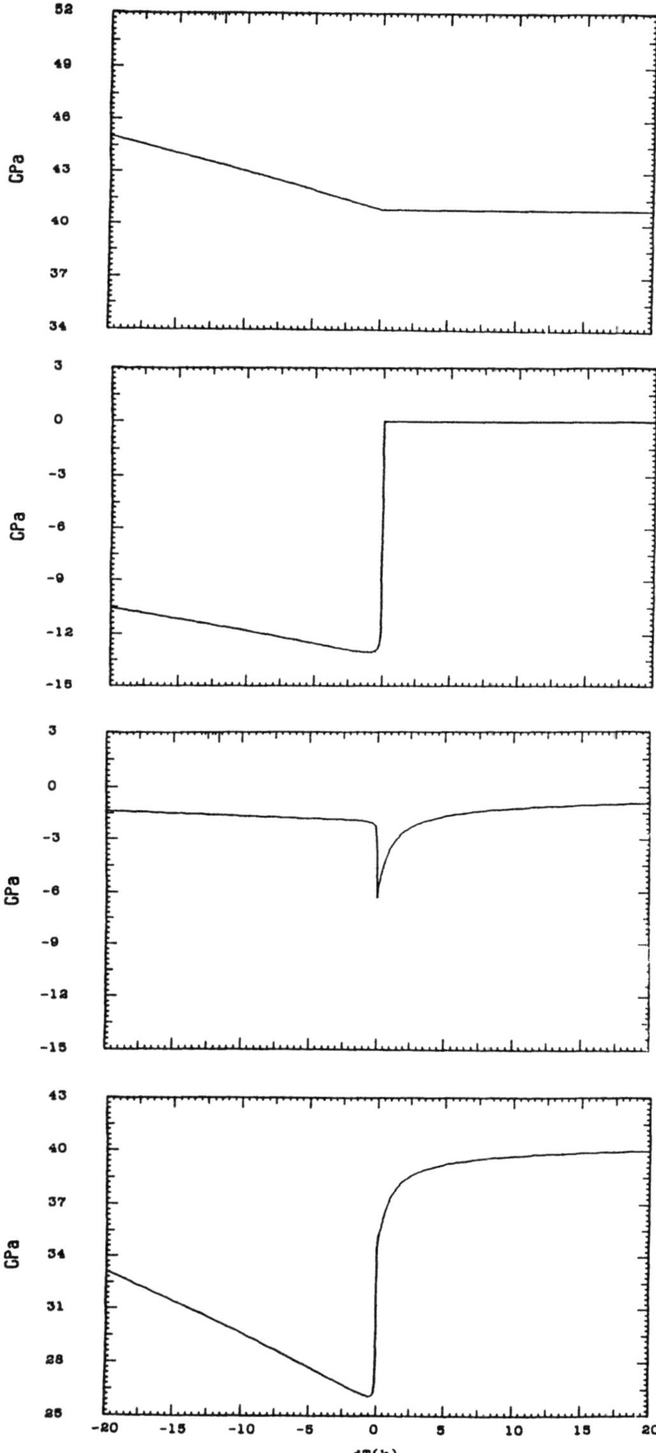

Figure 1. Theoretical acoustic anomaly in $C'(\omega)$ [left] and $\gamma(\omega)$ [right] predicted by Eqs. (16) and (17). Top to bottom: first, second and third terms and total values.

(continued)

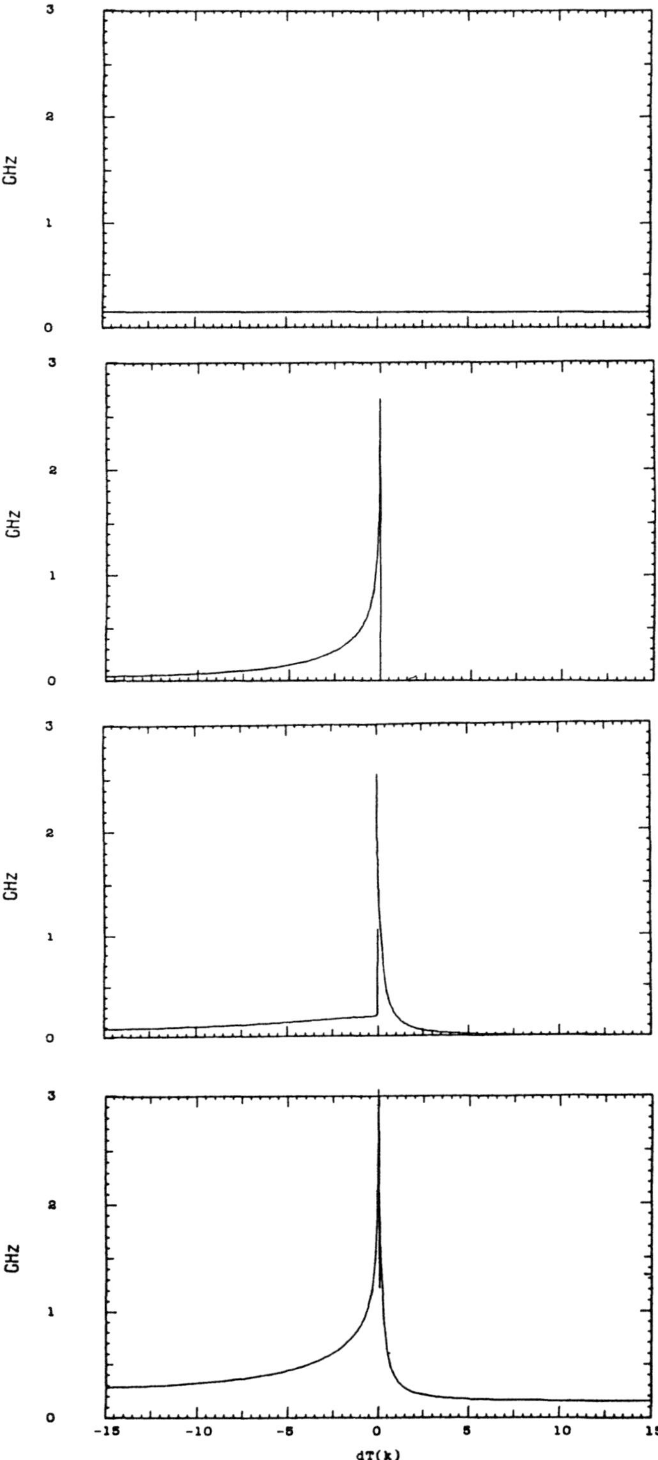

Figure 1 (Continued)

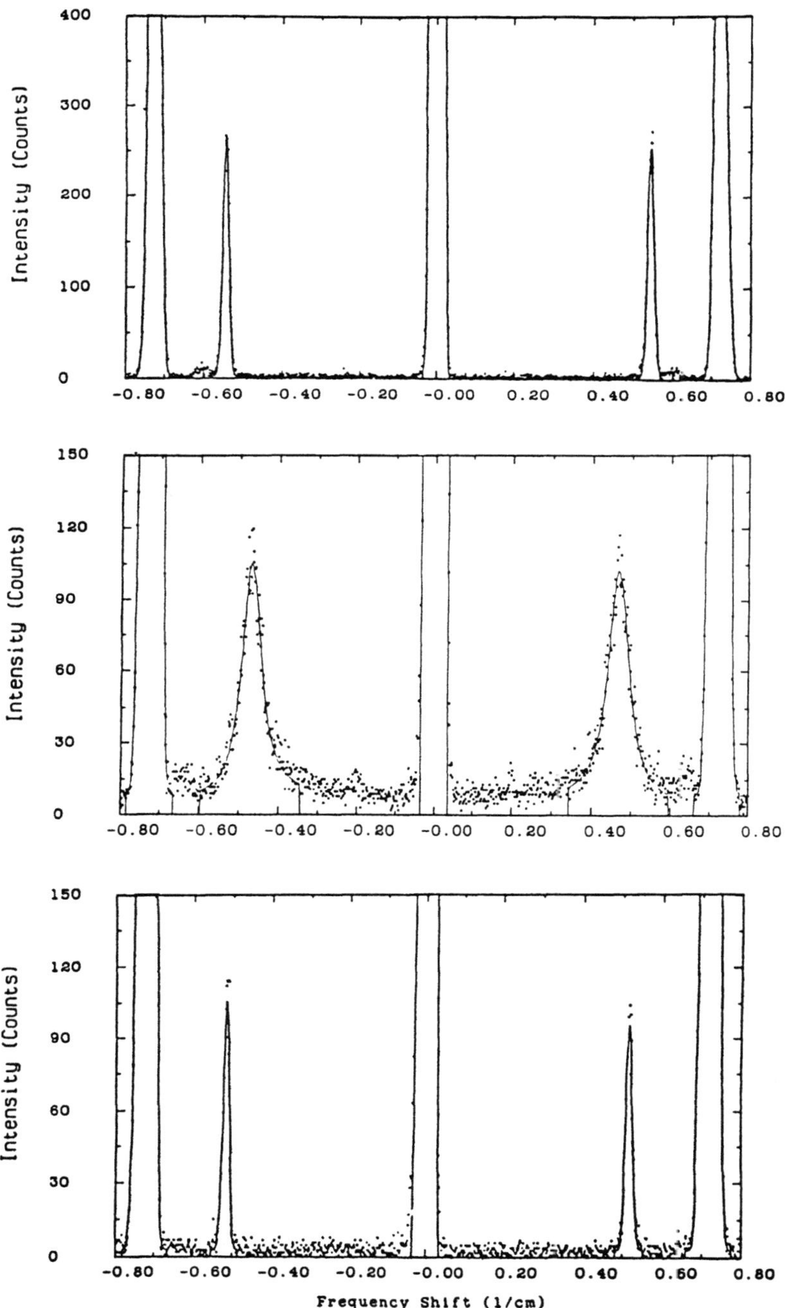

Figure 2. (a+c)[b,T](a-c) Brillouin spectra of K_2SeO_4 at (top to bottom) T = 300, 130 and 81 K.

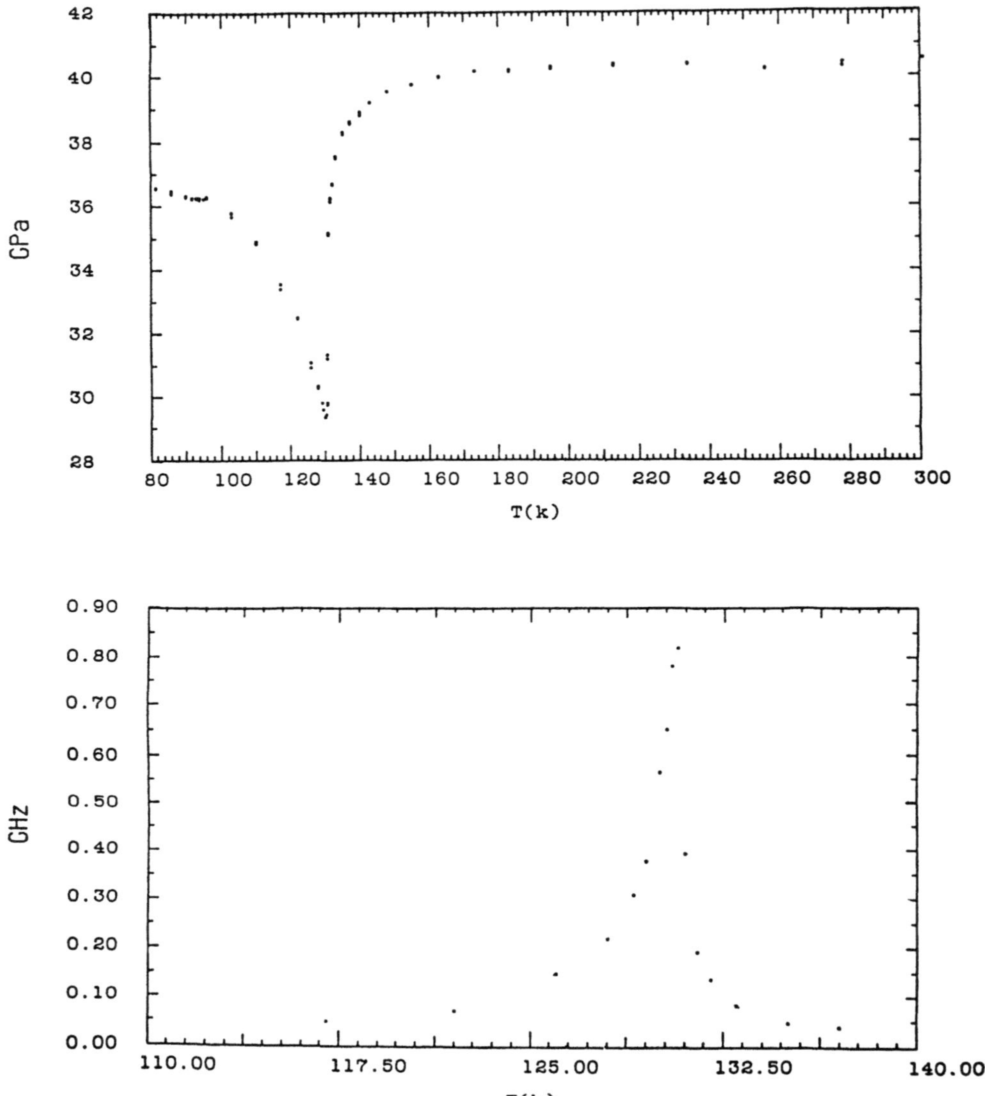

Figure 3. Temperature dependence of C_{33} (top) and γ (bottom) from Brillouin shifts and linewidths.

experiments were grown by G. Hauret at the Universite d'Orleans from 99%
stock material (Alfa Products, Danvers, Mass.) which was recrystallized
twice for purification.

For the 90° scattering experiments a crystal approximately 6 mm on
each side was cut with faces perpendicular to b, a+c and a-c. The
scattering geometry used was (a+c)[b, T](a-c).

The crystal was mounted on the coldfinger of a Cryotip continuous
flow cyrostat controlled by an Oxford ITC-4 temperature controller. A
Spectra-Physics 165 single-mode argon ion laser provided 100 mW of
incident v-polarized 4880Å light. Scattered light was analyzed with a
6-pass (3 x 2) Sandercock tandem Fabry Perot interferometer with 14mm
plate separation, scanned at two scans per second.

Data were acquired with an AT-type computer equipped with an EG&G
multiscalar board. 1024 data channels were stored, typically for 3,000
scans near T_I and 1,200 scans far from T_I where the Brillouin lines are
narrow. Data was subsequently transferred to a Vax 780 for analysis with
a nonlinear least squares fitting program.

Figure 2 shows three such spectra, at T=300, 130 and 81 K. Note the
evident broadening at 130 K. The elastic constant C_{33} and linewidth γ
obtained from Lorentzian fits (after deconvolution of the instrument
function) are shown in Fig. 3. The agreement with the theoretical
prediction of Fig. (1d) is qualitatively reasonable, but the quantitative
aspects of the fit turn out to depend critically on the q=0 phason gap.
As our data analysis proceeds, the value of this gap (whose value is
currently somewhat controversial) should be clarified by our results.

We wish to thank G. Hauret for providing the crystals used in this
experiment. We acknowledge the support of the NSF, CNRS, and NATO
Scientific Affairs Division for the work described here.

REFERENCES

1. c.f. _Incommensurate Phases in Dielectrics_, R. Blinc and A.P. Levanyuk,
eds. (N. Holland, Amsterdam, 1986); H.Z. Cummins, Physics Reports (in
press).

2. D.G. Sannikov and V.A. Golovko, Sov. Phys. Solid State 26, 678 (1984)

3. M. Iizumi, J.D. Axe, G. Shirane and K. Shimaoka, Phys. Rev. B15, 4392
(1977).

4. T. Atake, K. Nomoto, B.K. Chaudhuri and H. Chihara, J. Chem.
Thermodynamics 15, 383 (1983).

5. I.N. Flerov, L.A. Kot and A.I. Kriger, Sov. Phys. Solid State 24, 954
(1982).

6. S. Kudo and T. Ikeda, J. Phys. Soc. Jpn. 50, 3681 (1981).

7. R. Rehwald, A. Vonlanthen, J.K. Kruger, R. Wallerius and H.G. Unruh,
J. Phys. C 13, 3823 (1980).

8. L.D. Landau and I.M. Khalatnikov, Akademiia Nauk SSR Doklady 96, 469
(1954).

9. V.A. Golovko and A.P. Levanyuk, Sov. Phys. JETP **54**, 1217 (1981).

10. S. Kh. Esayan and V.V. Lemanov, Izv. Akad. Nauk SSR Ser. Fiz. **47**, 591 (1983).

11. A.P. Levanyuk, Sov. Phys. JETP **22**, 901 (1966).

12. E. Pytte, Phys. Rev. B **1**, 924 (1970); Structural Phase Transitions and Soft Modes, edited by E.J. Samuelsen, E. Andersen and J. Feder (Universitets Forlaget, Oslo, 1971) p. 151.

13. J.M. Courdille, R. Deroche and J. Dumas, J. Phys. (Paris) **36**, 891 (1975).

14. Y. Luspin and G. Hauret, Phys. Status Solidi B **76**, 551 (1976).

15. J. Hu, J.O. Fossum, C.W. Garland and P.W. Wallace, Phys. Rev. B **33**, 6331 (1986).

16. W. Yao, H.Z. Cummins and R.H. Bruce, Phys. Rev. B **24**, 424 (1981).

17. H.-G. Unruh, W. Eller and G. Kirf, Phys. Status Solidi (a) **55**, 173 (1979).

18. M. Quilichini and R. Currat, Solid State Commun. **48**, 1011 (1983).

19. H. Hoshizaki, A. Sawada, Y. Ishibashi, T. Matsuda and I. Hatta, Jpn. J. Appl. Phys. **19**, L324 (1980).

20. W. Rehwald and A. Vonlanthen, Solid State Commun. **38**, 209 (1981).

21. V.V. Lemanov, S. Kh. Esayan and A. Karaev, Sov. Phys. Solid State **28**, 931 (1986).

22. S. Kudo and T. Ikeda, Jpn. J. Appl. Phys. **19**, L45 (1980).

23. P.A. Fleury, S. Chiang and K.B. Lyons, Solid State Commun. **31**, 279 (1979).

24. T. Yagi, M. Cho and Y. Hidaka, J. Phys. Soc. Jpn. **46**, 1957 (1979).

25. M. Cho and T. Yagi, J. Phys. Soc. Jpn. **50**, 543 (1981).

26. G. Hauret and J.P. Benoit, Ferroelectrics **40**, 1 (1982).

27. Y. Luspin, G. Hauret, A.M. Robinet and J.P. Benoit, J. Phys. C **17**, 2203 (1984).

28. J.P. Benoit, G. Hauret, Y. Luspin, A.M. Gillet and J. Berger, Solid State Commun. **54**, 1095 (1985).

NMR AND NQR IN INCOMMENSURATE SYSTEMS

F. Milia and G. Papavassiliou

National Center for Scientific Research "Demokritos"
Ag.Paraskevi Attikis, 153 10 Athens, GREECE

INTRODUCTION

NQR and NMR are -as far as structurally incommensurate (I) systems
are concerned- among the methods that were extensively used in the
investigation of these systems. Their usefulness is based on the fact that
the resonance frequencies vary in space in such a way that reflects the
variation of the (I) modulation and give unique information about the
static and dynamic states of these systems. The phase solitons are studied
from the line-shape calculations and the elementary excitations were
clarified from extensive studies of the spin-lattice and spin-spin
relaxation times T_1 and T_2. In addition the investigation of the influence
of defects and impurities to the relaxation mechanism, allowes us to
clarify some questions on the pinning of the modulation wave.

It is well known that the pinning is still an open problem in solid
state physics. The questions to be answered are the following: The
starting mechanism of the pinning, how large it is and its behaviour as a
function of temperature and impurity concentration. Especially valuable is
the study of the Goldstone or phase fluctuation mode, which has been

Geometry and Thermodynamics
Edited by J.-C. Tolédano
Plenum Press, New York, 1990

proved to be very sensitive to even the slightest amount of impurities, affecting thus its energy gap Δ_ϕ.

In this article we analyze the NQR line-shape in the (I) phase as well as the T_1 and T_2 relaxation times of Rb_2ZnCl_4. The measurements were done on single crystals $[Rb_{1-x}(NH_4)_x]_2ZnCl_4$ where $(NH_4)^+$ acts as an impurity with concentrations x=0.00, 0.01 and 0.04. Throughout the measurements a pulse Fourier transform spectrometer was used. For the line shape and frequency measurements, the free induction decay signal has been used while for the T_1 and T_2 relaxation times the 90^0-τ-90^0 and 90^0-τ-180^0 pulse sequences respectively.

Since the first radiospectroscopic studies with NQR[1,2], Rb_2ZnCl_4 is the most extensively investigated structurally (I) system. It undergoes successive transitions at T_I = 302K, T_{c1}= 192K and T_{c2}= 75K. The prototype paraelectric (P) phase in the high temperature phase is orthorhombic D_{2h}^{16}- P_{cmn} with four formula units per unit cell. The intermediate phase is (I) and the first low temperature phase is commensurate (C) and ferroelectric.

NQR LINE-SHAPE OF Rb_2ZnCl_4

The Hamiltonian describing our system is equal to $\overline{\mathcal{H}} = \overline{\mathcal{H}}_Q + \overline{\mathcal{H}}_D$ and is the sum of a quadrupolar and a dipolar term. Usually $\overline{\mathcal{H}}_D \ll \overline{\mathcal{H}}_Q$, introduces negligible corrections to the energy spectrum of the unperturbed Hamiltonian and can be neglected.

<u>EFG Tensor Calculations</u>

The following EFG tensor calculations (and experimental data) are for the sake of simplicity carried out for the local case on only one Cl nuclear site -the Cl(1)- but it is valid and can be carried out in the same way for the rest nuclear sites.

It is known that in the (P) phase the Cl(1) nuclei lie on a m_y plane and similarly to the ^{87}Rb EFG tensor calculations[3,4] the EFG tensor elements $V_{xy} = -V_{xy}$ and $V_{yz} = V_{yz}$, are in view of the mirror symmetry, equal to zero. Therefore for $T>T_I$ the ^{35}Cl EFG tensor, in the crystal fixed $x||a$, $x||b$, $z||c$ frame, has the form:

$$V^{(P)} = \begin{vmatrix} V_{xx} & 0 & V_{xz} \\ 0 & V_{yy} & 0 \\ V_{xz} & 0 & V_{zz} \end{vmatrix} \qquad T > T_I \qquad (1)$$

Below T_I the frozen-in soft mode displacements destroy the mirror plane and as a consequence the tensor components V_{xy} and V_{yz} are different from zero. From neutron scattering experiments[5] on the isomorphous Rb_2ZnBr_4 we know that the modulation wave is polarized along the b-axis. From the above, it can be proved that in the (I) phase[4]:

$$V_\mu(\varphi+\pi) = - V_\mu(\varphi) \qquad \text{where} \quad \mu = xy,yz \qquad (2)$$

$$V_\mu(\varphi+\pi) = V_\mu(\varphi) \qquad \text{if} \quad \mu \neq xy,yz$$

As a result the expansion of the quadrupole tensor takes the following form: (equation 3)

$$V^{(I)} = \begin{vmatrix} V^{(0)}_{xx} & 0 & V^{(0)}_{xz} \\ 0 & V^{(0)}_{yy} & 0 \\ V^{(0)}_{xz} & 0 & V^{(0)}_{zz} \end{vmatrix} + \begin{vmatrix} 0 & V^{(1)}_{xy} & 0 \\ V^{(1)}_{xy} & 0 & V^{(1)}_{yz} \\ 0 & V^{(1)}_{yz} & 0 \end{vmatrix} A\cos\varphi + \begin{vmatrix} V^{(2)}_{xx} & 0 & V^{(2)}_{xz} \\ 0 & V^{(2)}_{yy} & 0 \\ V^{(2)}_{xz} & 0 & V^{(2)}_{zz} \end{vmatrix} A^2\cos^2\varphi$$

In expression (3) we have in mind that $u(\varphi(z)) = A\cos\varphi(z)$ represents the frozen-in (I) distortion wave in the plane wave limit (PWM) and:

$$u(\varphi(z)) = A\cos(q_s z+\varphi_0).$$

The matrix elements of the appropriate quadrupole Hamiltonian are:

$$\langle m|\mathcal{H}_Q m\rangle = E\left[3m^2-I(I+1)\right]V_0$$

$$\langle m\pm1|\mathcal{H}_Q|m\rangle = E(2m\pm1)\left[(I\mp m)(I\pm m+1)\right]^{1/2}V_{\mp 1} \qquad (4)$$

$$\langle m\pm2|\mathcal{H}_Q|m\rangle = E\left[(I\mp m)(I\mp m-1)(I\mp m+1)(1\mp m+2)\right]^{1/2}V_{\mp 2}$$

and if $|m'-m|>2$ then $\langle m'|\mathcal{H}|m\rangle=0$. Here

$$V_0 \equiv V_{zz}$$

$$V_{\pm 1} \equiv V_{xz} \pm iV_{yz} \qquad (5)$$

$$V_{\pm 2} \equiv 1/2\,(V_{xx}- V_{yy})\pm iV_{xy}$$

$$E \equiv \frac{e^2 Q}{4I(2I-1)}$$

In the case of a Cl nucleus with $I=3/2$ the Hamiltonian $\mathcal{H}_Q$ is:

$$\mathcal{H}_Q = \begin{array}{c|cccc} & |3/2\rangle & |1/2\rangle & |-1/2\rangle & |-3/2\rangle \\ \hline |3/2\rangle & 3EV_0 & 2\sqrt{3}EV_1 & 2\sqrt{3}EV_2 & 0 \\ |1/2\rangle & 2\sqrt{3}EV_1 & -3EV_0 & 0 & 2\sqrt{3}EV_2 \\ |-1/2\rangle & 2\sqrt{3}EV_2 & 0 & -3EV_0 & -2\sqrt{3}EV_1 \\ |-3/2\rangle & 0 & 2\sqrt{3}EV_2 & -2\sqrt{3}EV_1 & 3EV_0 \end{array} \qquad (6)$$

In order to calculate the energy levels ($\mathcal{E}$) we must diagonalize the above Hamiltonian getting finally the equation:

$$(9E^2V_0^2-\lambda^2)^2+24E^2(V_1V_{-1}+V_2V_{-2})(9E^2V_0^2-\lambda^2)+144E^2(V_1V_{-1}V_2V_{-2})^2= 0 \quad (7)$$

The solutions of equation (7) are the following:

$$\lambda_1= \mathcal{E}_{\pm3/2}= 3E\left[V_0^2+4/3(V_1V_{-1}+V_2V_{-2})\right]^{1/2} \quad (8)$$

$$\lambda_2= \mathcal{E}_{\pm1/2}=-3E\left[V_0^2+4/3(V_1V_{-1}+V_2V_{-2})\right]^{1/2}$$

and as a consequence the energy levels are doubly degenerated and the quadrupole frequency is given by the expression:

$$\nu_Q= \frac{\mathcal{E}_{\pm3/2}-\mathcal{E}_{\pm1/2}}{h} = \frac{6E}{h}\left[V_0^2+4/3(V_1V_{-1}+V_2V_{-2})\right]^{1/2} \quad (9)$$

With the help of relation (3,5) after reordering the terms we get:

$$\nu_Q=\frac{6E}{h}\left\{\left[(V_{zz}^{(0)})^2+4/3(V_{xz}^{(0)})^2+1/3(V_{xx}^{(0)})^2+1/3(V_{yy}^{(0)})^2-2/3V_{xx}^{(0)}V_{yy}^{(0)}\right]+\right.$$
$$+\left[(V_{yz}^{(1)})^2+{}_{xy}^{(1)})^2+2V_{zz}^{(0)}V_{zz}^{(2)}+2/3(V_{xx}^{(0)}V_{xx}^{(2)}+V_{yy}^{(0)}V_{yy}^{(2)})-2/3(V_{xx}^{(0)}V_{yy}^{(2)}+\right.$$
$$+V_{xx}^{(2)}V_{yy}^{(0)})A^2\cos^2\varphi+\left[(V_{zz}^{(2)})^2+1/3(V_{xx}^{(2)})^2+1/3(V_{yy}^{(2)})^2+4/3(V_{xz}^{(2)})^2-\right.$$
$$\left.-2/3V_{xx}^{(2)}V_{yy}^{(2)}\right]A^4\cos^4\varphi\bigg\}^{1/2} \quad (10)$$

So we may conclude that the expansion of the quadrupolar frequencies ν_Q in the (I) phase can be given as:

$$\nu_Q= (\nu_0+ \nu_1\cos^2\varphi + \nu_2\cos^4\varphi)^{1/2} \quad (11)$$

where

$$\nu_0^{1/2}= \text{The unperturbed paraelectric frequency}$$
$$\nu_1= a_1(T_I-T)^{2\beta}$$
$$\nu_2= a_2(T_I-T)^{4\beta} \quad (12)$$

and a_1, a_2,...., are according to relation (10) a function of the EFG tensor elements.

From equation (11) we can deduce that in the plane wave limit, peaks (edge singularities) will appear whenever[6]

$$\frac{d\nu_Q}{d\varphi} = 0 \quad (13)$$

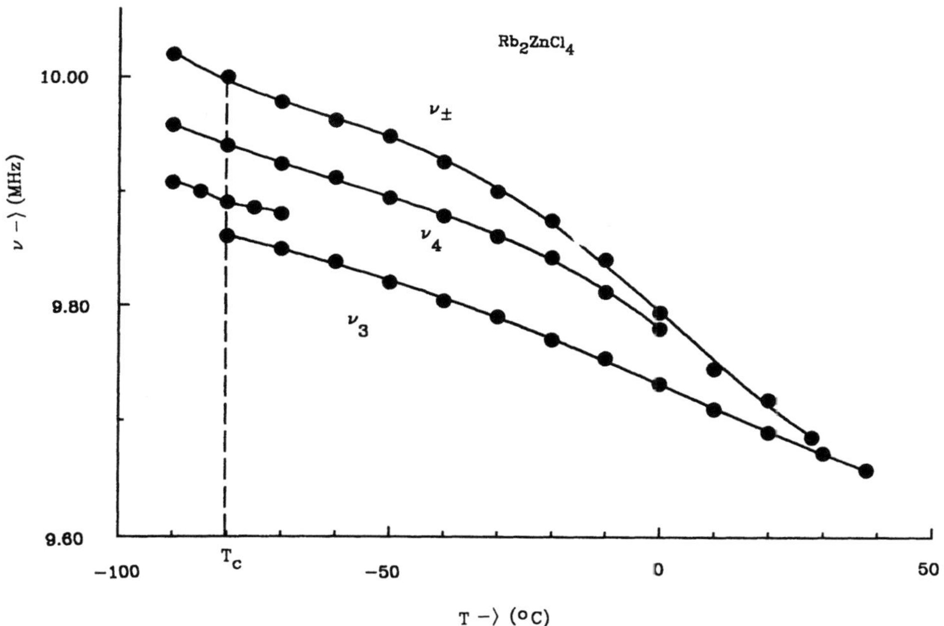

Figure 1. Temperature dependence of the experimental values of the NQR lines in pure Rb$_2$ZnCl$_4$

which in this case means that the edge singularities will appear at the following positions:

(i) $\qquad \cos\varphi = \pm 1 \qquad \nu_\pm = (\nu_0 + \nu_1 + \nu_2)^{1/2}$

(ii) $\qquad \cos\varphi = 0 \qquad \nu_3 = \nu_0^{1/2}$ $\hfill$ (14)

(iii) $\qquad \cos\varphi = \sqrt{|\nu_1|/2\nu_2} \qquad \nu_4 = (\nu_0 - |\nu_1|^2/4\nu_2)^{1/2} \qquad (\nu_1 < 0, \ |\nu_1|/2\nu_2 < 1)$

As we easily see from the above relations, at the extreme nuclear displacments $\cos\varphi = \pm 1$, we have only one frequency singularity $\nu_- = \nu_+$, a second one ν_3 for $\cos\varphi = 0$ (zero nuclear displacement) and under certain conditions a third one which we call ν_4. In Figure 1, we see the experimental points and the theoretical fit for Cl(1) in Rb$_2$ZnCl$_4$. Here we want to note that the NQR "line" designated as ν_4 has varying values for $\cos\varphi$ which lie between 1 and 0.6. The lower values are very close to T$_c$.

SPIN-LATTICE RELAXATION OF THE PHASE AND AMPLITUDE MODES OF FLUCTUATION

Close to the (P)-(I) transition temperature T$_I$, the PWM approximation describes very well the (I) phase[6,7]. The condensed soft mode splits into two branches. (i) The acoustic-like "phason" which is the Goldstone symmetry recovery mode, is gapless in the absence of any pinning of the modulation wave and has a linear dispersion :

$$\omega_\phi^2 = \varkappa^2 k^2 \qquad\qquad T < T_I \qquad (15)$$

Here ω_ϕ^2 is the phason frequency, $k = q - q_s$ and $\varkappa$ is a constant. (ii) The amplitude fluctuation branch (amplitudon) is optic-like and behaves as a normal soft mode

$$\omega_A^2 = \Delta_A^2 + \varkappa^2 k^2 \qquad\qquad T < T_I \qquad (16)$$

where the amplitudon gap $\Delta_A = \sqrt{2a(T_I - T)}$ and "a" is a constant.
We see that in the absence of any pinning the phason branch is temperature independent while the amplitudon in contrast is temperature dependent. In the case of pinning of the modulation wave, an energy gap is introduced in the excitation spectrum of the phason branch and relation (15) is now:

$$\omega_\phi^2 = \Delta_\phi^2 + \varkappa^2 k^2 \qquad\qquad (17)$$

Close to the transition temperature T_c (strong pinning case) the PWM is not valid. The phason spectrum splits into an acoustic-like and an optic-like branch. This splitting coincides with the appearance of the multi-soliton lattice (MSL) and the existence of a metastable chaotic phase[8]. Both branches are related to the intersoliton distance which goes to infinity as the temperature goes to T_c from above. The experimental results verify this model according to which the spin-lattice relaxation rate for the phason relaxed singularity is temperature independent (pure systems) for the main part of the (I) phase and temperature dependent in the MSL through the intersoliton distance[9,10].

In analyzing the inhomogeneous frequency distribution of the (I) phase, we consider that the contribution of the static dipolar part $\overline{\mathcal{H}}_D$ of the Hamiltonian is negligible in comparison to the quadrupolar part $\overline{\mathcal{H}}_Q$. Things are completely different if we examine the relaxation processes. Here the dipolar contribution is of equal importance to the quadrupolar one[11]. The main dipolar interaction at the Cl(1) site is produced by the coupling with the Cl(2), Cl(3) and Cl(4) in the $ZnCl_4$ tetrahedra. (The Zn nucleus being even-even has spin I=0). Since the spin-spin relaxation time T_2 behaves in exactly the same way as the spin-lattice relaxation time T_1, we will concentrate our current investigation only on T_1[11,12]. The spin-lattice relaxation rate is thus equal to:

$$\frac{1}{T_1} = \frac{1}{T_{1Q}} + \frac{1}{T_{1D}} \qquad\qquad (18a)$$

Here

$$\frac{1}{T_{1Q}} = 24E^2(J_1^Q(\omega)+J_2^Q(\omega)) \tag{18b}$$

$$J_1^Q(\omega)=\int_{-\infty}^{+\infty}\overline{(\Delta V_{xz}(0)+i\Delta V_{yz}(0))(\Delta V_{xz}(t)-i\Delta V_{yz}(t))}e^{i\omega t}dt \tag{18c}$$

$$J_2^Q(\omega)=\int_{-\infty}^{+\infty}\overline{1/4(\Delta V_{xx}(0)-\Delta V_{yy}(0)+i2\Delta V_{xy}(0))(\Delta V_{xx}(t)-\Delta V_{yy}(t)-i2\Delta V_{xy}(t))}e^{i\omega t}dt$$

($\Delta V(t)$ is the time dependent part of the EFG tensor, $\Delta V(t)=V(t)-\bar{V}$) and

$$\frac{1}{T_{1D}} = 3\gamma^4\hbar^2\sum_{k=2}^{4} J_k^D(\omega) \tag{18d}$$

$$J_k^D(\omega)=\int_{-\infty}^{+\infty}\overline{(\Delta F_{xz}(1,k)(0)+i\Delta F_{yz}(1,k)(0))(\Delta F_{xz}(1,k)(t)-i\Delta F_{yz}(1,k)(t))}e^{i\omega t}dt$$

where $\Delta F(t)$ is the time dependent part $\Delta F(t)=F(t)-\bar{F}$ of the tensor

$$F = \begin{cases} F_{zz}= \dfrac{r^2-3z^2}{r^5} \\[2ex] F_{ij}= \dfrac{ij}{r^5} \qquad i,j=x,y,z \quad \text{(for the rest of the elements)} \end{cases}$$

r is the distance between nuclear site Cl(1) and one of the neighbouring, Cl(2), Cl(3) and Cl(4). If we expand the time dependent part of both tensors $\Delta V(t)$ and $\Delta F(t)$ in a Taylor series up to second order terms of the fluctuating part of the nuclear displacements[11], we get:

$$\frac{1}{T_{1Q}} = 12E^2\Big\{1/2[(V_{xx}^{(2)}A\cos\varphi)^2+(V_{yy}^{(2)}A\cos\varphi)^2]+2[(V_{xz}^{(2)}A\cos\varphi)^2+(V_{xy}^{(1)})^2+$$
$$+(V_{yz}^{(1)})^2\Big\}\times\Big\{\cos^2\varphi\, J_A+\sin^2\varphi\, J_\phi\Big\} \tag{19a}$$

$$\frac{1}{T_{1D}} = 3\gamma^4\hbar^2\Big\{(F_{xz}^{(2)}(1,2)A\cos\varphi)^2+(F_{yz}^{(1)}(1,2))^2+(F_{xz}^{(1)}(1,3)+F_{xz}^{(2)}(1,3)A\cos\varphi)^2+$$
$$+(F_{yz}^{(1)}(1,3)+F_{yz}^{(2)}(1,3)A\cos\varphi)^2+(F_{xz}^{(1)}(1,4)+F_{xz}^{(2)}(1,4)A\cos\varphi)^2+$$
$$+(F_{yz}^{(1)}(1,4)+F_{yz}^{(2)}(1,4)A\cos\varphi)^2\Big\}\times\Big\{\cos^2\varphi\, J_A+\sin^2\varphi\, J_\phi\Big\} \tag{19b}$$

The total spin-lattice relaxation rate can thus be written as,

$$\frac{1}{T_1} = \Big\{C_0+ C_1A\cos\varphi + C_2A^2\cos^2\varphi\Big\}\times\Big\{\cos^2\varphi\, J_A+\sin^2\varphi\, J_\phi\Big\} \tag{20}$$

where J_A, J_ϕ are the spectral densities for the phase and amplitude fluctuations, which are given by the following relations[6]:

(i) $\quad J_\phi= kT_\phi/\Delta_\phi, \qquad J_A= kT_A/\Delta_A \qquad$ (PWM) $\tag{21a}$

(ii) $\quad (J_\phi)_{ac}= \Lambda^2\Gamma_\phi/8(\Delta_\phi)^4 b \qquad$ (MSL) $\tag{21b}$

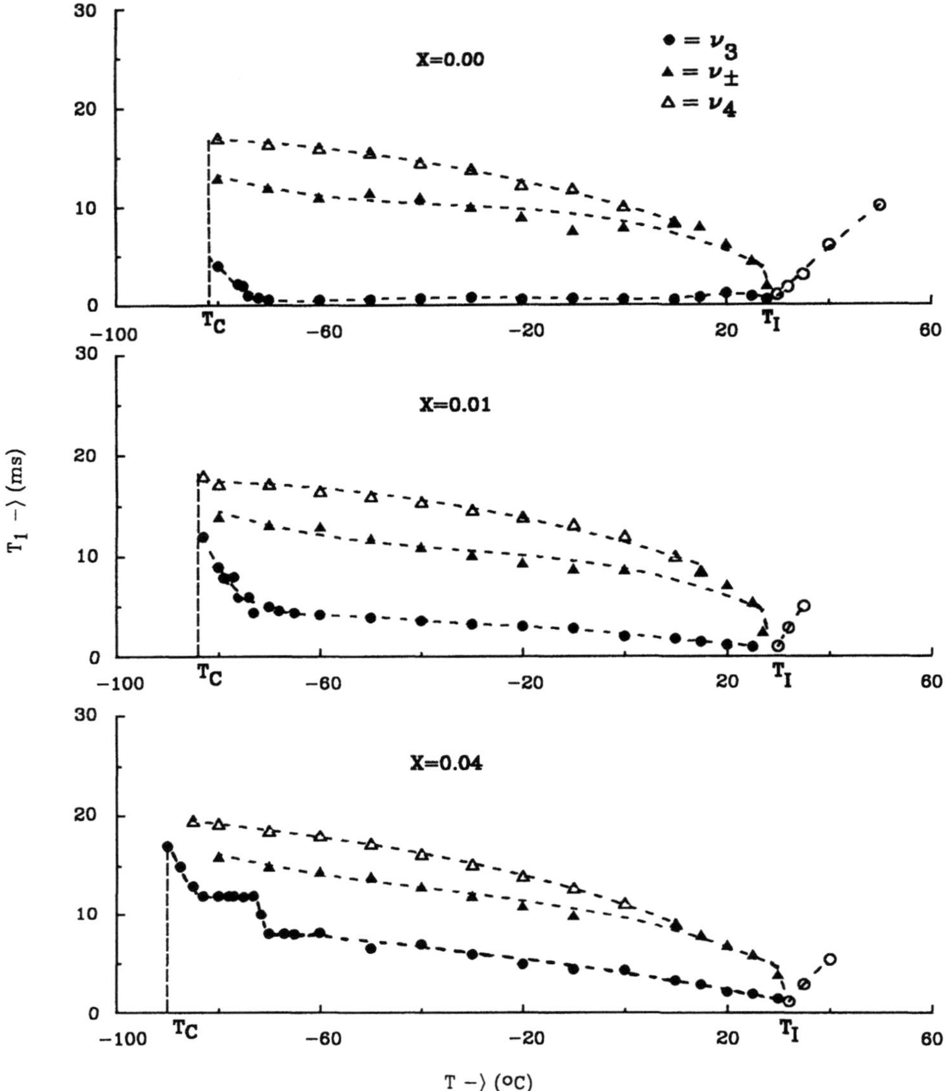

Figure 2. Temperature dependence of the ^{35}Cl spin-lattice relaxation time T_1 in incommensurate $[Rb_{1-x}(NH_4)_x]_2ZnCl_4$ for impurity concentration x=0.00, 0.01 and 0.04

Here K is a constant, Γ_ϕ and Γ_A are the phason and amplitudon damping constants, Λ is an effective maximal value of the wave vector q and b is the intersoliton distance, $b\to\infty$ as $T\to T_c$.

According now to relations (14) and (20) the edge singularity ν_3 of the inhomogeneous (I) frequency distribution which corresponds to $\cos\varphi=0$, is relaxing by phasons, the edge singularities $\nu_\pm$ that correspond to $|\cos\varphi|=1$ are relaxing by amplitudons and the third singularity ν_4 ($0.6\leq\cos\varphi\leq1$) which appears at $T\cong0^\circ C$ by a mixture of phason and amplitudons. However since the lower values of $\cos\phi$ are valid close to the transition temperature T_c, where the PWM approximation starts to break down and new commensurate lines appear coinsiding with the ν_4 and $\nu_\pm$ singularities, we may assume that in the PWM region also the ν_4 singularity relaxes mainly by amplitudons.

From relations (20), (21a) and (21b) we see that the spin-lattice relaxation time T_1 of the above edge singularities is proportional to the corresponding energy gap Δ_β (β = A, ϕ), showing thus the same as Δ_β temperature dependence. For the pure compounds in the PWM region, the phason gap Δ_ϕ is temperature independent while the amplitudon gap is according to relation (16) temperature dependent. For the mixed systems Δ_ϕ becomes temperature and impurity concentration dependent[13]. In the MSL region, the phason gap $(\Delta_\phi)_{ac}$ is temperature independent[14] and T_1 becomes temperature dependent via the intersoliton distance b. The above theoretical conclusions are in a good agreement with the expe- rimental data (Figure 2). The theoretical fit is better if we consinder 3rd and 4th order terms in the expansion of the tensors ΔV and ΔF. Here we want to note that in the region very close to T_c: (i) The pure and mixed compounds have a T_1 behaviour which agrees very well with the theory. (ii) The observed steps of the phason induced $(T_1)_\phi$ versus temperature , in the compound with impurity concentration x=0.04, indicate the locking of the intersoliton distance and consequently of the modulation wavevector, for certain temperature intervals. This effect must be also present in the x=0.01 compound, but the steps are probably too small to be identified. Generally it can be explained by the pinning of the narrow solitons to the impurities.

Acknowledgment: The authors would like to thank Mr. A.Anagnostopoulos for his valuable help during the preparation of the manuscript.

REFERENCES

1. A. K. Moskalev, I. A. Belobrova, I.P.Aleksandrova, Sh.Sawada and I. Shiroishi, Investigations of the Incommensurate Phase of Rb_2ZnCl_4 by NQR method, Phys. Status solidi 50a:K157 (1978)

2. F. Milia, ^{35}Cl NQR Study of the Incommensurate Structural Phase Transition in $Rb_2Zn Cl_4$, Phys. Lett. 70A:218 (1979); F. Milia and V. Rutar, Structure ofthe Incommensurate Phase of Rb_2ZnCl_4 as determined by ^{35}Cl Nuclear Quadrupole Resonance, Phys. Rev. B23:6061 (1981)

3. V. Rutar, J. Seliger, B. Topic, R. Blinc and I. P. Aleksandrova, ^{87}Rb Electric Field Gradient Tensors in Incommensurate Rb_2ZnCl_4, Phys. Rev. B24:24 (1981)

4. B. W. Van Beest, A. Janner and R. Blinc, ^{87}Rb Electric Field Gradient Tensors and the Symmetry of the Incommensurate Phase in Rb_2ZnBr_4 and Rb_2ZnCl_4, J. Phys. C 16:5409 (1983)

5. C. J. de Pater, Average Structure of Rb_2ZnBr_4, Acta Cryst. B35:299 (1979)

6. R.Blinc, Magnetic Resonance and Relaxation in Structurally Incommensurate Systems, Phys. Rep. 79:331 (1981) and references therein

7. R. Blinc, P. Prelovsek, V. Rutar, J. Seliger and S. Zumer, Experimental Observations of Incommensurate Phases, in: "Incommensurate Phases in Dielectrics", R. Blinc and A. P. Levanyuk, ed., North Holland, Amsterdam (1986), Vol.I

8. R. Blinc, P. Prelovsek, A. Levstic and C. Filipic, Metastable Chaotic State and the Soliton density in Incommensurate Rb_2ZnCl_4, Phys. Rev. B29:1508 (1984)

9. R. Blinc, F. Milia, V. Rutar and S. Zumer, Spin-Lattice Relaxation via Phasons in a Multisoliton Lattice, Phys. Rev. Lett. 48:47 (1982)

10. F. Milia, G. Papavassiliou and E. Giannakopoulos, ^{35}Cl Spin-Lattice Relaxation and Temperature-Dependent Phason Gaps in Substitutionally Disordered Incommensurate Systems, Phys. Rev. B39:12349 (1989)

11. G. Papavassiliou, F. Milia and E. Giannakopoulos, ^{35}Cl Spin-Spin Relaxation Time T_2 in Rb_2ZnCl_4 Studied by NQR, Phys. Rev. B (to be published)

12. R. Blinc, S. Zumer, O. Jarh and F. Milia, T_2 in Incommensurate Systems, Phys. Rev. B. (to be published)

13. R. Blinc, J. Dolinsek, P. Prelovsek and K. Hamano, Phason Gap in Substitutionally Disordered Incommensurate Systems, Phys. Rev. Lett. 56:2387 (1986)

14. P. Prelovsek, Theoretical Results concerning the Influence of Impurities on Incommensurate Systems, Phase Transitions 11:203 (1988)

DO WE UNDERSTAND OPTICAL ACTIVITY
IN THE INCOMMENSURATE PHASE OF THE A_2 BX_4 FAMILY ?

R.M. PICK

Département de Recherches Physiques, Univ. P. et M. Curie
4 place Jussieu - 75252 Paris cedex 05

ABSTRACT : We show that a microscopic theory of light propagation in incommensurate insulators cannot explain the present experimental results on optical activity in the A_2 BX_4 family.

In (macroscopic) classical optics, the basic equation which describes the electric properties of the material reads

$$D^\alpha(\vec{q}) = \sum_\beta \in^{\alpha\beta}(\vec{q})\ E^\beta(\vec{q}) \tag{1}$$

where $\vec{D}(\vec{q})$ (resp. $\vec{E}(\vec{q})$) is the Fourier transform of the electric induction (resp. **macroscopic** electric field) and $\overline{\overline{\in}}(\vec{q})$ the Fourier transform of the relative dielectric permittivity tensor. The $\vec{q}$ dependence of this tensor reflects the (weak) non local character of the relation between these two fields, so that it is physically meaningful to expand $\overline{\overline{\in}}(\vec{q})$ in increasing powers of $\vec{q}$.

Let us attach, for sake of simplicity the cartesian indices α, β to the crystal axes. The two electromagnetic waves which propagate, inside the bulk crystal, with a wave vector $\vec{q}$ may be defined through their optical index $n(\vec{q}, \omega)$, the latter being one of the two solutions of the 3x3 Fresnel equations :

Geometry and Thermodynamics
Edited by J.-C. Tolédano
Plenum Press, New York, 1990

$$\sum_{\beta} F^{\alpha\beta}(\vec{q})\ E^{\beta}(\vec{q}) \equiv \sum_{\beta} \left[n^2(\vec{q},\omega)\ M^{\alpha\beta}(\hat{q}) + \in^{\alpha\beta}(\vec{q}) \right] E^{\beta}(\vec{q}) = 0 \qquad (2)$$

with :

$$n(\vec{q},\omega) = c\ \frac{|\vec{q}|}{\omega} \quad ; \quad M^{\alpha\beta}(\hat{q}) = \hat{q}_{\alpha}\ \hat{q}_{\beta} - \delta_{\alpha\beta} \qquad (3)$$

- In the $\vec{q}_{\to 0}$ limit, to each optical index n_j ($j = 1,2$) corresponds an electromagnetic wave which may be characterized by the direction of its transverse, <u>linearly polarized</u>, electric induction field

$$\vec{D}_j(\vec{r},t) = \vec{d}_j(\hat{q})\ e^{i(\vec{q}\cdot\vec{r}-\omega t)} \qquad (4)$$

- The lowest term in the $\vec{q}$ expansion of $\overline{\overline{\in}}(\vec{q})$ may be written as $i\sum_{\gamma} \in^{\alpha\beta,\gamma} q^{\gamma}$ where $\in^{\alpha\beta,\gamma}$ is a real, antisymetric in $\alpha\beta$, tensor. To each optical index solution of (1) corresponds now a transverse but <u>elliptically polarized</u> electric induction field, and these two fields may be written as

$$\vec{d}_+(\hat{q}) = \vec{d}_a(\hat{q}) + i\lambda\ \vec{d}_b(\hat{q}) \quad ; \quad \vec{d}_-(\hat{q}) = i\lambda\ \vec{d}_a(\hat{q}) + \vec{d}_b(\hat{q}) \qquad (5.a)$$

with
$$\vec{d}_a(\hat{q})\ .\ \vec{d}_b(\hat{q}) = 0 \quad ; \qquad 0 < \lambda < 1 \qquad (5.b)$$

The value of λ depends on the ratio between the off diagonal imaginary terms in $\sum_{\gamma} \in^{\alpha\beta,\gamma} q_{\gamma}$ and $(n_+)^2 - (n_-)^2$, n_+ and n_- being the two optical indices corresponding to $\vec{d}_+(\hat{q})$ and $\vec{d}_-(\hat{q})$ respectively. In optically biaxial crystals, the usual situation corresponds to a ratio much smaller than one, (typically 10^{-2}) which leads to a similar value for λ. This (weak) ellipticity of the eigenmodes, called optical activity is difficult to discriminate from the (typically) 100 times larger birefringence, which explains the very small number of measurements of this quantity until the technical break through made by Kobayashi et al.[1,2].

In the preceding <u>macroscopic</u> theory, the existence of non zero $\in^{\alpha\beta,\gamma}$ coefficients depends only on the <u>point symmetry</u> of the crystal ; in particular, as, for any centered group, no odd rank tensor can exist, no optical activity can exist either. In that respect, the case of the A_2BX_4

crystals is of particular interest. Indeed, below a high temperature Pmcn (D_{2h}^{16}) phase without optical activity, these crystals may display a low temperature $P2_1cn$ (C_{2v}^9) phase in which $\in^{xy \cdot y}$ and $\in^{xz \cdot z}$ may be different from zero, the latter having been effectively measured for some of them. In that case, an incommensurate phase exists between these two phases, which is characterized by a wave vector $\vec{q}_i = \delta(T)\vec{c}^*$. $\delta(T)$ is frequently $(K_2SeO_4, Rb_2ZnCl_4, Rb_2ZnBr_4 \ldots)$ in the vicinity of $\frac{1}{3}$ with a lock-in at $\frac{1}{3}$, or, for e.g. $[N(CH_3)_4]_2 Zn Cl_4$ (in short TMAZC) in the vicinity of $\frac{2}{5}$ with a partial lock-in at 0.40. Following Landau theory which predicts equal amplitudes and opposite phases for the static deformations with wave vectors $\pm p\vec{q}_i$ (where p is an integer), the incommensurate phase should be macroscopically centered, i.e. with no optical activity. Nevertheless, this property has been reported by different authors[2,3,4,5,6] and, quite recently, J. Kobayashi et al. on the one hand[2], Dijkstra et al.[6] on the other hand, have reported a similar, and quite surprising thermal dependence of a coefficient g_{23}, proportional to $\in^{xz \cdot z}$, for TMAZC, (Fig. 1) using the same technique, but crystals obtained from different sources.

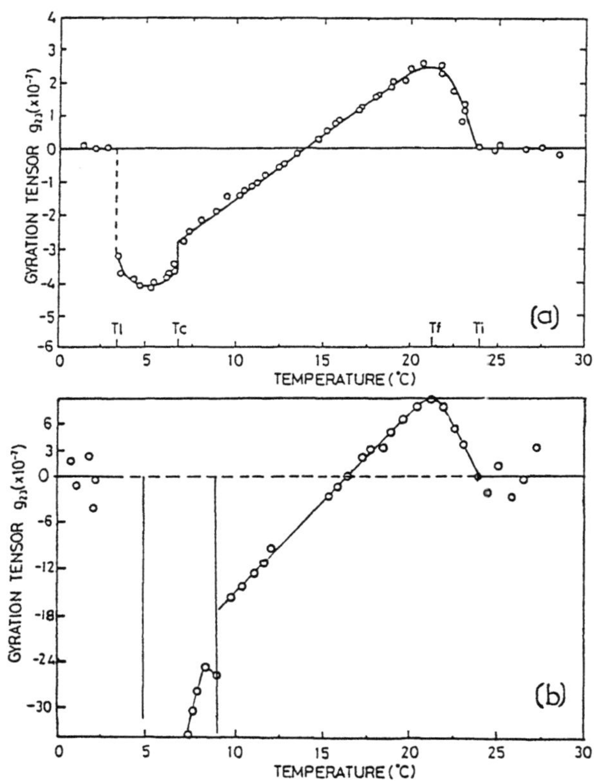

Figure 1 . Thermal dependence of g_{23} in $[N(CH_3)_4]_2 Zn Cl_4$

    a) after Kobayashi et al.[2]. *b)* after Dijkstra et al.[2].

In order to get more insight into eq. (1), it is worthwhile to derive it, following[7] , though a microscopic analysis, which we shall first performed for a normal crystal, characterized by its reciprocal lattice vectors $\vec{G}$. Eq. (1) may be generalized into

$$d^\alpha(\vec{q}+\vec{G}) = \sum_{\beta,\vec{G}'} \in^{\alpha\beta}(\vec{q}+\vec{G},\vec{q}+\vec{G}') \, e^\beta(\vec{q}+\vec{G}') \tag{6}$$

where, for an infinite crystal, $\vec{d}(\vec{q}+\vec{G})$ is the externally applied electric field and $\vec{e}(\vec{q}+\vec{G})$ the microscopic total electric field, the $\vec{G} = 0$ component of these two quantities being the macroscopic fields of eq. (1). The 3 macroscopic Fresnel equations (eq. (2)) have to be replaced by an infinite set of equations :

$$F^{\alpha\beta}(\vec{q}+\vec{G},\vec{q}+\vec{G}')e^\beta(\vec{q}+\vec{G}') \equiv \left[n^2(\vec{q}+\vec{G},\omega)M^{\alpha\beta}(\widehat{\vec{q}+\vec{G}})\delta_{\vec{G},\vec{G}'} + \in^{\alpha\beta}(\vec{q}+\vec{G},\vec{q}+\vec{G}') \right] e^\beta(\vec{q}+\vec{G}') = 0 \tag{7}$$

(where summation over repeated indices has been assumed).

As the matrix elements of the generalized dielectric tensor are, at most, of order unity, while, typically $[\,|\vec{q}|/|\vec{q}+\vec{G}|\,]^2 \sim 10^{-6}$, and as $\left\| M^{\alpha\beta}(\widehat{\vec{q}+\vec{G}}) \right\| = 0$, one can show that whatever are $\vec{G}$ and $\vec{G}$ <u>different from zero</u> :

$$F^{-1\,\alpha\beta}(\vec{q}+\vec{G},\vec{q}+\vec{G}') = (\widehat{\vec{q}+\vec{G}})_\alpha \, S^{-1}(\vec{q}+\vec{G},\ \vec{q}+\vec{G}')(\widehat{\vec{q}+\vec{G}'})_\beta + O\left(\frac{1}{n^2(\vec{q}+\vec{G},\omega)} \right) \tag{8}$$

with

$$S(\vec{q}+\vec{G},\vec{q}+\vec{G}') = (\widehat{\vec{q}+\vec{G}})_\alpha \in^{\alpha\beta}(\vec{q}+\vec{G},\vec{q}+\vec{G}')(\widehat{\vec{q}+\vec{G}'})_\beta \tag{9}$$

Eq. (8) implies that, up to terms typically of order 10^{-6} an electromagnetic wave which propagates with wave vector $\vec{q}$, and frequency ω may be <u>totally characterized</u> by its <u>macroscopic</u> electric <u>field</u> $\vec{e}(\vec{q})$, all its microscopic components being related to it through :

$$e^\alpha(\vec{q}+\vec{G}) = (-) \, F^{-1\,\alpha\gamma}(\vec{q}+\vec{G},\vec{q}+\vec{G}')(\widehat{\vec{q}+\vec{G}'})_\gamma \in^{\gamma\beta}(\vec{q}+\vec{G}',\vec{q}) \, e^\beta(\vec{q}) \tag{10}$$

(see eq.(8)) while the optical index related to this electric field is given by the usual set of 3 Fresnel equations in which the dielectric tensor is defined by

$$\in^{\alpha\beta}(\vec{q}) = \in^{\alpha\beta}(\vec{q},\vec{q}) - \in^{\alpha\beta}(\vec{q},\vec{q}+\vec{G}) \; F^{-1\gamma\delta}(\vec{q}+\vec{G},\vec{q}+\vec{G}') \; \in^{\delta\beta}(\vec{q}+\vec{G}',\vec{q}) \qquad (11)$$

the second term of the r.h.s. of (11) being defined as the local field correction to the dielectric tensor.

■ Clearly, the expansion of the first term of the r.h.s. of eq. (11) in powers of $\vec{q}$ is governed by the point group symmetry of the crystal. Conversly, the same expansion, for each element of the second term, is governed by the space group symmetry, but the sum over all the reciprocal lattice vectors (different from zero) has again the point group symmetry.

■ Finally, as shown by Pick[7], each element of the generalized dielectric tensor, $\in^{\alpha\beta}(\vec{q}+\vec{G},\vec{q}+\vec{G}')$ is easily related to the corresponding element of the current-current correlation function, which, in turn, is expressed through the sole knowledge of the occupied and empty electronic wave functions, and of their eigenvalues.

From this point of view, an incommensurate crystal does not differ from an usual 3-d crystal, except for the extra dependence of the generalized dielectric tensor on $p\vec{q}_i$. Indeed eq.(6) generalizes into

$$d^{\alpha}(\vec{q}+\vec{G}+p\vec{q}_i) = \sum_{\beta,\vec{G}',p'} \in^{\alpha\beta}(\vec{q}+\vec{G}+p\vec{q}_i,\vec{q}+\vec{G}'+p'\vec{q}_i) \; e^{\beta}(\vec{q}+\vec{G}'+p'\vec{q}_i) \qquad (6.b)$$

and this generalized tensor may be expressed with the help of the electronic wave functions of the incommensurate crystal in the same way as for a normal crystal, while the symmetry of these wave functions have been specified e.g. by Zak et al.[8], a result which is valid as long as the incommensurate potential is weak enough with respect to the interatomic potential to prevent any analyticity breakdown[9].

No optical activity should thus exist in such an incommensurate phase, except if, at least, one of the hypotheses used for the above derivation is incorrect. Let us review them in turn.

- The existence of a center of symmetry in the incommensurate phase of these crystals is supported by X Ray experiments. More precisely, specific searchs for a signature of the absence of such a center of inversion has been made for instance in Rb_2ZnBr_4 [10], in which optical activity has been reported[10], and lead to negative results.

- $\vec{q}_i$ being incommensurate with $\vec{G}$, $p\vec{q}_i + \vec{G}$ may be small enough to be of the same order of magnitude as $\vec{q}$. Nevertheless, the corresponding values of p are always large ($p \gtrsim 15$), and an analysis of the structure of eq. (7) and of the generalized dielectric tensor shows that in the vicinity of the normal-incommensurate phase transition temperature, T_i, the corresponding g_{23} coefficient should be proportional to η^p, while the order parameter η is proportional to $(T_i - T)^\beta$. As $\vec{q}_i = \delta(T)\ \vec{c}^{\,*}$ varies with temperature, one would expect a discontinuous thermal variation of g_{23}, and an overall $(T_i - T)^{p\beta}$ dependence of g_{23}, (with a large p) which does not fit with Fig. 1.

- The previous analysis may also be erroneous in the vicinity of T_c, the incommensurate—lock-in phase transition, due to the inhomogeneity brought it by discommensurations in this temperature region. The disclination walls[10] separate homogeneous domains of equal width with opposite polarization, so that they give a zero net optical activity, but the walls themselves will all display the same chirality. Nevertheless, the width of the walls does not vary with temperature while their number decrease when approaching T_c. One would thus expect a decrease of $|g_{23}|$ with $T - T_c$, which is in contradiction again with Fig. 1.

The explanation of the experimental results is thus a real challenge, and, even if the correlation between the maximum of Fig. 1 and the existence of a partial lock-in at the same temperature in TMAZC[11] suggests that impurities may play a role, the mechanism of the effect needs still to be unravelled.

ACKNOWLEDGEMENTS

I would like to thank E. Dijkstra for communicating his experimental data before publications and acknowledge long and fruitfull discussions with A. Janner. But most of the ideas exposed here owe much to T. Janssen and H. Poulet with whom I have constantly collaborated during this project.

REFERENCES

1. J. Kobayashi, T. Takahashi, T. Hosokawa and Y. Uesu, J. Appl. Phys. 49:809 (1978)
2. J. Kobayashi, H. Kumomi and K. Saito, J. Appl. Cryst., 19:377 (1986)
3. J. Kobayashi, Y. Uesu, Y. Yamada and H. Takehara, Ferroelectrics 36:371 (1981)
4. R.A. Sanctuary, Thesis, E.T.H. Zurich (1985)
5. H. Meekes and A. Janner, Phys. Rev. B38:8075 (1988)
6. E. Dijkstra and A. Janner, Ferroelectrics, to appear
7. R.M. Pick, Adv. in Phys. 19:269 (1970)
8. J. Zak and R.M. Pick, Phys. Lett. A126:33 (1987)
9. S. Aubry in "Solitons and Condensed Matter Physics", A.R. Bishop and T. Schneider ed., Springer Verlag, Berlin p. 264 (1978)
10. V. Janovec, Phys. Lett. A99:384 (1983)
11. M. Bizouet, R. Almayrac and P. Saint Gregoire, J. Phys. C20:2635. (1987).

THE COUPLING BETWEEN PHASON AND SOFT MODES

NEAR THE SMECTIC C* TO SMECTIC A PHASE TRANSITION

Harald Pleiner and Helmut R. Brand

Fachbereich Physik
Universität Essen
D 4300 Essen 1, West Germany

INTRODUCTION

In condensed matter physics incommensurate phases can occur, if a symmetry of the underlying Hamiltonian is spontaneously broken more than once. The usual situation in incommensurate crystalline systems is that translational symmetry in one direction is broken twice due to the occurence of the periodical order of two different kinds of atoms or ions with different wavelengths. In a truly incommensurate phase the two wavelengths are incommensurate, there is no interaction energy between the two kinds of atoms and the relative motion is described by an acoustic-like excitation (phason). In a commensurate phase, where the ratio of the two wavelengths is rational, there is an interaction energy and the relative motion is optic-like (phason with gap). The transition between these two situations is the commensurate-incommensurate phase transition. In the following we will discuss incommensurate situations in liquid crystal systems and their similarities and differences compared to the usual crystalline phases.

In liquid crystals rotational symmetry can be broken spontaneously twice. In smectic I, F, [1] and L phases rotational symmetry within the smectic planes is broken by the (homogeneous) tilt of the molecules as well as by the bond-orientational order [2]. The direction of the tilt is changed by rotating all molecules around the layer normal, while the preferred directions due to bond-orientational order are changed by rearranging the centers of mass of the molecules. Although these two operations are completely independent, in smectic I anf F the tilt direction is energetically fixed [3, 4, 5] with respect to (one of) the preferred bond directions (with a relative equilibrium angle of 0 or 30 degrees). If the current interpretation of the smectic L phase [6] is correct, this also is a commensurate phase with a fixed relative equilibrium angle that is arbitrary between 0 and 30 degrees. Thus, the phason in all these phases has a large gap and no commensurate to incommensurate phase transition has yet been found.

Quite the opposite situations occurs for the smectic C* (Sm-C^*) phase [7, 8]. There, translational symmetry along the layer normal is broken twice: by the layering and by the helical ordering of the long molecular axes (which are all tilted with respect to the layer normal with the tilt direction changing helicoidally from one layer to the next). The two wavelengths according to the two types of translational order are incommensurate and a (homogeneous) translation of the helix (along the layer normal) relative to the layers does not cost energy and constitutes a true gapless

Geometry and Thermodynamics
Edited by J.-C. Tolédano
Plenum Press, New York, 1990

phason mode [9]. A commensurate phase, where the layer spacing and the helical pitch are energetically fixed at a rational ratio, is impossible for symmetry reasons. The existence of the helix not only breaks translational symmetry but also rotational symmetry (a translation of the helical structure can be compensated by a translation). Thus, the hydrodynamic variable connected with helix displacements (called u_C in [9]) is not invariant under rotations around the helix axis in contrast to the hydrodynamic variable describing layer displacements (u_A). This implies that homogeneous coupling terms between u_C and u_A are not possible in the free energy (but inhomogeneous ones are possible [9]). This situation is therefore qualitatively different from that arising in incommensurate crystalline phases, where such homogeneous coupling terms between the two displacement felds are allowed by symmetry [10]. We conclude that there is no incommensurate to commensurate phase transition of the usual type in the Sm-C^* phase, since the two broken symmetries connected with layering and helical ordering are not identical.

On the other hand, at the Sm-C^* to smectic A (Sm-A) phase transition the macroscopic tilt angle vanishes and the helix ceases to exist. However, approaching this phase transition from the Sm-C^* phase, neither the layer spacing nor the helical pitch change very much (except in the critical region [11]) and their ratio does not lock to a commensurate value. Thus, although the incommensurability vanishes in the Sm-A phase (because there does not exist a helical order), the Sm-C^* to Sm-A phase transition is of a completely different nature compared to the usual incommensurate to commensurate phase transitions in crystalline systems. In the following we will deal with one special feature of the Sm-C^* to Sm-A phase transition, namely the coupling of the phase variable (rotation or translation of the helix) to the soft variable (change of the tilt angle) [12].

HYDRODYNAMIC EQUATIONS

In order to describe the hydrodynamics of the Sm-C^* phase near the phase transition to the Sm-A phase one has to add changes of the order parameter modulus, δS [13, 14], to the list of hydrodynamic variables [9]. We restrict ourselves to a linearized description, $\delta S \ll S$, where the equilibrium order parameter modulus, S is related to the equilibrium tilt angle, θ by $S = \sin\theta$. Of course, δS is related to tilt angle changes. Since we will use hydrodynamic-like power and gradient expansions, we have to stay outside the critical region and S is always finite. Concentrating on the main point of this communication we will consider explicitly only helix rotations, $\delta\phi$ (called $-n_2$ in ref.[15]), in addition to tilt angle changes and suppress all other hydrodynamic degrees of freedom. Of course, a helix rotation, $\delta\phi$ (or the appropriate helix translation, $q_0\delta\phi$ along the layer normal, $\hat{p}_0$, with q_0 the helical wavevector) is defined with respect to some arbitrary equilibrium direction (position) only. This arbitrariness of the reference system reflects the hydrodynamic character of $\delta\phi$, i.e. homogeneous changes of $\delta\phi$ do not cost energy, and constitutes the main difference between $\delta\phi$ and δS.

The relevant part of the free energy density then reads

$$f = a(\delta S)^2 + \lambda_i q_0(\delta S)\nabla_i\delta\phi + \frac{1}{2}K_{ij}(\nabla_i\delta\phi)(\nabla_j\delta\phi) \tag{1}$$

where the material tensors λ_i and K_{ij} are of the general biaxial form [12], which we do not need to repeat here. Since we use a local description of the Sm-C^* phase [15, 16], they are position dependent according to the helical ground state. The crosscoupling term in (1) is specific for Sm-C^* and has to vanish in a smectic C (Sm-C) phase, because of the spatial inversion symmetry present there. The coefficients in (1) also vanish at the phase transition to the Sm-A phase and comparing with a nonlinear Ginzburg-Landau expansion in S (valid close to the appropriate second order phase transition temperature T_c^* but only outside the critical region) one finds $K_{ij} \sim S^2$, $\lambda_i \sim S^3$ and $a \sim S^2$ or $\sim S^4$, depending on whether the fourth order Ginzburg-Landau parameter is considered more important than the sixth order one or not [17]. With the assumption $a \sim (T_c^* - T)/T_c^*$ this defines the temperature dependence of the parameters.

Eq.(1) gives the thermodynamic forces h_S and $\vec{h}_\phi$, defined by the (truncated) Gibbs relation $df = h_S \delta S + \vec{h}_\phi d\vec{\nabla}\delta\phi$, via partial derivation. Neglecting flow the dynamics of δS and $\delta\phi$ is purely dissipative [12]

$$\frac{\partial \delta\phi}{\partial t} = \frac{\partial R}{\partial(\mathrm{div}\,\vec{h}_\phi)} \qquad \text{and} \qquad \frac{\partial \delta S}{\partial t} = -\frac{\partial R}{\partial h_S} \tag{2}$$

where the relevant part of the dissipation function reads

$$R = \frac{\zeta_1}{2}h_S^2 + \frac{\zeta_2}{2}(\mathrm{div}\,\vec{h}_\phi)^2 + \zeta_3 h_S\,\mathrm{div}\,\vec{h}_\phi \tag{3}$$

Eqs. (1) – (3) describe the combined relaxation and diffusion of any non-zero initial value of $\delta\phi$ and δS to their stationary values $\delta\phi = 0 = \delta S$. The new crosscoupling ($\sim \zeta_3$), which has not been given before [18 – 22], is specific for the Sm-C^* phase and does not exist in the Sm-C phase, because of the presence of a vertical symmetry plane ($\phi \to -\phi$ symmetry). Comparing eq.(2) and (3) with the usual nematodynamic equations for the director and assuming that the kinetic coefficients there do not depend on T_c^* one finds that ζ_1 is independent of S, while $\zeta_3 \sim S^{-1}$ and $\zeta_2 \sim S^{-2}$, the latter relation being already confirmed experimentally [23].

We will discuss the dynamics in the homogeneous limit only. The two relaxation rates, γ_1 and γ_2 are

$$\gamma_1 = 0 \qquad \text{and} \qquad \gamma_2 = 2a\zeta_1 \tag{4}$$

characterizing the hydrodynamic mode and the soft mode, respectively. Apparently the eigenfrequencies are not affected by the new crosscoupling term ($\sim \zeta_3$), nor is the eigenvector of the hydrodynamic mode ($\delta\phi \neq 0$, $\delta S = 0$) in this limit. However, in the soft mode $\delta\phi$ and δS are coupled

$$\delta\phi = (\zeta_3/\zeta_1)\delta S \sim S^{-1}\delta S \tag{5}$$

and the phason variable takes part in the soft mode relaxation, where $\omega(k \to 0) \neq 0$. This can also be seen in the correlation function

$$< \delta\phi(\omega)\delta\phi^*(\omega') > = 2k_B T \left(\frac{\zeta_2 - \zeta_3^2\zeta_1^{-1}}{\omega^2} + \frac{\zeta_3^2\zeta_1^{-1}}{\omega^2 + 4a^2\zeta_1^2} \right) \delta(\omega - \omega') \tag{6}$$

The Sm-C^* phase seems to be the first one to show this phenomenon. In the following we will discuss some experimental consequences of eq.(5).

EXPERIMENTS

The most direct way of checking eq.(5) experimentally is to excite a relaxation of δS and to look for the accompaning rotation of the helix. This can be done as follows: Near the phase transition a rapid (homogeneous) temperature quench ΔT (with $\Delta T < T_c^* - T$) at time $t = 0$ brings the system into a nonequilibrium situation, where the (old) order parameter modulus S differs from its (new) equilibrium value S_{eq} by ΔS. According to eq.(5) the relaxation of $\delta S(t) = S(t) - S_{eq} = \Delta S\,exp(-\gamma_2 t)$, is then accompanied by a helix rotation $\delta\phi(t) = (\zeta_3/\zeta_1)\Delta S\,exp(-\gamma_2 t)$, which finally rotates the helix by $\phi(t=0) - \phi(\infty) = (\zeta_3/\zeta_1)\Delta S$. Thus, in Sm-$C^*$ (but not in Sm-C phases) a homogeneous change in the tilt angle (i.e. a change in S) is accompanied by a rotation of the tilt direction by the same angle in every plane. The latter can be detected by using a well oriented thin film (thin compared to the helical pitch), since there the average over the in-plane director is nonvanishing and the film is birefringent. If the tilt direction is not clamped at the boundaries, the rotation of the averaged tilt direction can take place homogeneously and can be detected optically.

A second possibility for experimentally testing eq.(5) is optical four-wave-mixing and we believe [24] that the coupling between relaxation of the tilt angle and of the tilt direction has actually already been seen in such an experiment [25]. The degenerate four-wave-mixing technique is sensitive to slow relaxational processes and the applied fields couple directly to in-plane fluctuations of the tilt direction. In the experiment [25] no slow relaxational process connected with fluctuations of the tilt direction was found in the Sm-C phase, while for the Sm-C^* phase, there was a strong effect. We interpret this result as evidence for the coupling, given in eq.(5), to the slow order parameter relaxation in a Sm-C^* phase, which does not exist in a Sm-C phase. This explanation seems to be simpler and more natural than the one given in [25], which employs, besides some peculiar time scale arguments, a nonlinear coupling to molecular rotations around their long axes, which is assumed to be slow.

Although the experiments are done on the A side of the Sm-A to Sm-C^* phase transition, we believe that our analysis given above is qualitatively valid for that case, too. Even in the Sm-A phase the molecules (or cluster of molecules) have a finite tilt angle [25], although the tilt direction point into any azimuthal directions. The variable $\delta\phi$ is then a non-hydrodynamic variable, which relaxes fast. The order parameter modulus (the macroscopic tilt averaged over all tilt directions) is zero in the Sm-A phase, but shows strong fluctuations, which relax rather slow near T_c^*. Thus, δS exists as a soft relaxational variable also at the A side of the phase transition. Since $\delta\phi$ is a non-hydrodynamic variable in the Sm-A phase, eqs. (1) – (3) have to be changed accordingly and we find instead [24] for the free energy density

$$f = a(\delta S)^2 + b(\delta\phi)^2 + \lambda q_0 (\delta S)\nabla_z \delta\phi \tag{1'}$$

where the z-direction is parallel to the layer normal. The thermodynamic forces are defined by $df = h_S\, d\delta S + h_\phi d\delta\phi$ and the dissipative dynamics is given by

$$\frac{\partial\delta\phi}{\partial t} = -\frac{\partial R}{\partial h_\phi} \quad \text{and} \quad \frac{\partial\delta S}{\partial t} = -\frac{\partial R}{\partial h_S} \tag{2'}$$

with the dissipation function

$$R = \frac{\zeta_1}{2}h_S^2 + \frac{\zeta_2}{2}h_\phi^2 + \zeta_3 q_0\, h_S \nabla_z h_\phi \tag{3'}$$

Eqs. (1') – (3') show that near a transition to a Sm-C phase $\delta\phi$ and δS are uncoupled, while near a transition to a Sm-C^* phase they are coupled. The wavelength of typical fluctuations (in the z-direction) is comparable to the pitch in the Sm-C^* phase and ∇_z is of the order of q_0. This argument parallels that given in [26] for the cholesteric-isotropic phase transition. In the soft mode the coupling between δS and $\delta\phi$ is then given by

$$\delta\phi \sim q_0^2 \delta S \tag{5'}$$

which can explain the experiment [25]. Thus we conclude that on both sides of the Sm-A to Sm-C^* phase transition crosscoupling terms between the phason and the soft variable are crucial for the understanding of the dynamics.

ACKNOWLEDGEMENTS

H.P. acknowledges interesting discussions at the workshop with G. Durand, D. Litster, and J. Prost. Financial support by the Deutsche Forschungsgemeinschaft is gratefully acknowledged.

REFERENCES

1 J.J. Benattar, F, Moussa, and M. Lambert, *J.Chim.Phys.* **80**, 99 (1983)
2 R.J. Birgeneau and J.D. Litster, *J.PhysiqueLett.* **39**, 399 (1978)
3 D.R. Nelson and B.I. Halperin, *Phys.Rev.* **B21**, 5312 (1980)
4 H. Pleiner and H.R. Brand, *Phys.Rev.* **A29**, 911 (1984)
5 H. Pleiner, *Mol.Cryst.Liq.Cryst.* **114**, 103 (1984)
6 G.S. Smith, E.B. Sirota, C.R. Safinya, and N.A. Clark, *Phys.Rev.Lett.* **60**, 813 (1988)
7 P.G. de Gennes, *The Physics of Liquid Crystals* (Clarendon, Oxford, 1975)
8 R.B. Meyer, L. Liébert, L. Strzelecki, and P. Keller, *J.Phys.(France)* **36**, 69 (1975).
9 H.R. Brand and H. Pleiner, *J.Phys.(France)* **45**, 563 (1984)
10 H. Brand and P. Bak, *Phys.Rev.* **A27**, 1062 (1983)
11 L. Musevic, B. Zeks, R. Blinc, L. Jansen, A. Seppen, and P. Wyder, *Ferroelectrics* **58**, 71 (1984)
12 H. Pleiner and H.R. Brand, *J.Phys.(France)* **50**, 851 (1989)
13 I.M. Khalatnikov, *An Introduction to the Theory of Superfluidity* (Benjamin, New York, 1965).
14 H. Pleiner, in *Incommensurate Crystals, Liquid Crystals, and Quasi-Crystals* eds. J.F. Scott and N.A. Clark, Plenum New York, p. 241 (1987)
15 H.R. Brand and H. Pleiner, *Mol.Cryst.Liq.Cryst.Lett.* **5**, 53 (1987)
16 J. Prost, *Ferroelectrics* **84**, 261 (1988)
17 C.C. Huang and J.M. Viner, *Phys.Rev.* **A25**, 3385 (1982)
18 E.P. Pozhidayev, L.M. Blinov, L.A. Beresnev, and V.V. Belyayev, *Mol.Cryst. Liq.Cryst.* **124**, 359 (1985)
19 K. Skarp, *Ferroelectrics* **84**,119 (1988)
20 T. Carlsson, B. Zeks, C. Filipic, and A. Levstik, *ibid.* 223 (1988)
21 G. Andersson, I. Dahl, W. Kuczynski, S.T. Lagerwall, K. Skarp, and B. Stibler, *ibid.* 285 (1988)
22 R. Blinc and B. Zeks, *Phys.Rev.* **A18**, 740 (1987)
23 E.P. Pozidaev, M.A. Osipov, V.G. Chigrinov, V.A. Baikalov, L.M. Blinov, and L.A. Beresnev, *Sov.-Phys. JETP* **67**, 283 (1988)
24 H.R. Brand and H. Pleiner, submitted for publication
25 J.R. Lalanne, J. Burchert, C. Destrade, H.T. Nguyen, and J.P. Marcerou, *Phys.Rev.Lett.* **62**, 3046 (1989)
26 P.G. de Gennes, *Mol.Cryst.Liq.Cryst.* **12**, 193 (1971)

OPTICAL MEASUREMENTS OF DIFFUSIVITIES

IN INCOMMENSURATE BARIUM SODIUM NIOBATE

J. F. Scott[*] and W. F. Oliver[+]

Departments of Physics
University of Colorado,[*] Boulder, CO 80309-0390 and
Arizona State University,[+] Tempe, AZ 85287

Abstract   Three different optical techniques have been
used for measuring diffusivities in $Ba_2NaNb_5O_{15}$: The
original method of Burkhart and Rice (1977) yielded a
thermal diffusivity of 0.02 cm^2/s at 373K in the quasi-
commensurate phase; dynamic central mode measurements of
Oliver et al. (1988, 1989) yield a value 20x larger near
565K, at the hypothesized 1q-2q transition in the middle
of the incommensurate phase from 543-582K; and two quite
different diffusivities are measured at 855K=T_C -- a
thermal diffusivity of 0.03 cm^2/s from a new thermal lens
effect recently discovered (Scott et al., 1989), and a
much larger value of 1.2 cm^2/s from dynamic central mode
light scattering. This shows that at both T_C and T_I at
least two separate diffusion processes occur; the light
scattering studies are sensitive to the faster process,
which at 565K may be related to the roughening transition
of anti-phase boundaries originally predicted by Rice,
Whitehouse and Littlewood (1981).

INTRODUCTION

Optical techniques have been utilized to measure thermal
diffusivities in ferroelectric insulators for more than a
decade. In 1977 Burkhart and Rice reported measurements [1]
on both lithium niobate and barium sodium niobate, using a
technique that involved counting phase fringes. Their
experimental values correspond to D_T = 0.02 cm^2/s at 373K for

Geometry and Thermodynamics
Edited by J.-C. Tolédano
Plenum Press, New York, 1990

$Ba_2NaNb_5O_{15}$, a diffusion constant found to be quite isotropic
(in contrast to the 6% anisotropy determined in lithium
niobate). More recently Lyons et al. have compared diffusivities inferred from the linewidths of dynamic central mode [2]
light scattering in incommensurate $BaMnF_4$ with thermal
diffusion measured conventionally and have inferred that there
are at least two separate diffusion processes occurring near
T_I, the commensurate-incommensurate transition temperature (ca.
254K in that material), with a phason-like diffusion having a
diffusivity an order of magnitude larger than D_T. This faster
diffusivity for kink-like diffusion agrees with a theoretical
model calculation of Hassold et al. [3] that predicted for a
2D model faster diffusion of kinks than of entropy fluctuations.

In the present paper we compare different optical techniques for measurement of diffusion in incommensurate barium
sodium niobate. We have measured diffusivities from central
mode light scattering from ca. 400K (in the quasicommensurate
phase of $Ba_2NaNb_5O_{15}$ where residual discommensurations are
thought to arise from imperfect stoichiometry) to 600K.
Special emphasis was placed upon the transition temperatures
543K, 565K, and 582K, which correspond, for heating cycles
(substantial hysteresis occurs upon cooling) to the transitions:
C/1q; 1q/2q; and 2q/C, as shown in Fig. 1. Here C designates
a commensurate phase (space group $Ccm2_1$ from 105-543K; P4bm
from 582-855K); 1q designates an orthorhombic incommensurate
phase with modulation along one axis, corresponding to an
averaged mm2 structure; and 2q designates a tetragonal
incommensurate phase with modulation along two orthogonal axes,
corresponding to an averaged 4mm structure.

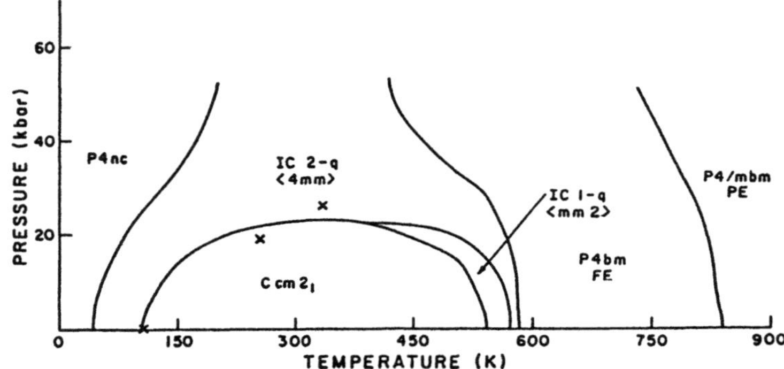

Fig. 1. Hypothetical phase diagram for $Ba_2NaNb_5O_{15}$. x's are
data points from S. Kojima et al., [4].

EXPERIMENTAL

We describe below the results [5] of dynamic central mode
light scattering near T_C = 855K and also through the three IC
phase transitions at 543, 565, and 582K. In each case we
observe a dynamic central mode whose linewidth exhibits the q^2
dependence expected for a diffusion process, as determined
experimentally by varying the scattering angle (we employed
angles of approximately 5°, 90°, and 178° arc). q could be
varied in magnitude by x2 by going from right-angle scattering
to backscattering, and in some cases near-forward scattering
geometries were also successful, at least in obtaining upper
limits on linewidths. In these experiments we were careful to
keep momentum transfer confined within the isotropic ab-plane,
so that q^2 effects were seen and no corrections for index
of refraction variation were required. q was varied in this
way from 5.52 x 10^7 m^{-1} for backscattering to 3.90 x 10^{-7} m^{-1}
for 90° geometries.

The diffusivity corresponding to

$$\Gamma = D_k\, q^2 \qquad\qquad (1)$$

in this experiment is 0.8 $\pm$ 0.1 cm^2/s and peaks at 565K, a
temperature independently assigned on the basis of electron
diffraction experiments as the 1q-2q transformation within the
IC phase(s). [6]

It is worth pointing out that the intensity of this
dynamic central mode, as shown in Fig. 2, peaks not at the
lock-in transition at 543K = T_L nor at the C-IC transition at
582K = T_I, but at the hypothesized [6] 1q-2q transition at 565K.
In this sense it is somewhat analogous to the anti-phase
boundary (APB) roughening transition initially predicted for
incommensurates by Rice, Whitehouse, and Littlewood [7].

We show in the section that follows that this fast
diffusion is not due to thermal diffusion (of entropy
fluctuations); D_T is in fact rather independent of temperature
from 373K to 855K, as shown by our new thermal lens data, to
be discussed below.

We also note that a detailed deconvolution of asymmetric
lineshapes near 565K shows that strong coupling occurs between
this dynamic central mode and the longitudinal acoustic (LA)
phonon near 40 GHz. It is interesting to note that there is no

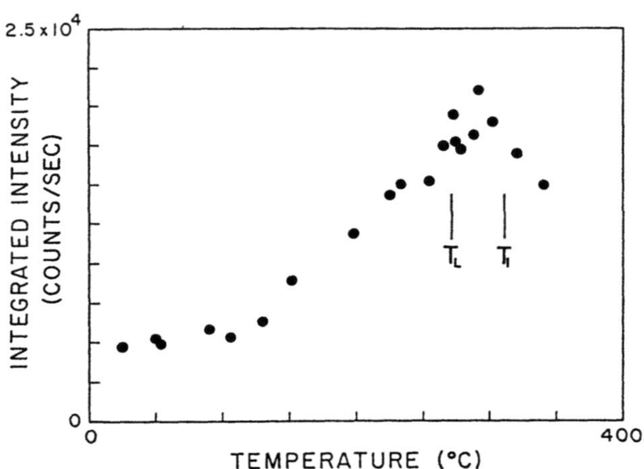

Fig. 2.   Central mode light scattering intensity versus
            frequency shift, showing a maximum at 565K (292°C)
            where the 1q-2q transition has been reported [6],
            and not at the lock-in transition T_L =543K (270°C)
            nor the commensurate-incommensurate transition T_I =
            582K (309°C). All data were taken upon heating.

coupling with the TA phonons and that for the (different)
central mode near T_C = 855K there is no coupling with either
LA or TA phonons. This shows that the dynamic central modes
near T_I and near T_C have quite different microscopic origins.
The coupling of central mode and LA phonon in the incommensurate phase permits the LA frequency to renormalize by ca. 4% by
relaxing into the diffusive mode; and it also results in the
strong central mode scattering intensity -- we find that at
565K 89% of the central mode cross section arises from coupling.
This helps reconcile the intense central mode scattering we see
with phason scattering theories, which often predict extremely
small cross sections for phason-like processes. [8]

    There are some difficulties in associating the diverging
cross section for the dynamical central mode near 565K with a
1q-2q phase transition: Firstly, the divergence in intensity
graphed above is of the smooth form normally associated with a
second-order or at least nearly continuous phase transition,
whereas the 1q-2q transition is reported to be extremely first
order, with large temperature regions over which both phases
coexist metastably. Second, there is some very recent evidence
(Xiaoqing Pan, private communication, April 1989) that the
hypothesized 2q phase may not even exist, but perhaps is only

a coexistence of two micro-domains, each of 1q structure but oriented at 90° to each other. How then can one interpret the diverging central mode intensity near 565K shown in Fig. 2 above, or the electron microscopy data of Ref. [6]? We think that some of the phenomena near 565K may result from a true roughening transition. In general, it is not expected that ferroelectric incommensurate insulators will undergo APB wall roughening transitions [Arkadiy P. Levanyuk, private communication, July 1989]. The reason is that the macroscopic strain that is associated with the antiphase boundary curvature carries a long-range Coulombic field. However, $Ba_2NaNb_5O_{15}$ is an unusual case in which ferroelectricity is independent of the incommensurate structure and in fact occurs perpendicular to the direction of incommensurate modulation. In such a case, roughening transitions of the type hypothesized by Rice et al. [7] seem quite possible. Because the hypothesized 1q-2q transition in barium sodium niobate is so discontinuous (if it exists!), it does not seem likely that this transition _per se_ can be directly responsible for the divergence in central mode intensity at 565K; rather, a second-order roughening transition at this temperature is a plausible alternative mechanism for these spectroscopic phenomena.

THERMAL LENS EFFECT

In the ordinary thermal lens effect in fluids [9] thermal focussing occurs because of thermal expansion in the illuminated cylinder of a sample and the fact that density and hence index of refraction decrease upon heating. The effect is observed to be isotropic and to produce a negative (concave) lens with effective focal length of order -100 cm. Strong spherical aberration and astigmatism can be present.[10] We have recently discovered an analogous effect in ferroelectrics near T_C.[11] As shown in Fig. 3, the strong dependence of dn/dT for certain crystal axes produces a very strong thermal focussing. The focal length may be positive or negative and is of order 3 cm. The formula given below in Eq.(2) is accurate to about 50% (neglecting curvature of field and spherical aberration):

$$F = (\pi K_T n\, w_o^2)\,(0.24\ PbL\ dn/dT)^{-1} \qquad (2)$$

where K_T is the thermal conductivity (and is generally not isotropic); n, the index of refraction; w_o, the beam waist; P,

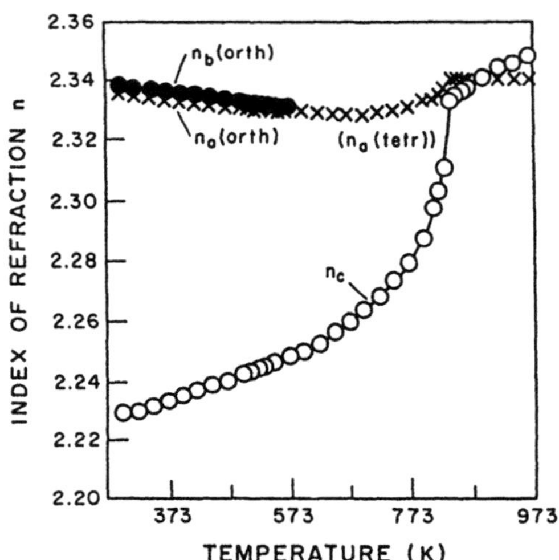

Fig. 3.   Index of refraction variation in barium sodium niobate near T_C = 855K and T_I = 582K. The large value of dn_c/dT near T_C produces positive thermal focussing (convex thermal lens effect). From T. Yamada et al. [12]

the laser power;b, the optical absorption coefficient; and L, the sample length.

A simpler geometric relationship between focal length F and beam divergence θ is given by

$$F = w_o/(1.6\ \theta).\qquad(3)$$

In $Ba_2NaNb_5O_{15}$ Eqs.(2) and (3) agree within 50%, which is the uncertainty in the absorption coefficient b (which varies by x2 from sample to sample).

Measurement of the thermal lens effect in barium sodium niobate shows that K_T (and hence D_T) changes very little from room temperature to 855K and is between 0.06 W/K.cm^2 and 0.09 W/K.cm^2 (i.e., D_T between 0.02 and 0.03 cm^2/s). Thus, the diffusivity inferred from central mode light scattering of 0.8 cm^2/s at 565K and 1.2 cm^2/s at 855K must arise from additional, non-thermal diffusion.

We have suggested above that the diffusion near 565K arises from APB wall roughening. To what can we attribute the slightly faster diffusion near T_C? First, we observe that the latter diffusion is highly anisotropic; it arises only for cc-polarizability tensor component (i.e. light polarized along the optic polar axis). It probably consists of polarization

fluctuations involving mobile charged defects being driven up
and down the polar axis.

At higher laser powers (greater than 1 W) several more
complex phenomena are observed with the thermal lens effect in
barium sodium niobate. These include self-induced phase
modulation and some detailed anisotropic pattern formation.

At powers below 1W the thermal lens pattern in far-field
is usually a simple elliptical ring. The ellipse is eccentric
because the thermal conductivity is slightly greater along the
[001] direction than along [100] or [010] in $Ba_2NaNb_5O_{15}$. And
a single ring results from the effect of spherical aberration.
In this low power regime we have used the beam divergence $\theta(T)$
in Eqs. (2,3) above to determine the critical exponent β,
or more precisely, $2\beta -1$, which comes from the term dn/dT in
Eq.(2). A preliminary result gave $2\beta = 0.5\pm0.1$ [12] However,
more recent measurements in our laboratory have given higher
values (ca. 0.7 - 0.8). A possible explanation of these
exponents is that in addition to the divergence in $n(T)$ near
T_C, which varies as $t^{2\beta}$, there is also a possible divergence
in thermal conductivity $K_T(T)$; the combination of these
two terms is proportional to $K_T (dn_C/dT)^{-1}$, which is
predicted to vary as
(tricritical point)

$$t^{2\beta - 1 - \nu + \gamma} = t^0 \tag{4}$$

which is logarithmic in mean field. More precise experiments
will be required to determine the actual dependence.

The high-power thermal lens patterns observed in
incommensurate barium sodium niobate will be analyzed in detail
elsewhere. Here we mention only two effects which are not
observed at lower powers: Above 1W near T_C we observe an
extremely complicated elliptical pattern consisting of as many
as a dozen concentric rings, the intensities of which are
described by higher order Bessel functions [13-17]. The
patterns are quite similar to those observed in light trapped
in filaments [18]. We have observed several effects, both
spatial and temporal, in barium sodium niobate analogous to
those in fluids, including chirping. The second effect we
observe above 1W is shown in Fig. 4. It is the appearance in
the far-field pattern of transmitted light of an image that
displays the crystalline symmetry (approximately four-fold).
Such an effect arises from the directional anisotropy of the

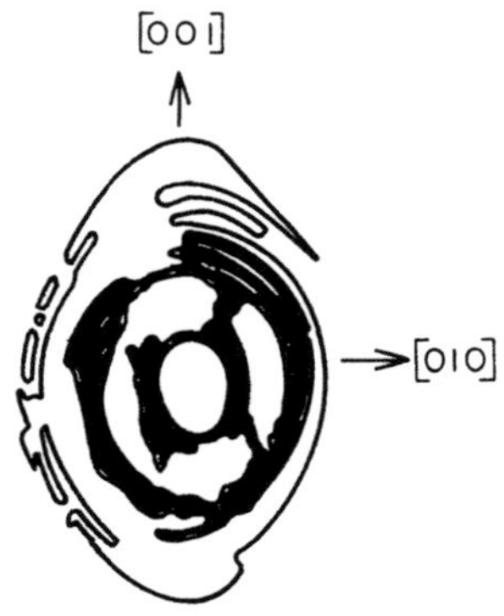

Fig. 4.   Intensity pattern (far-field) of thermal lens effect
in barium sodium niobate near T_C, showing
approximate fourfold crystalline symmetry due to
anisotropy in thermal conductivity K_T.

thermal conductivity K_T and has not been reported before. Note
that any anisotropy in K_T in barium sodium niobate is small at
373K or below [7] but may become large near T_C.

CONCLUSION

We conclude that at 565K, the temperature inferred to be
the transition temperature for a 1q to 2q transformation within
the IC phase of $Ba_2NaNb_5O_{15}$, there is a diffusional process
that is very fast -- approximately 20x that of thermal diffusion. We suggest that this physical process resembles the APB
roughening hypothesized by Rice et al. The light scattering
cross-section for this diffusive mode is intrinsically very
small, but it is enhanced one order of magnitude by bilinear
coupling with the LA phonon.

Near T_C = 855K we see an even more intense dynamic central
mode (Fig. 5 below) in our light scattering spectra. A low-
resolution (12 GHz) grating spectrometer study first detected
this mode several years ago. [19] This mode is qualitatively
different from that in the IC phase in that it exhibits
absolutely no coupling with either TA or LA phonons.

The diffusivity of the mode at 565K agrees rather well
with earlier estimates [20] based upon LA phonon sound velocity
anomalies and dispersion. The somewhat faster diffusion
observed near T_C = 855K is not well understood at present, nor

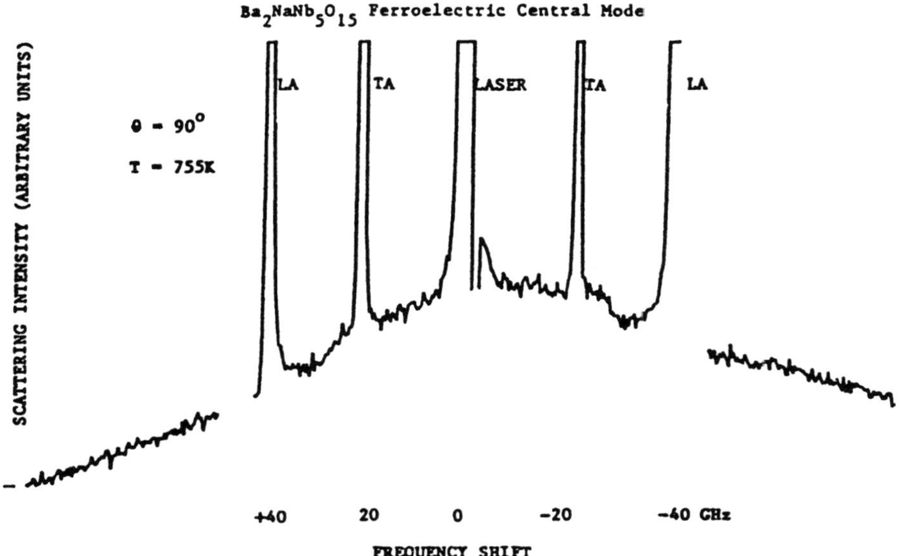

Fig. 5.   Dynamic central mode light scattering intensity
versus frequency shift 100K below T_C in Ba$_2$NaNb$_5$O$_{15}$.

is its very strong light scattering intensity. The system
appears very similar to lead germanate [21], where Lyons and
Fleury suggested phonon density fluctuations as the origin of
the observed diffusion. The c-axis anisotropy of the
scattering is compatible with polarization fluctuations near
T_C.

Finally, the technique of thermal focussing near T_C in
ferroelectrics is established as a contactless method of
determining thermal diffusivity at elevated temperatures.
The utility of this technique for the evaluation of critical
exponents has been introduced. And several new effects have
been observed, including those due to anisotropy of the thermal
conductivity. Most important, from the context of the present
paper, the use of thermal lens effects at the same temperature
at which dynamic central mode light scattering is carried out
establishes that at least two different diffusion processes,
differing in speed by an order of magnitude, exist simultaneously near T_I and near T_C in barium sodium niobate.

ACKNOWLEDGMENT

This work was supported in part by NSF grant DMR86-0666
and by ARO grant DAAL-86-K-0053.

REFERENCES

1.) G. H. Burkhart and R. R. Rice, J. Appl. Phys. $\underline{48}$, 4817 (1977).

2.) K. B. Lyons, T. J. Negran, and H. J. Guggenheim, Phys. Rev. $\underline{B25}$, 1791 (1982); J. Phys. (Sol. St.) $\underline{C13}$, L415 (1980); Sol. St. Commun. $\underline{31}$, 285 (1979).

3.) G. N. Hassold, J. F. Dreitlein, P. D. Beale, and J. F. Scott, Phys. Rev. $\underline{B33}$, 3581 (1986).

4.) S. Kojima, K. Asaumi, T. Nakamura, and S. Minomura, J. Phys. Soc. Jpn. $\underline{45}$, 1433 (1978).

5.) W. F. Oliver, J. F. Scott, S. A. Lee, and S. M. Lindsay, Mater. Sci. & Engineering, $\underline{A150}$ (in press, 1989), and in Laser Optics of Condensed Matter, edited by J. L. Birman, H. Z. Cummins, and A. A. Kaplyanskii (Plenum, New York, 1988), p.263.

6.) Feng Duan and Pan Xiaoqing, Phys. Stat. Sol. $\underline{A106}$, K117 (1988) and in Proc. XIth Int. Congress Electron Microscopy (Kyoto, 1986), p.1239.

7.) T. M. Rice, S. Whitehouse, and P. Littlewood, Phys. Rev. $\underline{B24}$, 2751 (1981).

8.) N. I. Lebedev, A. P. Levanyuk, and A. S. Sigov, Zh. Eksp. Teor. Fiz. $\underline{92}$, 248 (1987); K. B. Efetov and A. I. Larkin, Zh. Eksp. Teor. Fiz. $\underline{72}$, 2350 (1977); V. A. Golovko and A. P. Levanyuk, Zh. Eksp. Teor. Fiz. $\underline{81}$, 2296 (1981); R. Zeyher and W. Finger, Phys. Rev. Lett. $\underline{49}$, 1833 (1982); see also the review by J. C. Toledano, CNET Report #576 (Bagneux, France, 1985).

9.) J. P. Gordon, R. C. C. Leite, R. S. Moore, S. P. S. Porto, and J. R. Whinnery, J. Appl. Phys. $\underline{36}$, 3 (1965).

10.) S. J. Sheldon, L. V. Knight, and J. M. Thorne, Appl. Opt. $\underline{21}$, 1663 (1982); C. A. Carter and J. M. Harris, Appl. Opt. $\underline{23}$, 476 (1984); C. Hu and J. R. Whinnery, Appl. Opt. $\underline{12}$, 72 (1973); P. Calmettes and C. Laj, Phys. Rev. Lett. $\underline{27}$, 239 (1971).

11.) J. F. Scott, Shou-Jong Sheih, K. R. Furer, N. A. Clark, W. F. Oliver, and S. A. Lee, Phys. Rev. B (submitted).

12.) T. Yamada, H. Iwasaki, and N. Niizeki, J. Appl. Phys. $\underline{41}$, 4141 (1970).

13.) S. A. Akhmanov, D. P. Krindach, A. P. Sukhorukov, and R. V. Khokhlov, Pis'ma Zh. Eksp. Teor. Fiz. $\underline{6}$, 509 (1967); translation: JETP Lett. $\underline{6}$, 38 (1967).

14.) F. W. Dabby, T. K. Gustafson, J. R. Whinnery, Y. Kohanzadeh, and P. L. Kelley, Appl. Phys. Lett. $\underline{16}$, 362 (1970).

15.) S. A. Akhmanov, D. P. Krindach, A. V. Migulin, A. P. Sukhorukov, and R. V. Khokhlov, J. Quant. Electron. $\underline{QE-4}$, 568 (1968).

16.) S. A. Akhmanov, A. P. Sukhorukov, and R. V. Khokhlov, Zh. Eksp. Teor. Fiz. $\underline{50}$, 1537 (1966); translation: Sov. Phys. JETP $\underline{23}$, 1025 (1966).

17.) J. R. Whinnery, D. T. Miller, and F. W. Dabby, J. Quant. Electron. $\underline{QE-3}$, 382 (1967).

18.) F. Shimizu, Phys. Rev. Lett. $\underline{19}$, 1097 (1967); A. C. Chung, D. M. Rank, R. Y. Chiao, and C. H. Townes, Phys. Rev. Lett. $\underline{20}$, 786 (1968).

19.) G. Errandonea, M. Hebbache, and F. Bonnouvrier, Phys. Rev. $\underline{B32}$, 169 (1985).

20.) P. W. Young and J. F. Scott, Phase Trans. $\underline{6}$, 175 (1986).

21.) K. B. Lyons and P. A. Fleury, Phys. Rev. $\underline{B17}$, 2403 (1978); Ferroelectrics $\underline{35}$, 37 (1981); Structural Phase Transitions (ed. K. Muller and H. Thomas, Springer, Berlin, 1981), p.9.

INDEX